Relativistic Quantum Mechanics and Quantum Fields

Second Edition for the 21st Century

Relativistic Quantum Mechanics and Quantum Fields

Second Edition for the 21st Century

W-Y Pauchy Hwang
National Taiwan University, Taiwan

Ta-You Wu

World Scientific

NEW JERSEY • LONDON • SINGAPORE • BEIJING • SHANGHAI • HONG KONG • TAIPEI • CHENNAI • TOKYO

Published by

World Scientific Publishing Co. Pte. Ltd.
5 Toh Tuck Link, Singapore 596224
USA office: 27 Warren Street, Suite 401-402, Hackensack, NJ 07601
UK office: 57 Shelton Street, Covent Garden, London WC2H 9HE

British Library Cataloguing-in-Publication Data
A catalogue record for this book is available from the British Library.

RELATIVISTIC QUANTUM MECHANICS AND QUANTUM FIELDS
Second Edition for the 21st Century

ISBN 978-981-3270-02-2

For any available supplementary material, please visit
http://www.worldscientific.com/worldscibooks/10.1142/10984#t=suppl

Desk Editor: Christopher Teo

Typeset by Stallion Press
Email: enquiries@stallionpress.com

Printed in Singapore

PREFACE (for the 21st-Century Edition)

Several years ago, I realized that, in my style, the 1st Edition was already out of date, but at that time I was not ready to write the edition of my like. After a few years of intensive researches on the Standard Model, I finally felt that it is the time to have a textbook for the kids of the 21st Century.

I think that "relativistic quantum mechanics and quantum fields" should be the language of our world - to replace the thoughts in Newton's classic era (which was there for the last four centuries since Newton). I would be much surprised if Einstein's relativity principle and the quantum principle, together with the existence of the *smallest* units of matter (such as electrons, neutrinos, quarks, etc.), was not there a thousand years from now (maybe much longer). Relativistic quantum mechanics and quantum fields is the language of the description of these smallest units of matter on the basis of Einstein's relativity principle and the quantum principle.

For me, the research that enables me to touch the truth is full of joys. For my wife, Jane Su-Chen, had to find something else to entertain herself, occasionally. I also realize that my secretary Chih-Hsin Huang has had the superior typing abilities to put things together in beautiful forms. As a private and quiet person, I just don't have the patience to dig out most of typo errors - so, the smart students should realize that. I think that the Truth overcomes everything else.

W-Y. Pauchy Hwang
National Taiwan University

PREFACE (for the 1st Edition)

There exists already an ample literature on relativistic quantum mechanics, quantum electrodynamics, and even on the more recent developments on the so-called "standard model" which consists of quantum chromodynamics for strong interactions and the Glashow-Weinberg-Salam $SU(2)\times U(1)$ theory for electroweak interactions. A few words may perhaps be in order for the addition of the present volume to the literature.

When in 1986 one of the present authors (Wu) published Quantum Mechanics ("Q.M."), the manuscript originally contained chapters on the theory of Klein-Gordon and of Dirac, classical fields and the quantization of free fields. It stopped just before the important development of quantum electrodynamics. This would be very unsatisfactory, and Wu decided to publish the non-relativistic part as "Quantum Mechanics" and to hold the rest of the manuscript, hoping to interest a colleague in completing it into a sequel volume that will cover relativistic quantum mechanics and an introduction to quantum electrodynamics and the standard model. In 1988, the present authors discussed and agreed upon a joint effort to prepare this sequel to "Q.M.". The first eight chapters come from the earlier manuscript as mentioned above, and follow the generally pedagogic style of "Q.M.", and the last six chapters are based on the notes of lectures given by Hwang at Indiana University and at National Taiwan University. Attempt has been made to present the latter chapters at the same level of ease (or difficulty) as the earlier ones, but an entirely smooth integration is difficult because of the nature of the subject matter itself and of the difference in style of two co-authors.

The scope of the present volume may be seen from the table of content. It ranges from relativistic quantum mechanics where the Klein-Gordon and Dirac equations are treated in detail, to an introduction to quantum field theory with quantum electrodynamics (QED) as our basic example, and to a pedagogic exposition of important issues related to the standard model. As our primary objective is to expose the students to the themes and ideas underlying the contemporary developments, we attempt to avoid using excessively mathematical tools such as the method of path integrals, which may add to mathematical elegance pertaining to the subject but without enhancing significantly the global understanding of the subject matter in terms of basic and easy-to-grasp notions. Thus, we hope that this volume will serve as a useful textbook or reference material in the area of relativistic quantum mechanics and quantum electrodynamics and an introduction to the more recent developments on gauge field theories.

The convention of choosing units such that $\hbar = c = 1$ is almost universal in the field theory literature but is less so in quantum mechanics. For the convenience of some readers, certain equations are written out also in ordinary units in the footnotes (with primed equation numbers).

In preparing the final version of the manuscript, we have generally followed closely the line of presentation given by authors in the original papers since, after all, many of these things have become so standard that they may as well appear without much distortion in a textbook on the subject. For example, the section on dimensional regularization is based on the original paper of 't Hooft and Veltman; most materials for the chapter on experimental tests of the standard model are from the 1988 publication of the Particle Data Group; and, since Hwang used in his lectures a couple of very nice books, notably the one by E.D.

Commins and P.H. Bucksbaum ("Weak Interactions of Leptons and Quarks", Cambridge University Press, 1983) and the other one by F. Halzen and A. D. Martin ("Quarks and Leptons", John Wiley & Sons, 1984), it should not be a surprise that footprints of these authors may be found occasionally in the last few chapters. Wu first learned from Professor George Uhlenbeck, in 1932, about the Zitterbewegung, Klein paradox, and the "divorce" of velocity and momentum in Dirac's theory, and later studied Pauli's article. The treatments in many sections in Chapters 7 and 8 have followed the memeographed notes of lectures by Professor E. Mishkin at the Polytechnique Institute of Brooklyn in 1964. To these and many other authors from whose works we have drawn freely, we gratefully express our indebtedness.

Finally, we wish to acknowledge Ms. Bi-Lien Lin and Ms. Yu-Ru Hung for helping to prepare the entire camera-ready manuscript in TeX and Professor Wen-Kwei Cheng at State University of New York at Old Westbury and Mr. Yen-Sheng Su at National Taiwan University for offering useful remarks upon reading the manuscript.

September 1989

Ta-You Wu, Academia Sinica

W-Y. Pauchy Hwang, National Taiwan University

Foreword (in the 1st Edition)

In my Foreword to Professor Wu's well-known volume on Quantum Mechanics, I expressed the hope that there would be a sequel which would cover, in addition to Relativistic Quantum Mechanics, also Quantum Field Theory. This wish has been amply fulfilled in the present book by Professor Wu in collaboration with Professor W-Y. Pauchy Hwang.

In reading this treatise, I was reminded of the period 1945-46 in Kunming when Professor Ta-You Wu gave us extra lessons on Quantum Mechanics and Relativistic Quantum Mechanics. It was my good fortune tobe his student. In that brief time of one year, he gave me a solid foundation in physics and immediately start doing graduate work. Other students in that special enrichment class in Kunming were J.S. Wang, G.Y. Zhu, A.Q. Tang, and B.H. Sun; all of them benefited greatly from Professor Wu's lectures and made important contributions to science later on. The publication of Professor Wu's Quantum Mechanics and the present sequel will make it possible for serious science students everywhere to share the insight and the wisdom that were instrumental in launching our careers.

This is indeed an important event for the entire physics community.

T.D. Lee
April 10, 1990
New York

Table of Contents

Part B. Introduction to Quantum Field Theory

Part C. The Standard Model

Chapter 0: We live in the quantum 4-dimensional Minkowski space-time

Perhaps it might be a little bit too subjective to introduce the subject "Relativistic Quantum Mechanics and Quantum Fields" in this personal way. But as we see it, we should declare that we are in fact living in the quantum 4-dimensional Minkowski space-time with force-fields gauge-group structure built-in from the very beginning — why we have to study relativistic quantum mechanics and quantum fields. What we intend to show is the physics of where we live — we should not pretend to live the Newtonian classic world, not from the 21st Century onward.

On 28 March 2016 (2016.3.28), I (Pauchy) woke up at around two o'clock in the early morning and realized that what we have been doing these days is in fact "the Standard Model of all centuries". Newton's classic frame of physics is now being replaced by Einstein's relativity principle and the quantum principle, the two pillars established during the 20th Century, basing on these two pillars to describe the smallest units of matter, including electrons, neutrinos, and quarks (a finite number of them) as called "the Standard Model". The existence of the smallest units of matter, such as electrons, neutrinos, quarks, etc., is now firmly established as the basic experimental fact — the theory, the Standard Model, to describe these smallest units of matter is so fundamental, having the character to replace Newton's classic era of four or five centuries.

At this beginning of the 21st Century, we should declare that we are living in the quantum 4-dimensional Minkowski space-time with some force-fields gauge-group structure built-in from the very beginning. For this "overall background" of our world, we should be able to name all the relevant force-fields gauge-groups structure beyond the Lorentz group. We used the entire 20th Century to make certain that both Einstein's relative principle and the quantum principle are two pillars of our world, but some of us still hesitate to view the world as such, still using the thinking in the old Newton's classic era.

To really think in the modern way (or, after the 21st-Century way), we thus claim that we already live in the quantum 4-dimensional Minkowski space-time — a world that satisfies Einstein's relativity principle and the quantum principle. Everything goes from there on.

§. 0.0. The Space-Time

We should stress that in fact we live in the quantum 4-dimensional Minkowski space-time — a space-time that satisfies both the Einstein's relativity principle and the quantum principle. The easiest way of seeing this is to go to the small world, called "the atomic world (of sizes of anstrons)", or to go to even a smaller world such as the nuclear world (of the scales of Fermi's), where elementary particles or nuclei play the dominant roles. The era of Newton's classical physics, of daily sizes of centimeters (cm's), established since Newton, could be declared to be completely irrelevant, or dead.

We assume that you know about the length measurement and the standard clock. Here we assume that the operations of sticks and clocks do not affect anything except giving us the length and the time. Further. we can make sure that the local space-time which we are living is flat. If we label an event or point as (x, y, z, ct) or simply (x, y, z, t), then the interval between two points, (x_1, y_1, z_1, ct_1) and (x_2, y_2, z_2, ct_2), would be an invariant quantity, according to Einstein's relativity principle. Thus, we, or you and I, could begin to talk about something and in fact set up some common units for space and time.

So, we have, with $\Delta x = x_2 - x_1$, $\Delta y = y_2 - y_1$, $\Delta z = z_2 - z_1$, and $\Delta t = t_2 - t_1$,

$$(\Delta s)^2 = (\Delta x)^2 + (\Delta y)^2 + (\Delta z)^2 - (c\Delta t)^2, \tag{1}$$

an invariant quantity among all the observers. Or, in infinitesimal forms,

$$ds^2 = dx^2 + dy^2 + dz^2 - c^2dt^2 \equiv g_{\mu\nu}dx^\mu dx^\nu = \eta_{\mu\nu}dx^\mu dx^\nu. \tag{2}$$

Here ds is called "the proper length", or "the proper distance", $g_{\mu\nu}$ is the second-rank metric tensor, and $\eta_{\mu\nu} = diag(1, 1, 1, -1)$. Similarly, the $d\tau$ (with $d\tau^2 \equiv -ds^2$) is "the proper time". We shall use "the invariant distance" or "the invariant time" for the sake of being precise.

From the measurements of the space and of the time which are rather independent, there is no reason why the Minkowski constraint, the equation specified above, should be valid. Thus, we live in the 4-dimensional Minkowski space-time. The independent nature of the four measurements on space-time should be seriously questioned, in view of the invariance of $(ds)^2$.

It is the quantum 4-dimensional Minkowski space-time - the measurements of space and time should strictly take the meaning of the quantum mechanics sense (e.g., see the discussion about Snyder's as below), or, satisfying the quantum principle, even though, so far, we do not understand exactly what the meaning of the quantum principle really is.

To make sure that the quantum principle is meaningful, for example, we note that the differences Δx, Δy, Δz, and Δt are the quantum observables, but Δs^2 is not always the quantum observable (cf. such as part in the Schwarzschild metric). The quantum observable assumes a value in real numbers, i.e., in $(-\infty, +\infty)$.

Let's come to think about it. The so-called "intuition" (classically) could be completely wrong. Because of "Lorentz contraction" and "time dilation" already built in the above Minkowski's metric, the world of an elementary particle differs from our limited intuitive one, based on classical physics or Newton's. And, for this book on "Relativistic Quantum Mechanics and Quantum Fields", we suggest that we should imagine that our "intuition" of the world should be identical with that for an elementary particle. Thus, in this 21st Century, we would try to downplay the importance of the non-relativistic, or, Newton's, descriptions.

As another important example of the 21st-century post-modern physics, let's start with the Schwarzschild metric,

$$d^2s = -(1 - \frac{2GM}{r})dt^2 + (1 - \frac{2GM}{r})^{-1}dr^2 + r^2(d\theta^2 + sin^2\theta d\phi^2). \tag{3}$$

We should try to establish our intuitive explanation of the time and the space, at all r and all t. Sign switch may be identified as the important source of confusion. In this example, we use the natural units. We should eventually establish a decent 21st-century physics, with this "black hole" solution as one of the beginning points.

To this end, we identify t as our time and (x_1, x_2, x_3) as our space. The problem with the Schwarzschild metric is the coefficient of dr^2 which goes over to ∞ at the horizon, i.e., $\frac{2GM}{r} \to 1$. The coefficient of dt^2 would go to zero simultaneously. So, in our space and our time, maybe we never get to this point — that is, black holes in our space-time never form and all the black holes are pre-formed. Einstein equation tells us so.

The other way of saying it: We try to write the Schwarzschild metric as follows:

$$(ds)^2 = -c^2(dt^*)^2 + (dr^*)^2 + r^2(d^2\theta + sin^2\theta d\phi^2). \tag{4}$$

Here we have

$$\frac{dr^*}{dr} = \pm\sqrt{\frac{r}{r - r_h}}, \quad \frac{dt^*}{dt} = \pm\sqrt{\frac{r - r_h}{r}}; \qquad r_h = 2GM. \tag{5}$$

So, it becomes pure imaginary beyond the horizon ($r = r_h$) rather than the switching between time and space, which might be wrong at the first place. Our space variables and our time variable are real numbers, i.e., $(-\infty, +\infty)$, after all. This may offer an entirely new road for the knowledge of black holes.

We thus suggest that we may stick to the interpretation of viewing t as the time and (x_1, x_2, x_3) as the space. Let these variables run from $-\infty$ to $+\infty$ to define the observable space-time for us. We need to talk about the events beyond us — not within the $\pm\infty$. We will be "conservative" in this regard.

There is an important sideline solution to the Schwarzschild metric - arising from the fact that there is the 25% dark matter while there is only 5% visual ordinary matter. According to the Standard Model [1]*, the 25% dark matter must be the cosmic background (CB) ν's (i.e., abbreviated as "neutrinos of three flavors, including antineutrinos"). But neutrinos have tiny masses and so were clustered, to form neutrino halos, each of them followed the visual ordinary-matter object, such as the Earth, the Venus, the Sun, and the stars, five times in weight. These neutrino halos, being the Fermi gases, must have the incompressible Fermi-Dirac spheres that resist the final collapse into the presumed black holes* [2]*. The known neutrino tiny mass indicates that the neutrinos near the surface of the Fermi-Dirac sphere (say, of the Earth) have the Fermi momentum $k_F^\nu \sim 1\,PeV$ $(10^{15}\,eV)$ — the "movable" surface neutrinos/antineutrinos.*

We may choose the *cgs* unit system, the length in *cm*, the time in *sec*, and the mass in *gm*, so that the light velocity $c = 3.00 \times 10^{10}\,cm/sec$. Or, for this textbook, we choose the natural units such that $\hbar = c = 1$.

We have to look at the space and the time in particular not too "ideologically". First of all, the space-time changes according to the matter distribution — if we adopt the view from Einstein's general relativity. In other words, the "definition" of the space-time varies with the matter distribution, which could be anything. Then, how do we "define" the matter in this context? It is so easy to fall into arguing in a vicious circle, and we don't get anywhere.

In the Schwarzschild metric given above which is the solution to Einstein general-relativity equation of a "point" mass, the rationale of the "black-hole" solution still stirs up so many puzzles, even until today (for about a hundred years later). So, the mix-up of the "simple" space-time notion with the matter notion sort of drives these concepts with a lot of ambiguities. *Meanwhile, regarding the five-times-in-weight dark matter as neutrino halos while*

[1] *W-Y. Pauchy Hwang, arXiv: 1304.4705v2 [hep.ph] 25 August 2015; "The Standard Model".*

[2] *W-Y. Pauchy Hwang, The Universe,* **3-4***, 7 (2015); ibid,* **4-1***, 7 (2016).*

the neutrino halo, as a Fermi gas obeying Pauli's exclusion principle, cannot collapse into a presumed black hole, we get a side-way outlet to the heaven.

It is clear that we might regard the "space-time" as a whole as the "physical system" of some sort. In the simplest case, the physical system possesses the Lorentz invariance and other symmetries. When we talk about a "point-like" particle in the space-time, we could try to "define" what the "point-like" means in the physical system of the space-time. In other words, the space-time is more physical than mathematical.

In the last part of the book, we shall study the Standard Model of particle physics. Electrons, neutrinos, quarks, etc., the so-called "building block of matter", all are point-like particles - up to the size of $10^{-20}\,cm$ (i.e., the known resolutions), these particles are still point-like.

So, we are studying the behaviors of these "points" in our space-time, physically rather than mathematically, at the scale far small than the atomic size of 10^{-8} centimeter.

§. 0.1. The Point in the Quantum Sense

Everything would begin with the quantum variables $(x, y, z,\ ict)$, a point in the quantum 4-dimensional Minkowski space-time. Because of the quantum principle, $(x, y, z,\ ict)$ is no longer the point in the geometric sense. Because we in fact live in this quantum 4-dimensional Minkowski space-time, we will have enough time to learn this "quantum" "4-dimensional Minkowski space-time", starting from the beginning of the 21st century.

Mathematically, a "point" does not have any size or volume, according to the geometry. When we talk about a "point-like" particle in the physical sense; it should be different from the "point" in the mathematical sense. When we describe a particle by Dirac equation or by Klein-Gordon equation, that does not have the size parameter, we call it "point-like", or "point-like-ness" in the quantum sense.

Quantum mechanics, or more precisely the quantum principle, comes to rescue in a mysterious way. At the scales around $1\,cm$, we deal with the system macroscopically and we developed the classical physics. At the scales of about $10^{-8}\,cm$, we know that the quantum principle has to be there. The uncertainty principle and others, i.e., the manifestations of the quantum principle, are working there, down to the scale of $1\,fm$, or $10^{-13}cm$, or much smaller.

The quantum principle states that the position (x, y, z) does not commute with the momentum $m\frac{d}{dt}(x, y, z)$, with the commutator equaling $i\hbar$. Thus, we have to treat the position (x, y, z) and the momentum $m\frac{d}{dt}(x, y, z)$ as operators when we speak of them simultaneously. $\hbar$ sets the scale when the quantum effects are "visible".

So, what is the concept of "point" in this quantum regime? It is relevant for the atomic size (i.e. $10^{-8}cm$) or for the size of a nucleus (i.e. $10^{-13}cm$). Eventually, it has become rather murky at the distance of $10^{-20}cm$ or shorter. $10^{-20}cm$ is sightly smaller than the resolution set by the Large Hadron Collider (LHC) at Geneva.

We should stress the following basic thinking. When we define the point (x, y, z, ct) via the due measurements, we put in the Lorentz invariance, or the Lorentz group, in the game. We in fact did not insist on the commuting properties of the various variables, such as those among (x, y, z, ct). We do insist the Lorentz symmetry, the property under the Lorentz group, and so on.

When it becomes much smaller, such as $10^{-25} cm$ or smaller, which are so small that we never get there, the coordinates (x, y, z, ct) might not commute among themselves – the so-called "non-commutative geometry". This phenomenon, if it exists, which we call it the "super-quantum regime" – it is rather natural in the line of "reasoning".

This noncommutative aspect among the coordinates was raised by Hartland S. Snyder in as early as 1947 [Phys. Rev. **71**, 38 (1947)]. This aspect was discussed later on in a variety of contexts. The idea could be relevant when we discuss the alternatives of "the point-like structure" or point-like particles.

In any event, it is quite clear that the "point-like" in the physical sense could be so different from the "point" in the mathematical sense. As a physicist, we'd better keep in mind these subtleties, particularly when we "polish" our theories to eventually describe lot's much more peculiar systems or objects.

In what follows, we try to describe "relativistic quantum mechanics and quantum fields" following the tone given above - though a little non-orthodox but maybe more appropriate. We hope that students in the 21st century could benefit from our presentations. In fact we are waiting for these next generations of scientists to come up with the clear picture of the 21st-century physics. We are updating and revising this textbook with an eye on that.

For the students majoring in theoretical physics, the present "advanced quantum mechanics" would be appropriate for their learning, even though we use the examples from particle physics and cosmology. Here we try to define "advanced quantum mechanics" as those concepts which would be mentioned or used beyond "quantum mechanics" — as we will see, quantum fields are "quantum mechanics with an infinite or continuous number of degrees of freedom" while quantum mechanics is the "quantum fields in the zeroth dimension". Today's theoretical physicists should treat these terms in a more precise way.

§. 0.2. The Standard Model of the 20th Century

Quantum mechanics and Einstein's theory of special relativity constitute the basis for the main body of modern physics developed during the first half of the twentieth century (that is, the two pillars of the 20th-Century modern physics). Perhaps, in the 21st-century physics, we might, carefully, introduce the third pillar, the fourth pillar, and so on. Alternatively, following the experience of the Standard Model, it comes to recognize that there exist the so-called "smallest units of matter", such as electrons, neutrinos, quarks, etc. — thus, it is the end of the era to introduce some new principle(s).

Physical systems that can be described by relativistic wave equations include electrons, quarks, and many other elementary particles in the subatomic world. The presentation on the Klein-Gordon equation could be simple since it has the same symmetry as the background, i.e., the 4-dimensional Minkowski space-time. It describes spinless particles such as the Higgs particles, for which the example for its existence was in 2012, and the pions, the composite systems which were discovered in 1940's. So after we explain the axiom box for the meaning of "quantization" on the language, we shall present the Dirac relativistic equation in great details. The Dirac theory for the electron has many successes: It is Lorentz covariant; it contains the electron spin with the gyromagnetic ratio $g = 2$; it predicts the anti-particle positron which is experimentally discovered; it gives the correct fine structure of the hydrogenic atom levels. Application of the Dirac equation to other leptons such as muons and neutrinos, and to quarks in the context of bag models, also leads to quantitative successes. It is clear that, unlike atomic or molecular physics where relativistic effects can often be treated as small perturbations, a suitable introduction to elementary particle physics and field theories must commence with a description of relativistic wave equations.

Early on, there seemed to have "difficulties" too with the Dirac equation. One was that of extending the theory for a single electron to a system of many electrons. Another was of an even more basic nature, namely, a theory started out to represent a single electron ends up being inseparably bound with a many-body effect on account of the infinite sea of electrons in negative energy states — as most physicists thought that way. We resolve, or try to resolve, the first puzzle by realizing that we are dealing with the innumerable infinite degrees of freedom. On resolving the second puzzle, the axiom box of the language, in Chapter One as follows, does not call for the existence of the so-called "electron sea" and especially the so-called "electron sea" does not stop at the minute infinite so that the concept is ill-defined.

The antiparticle of the electron, or the positron, was discovered. It explains why the Dirac theory must have four components — two for the electron and another two for the positron. The electron and the positron are born to be described by the same Dirac equation. There is no such thing which are called as "negative-energy states". The entire things could have been clear from there, then. The notion of "negative-energy states" should not appear, and in fact no need to appear — we could call it "the Conceptual Mistake of the 20th Century Physics" but it was due to the fact that the discovery of the positron happens after the discovery of the Dirac equation for the electron.

Historically, the story is as follows: The theory of quantized electromagnetic fields begins with the work of Dirac in 1927, followed immediately by the work of Jordan and Wigner, and by Fermi in 1930. A breakthrough comes only in the mid 1940's with the work of Tomonaga in Japan and Schwinger, Feynaman, and Dyson in the United States. Although infinities still remain, the theory succeeds in "subtracting" them away in a definite, covariant way so that finite results can be obtained, which have been found to be in excellent agreement with the observed Lamb shifts and the "$g-2$ anomaly". The decade from the mid 1940's to the mid 1950's is a period of fervent studies of quantum electrodynamics, both in further calculations on this "renormalization" theory and in attempts to rid the theory of the infinities (not just to isolate and bury them, so to speak). Till now (2016), there seems not yet a satisfactory solution of this problem and in the meantime physicists have turned their interest to other areas — elementary particles and the nature of their interactions and their unification.[3]

During the decade from the mid 1950's to the mid 1960's, serious attempts were made in searching for an alternative means of describing interactions among elementary particles in terms of the so-called "S-matrix theory", in which one tries to determine the scattering amplitudes, or the S-matrix elements, using general principles such as unitarity and microscopic causality (which lead to dispersion relations) and a minimal set of dynamical

[3] *Dirac believes that in a completely satisfactory theory, infinities should not appear. Since the late 1940's, he set out to re-examine and reformulate classical dynamics and electromagnetic theory with a view to find a different basic theory for quantization. He believes even some new mathematics not yet known may be needed. He expressed his views only occasionally in writing, but much more freely in private discussions.*

assumptions. Altogether in the S-matrix approach, the question concerning the underlying dynamics must be answered, or postulated, before any quantitative predictions can be consistently made. Therefore, the notion of using "gauge field theories"[4] to describe interactions among "building blocks" of mater has scored an amazingly successful comeback since the late 1960's, while progresses in the S-matrix approach, which have since been relatively limited, have become things of the past,at least for the time being.

The development of particle physics of the last half century (since the late 1960's till the present — 2015) consists of several major breakthroughs which culminate in the general acceptance of the $SU(3)_{color} \times SU(2)_{weak} \times U(1)$ gauge field theory of strong, electromagnetic, and weak interactions as the "standard model". The Glashow-Salam-Weinberg (GSW) $SU(2)_{weak} \times U(1)$ gauge theory provides a unified description of electromagnetic and weak interactions. It predicts the existence of neutral weak interactions, the existence of the charm quark, and the existence of weak bosons $W^{\pm}$ and Z^0 (which mediate weak forces), all of which have been substantiated experimentally.[5] In the meantime, the quantized $SU(3)_{color}$ gauge theory, or quantum chromodynamics (QCD), has been established as the candidate theory of strong interactions among quarks and gluons, which are believed to be the building blocks of all observed strongly interacting elementary particles or hadrons. QCD supports the dual picture of considering, e.g., a proton as a collection of almost non-interacting quarks, antiquarks, and gluons at high energies (because of the asymptotically free nature of QCD) and as a system of confined, dressed valence quarks (because QCD is consistent with color confinement).

What is troubling the theoretical physicists over almost a whole century is the occurrence of ultraviolet divergences — see Chapter 10 as example. If we "believe" our Standard Model, the infinite parts should cancel out if all ultraviolet divergences of the same characteristics all are taken account. Though there remain some basic progresses to be made, the answer to this basic question seems to be on the positive side.

In Part B of the present volume, we shall present in some detail quantum electrodynamics (QED), the simplest prototype of all quantized gauge field theories. We also describe the conventional standard model, i.e., QCD and the GSW electroweak theory, and the experimental tests of it. In Part C, we move on to describe the Standard Model of the 21st century. We hope to conclude the book with the real Standard Model, the real final chapter.

§. 0.3. From building blocks of matter to the smallest units of matter

We shall for convenient reference begin with a qualitative summary concerning the subatomic and atomic world. In the minimal Standard Model, building blocks of matter are known to include (a) three generations of fermions, (b) mediators of fundamental interactions, and (c) scalar particles which are responsible for spontaneous symmetry breaking

[4] *QED is the simplest prototype of gauge field theories, in which there exist a class of so-called "gauge transformations" which do not affect physical observables classically. The notion of a gauge transformation will be introduced later and it will become the major theme in the second and third parts of this book.*

[5] *In a short time span of ten years starting from 1976, Nobel prizes have been awarded three times to discoveries related to the GSW electroweak theory: for the discovery of the charm quark, for the successes of the GSW theory, and for the experimental observation of $W^{\pm}$ and Z^0.*

related to the physical vacuum. Specifically, fermions consist of leptons, quarks, and their antiparticles:

Leptons:

$$(e^-,\ \nu_e),\quad (\mu^-,\ \nu_\mu),\quad (\tau^-,\ \nu_\tau),\quad (columns). \tag{6}$$

Quarks:

$$\begin{pmatrix} (u_R,\ d_R), & (c_R,\ s_R), & (t_R,\ b_R), \\ (u_Y,\ d_Y), & (c_Y,\ s_Y), & (t_Y,\ b_Y), \\ (u_B,\ d_B), & (c_B,\ s_B), & (t_B,\ b_B). \end{pmatrix} \quad (columns) \tag{7}$$

Note that leptons include electrons (e^-), muons (μ^-), tau-leptons (τ^-), electronlike neutrino (ν_e), and so on. Quarks come in with six possible flavors: up (u), down (d), charm (c), strange (s), bottom (b), and top (t). A quark of any given flavor is assumed to carry one of three possible colors: red (R), yellow (Y), blue (B), or, x, y, and z with

$$x = (1,\ 0,\ 0),\quad y = (0,\ 1,\ 0),\quad z = (0,\ 0,\ 1),\quad (columns). \tag{8}$$

Quarks are not observed in isolation presumably because color is strictly confined, a property consistent with the conjectured two-phase picture of QCD.

Mediators of fundamental interactions include (1) the photon (γ), which mediates the well-known electromagnetic interaction, (2) three weak bosons ($W^\pm, Z^0$), which mediate charged and neutral weak interactions, and (3) eight gluons, which mediate strong interactions among quarks and antiquarks. Gluons carry one of eight possible octet colors and cannot exist in isolation because of color confinement.

Hadrons, or strongly interacting elementary particles, are by assumption color-singlet, or colorless, composites of quarks, antiquarks, and gluons. The hadrons include mesons, baryons, glueballs, and so on. Pions ($\pi^\pm, \pi^0$), kaons ($K^\pm$, K^0, $\bar{K}^0$), etas (η, η'), rho-mesons ($\rho^\pm, \rho^0$), ψ/J, and upsilons ($\Upsilon, \Upsilon', ...$) all are *mesons* which, at low energies, are believed to be quark-antiquark pairs confined to within the region defined by the meson size. Nucleons (p, n), lambda (Λ), sigmas ($\Sigma^\pm, \Sigma^0$), xi's (Ξ^-, Ξ^0), deltas (Δ^-, Δ^0, Δ^+, Δ^{++}), charmed lambda (Λ_c), and bottomed lambda (Λ_b) all are *baryons* which, at low energies, look like systems of three quarks confined to within the region defined by the baryon size. *Glueballs* are colorless objects consisting of gluons only.

Nucleons, i.e., protons (p) and neutrons (n), are believed to be primary building blocks of the *nuclei* of the various atoms, ranging from the proton itself [as the simplest nucleus], to the deuteron, α, ^{12}C, ^{56}Fe, ^{208}Pb, and even to neutron stars [$A = \infty$ nuclei]. Replacement of a nucleon in an ordinary nucleus by a lambda (Λ) or by a sigma-baryon ($\Sigma^\pm, \Sigma^0$) results in a *hypernucleus*. An *atom* is a system of electrons around an ordinary nucleus, the whole system being often electrically neutral. *Molecules* are built from atoms. All matter observed terrestrially or celestially are believed to be composed of building blocks conjectured in the standard model.

The GSW $SU(2)_W \times U(1)$ electroweak theory invokes the so-called "Higgs mechanism", in which the physical vacuum, or the true ground state, differs from the trivial vacuum [where expectation values of all fields vanish identically]. This theory predicts, among other things, the existence of a Higgs particle, which has not been observed so far. However, many other important predictions of the GSW electroweak theory have been substantiated experimentally, including the existence of: (1) neutral weak interactions, (2) the charm quark, and (3) weak bosons $W^{\pm}$ and Z^0. It is very likely that any new theory which is beyond the standard model must reproduce or explain the successes of the (old) standard model and so will contain the (old) standard model as a limiting case.

Up to the moment of writing, building blocks of matter, as conjectured in the (old) standard model, appear to be structureless (or, less precisely, pointlike) at the highest energy scale (or the smallest distance scale) which we are capable of probing. Qualitatively speaking, quarks and leptons can be described by Dirac equations of some sort while mediators of fundamental interactions (spin-1 particles) and the Higgs particle (spin-0 particle) are described, respectively, by gauge field theories and a generalized Klein-Gordon equation. It is clear that the presentation of relativistic quantum mechanics in Part A is most relevant in the subatomic world. Indeed, it is not clear at all whether a Dirac equation will ever be relevant in the description of a composite spin-$\frac{1}{2}$ system such as a proton or a neutron.

We move on to describe qualitatively the Standard Model after the 21st Century.

§. 0.4. The Standard Model of All Centuries

Our world, or our Universe, is described by the two pillars of the 20th Century, Einstein's relativity principle and the quantum principle. In this world, there exist the smallest units of matter, such as electrons, neutrinos, quarks, etc., a finite number of them.

In the year of 2012, the Standard Higgs particle is finally discovered at CERN, Geneva, Switzerland. Subsequently in 2013, the Nobel prize was given to Englert and Higgs for their realization of the Higgs mechanism (or, the BEH mechanism).[6]

There are two basic implications involved here. After searching for the Higgs or point-like scalar particles for forty years (i.e., almost half a century), the only thing found is the Standard-Model Higgs particle, which represents some constrained existence of the scalar particles, as described by the Klein-Gordon equation. The constraint for the real world is as follows: The general permissible renormalizable lagrangian for a complex scalar field $\phi(x)$ should be $-(\partial_\mu\phi^\dagger)(\partial_\mu\phi) - m^2\phi^\dagger\phi - \lambda(\phi^\dagger\phi)^2$. Here the system would collapse if the dimensionless coupling λ (in the 4-dimensional Minkowski space-time) is negative; and it would be self-repulsive if λ is positive; and it would be meta-stable if λ is zero. In the Standard Model, we advocate that in addition to the Standard-Model Higgs $\Phi(1,2)$ there are also the mixed family Higgs $\Phi(3,2)$ and the purely family Higgs $\Phi(3,1)$ such that each Higgs field is self-repulsive and the mutual interactions are maximally attractive.[7]

The other important implication is that the story should end at the "final" Standard Model — particularly, why there are three generations? This is the task of the real Standard Model.[8] The joke regarding why there is the second electron (the muon) is indeed there.

We could try to summarize the (real) Standard Model as follows: Under two pillars of the 20th Century, i.e., Einstein's relativity principle and the quantum principle, there are the

[6] *The Universe, Vol. 1, No. 4, the Nobel Issue.*

[7] *Thus, we put off or delete the previous presentation of the Klein-Gordon equation, the old Chapter One, noting that we start with Chapter Zero (this chapter).*

[8] *W-Y. Pauchy Hwang, arXiv:1304.4705v2 [hep-ph] 25 Aug 2015; version-1 [hep-ph] 17 April 2013.*

smallest units of matter, such as electrons, neutrinos, and quarks, only a limited number of them, and the smallest units of matter are described by the Standard Model.

Thus, we will use "the basic units of motion for matter" (or, the basic units of matter, for simplicity) when we get to the very details in our description of the "Standard Model".

Instead of "the Standard Model of the 21st Century", we are in fact in the search of "the Standard Model of All Century" — the theory that uses Einstein's relativity principle and the quantum principle as the two basic pillars. The language should replace those in the Newton's classic era of the last three or four centuries.

§. 0.5. The Origin of Mass

Suppose that, before the spontaneous symmetry breaking (SSB), the Standard Model does not contain any parameter that is pertaining to "mass", but, after SSB, all particles in the Standard Model acquire the mass terms as it should — we call it "the origin of mass".[9]

This turns out to be a true statement — a statement that carries many underlying basic meanings.

Note that, at high enough temperature such as in the early Universe, all the mass terms are negligible and could thus be set to zero, but it is different from the stage before the turn-on of the spontaneous symmetry breaking (*SSB*). In terms of the temperature, the *SSB* term is some sort of pure imaginary temperature; in fact, the real world never switches on the "phase transition" which generates the various masses — only the hand of the God can do that for all of us.

The concept of "mass" is rather tricky — we know nothing from Newton's classical mechanics, while, turning to quantum field theory, it is a scalar parameter for the Dirac equation or the mass-squared parameter m^2 for the complex scalar field, etc. In the Standard Model the spontaneous symmetry breaking (*SSB*) is doing everything for us — all the mass parameters are generated through the *SSB* via the dimensionless couplings. Here the "dimensionless-ness" means that it should be determined by the 4-dimensional Minkowski space-time *globally*, rather than by the details of the individual fields.

We can divide the matter world into "the quark world" and "the lepton world". The quark world operates *effectively* in the fermi scale, with one fermi equalling to $10^{-13}\,cm$. It sees the force-fields gauge group $SU_c(3) \times SU_L(2) \times U(1)$. On the other hand, the lepton world operates in the atomic scale, or at the $10^{-8}\,cm$ scale. We believe that it sees the force-fields gauge group $SU_L(2) \times U(1) \times SU_f(3)$, although the family group $SU_f(3)$ is feeble and murky (for seeing). We believe the (123) symmetry since it guarantees the smooth behaviors as the particle distance $r \to 0$ or as the momentum-transfer squared $Q^2 \to \infty$.

In either of the quark world and the lepton world, all the objects are massless before the *SSB* turns on. All the couplings are dimensionless in the quantum 4-dimensional Minkowski space-time. This dimensionless-ness is the basic signature of the (quantum) 4-dimensional Minkowski space-time, implying that everything should be settled from there on.

After the spontaneous symmetry breaking (*SSB*) in the (right) family channel, everything is entitled a mass if necessary. This is so-called "the origin of mass".

[9] *W-Y. Pauchy Hwang, The Universe*, **2-2**, *47 (2014). "The Origin of Mass"*

The *SSB* requires a negative-energy term as the input; besides the *SSB* term, there is nothing which has the dimension in the 4-dimensional Minkowski space-time. It is quite remarkable; the "phase transition" happens in the pure imaginary temperature — only the hand of the God could do that for all of us. In the real world, the dimensionless theory means that there is no phase transition whatsoever.

In fact, to answer the "origin" of mass, it is one basic question of physics, that we should try to answer whenever we could (since we learnt "general physics"). We think that the developments of the Standard Model of particle physics are reaching at the point really close to actually answering this basic question, i.e., what the "mass" is all about.

§. 0.6. The Origin of Fields (Point-Like Particles)

Why do you have to worry about *only three* fields — complex scalar fields, Dirac fields, and gauge fields? It seems to be that nothing other than the three fields would be relevant in the quantum 4-dimensional Minkowski space-time. It is indeed true, as a matter of fact, in the 4-dimensional Minkowski space-time.

In other words, our world, or our Universe, is already *closed* — using the three fields to describe the behaviors of the smallest units of matter, such as electrons, neutrinos, quarks, etc., via the Standard Model after the 21st Century. These all happen in the (quantum) 4-dimensional Minkowski space-time. No more fields (particles) are needed for the closure of our Universe. Our language, logical and mathematical, turns out to be consistent and complete.

We can describe our matter world, the physics of all point-like particles, in terms of quantum fields, the complex scalar fields, the spin-1/2 Dirac fields, and the spin-1 force-fields gauge fields, and nothing more.[10] All these happen in the quantum 4-dimensional Minkowski space-time with the force-fields gauge-structure $SU_c(3)\times SU_L(2)\times U(1)\times SU_f(3)$ built-in from the very beginning. The 3 K cosmic microwave background (CMB) is a good evidence of this force-fields structure built-in at the outset.

Basically, a potential $\lambda(\phi^\dagger\phi)^2$ with a positive λ is repulsive where ϕ is a complex scalar field, something quite close in nature to the 4-dimensional Minkowski space-time. Here λ is dimensionless and is determined solely by the 4-dimensional Minkowski space-time, *not* by the complex scalar field. (We suspect that $\lambda = \frac{1}{8}$.) It is something topological.

Therefore, a single complex scalar field cannot exist by itself. When there are two, or more, such fields which are related, we can introduce the attractive terms just enough to overcome the self-repulsive terms. Among themselves, they are "arranged" so elegantly that all gauge bosons, if necessary, become massive and the remainders of the scalar fields become the Higgs particles. We are talking about the physics among three related Higgs fields: $\Phi(1,2)$ (the Standard-Model Higgs), $\Phi(3,2)$ (the mixed family Higgs), and $\Phi(3,1)$ (the purely family Higgs), with the first label for $SU_f(3)$ while the second for $SU_L(2)$.

In fact, $\lambda(\phi^\dagger\phi)^2$ is the *only* extra "renormalizable" term in the world that consists of the complex scalar fields, the Dirac fields, and the spin-1 force-fields gauge fields.

For the matter world, i.e., the quark world or the lepton world, the Einstein relation $E^2 = \vec{p}^{\,2} + m^2$ has to be linearized $E = \vec{\alpha}\cdot\vec{p} + \beta m$, resulting in the Dirac equation. So far,

[10] *W-Y. Pauchy Hwang, The Universe,* **3-1**, *3 (2015). "The Origin of Fields (point-like Particles)"*

we know that it is unique. We have use the linearized version to describe the motions of matter.

Thus, there is a one-to-one correspondence between the quantum 4-dimensional Minkowski space-time with the observed point-like particles, together with some mathematics.

The word "quantum" in the quantum 4-dimensional Minkowski space-time needs to be further clarified or to be quantitatively "defined". The complex scalar fields, the Dirac fields, or the force-fields gauge fields of the quantum variables in the quantum 4-dimensional Minkowski space-time are all what we need in this world, nothing more (strangely enough, nothing more is needed for our world).

Thus, it seems that we are on the right track in approaching the "truth". It seems that the language consisting of "relativistic quantum mechanics and/or quantum fields" is chosen by the Nature as the final language.

Eventually we shall establish, in this book, that the Standard Model of particle physics is in fact the **direct** consequence of the quantum 4-dimensional Minkowski space-time. We hope that this book provides, for you as the scientist of the 21st Century, a rather convincing proof of that.

§. 0.7. A Few Words About Gravity

The students of this textbook are often curious about the gravitational force, as expressed by the Newton's universal gravitational law or, implicitly, by Einstein's general relativity.

As mentioned earlier, the two basic principles, the two pillars established in the 20th Century, are Einstein's relativity principle and the quantum principle. In addition, there exist the smallest units of matter, such as electrons, neutrinos, quarks, etc. (a finite number of them) and the description of these smallest units of matter is just "the Standard Model".

Conceptually, the recognition of the existence of the smallest units of matter is of fundamental importance. For instance, it is easy to argue that the gravity, the approximate Newton's gravitational law, deals with the force between the two aggregates of 10^{60}, or so, smallest units of matter — thus, complexities and symmetries would dominate the question concerning the origin of gravity.

Looking at a star of five solar mass, we are in fact talking about an aggregate of $10^{24} \cdot 10^{36}$, or 10^{60}, smallest units of matter. It is a gigantic number. So, when we describe the motion of the Earth around the Sun, we deal with the gravitational force between two aggregates of order 10^{60} smallest units of matter. That is what Newton's universal gravitational force involves.

Besides, the dark matter of five times in weight, now thought-to-be neutrino halos, one for each ordinary-matter object (such as the Earth, the Sun, and stars), sneaks in as a re-normalized effect — in modifying the Newton's gravitational constant G_N.

In short, as the students of this textbook, we learn something very important but we have to realize that they used to be the permanent "truths" but they now seem no longer to be the "truth". However, it would be delightful to learn the truth if you realize what the "false" truth should be discarded, from your brain or from the bookshelves.

§. 0.8. Natural Units

The brief discussions in this chapter emphasize that, in this 21st century, we would safely declare that we live in the quantum 4-dimensional Minkowski space-time with the force-fields gauge-group structure $SU_c(3) \times SU_L(2) \times U(1) \times SU_f(3)$ built-in from the very beginning. The 3 K cosmic microwave background (CMB) observed in our Universe is the evidence of this "built-in from the very beginning" assertion.

Thus, the knowledge of relativistic quantum mechanics and quantum fields is what we have to learnt; it would of course be the most relevant in particle physics as well as in nuclear physics. Since the kinetic energy of a particle under investigation is often more important than its rest mass in a typical particle or nuclear physics problem, it is convenient to measure a given velocity in units of the light velocity c.[11]

$$c = 2.9979 \times 10^{10} cm/sec. \tag{9}$$

Similarly, it is convenient to express the action in units of $\hbar c$:

$$\hbar c = 197.33\, MeV - fm, \tag{10}$$

where $1fm \equiv 10^{-13}cm$. The remaining units can be chosen as powers of MeV (or GeV) or as powers of fm or as powers of $seconds$. For instance, a delta width of $110\, MeV$ corresponds to

$$\begin{aligned} & 110MeV \cdot \tfrac{1}{\hbar c} \cdot 10^{13} fm/cm \cdot c \\ = \; & 110MeV \cdot \tfrac{1}{197.33MeV - fm} \cdot 10^{13} fm/cm \cdot 2.9979 \times 10^{10} cm/sec \\ = \; & \{5.98 \times 10^{-24} sec\}^{-1}, \end{aligned} \tag{11}$$

or to a very short lifetime of $5.98 \times 10^{-24} sec$. The system in which quantities are measured in units of c and $\hbar c$ is referred to as "natural units". Customarily, one writes

$$\hbar = c = 1. \tag{12}$$

For problems of present-day particle and nuclear physics, adoption of natural units leads to equations which look simpler than those obtained in ordinary units, although the physical content remains the same. Natural units are used in the present volume. The situation may be considered as different from ordinary quantum mechanics where the role played by $\hbar$ should always be emphasized.

Finally, we note that the metric used by us (Wu and Hwang) is the same as that by W. Pauli, T. D. Lee, H. Primakoff, and others. R. P. Feynman used both notations and vice versa. This is the so-called "old-fashioned" notations. This choice involves the hermitian gamma matrices on the Dirac equation. In fact, one of us (Hwang) began on the notations by Bogoliubov but, later, switched to (the mentor) Henry Primakoff — and fixed from there on. The choice of notations does not affect the physics which we are learning. When there is a need to differentiate the upper and lower indices (such as equations in general relativity), we will make it clear when necessary. (mainly in a couple of appendices throughout this textbook)

$$g_{\mu\nu} = \delta_{\mu\nu}; \quad \mu, \nu \in (1, 2, 3, 4). \tag{13}$$

---

[11] *Note that we do not quote in this book all the significant figures of many basic constants which have been measured with great precision.*

For instance, the four-momentum of a particle is specified by

$$p_\mu = (\mathbf{p}, \mathrm{iE}), \quad \text{with} \quad \mathrm{p}_4 = \mathrm{ip}_0 = \mathrm{iE}. \tag{14}$$

The inner product of two four-vectors A_μ and B_μ is given by[12]

$$A \cdot B = A_\mu B_\mu = \mathbf{A} \cdot \mathbf{B} - \mathrm{A}_0 \mathrm{B}_0. \tag{15}$$

The only complication arises when one tries to take the complex conjugate of a complex four-vector. For instance, we have

$$\xi_\mu^* = (\xi^*, i\xi_0^*), \tag{16}$$

for a complex polarization four-vector $\xi_\mu = (\xi, i\xi_0)$, so that $\xi^* \cdot p = \xi^* \cdot \mathbf{p} - \xi_0^* \mathrm{p}_0$. We shall write out the expressions explicitly in the case that some confusion may arise.

As the other final remark, we note that[13] equations and results in the previous volume entitled "Quantum Mechanics"[14] will be used occasionally in this book. As such referencing occurs, we shall use notations such as "(VIII-58), Vol. I" or "p. 12, Vol. I" where "(VIII-58)" is the equation number and "p.12" indicates the page number.

[12] *As a convention, we use a bold-faced Roman letter or a Greek letter to denote a three-vector in configuration space. The arrow is reserved only for a vector in isospin space or in other internal space.*

[13] *To benefit readers in the area of atomic physics, certain equations (up to Chapter 8) are written in ordinary (m.k.s.a.) units as footnotes with primed equation numbers.*

[14] *Henceforth referred to as "Vol. I", authored by one of us, T.-Y. Wu, and published by World Scientific, 1986.*

Part. A. Relativistic Quantum Mechanics

Chapter 1. The Language

Chapter 2. The Dirac Theory for Free Electrons

Chapter 3. γ-matrices; Helicity; Charge Conjugation

Chapter 4. Transformations of Dirac Equations

Chapter 5. The Dirac Electron in an Electromagnetic Field

1. The Axioms Leading to Quantum Fields

The axiom box illustrating quantum fields is given in Fig. 1. From the mechanics to the field, the degrees of freedom go from the discreteness to the continuum. The jump from the classical systems to the quantum system, as observed by Dirac, is described by the Dirac correspondence principle, in which the algebra of all the classical Poisson brackets can be mapped one-to-one into the algebra of all the commutators (modulated by $i\hbar$). This correspondence is the consistency statement that classical physics and quantum physics are both consistent at the language level, or both are wrong.

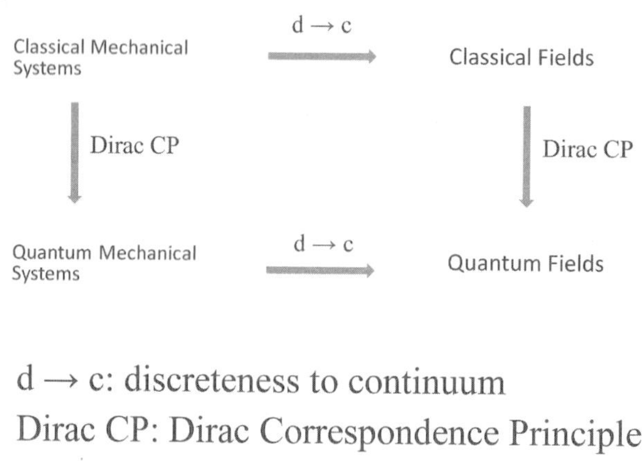

Figure 1: The axiom box.

1.1. The Axioms

We try to briefly summarize the axioms for the four corners of Fig. 1.
Classical Mechanical System: "For a given system, we can find a function (lagrangian) of the coordinates and velocities such that the integral (action) between two instants is an extremum for the real motion."
Quantum Mechanical System: "For the coordinates we can find the conjugate momenta such that the basic (elementary) commutation relations hold." — Now, they are operators.
Classical Fields: "For a given system, we can find a function (lagrangian) of the coordinates and velocities such that the integral (action) between two instants is an extremum for the real motion." (except that quantities take continuum meaning).
Quantum Fields: "For the coordinates we can find the conjugate momenta such that the basic (elementary) commutation relations hold." (except that quantities take continuum

meaning and we also generalize the notion to include fermions i.e., anti-commutation relations).

It is well-known that we have different ways to write these axioms. "For a given system, we can find a function (lagrangian) of the coordinates and velocities such that the integral (action) between two instants is an extremum for the real motion." Here we adopt the lagrangian formalism (of the classical mechanical systems or the classical fields).

Quantum Principle — the most mysterious conjecture. "For the coordinates we can find the conjugate momenta such that the basic (elementary) commutation relations hold." We elevate the coordinates and the momenta from the c-numbers to operators.

We understand that the advanced physics students might have some difficulty in understanding these axioms. In making these axioms including their connections, we have to be precise (and so brief) in the wording. You have to keep these axioms in mind during the entire course and for the future, it is one of our suggestions.

1.2. The Quantum Principle

The state of a quantum mechanical system, as shown by the left bottom corner in Fig. 1 (the axiom box), is described by the state function $\Psi(\vec{x},t)$.

$$i\frac{h}{2\pi}\frac{\partial}{\partial t}\Psi(\vec{x},t) = \hat{H}\psi(\vec{x},t) \tag{1}$$

Observable quantity displays the operator $\hat{Q}$ and the observable expectation is $\langle\hat{Q}\rangle = \int d^3x\Psi^*(\vec{x},t)\hat{Q}\Psi(\vec{x},t)$. The distribution function is $|\Psi(\vec{x},t)|^2$.

Quantum Principle: When we elevate the coordinates and the momenta to operators, the states become c-number functions. Or, all the states form a functional space. This has a basic implication, that the energy spectra must be bounded below.

The system, with the energy spectrum unbound from below, that is, going to minus infinite, cannot be justified as a physical system (and would collapse).

There were so many 20th-century textbooks that introduced the so-called "second quantization", a major misnomer in the 20th century.

Basically, there is a conceptual "paradox" or "contradiction": Einstein's basic relation: $E^2 = m^2 + p^2$, with ($c = 1$). If treated as an operator, it cannot be combined easily with the quantum principle (especially that the energy spectra must be bounded from below to achieve the stability). In other words, Einstein's basic relation should not be regarded as an operator relation.

However, the puzzle of the quadratic nature of Einstein's relation (1905) is in fact resolved in 1928 by Dirac's linearization, $E = \vec{\alpha}\cdot\vec{p} + \beta m$. As used by Newton in classical dynamics, the description of motion of the matter should begin with the linear form in the energy E. Thus, the quadratic nature of Einstein's relation precludes it to serve as the beginning of the description of motion.

The quantum principle sneaks into the game when we agree to adopt the Dirac algebra in our language, noting that they satisfy the anti-commuting relations. This is the quantum principle beyond those implied by the axiom box at the beginning of the Chapter.

The axiom box is of upmost importance, as this is the first time to realize the quantum principle — how the quantum concept is connected with the classical concept. On that, we should acknowledge Paul A.M. Dirac for helping us to begin understanding the mystery of quantization. You should make sure of that you understand the details of the box. To some extent, this axiom box constitutes one of the basic knowledge of physics.

Nevertheless, the introduction of the Dirac equation to describe the electron seems to be something coming from nowhere. It is certainly the quantum principle in operation - but indeed it is something coming from nowhere.

2. Our Universe

In the 20th Century, we have established the validity of the two pillars of modern science, Einstein's relativity principle and the quantum principle. As a matter of fact, we're supposed to give up the language of the Newton's classic era.

At the beginning of the 21st Century, we are firming up the other basic discovery regarding the existence of the *smallest* units of matter, such as electrons, neutrinos, quarks, etc., a finite number of them.

Thus, our Universe is basically the description of the smallest units of matter using the two pillars which we learned in the 20th Century, i.e., Einstein's relativity principle and the quantum principle.

In more precise terms, we declare that we are living in the quantum 4-dimensional Minkowski space-time with the force-fields gauge-group structure $SU_c(3) \times SU_L(2) \times U(1) \times SU_f(3)$ built-in from the very beginning. From our Universe, we can see the quark world, of nuclear sizes, which respects the $SU_c(3) \times SU_L(2) \times U(1)$, or (123), symmetry. From our Universe, we can also see the lepton world, of atomic sizes, which respects the other (123) symmetry $(SU_L(2) \times U(1) \times SU_f(3))$.

The wording follows precisely the Standard Model[1], as the Standard Model was invented to describe our Universe.

A few words are in order. The group characterization, in the description of our Universe, implies that the neutrinos, since they enter the same member of the group as the electrons, must be described by the Dirac equation, like electrons.

The other remark has to with the role of relativistic quantum mechanics, they should be used in the quark world, or in the lepton world. Or, the entire body of nuclear physics, or that of atomic physics, should be described by the relativistic quantum mechanics, rather than by ordinary, or non-relativistic, quantum mechanics. We've made the basic mistake because the predictions of the non-relativistic quantum mechanics are, in most cases, are the same as the approximate relativistic quantum mechanics.

Our Universe is described by the Standard Model; thus, the smallest units of matter, including electrons, neutrinos, quarks, etc., should exhaust all possible smallest units of matter — it seems to be valid in our Standard Model by the reason of mathematical consistency and completeness. Thus, we need to learn how to describe the various smallest units of matter, including electrons, neutrinos, quarks, etc.

[1]W-Y. Pauchy Hwang, arXiv:1304.4705v2 [hep-ph] 25 August 2015; "The Standard Model".

3. The Concepts Developed in the 20th Century

At the classical level, the specific way to determine what field equations are to be used can be summarized by the knowledge on the lagrangian density $\mathcal{L}(x)$ (or the Hamiltonian density $\mathcal{H}(x)$), since all relevant field equations should follow from the principle of least action, Eq. (2), or the variational principle.

$$\delta \int d^4x \mathcal{L}(x) = 0. \tag{2}$$

$$\mathcal{L}(x) = L(u^i(x), \partial_\mu u^i(x)). \tag{3}$$

which depends on x through its dependence on the field functions $u^i(x)$ and on derivatives of the field functions $\partial_\mu u^i(x)$.

$$\begin{aligned} &\delta \int d^4x \mathcal{L}(u^i(x), \partial_\mu u^i(x)) \\ = &\int d^4x \left\{ \frac{\partial \mathcal{L}}{\partial u^i} \delta u^i + \frac{\partial \mathcal{L}}{\partial \partial_\mu u^i(x)} \delta \partial_\mu u^i(x) \right\} \\ = &\int d^4x \left\{ \frac{\partial \mathcal{L}}{\partial u^i} \delta u^i + \frac{\partial \mathcal{L}}{\partial \partial_\mu u^i(x)} \partial_\mu \delta u^i(x) \right\} \\ = &\int d^4x \left\{ \frac{\partial \mathcal{L}}{\partial u^i} - \partial_\mu \frac{\partial \mathcal{L}}{\partial \partial_\mu u^i(x)} \right\} \delta u^i(x) = 0, \end{aligned}$$

or,

$$\frac{\partial \mathcal{L}}{\partial u^i} - \frac{\partial}{\partial x_\mu} \left(\frac{\partial \mathcal{L}}{\partial \partial_\mu u^i(x)} \right) = 0 \tag{4}$$

Eq. (4) is the so-called Euler-Lagrange equation.

3.1. Classical Mechanical Systems Versus Classical Fields

Complex Scalar Field $\phi(x)$

$$\mathcal{L}(x) = -(\partial_\mu \phi^*(x))(\partial_\mu \phi(x)) - m^2 \phi^*(x)\phi(x) \tag{5}$$

For a complex field $\phi(x)$ we may treat $\phi(x)$ and $\phi^*(x)$ as two "independent" field functions.

$$\begin{aligned} \frac{\partial \mathcal{L}}{\partial \phi^*(x)} &= -m^2 \phi(x) \\ \frac{\partial \mathcal{L}}{\partial (\partial_\mu \phi^*)} &= -\partial_\mu \phi(x) \end{aligned}$$

$$\partial_\mu \partial_\mu \phi(x) - m^2 \phi(x) = 0. \tag{6}$$

Classical Mechanics

$$\delta \int dt L(t) = 0, \tag{1'}$$

$$L(t) = L(x_i, \dot{x}_i), \tag{2'}$$

$$\frac{\partial L}{\partial x_i} - \frac{d}{dt}\left(\frac{\partial L}{\partial \dot{x}_i}\right) = 0, \tag{3'}$$

$$L(t) = \Sigma_i \frac{1}{2} m_i \dot{x}_i^2 - V(\{x_i\}), \tag{4'}$$

$$m_i \ddot{x}_i + \frac{\partial V}{\partial x_i} = 0. \tag{5'}$$

$$q_i(t) \to \phi(\mathbf{x}, t) \tag{7}$$

Here $q_i(t)$ ($\equiv x_i(t)$ in the above example) is the generalized coordinates with discrete index i while the field function $\phi(\mathbf{x}, t)$ may be regarded as the generalized coordinates with continuous index $\mathbf{x}$.

So far, we have explained briefly what the classical mechanics system and the classical field are about and their correspondence Eq. (7). That is, we have illustrated the top section of the "language" (in Fig. 1).

The Klein-Gordon equation reads

$$\left(i\frac{\partial}{\partial t}\right)^2 \phi(\mathbf{x}, t) = ((-i\nabla)^2 + m^2)\phi(\mathbf{x}, t) \tag{8}$$

Some people in the 20th century tried to "interpret" the procedure as follows:

$E \to i\frac{\partial}{\partial t}$ and $\mathbf{p} \to -i\nabla$

$E^2 = \mathbf{p}^2 + m^2$ (Einstein's relation)

Here the first equations look like the first "quantization" but, for the consistency of the language (Fig. 1), the Klein-Gordon equation is the field equation (for the classical fields). Similarly the field equation for the Dirac equation for the upper right of Fig. 1.[2]

We could simply introduce the Dirac equation, as a field equation, as follows: $E = \alpha \cdot \mathbf{p} + \beta m$ with α_i and β some generalized numbers. Whether these would be the 4×4 matrices is an irrelevant story, they have to satisfy the "Dirac algebra" (in what follows).

Dirac algebra: $\alpha_i \alpha_j + \alpha_j \alpha_i = 2\delta_{ij}$, $\alpha_i \beta + \beta \alpha_i = 0$, $\beta^2 = 1$, $\alpha^\dagger = \alpha$ and $\beta^\dagger = \beta$

By introducing $\gamma_i \equiv -i\beta\alpha_i$ and $\gamma_4 \equiv \beta$, we may write the Dirac equation as follows:

$$\{\gamma_\mu \partial_\mu + m\}\psi(x) = 0. \tag{9}$$

the Dirac algebra becomes

$$\gamma_\mu \gamma_\nu + \gamma_\nu \gamma_\mu = 2\delta_{\mu\nu}; \qquad \gamma_\mu^\dagger = \gamma_\mu. \tag{10}$$

It may be useful to recall that Pauli matrices $\{\sigma_x, \sigma_y, \sigma_z\}$ satisfy

$$\sigma_i \sigma_j + \sigma_j \sigma_i = 2\delta_{ij}, \tag{11a}$$

[2] *This is one of the most famous misnomers, the so-called "second quantization". As a consistent language, we reject this notion.*

$$\sigma_x = \begin{pmatrix} 0 & 1 \\ 1 & 0 \end{pmatrix}, \sigma_y = \begin{pmatrix} 0 & -i \\ i & 0 \end{pmatrix}, \sigma_z = \begin{pmatrix} 1 & 0 \\ 0 & -1 \end{pmatrix}. \tag{11b}$$

$$\gamma_1 = \begin{pmatrix} 0 & 0 & 0 & -i \\ 0 & 0 & -i & 0 \\ 0 & i & 0 & 0 \\ i & 0 & 0 & 0 \end{pmatrix}, \quad \gamma_2 = \begin{pmatrix} 0 & 0 & 0 & -1 \\ 0 & 0 & 1 & 0 \\ 0 & 1 & 0 & 0 \\ -1 & 0 & 0 & 0 \end{pmatrix},$$

$$\gamma_3 = \begin{pmatrix} 0 & 0 & -i & 0 \\ 0 & 0 & 0 & i \\ i & 0 & 0 & 0 \\ 0 & -i & 0 & 0 \end{pmatrix}, \quad \gamma_4 = \begin{pmatrix} 1 & 0 & 0 & 0 \\ 0 & 1 & 0 & 0 \\ 0 & 0 & -1 & 0 \\ 0 & 0 & 0 & -1 \end{pmatrix}. \tag{12}$$

$$\vec{\gamma} = \begin{pmatrix} 0 & -i\vec{\sigma} \\ i\vec{\sigma} & 0 \end{pmatrix}, \qquad \gamma_4 = \begin{pmatrix} 1 & 0 \\ 0 & -1 \end{pmatrix}. \tag{12'}$$

$$\gamma_5 \equiv \gamma_1\gamma_2\gamma_3\gamma_4 = \begin{pmatrix} 0 & -1 \\ -1 & 0 \end{pmatrix}, \tag{13a}$$

$$\sigma_{\mu\nu} \equiv \frac{1}{2i}(\gamma_\mu\gamma_\nu - \gamma_\nu\gamma_\mu), \tag{13b}$$

$$\sigma_{ij} = \epsilon_{ijk}\begin{pmatrix} \sigma_k & 0 \\ 0 & \sigma_k \end{pmatrix}, \qquad \sigma_{i4} = \begin{pmatrix} 1 & \sigma_i \\ \sigma_i & 0 \end{pmatrix}; \tag{13c}$$

(i, j, k) running from 1 to 3.

$$\psi(x) = u_s(\mathbf{p})e^{ip\cdot x}, \tag{14a}$$

$$\psi(x) = \upsilon_s(\mathbf{p})e^{-ip\cdot x}, \tag{14b}$$

$$(i\gamma \cdot p + m)u_s(\mathbf{p}) = 0, \tag{15a}$$

$$(-i\gamma \cdot p + m)\upsilon_s(\mathbf{p}) = 0. \tag{15b}$$

$$u_s(\mathbf{p}) = N\begin{pmatrix} \chi_s \\ (E+m)^{-1}\vec{\sigma}\cdot\mathbf{p}\chi_s \end{pmatrix}, \tag{16a}$$

$$\upsilon_s(\mathbf{p}) = N\begin{pmatrix} -(|E|+m)^{-1}\vec{\sigma}\cdot\mathbf{p}\chi_s \\ \chi_s \end{pmatrix}, \tag{16b}$$

$\begin{pmatrix} 1 \\ 0 \end{pmatrix}$ (for $s = +1$) or $\begin{pmatrix} 0 \\ 1 \end{pmatrix}$ (for $s = -1$), and N is a normalization constant.

$$N = \left(\frac{|E|+m}{2|E|}\right)^{\frac{1}{2}}, \tag{17a}$$

$$u^\dagger(\mathbf{p})u(\mathbf{p}) = \upsilon^\dagger(\mathbf{p})\upsilon(\mathbf{p}) = 1. \tag{17b}$$

$$\bar{\psi} \equiv \psi^\dagger\gamma_4, \tag{18a}$$

$$\partial_\mu \bar{\psi}\gamma_\mu - m\bar{\psi} \equiv \bar{\psi}(\overleftarrow{\partial} \cdot \gamma - m) = 0. \tag{18b}$$

$$\mathcal{L}(x) = -\bar{\psi}\gamma_\mu \partial_\mu \psi - m\bar{\psi}\psi. \tag{19}$$

$$\frac{\partial \mathcal{L}}{\partial \bar{\psi}} - \frac{\partial}{\partial x_\mu}\left(\frac{\partial \mathcal{L}}{\partial \partial_\mu \bar{\psi}}\right) = -(\gamma_\mu \partial_\mu + m)\psi = 0,$$

$$\frac{\partial \mathcal{L}}{\partial \psi} - \frac{\partial}{\partial x_\mu}\left(\frac{\partial \mathcal{L}}{\partial \partial_\mu \psi}\right) = -m\bar{\psi} + \partial_\mu \bar{\psi}\gamma_\mu = 0,$$

which are just Eqs. (9) and (18b), respectively.

Lorentz four-vector field $A_\mu(x)$ subject to an auxiliary Lorentz condition, $\partial_\mu A_\mu(x) = 0$.

$$\mathcal{L}(x) = -\frac{1}{4}F_{\mu\nu}F_{\mu\nu}, \tag{20}$$

with

$$F_{\mu\nu}(x) \equiv \partial_\mu A_\nu(x) - \partial_\nu A_\mu(x), \tag{21}$$

and

$$\partial_\mu A_\mu(x) = 0. \tag{22}$$

$$\mathcal{L}(x) = -\frac{1}{4}B_{\mu\nu}B_{\mu\nu} - \frac{1}{2}m^2 B_\mu B_\mu, \tag{23}$$

with $B_{\mu\nu} \equiv \partial_\mu B_\nu - \partial_\nu B_\mu$.

$$A_\mu(x) \to A'_\mu = A_\mu(x) - \partial_\mu \Lambda(x), \tag{24}$$

where $\Lambda(x)$ satisfies $\partial_\mu \partial_\mu \Lambda(x) = 0$. It is clear that $\mathcal{L}(x)$ of Eq. (20) is invariant under such transformation. In classical electrodynamics, such (gauge) invariance means that $A_\mu(x)$ can be replaced everywhere by $A'_\mu(x)$ without leaving any trace on the physical observables $\{\mathbf{E}, \mathbf{B}\}$.

We note that a similar gauge transformation for $B_\mu(x)$ will definitively change $\mathcal{L}(x)$ of Eq. (23) into something different. It is the mass term i.e., $-\frac{1}{2}m^2 B_\mu B_\mu$ that destroys gauge invariance.

It is very interesting to note that a massless gauge field has two degrees of freedom (DOF), the spin-one particle has three DOF, while the real Lorentz 4-vector have four — the mismatch is resolved in the way that the massless particle is described by a simple gauge theory while the massive particle is described by the gauge field with spontaneous symmetry breaking (SSB). The gauge theory with SSB, or with the Higgs mechanism, will be described in Part B and Part C of this book.

3.2. Classical Systems Versus Quantized Systems

The conjugate momenta

$$p_i \equiv \frac{\partial L}{\partial \dot{x}_i} \tag{25}$$

and impose the commutation relations:

$$[x_i, x_j] = 0, \qquad [p_i, p_j] = 0 \qquad [x_i, p_j] = i\delta_{ij} \tag{26}$$

3.3. Classical Poisson Brackets Versus Quantum Commutation Relations

Poisson brackets in classical dynamics:

$$\{A, B\}_{P.B.} = \sum_i \left\{ \frac{\partial A}{\partial q_i}\frac{\partial B}{\partial p_i} - \frac{\partial B}{\partial q_i}\frac{\partial A}{\partial p_i} \right\}$$

The elementary commutation relations in QM:

$$[x_i, x_j] = 0, \qquad [p_i, p_j] = 0 \qquad [x_i, p_j] = i\delta_{ij}$$

Dirac Correspondence Principle:

$$\{A, B\}_{P.B.} \to [A, B] \qquad (i\hbar).$$

This is one of the most basic principles.

Eq. (26) is a specific application of the Dirac correspondence principle which defines a quantized physical system out of a classical physical system. Dirac observed that the algebra satisfied by all Poisson brackets for a classical system can be mapped in one-to-one correspondence onto the algebra satisfied by all commutators for the corresponding quantized system. More specifically, the value of the commutator between two quantum observables can be obtained by multiplying that of the corresponding Poisson bracket by the famous factor $i\hbar$. Since all commutators (Poisson brackets) can be expressed in terms of the so-called elementary commutators (elementary Poisson brackets), quantization of a classical system is determined completely by the set of elementary commutators, such as Eq. (26).

$$\begin{aligned} [\phi(\mathbf{x}, t), \phi(\mathbf{y}, t)] &= 0, \\ [\dot{\phi}^*(\mathbf{x}, t), \dot{\phi}^*(\mathbf{y}, t)] &= 0, \\ [\phi(\mathbf{x}, t), \dot{\phi}^*(\mathbf{y}, t)] &= i\delta^3(\mathbf{x} - \mathbf{y}). \end{aligned} \tag{27}$$

$$\pi(\mathbf{x}, t) \equiv \frac{\partial \mathcal{L}}{\partial(\partial_0 \phi(\mathbf{x}, t))} = \dot{\phi}^*(\mathbf{x}, t), \tag{28}$$

$$p^i \equiv \frac{\partial L}{\partial \dot{x}^i} = m_i \dot{x}_i. \tag{28$'$}$$

$$\phi(x) = \int \frac{d^3k}{(2\pi)^{\frac{3}{2}}} \cdot \frac{1}{\sqrt{2k_0}} \{a_+(\mathbf{k}) e^{ik\cdot x} + a_-^*(\mathbf{k}) e^{-ik\cdot x}\}, \tag{29}$$

$$\dot{\phi}^*(x) = \int \frac{d^3k}{(2\pi)^{\frac{3}{2}}} \cdot \sqrt{\frac{k_0}{2}} (+i) \{a_+^*(\mathbf{k}) e^{-ik\cdot x} - a_-(\mathbf{k}) e^{ik\cdot x}\}. \tag{30}$$

$$\begin{aligned} [a_+(\mathbf{k}), a_+^*(\mathbf{k}')] &= \delta^3(\mathbf{k} - \mathbf{k}'), \\ [a_-(\mathbf{k}), a_-^*(\mathbf{k}')] &= \delta^3(\mathbf{k} - \mathbf{k}'), \\ \text{Other elementary commutators} &= 0. \end{aligned} \tag{31}$$

The equivalence between field quantization in configuration space and that in momentum space is dictated by the so-called "canonical transformation" (which preserves the elementary Poisson brackets).

To understand the meaning of the various field operators, we first introduce the vacuum state $|0\rangle$:

$$\begin{aligned} a_+(\mathbf{k})|0\rangle &= 0, \\ a_-(\mathbf{k})|0\rangle &= 0, \\ \langle 0|0\rangle &= 1. \end{aligned} \tag{32}$$

Note that

$$\begin{aligned} &\langle 0|a^+(\mathbf{k}).a_+^*(\mathbf{k}')|0\rangle \\ &= \langle 0|[a_+(\mathbf{k}), a_+^*(\mathbf{k}')]|0\rangle = \delta^3(\mathbf{k}-\mathbf{k}'), \end{aligned}$$

which suggests that $a_+^*(\mathbf{k})|0\rangle$ may be interpreted as a single-particle plane-wave state $(2\pi)^{-\frac{3}{2}}$ $exp(i\mathbf{k}\cdot\mathbf{x} - ik_0t)$. Indeed, we find that, upon quantization of the classical field $\phi(x)$, $a_+^*(\mathbf{k})$ may be interpreted as the creation operator for a particle of three-momentum $\mathbf{k}$ and energy $k_0 = (\mathbf{k}^2+m^2)^{\frac{1}{2}}$ while $a_-^*(\mathbf{k})$ is the creation operator for the corresponding antiparticle of three-momentum $\mathbf{k}$ and $k_0 = +(\mathbf{k}^2+m^2)^{\frac{1}{2}}$. Such interpretation of these operators can be made more explicit by letting the conserved dynamical operators such as the three-momentum $\mathbf{p}$, the energy H, or the charge Q act on the states $a_+^*(\mathbf{k})|0\rangle$ and $a_-^*(\mathbf{k})|0\rangle$.

$$\pi(\mathbf{x},t) = \frac{\partial\mathcal{L}}{\partial(\partial_0\psi)} = +i\bar\psi\gamma_4 = i\psi^\dagger(\mathbf{x},t). \tag{33}$$

$$\begin{aligned} \{\psi_i(\mathbf{x},t), \psi_j(\mathbf{y},t)\} &= 0, \\ \{\psi_i^\dagger(\mathbf{x},t), \psi_j^\dagger(\mathbf{y},t)\} &= 0, \\ \{\psi_i(\mathbf{x},t), i\psi_j^\dagger(\mathbf{y},t)\} &= i\delta_{ij}\delta^3(\mathbf{x}-\mathbf{y}). \end{aligned} \tag{34}$$

$$\psi(\mathbf{x},t) = \int \frac{d^3p}{(2\pi)^{\frac{3}{2}}} \sum_s \{a_s^{(+)}(\mathbf{p})u(\mathbf{p},s)e^{ip\cdot x} + a_s^{(-)*}(\mathbf{p})v(\mathbf{p},s)e^{-ip\cdot x}\}. \tag{35}$$

$$\begin{aligned} \{a_s^{(+)}(\mathbf{p}), a_s^{(+)*}(\mathbf{p}')\} &= \delta_{ss'}\delta^3(\mathbf{p}-\mathbf{p}'), \\ \{a_s^{(-)}(\mathbf{p}), a_s^{(-)*}(\mathbf{p}')\} &= \delta_{ss'}\delta^3(\mathbf{p}-\mathbf{p}'), \\ \text{all other elementary anticommutators} &= 0. \end{aligned} \tag{36}$$

The fact that anti-commutation relations are adopted for quantization of Dirac fields is tied to the Fermi-Dirac statistics assumed for spin-$\frac{1}{2}$ particles [which is reflected to some extent by the anti-commuting properties of γ-matrices, Eq. (10)]. It turns out that the Dirac correspondence principle is applicable even in this case if all Poisson brackets are defined in the anticommuting sense for classical Dirac fields.

As a footnote remark, we note that $a_s^{(+)*}(\mathbf{p}')[a_s^{(-)*}(\mathbf{p}')]$ are to be interpreted as creation operators for particles [antiparticles] of spin s, three-momentum $\mathbf{p}$, and energy E

($\equiv \sqrt{\mathbf{p}^2+m^2}$) while $a_s^{(+)}(\mathbf{p}')$ and $a_s^{(-)}(\mathbf{p}')$ are the corresponding annihilation operators. We wish to emphasize that the concept of a state of negative energy invented for $v-$spinors in relativistic quantum mechanics is somewhat misleading and can be avoided altogether once quantization of classical fields has been carried out.

We wish to remind all the students that the notion of the negative-energy states never appears in our presentations. Dirac was troubled by the fact of his time in obtaining the (Dirac) equation that the antiparticle (positron) was yet to be discovered, so he had the burden to interpret "away" the extra stuffs in his equation.

It is well-known that covariant quantization of the electromagnetic field $A_\mu(\mathrm{x})$ is a non-trivial task. This is due to *both* the presence of a constraint, the Lorentz condition of Eq. (22), *and* the gauge symmetry. The presence of a constraint such as the Lorentz condition implies that there is a redundant degree of freedom. The "generalized" momentum conjugate to the constrained coordinate must vanish identically. Implementation of the constraint at the quantum level often violates manifest Lorentz covariance. On the other hand, gauge symmetry means that two different coordinates $A_\mu(x)$ and $A'_\mu(x)$ are completely equivalent if they are related by a gauge transformation. This causes problem for quantization because we need to select out a set of independent generalized coordinates as the starting point.

The "normal-mode" expansion is given by

$$A_\mu(x) = \int \frac{d^3k}{(2\pi)^{\frac{3}{2}}} \frac{1}{\sqrt{2k_0}} \sum_{\lambda=1}^{2} \{b_\lambda(\mathbf{k})\epsilon_\mu^\lambda(\mathbf{k})e^{ik\cdot x} + b_\lambda^\dagger(\mathbf{k})\epsilon_\mu^{\lambda^*}(\mathbf{k})e^{-ik\cdot x}\}, \tag{37}$$

where λ (=1 or 2) refers to two possible transverse polarizations:

$$\epsilon^\lambda \cdot \epsilon^{\lambda^\dagger} = \epsilon^\lambda \cdot \epsilon^{\lambda *} - \epsilon_0^\lambda \cdot \epsilon_0^{\lambda^*} = 1, \tag{38a}$$

$$k \cdot \epsilon^\lambda = \mathbf{k} \cdot \epsilon - k_0\epsilon_0 = 0. \tag{38b}$$

So, if we have

$$\mathbf{k} = k_0\hat{z}, \tag{39a}$$

then we may choose [with $\pm$ standing for the polarization index λ]

$$\epsilon^{(+)} = -\frac{1}{\sqrt{2}}(\hat{x}+i\hat{y}), \qquad \epsilon_0^{(+)} = 0, \tag{39b}$$

$$\epsilon^{(-)} = +\frac{1}{\sqrt{2}}(\hat{x}-i\hat{y}), \qquad \epsilon_0^{(-)} = 0. \tag{39c}$$

$$\epsilon_\mu \to \epsilon'_\mu = \epsilon_\mu - ck_\mu, \tag{40}$$

$$[b_\lambda(\mathbf{k}), b_\lambda^\dagger(\mathbf{k}')] = \delta_{\lambda\lambda'}\delta^3(\mathbf{k}-\mathbf{k}'),$$
$$\text{all other elementary commutators} = 0. \tag{41}$$

$$b_\lambda(\mathbf{k})|0\rangle = 0. \tag{42}$$

The T-product, or the chronological product, is specified by

$$\begin{aligned} T(\varphi_i(x)\varphi_j(y)) \equiv \quad & \varphi_i(x)\varphi_j(y), \qquad if \quad x_0 > y_0; \\ & \pm\varphi_j(y)\varphi_i(x), \qquad if \quad y_0 > x_0. \\ & (+ \text{ for bosons \& - for fermions.}) \end{aligned} \tag{43}$$

Accordingly, the T-pairing, or the chronological pairing, is defined by

$$\begin{aligned} A_\mu(x;pair)A_\nu(y;pair) &\equiv \langle 0 \mid T(A_\mu(x)A_\nu(y)) \mid 0\rangle \\ &\equiv -i\delta_{\mu\nu}D_0^c(x-y) \\ &= \frac{\delta_{\mu\nu}}{(2\pi)^4 i}\int d^4k e^{-ik\cdot(x-y)}\frac{1}{k^2-i\varepsilon}. \end{aligned} \tag{44}$$

$$\begin{aligned} \psi(x;pair)\bar{\psi}(y;pair) &\equiv \langle 0 \mid T(\psi(x)\bar{\psi}(y) \mid 0\rangle \\ &\equiv -iS^c(x-y) \\ &= \frac{1}{(2\pi)^4 i}\int d^4p e^{ip\cdot(x-y)}\frac{m-i\gamma\cdot p}{m^2+p^2-i\varepsilon}. \end{aligned} \tag{45}$$

Here we use two "pair" to indicate the T-pairing between the two, to replace the standard drawing notation, which is more difficult for the drawing (using the file simplewick.sty).

3.4. Brief Explanation of the Scattering Matrix

$$H = H_0 + H', \tag{46}$$

$$\mid t\rangle = U(t,t_0) \mid t_0\rangle, \tag{47}$$

where $U(t,t_0)$ is referred to as the "evolution operator".

i

$$U(t_2,t_0) = U(t_2,t_1)U(t_1,t_0), \qquad \text{for } \; t_2 > t_1 > t_0. \tag{48a}$$

ii

$$U^{-1}(t_1,t_0) = U(t_0,t_1), \qquad \text{(Existence of the inverse operator)}. \tag{48b}$$

iii

$$U^\dagger(t_1,t_0)U(t_1,t_0) = 1, \qquad \text{(unitarity)}. \tag{48c}$$

iv

$$U(t,t) = 1, \qquad \text{(Existence of the identity operator)}. \tag{48d}$$

We find

$$
\begin{aligned}
i\frac{d}{dt}U(t,t_0) &\equiv i\lim_{\triangle t\to 0}\frac{U(t+\triangle t,t_0)-U(t,t_0)}{\triangle t}\\
&= i\lim_{\triangle t\to 0}\frac{U(t+\triangle t,t)U(t,t_0)-U(t,t_0)}{\triangle t}\\
&= i\left\{\lim_{\triangle t\to 0}\frac{U(t+\triangle t,t)-1}{\triangle t}\right\}U(t,t_0).
\end{aligned}
\tag{49}
$$

We define the interaction Hamiltonian:

$$
H_{int}(t)\equiv i\lim_{\triangle t\to 0}\frac{U(t+\triangle t,t)-1}{\triangle t},
\tag{50}
$$

so that $H_{int}(t)$ characterizes how the system evolves with time at t. It follows that

$$
i\frac{d}{dt}U(t,t_0)=H_{int}(t)U(t,t_0).
\tag{51}
$$

Note that

$$
i\frac{d}{dt}U(t,t_0)\mid t_0\rangle = H_{int}(t)U(t,t_0)\mid t_0\rangle,
$$

or,

$$
i\frac{d}{dt}\mid t\rangle = H_{int}(t)\mid t\rangle,
\tag{52}
$$

which suggests that $H_{int}(t)=H'$ with H' given by Eq. (46).

Eq. (51) can readily be solved:

$$
\begin{aligned}
U(t,t_0) &= 1-i\int_{t_0}^{t}dt_1H_{int}(t_1)U(t_1,t_0)\\
&= 1-i\int_{t_0}^{t}dt_1H_{int}(t_1)\\
&\quad +(-i)^2\int_{t_0}^{t}dt_1\int_{t_0}^{t_1}dt_2H_{int}(t_1)H_{int}(t_2)\\
&\quad +...
\end{aligned}
\tag{53}
$$

Define the chronological or T-product:

$$
T(A(t_1)B(t_2))=\begin{cases} A(t_1)B(t_2) & \text{for } t_1>t_2;\\ B(t_2)A(t_1) & \text{for } t_1<t_2.\end{cases}
\tag{54}
$$

Note that

$$
\begin{aligned}
&\int_{t_0}^{t}dt_1\int_{t_0}^{t_1}dt_2H_{int}(t_1)H_{int}(t_2)\\
=\ &\frac{1}{2}\int_{t_0}^{t}dt_1\int_{t_0}^{t}dt_2T(H_{int}(t_1)H_{int}(t_2)),
\end{aligned}
\tag{55a}
$$

and, more generally,

$$
\begin{aligned}
&\int_{t_0}^{t}dt_1\int_{t_0}^{t_1}dt_2...\int_{t_0}^{t_{n-1}}dt_nH_{int}(t_1)H_{int}(t_2)...H_{int}(t_n)\\
=\ &\frac{1}{n!}\int_{t_0}^{t}dt_1\int_{t_0}^{t}dt_2...\int_{t_0}^{t}dt_nT(H_{int}(t_1)H_{int}(t_2)...H_{int}(t_n)).
\end{aligned}
\tag{55b}
$$

Therefore, Eq. (9) becomes

$$\begin{aligned} & U(t,t_0) \\ = \ & T(\exp(-i)\int_{t_0}^{t} dt H_{int}(t)) \\ \equiv \ & 1+\sum_{n=1}^{\infty}\frac{(-i)^n}{n!}\int_{t_0}^{t} dt_1 \int_{t_0}^{t} dt_2 ... \int_{t_0}^{t} dt_n T(H_{int}(t_1)H_{int}(t_2)...H_{int}(t_n)). \end{aligned} \quad (56)$$

$$S=\lim_{t\to\infty}\ \lim_{t_0\to-\infty} U(t,t_0), \quad (57)$$

$$\begin{aligned} H_{int}(t) &= \int d^3x \mathcal{H}_{int}(x,t) \\ &\equiv -\int d^3x \mathcal{L}_{int}(x,t), \end{aligned} \quad (58)$$

$$S=T(\exp i\int d^4x \mathcal{L}_{int}(x)). \quad (59)$$

$$\langle f \mid S \mid i\rangle = \delta_{fi} + i(2\pi)^4\delta^4(\sum_f p_f - \sum_i p_i)T_{fi}. \quad (60)$$

$$\mid i\rangle = b^+_{s_1}(\mathbf{p}_1)a^+_{s_2}(\mathbf{p}_2)\mid 0\rangle, \quad (61a)$$

$$\mid f\rangle = b^+_{s'_1}(\mathbf{p}'_1)a^+_{s'_2}(\mathbf{p}'_2)\mid 0\rangle. \quad (61b)$$

$$\sigma = \frac{1}{f}\int\frac{d^3p'_1}{P_0}\int\frac{d^3p'_2}{P_0}\bar{\sum}\mid\langle f\mid S\mid i\rangle\mid^2\cdot\frac{1}{VT}, \quad (62)$$

$$\begin{aligned} & \mid (2\pi)^4\delta^4(\sum p_f - \sum p_i)\mid^2 \\ = \ & \mid \int d^4x \exp ix\cdot(\sum p_f - \sum p_i)\mid^2 \\ = \ & \int d^4x \exp ix\cdot(\sum p_f - \sum p_i)\cdot(2\pi)^4\delta^4(\sum p_f - \sum p_i) \\ = \ & VT(2\pi)^4\delta^4(\sum p_f - \sum p_i). \end{aligned} \quad (63)$$

$$d\sigma = \frac{1}{f}(\frac{d^3p'_1}{N_0})(\frac{d^3p'_2}{N_0})(2\pi)^4\delta^4(p'_1+p'_2-p_1-p_2)\bar{\sum}\mid T_{fi}\mid^2. \quad (64)$$

$$N_0=(2\pi)^3. \quad (65)$$

On the flux factor, $f=1$ for a high energy fixed target scattering experiment while $f=2$ for a high-energy collider experiment.

The scattering matrix, or the S-matrix, and the differential cross section $d\sigma$ in the scattering will be used in, e.g., Chapter nine and others in this book. These are basic entities of learning in particle physics (and others).

The lessons in the 20th Century allow us to make predictions from quantum field theory on the scattering problems and from quantum mechanics on the bound-states problems. There are ripples here and there, to be remained to further clarified, mainly on the infinities (i.e., the various ultraviolet divergences). These ripples are the source of why some good theorists lose the confidence in the language. For that reason, we have an expanded section on the material in the 20th Century, such that we could come back to examine if necessary. The newcomer students should be afraid if you gets lost at this point, you would pick up to eventually form a solid knowledge.

4. Renormalizability

If we expand the S-matrix elements and examine the potentially infinities (divergences), we could conclude that, in the 4-dimensional Minkowski space-time, those terms of dimension five or higher would be "non-renormalizable", or making divergences even worst by going to next order in the expansion.

For instance, for the complex scalar field ϕ, we could write the lagrangian,

$$L = -\partial_\mu \phi^\dagger \partial_\mu \phi - M^2 \phi^\dagger \phi - \lambda(\phi^\dagger \phi)^2. \tag{66}$$

In the 4-dimensional Minkowski space-time, this is the only Lagrangian that contains all renormalizable terms.

Maybe we could add parenthetical remarks: M^2 has the dimension mass squared and the field ϕ itself could determine its value. The kinetic energy term and the self-interaction λ term are dimensionless — the kinetic energy term sets the scale and the dimensionless coupling λ may be determined by the 4-dimensional Minkowski space-time *globally*. For $\lambda > 0$, the λ potential-term is repulsive (the system will increase the total energy). The uniqueness of λ should be investigated further.

Thus, the $\phi(x)$ system, if alone, will be repulsive for $\lambda > 0$ and cannot exist, and should collapse if $\lambda < 0$. It would be meta-stable for $\lambda = 0$. Since a field represents a particle in our language, we have to realize that all particles exist in nature.

The group terminology is a little complicated but more clear and precise in what we're talking about. For our Universe, we are living in the quantum 4-dimensional Minkowski space-time with the force-fields gauge-group structure $SU_c(3) \times SU_L(2) \times U(1) \times SU_f(3)$ built-in from the very beginning. We introduce three complex scalar fields $\Phi(1,2)$ (Standard-Model Higgs), $\Phi(3,2)$ (mixed family Higgs), and $\Phi(3,1)$ (pure family Higgs). The dimensionless coupling would be

$$\begin{aligned}
V_{Higgs} = \mu_2^2 \Phi^\dagger(3,1)\Phi(3,1) + \quad & \lambda(\Phi^\dagger(1,2)\Phi(1,2) + cos\theta_P \Phi^\dagger(3,2)\Phi(3,2))^2 \\
& + \lambda(-4cos\theta_P)(\Phi^\dagger(3,2)\Phi(1,2))(\Phi^\dagger(1,2)\Phi(3,2)) \\
& + \lambda(\Phi^\dagger(3,1)\Phi(3,1) + sin\theta_P \Phi^\dagger(3,2)\Phi(3,2))^2 \\
& + \lambda(-4sin\theta_P)(\Phi^\dagger(3,2)\Phi(3,1))(\Phi^\dagger(3,1)\Phi(3,2)),
\end{aligned} \tag{67}$$

under the assumption that there is *only one* dimensionless coupling λ. Here the μ_2^2 term is the ignition term for the spontaneous symmetry breaking (*SSB*).

Even though a particular Higgs field, repulsive in nature, cannot exist, the three Higgs fields could co-exist in view of the mutual maximal attractions. The story behind the various Higgs Higgs (complex scalar) fields is very interesting and of fundamental importance.

As resulting from *SSB* (with $\mu_2^2 < 0$), the family gauge bosons all become massive. Cross-generation neutrino oscillations through $SU_f(3)$ become possible. All other Higgs are massive. The negative-energy input on *SSB* is rather amazing (maybe only through the hand of the God!!)

All these will be discussed in detail near the end of this textbook.

For the spin-1/2 Dirac field, the lagrangian is already saturated - no more renormalizable term could be add to the overall lagrangian. For the spin-1 gauge fields, gauge invariance makes the room even tighter. To sum up, except the terms from the gauge principle (∂_μ replaced by D_μ), no more renormaizable terms are allowed.

Thus, for the three fields, only the $\lambda(\phi^\dagger\phi)^2$ term survives for the game of "renormalizability", no other renormalizable terms.

Moreover, the λ in the 4-dimensional Minkowski space-time is dimensionless; that is, it is a pure number; trying to think of it more, it is the number determined *globally* by the 4-dimensional Minkowski space-time, *not* by the field(s) ϕ itself. Thus, the λ term is something belong to a non-renormalization theorem.

We note that the $\lambda(\phi^\dagger\phi)^2$ interaction is self-repulsive and, by itself, the field *cannot exist* at all.[3] We think that the existence of the renormalizable dimensionless λ term, and its fundamental importance, has been overlooked by the theoretical physicists of the 20th Century.

5. Relativistic Quantum Mechanics

Our book is entitled as "Relativistic Quantum Mechanics and Quantum Fields", promoting relativistic quantum mechanics to the level of quantum field theory. Our Universe is the quantum 4-dimensional Minkowski space-time with the force-fields gauge group structure $SU_c(3) \times SU_L(2) \times U(1) \times SU_f(3)$ built-in from the very beginning. The two pillars of the 20th Century, Einstein's relativity principle and the quantum principle, should be there. So, relativistic quantum mechanics, used to describe the bound-state problems in such as atomic physics, nuclear physics, etc., is the language which we should adopt (though this is at odd with the historical trend).

The reason is rather simple. Einstein's relation of 1905 is quadratic while Dirac's linearization of 1928, a nontrivial discovery, came much later. In order to do the motion problem linearly, as in Newton's classic era, it involves the Dirac linearization of which the algebra is anti-commutative (completely new in history) and in fact it is inherently in accord with the quantum principle.

What we have learnt from the 20th century is the two pillars of modern physics, i.e., Einstein's relativity principle and the quantum principle. As we move into the 21st century,

[3] *W-Y. Pauchy Hwang, The Universe,* **3-1**, *3 (2015).*

we stress that in fact we live in the quantum 4-dimensional Minkowski space-time — a space-time that respects both Einstein's relativity principle and the quantum principle.

How does the 4-dimensional Minkowski space-time satisfy the quantum principle, in loose or precise terms? Since we in fact do not quite understand the quantum principle so far, we have to learnt how the quantum principle is working in details in the specific physical systems, from one system to the others.

Maybe we should cut in by introducing briefly the Standard Model.[4] We propose that we are living in the quantum 4-dimensional Minkowski space-time with the force-fields gauge-group structure $SU_c(3) \times SU_L(2) \times U(1) \times SU_f(3)$ built-in from the very beginning. From this overall background, we can see the quark world of the nuclear sizes satisfying the $SU_c(3) \times SU_L(2) \times U(1)$ symmetry (or, the (123) symmetry), and we can also see the lepton world of the atomic sizes respecting the $SU_L(2) \times U(1) \times SU_f(3)$ symmetry or the another (123) symmetry.

The Standard Model describes electrons, neutrinos, quarks, etc., the so-called "the smallest units of matter". Our world, or our Universe, do have the smallest units of matter; these units become the "building blocks" of matter. So, the smallest units of matter and how they interact with one another (via the Standard Model) is something which we should add to the two pillars of the 20th-Century physics — Einstein's relativity principle and the quantum principle.

The force-fields gauge-group structure is built in in the very beginning (at the same time as the quantum 4-dimensional Minkowski space-time). This is why we observe the almost-uniform 3 K cosmic microwave background (CMB). Thus, we should anticipate to observe cosmic background (CB) ν's since they were manufactured abundantly during our early Universe.

Thus, in our Universe, there are a lot of CMB and CB ν's. We already studied about CMB, which are Bosons (massless photons). The CB ν's are Fermions, of some tiny masses, which should be clustered, in many neutrino halos. Each neutrino halo should have a Fermi-Dirac sphere, which is incompressible due to Pauli's exclusion principle.

So, our Universe is incompressible due to the neutrino halos. These neutrino halos are identified as the 25% dark matter. The incompressibility of the neutrino halos should be regarded as an application of the quantum principle.

We could add a little more on the quantum principle: We could argue that the observable such as Δx, Δy, Δz, or $c\Delta t$ is obtained from the "quantum" measurement or, it is a "quantum" observable. This measurement might use the "classical" instrument, where we could quantify the accuracy in some way (which we believe). Once we have (x, y, z, ict), we could proceed to write down the field functions, etc. In this way, we define the quantum measurements in the "operational" point of view, though there would be some loose ends to tie up. Is this "classical instrument" for the quantum measurement a necessity or not? The quantum principle provides us some murky guideline.

Atoms, nuclei, ions, the positronium, muonic atoms, and others are systems that satisfy both Einstein's relativity principle and the quantum principle, in fact, all should be so since we live in the quantum 4-dimensional Minkowski space-time, although historically we started with the Newton's classical concepts in this thinking.

---

[4] *W-Y. Pauchy Hwang, arXiv:1304.4705v2 [hep.ph] 25 August 2015; "The Standard Model".*

In atoms, with the hydrogen as the simplest example, we can easily separate the center-of-mass (CM) coordinates from the (internal) relative coordinates, the size of the relative coordinates of about $10^{-8}\,cm$ or one anstron. We do not know if this separation of the CM and relative coordinates is legitimate in the Minkowski space-time, we just assume that it is so.

In nuclei, with the proton the simplest example and the deuteron the simplest complex-nucleus example, we again could separate the CM coordinates from the internal degrees of freedom, the sizes of the internal degrees of freedom of about $10^{-13}\,cm$, or one Fermi that is much smaller than the typical atomic size.

For any of these systems, we separate the center-of-mass (CM) coordinates from the internal coordinates, including the (internal) relative time in the quantum 4-dimensional Minkowski space-time. The internal part would be in the atomic size for an atom, or, in the $fermi^3$ size for a nucleus. Thus, we could define a system of "relativistic quantum mechanics" in order to describe the internal motion of the system.

For these systems, the Lorentz symmetry, or symmetries under Lorentz transformations, applies on the CM coordinates. The internal coordinates are also of the relativistic origin and of the quantum origin. We somehow were educated in the non-relativistic way, so we have gained some non-relativistic intuitions that might need some major revisions.

The physics related to the internal coordinates can be referred to as "relativistic quantum mechanics (RQM)", a branch of the knowledge which is often mistaken as the sideline of ordinary quantum mechanics. We should argue that "relativistic quantum mechanics" should be the orthodox, with its approximation which we call "non-relativistic quantum mechanics" or "ordinary quantum mechanics".

We could talk about the two-body system a lot more. Let the space-time coordinates be (x_1, y_1, z_1, ct_1) and (x_2, y_2, z_2, ct_2). The separation of the center-of-mass (CM) coordinates and the internal coordinates is carried as follows:

$$\begin{aligned} &R_x = \tfrac{Mx_2+mx_1}{M+m}, \qquad R_y = ..., \qquad T = \frac{Mt_2+mt_1}{M+m}; \\ &r_x = x_1 - R_x, \qquad r_y = ..., \qquad t = t_1 - T. \end{aligned} \tag{68}$$

For example, in a hydrogen atom, M (the nucleus) is big (heavy) and m (the electron) is small. So, the CM coordinates would be pretty much fixed (in the spatial position) and the electron is circling around the nucleus.

As another example, in a positronium (e^+e^-) system, the CM coordinates would be in the center of the system and the electron would be in one side of the system.

These two-body systems constitute the entire body of atomic physics, of nuclear physics, of many others (we cannot ignore these branches of physics, even though they seem to be slightly different from the standard description of quantum field theory (QFT). We think that "relativistic quantum mechanics" and "quantum fields" is the two sides of a single coin).

The entire body of atomic physics, and of chemistry, were written in non-relativistic (ordinary) quantum mechanics — basically, a conceptual mistake because of the Newton's classic era. If we have chance in understanding the two pillars of modern physics, relativistic quantum mechanics should be there as the proper language. Even though most of time relativistic quantum mechanics differs slightly from ordinary quantum mechanics, we should try to write up the 21st century physics as such, everything should be understood from the two pillars of modern physics, Einstein's relativity principle and the quantum principle.

6. The Declaration

We declare that in fact we are living in the quantum 4-dimensional Minkowski space-time with the force-fields gauge-group $SU_c(3) \times SU_L(2) \times U(1) \times SU_f(3)$ structure built-in from the very beginning.

First of all, these things including us, being living in the quantum 4-dimensional Minkows space-time, should in fact be described, in the language, in terms of those in the quantum 4-dimensional Minkowski space-time. Note that, first, why is the reason that we use non-relativistic descriptions is purely historical, that we started from Newton's classic era, etc., we took the wrong path. Second, all of us do not quite understand the quantum principle.

We would like to start the 21st-century textbook on "Relativistic Quantum Mechanics and Quantum Fields" by saying that we live in the quantum 4-dimensional Minkowski space-time and organizing everything according to that. The description of electrons in bound states using the Dirac equation and, if not, we should check the "non-relativistic approximations" that would give the relativistic results, should be something that the 21st-century students are supposed to do.

As a matter of fact, we believe Einstein's relativity principle and the quantum principle, in the 21st Century and many centuries to follow up. The same is true for the existence of the smallest units of matter. Occasionally, we try to emphasize the permanence of the Standard Model by saying it of All Centuries. We think that this is something to replace Newton's classic doctrine, though part of the truth still remains to be slightly vague.

To say one more thing, the separation of the center-of-mass (CM) coordinates in fact might not mingle well with the idea of the 4-dimensional Minkowski space-time. The positive side is that we have a lot to think about. The negative side is that it is easy to trap ourselves into the deep hole in the concepts.

Chapters 2-5 are relativistic quantum mechanics (RQM) examples which we could learn from the 20th-century textbooks, mainly about the Dirac equation. We keep most of these chapters, from the previous edition by Wu and Hwang. Note that thoughts following Dirac, Einstein, etc. deserve a lot of second thoughts. Some of these should not be discarded until the theory is clearly written (and is invented by a lot of future Dirac's).

In fact, Chapters 2-5 are elementary treatments of the Dirac equation, Chapters 2-4 starting from the plain-wave solutions to the general concepts and Chapter 5 for the hydrogen-like atoms. Our suggestion is that we should avoid using the old terminology "negative-energy solutions", because it confuses the concepts and, if we are careful enough, the "negative-energy solutions" can be replaced everywhere by the solutions of the antiparticles. In fact, it is consistent to do so, as the language is supposed to be consistent. Typo-wise, I might call it as "the 20th-century typo mistake". Only when you encountered the Dirac equation but you did not observe the antiparticle, that you, like Dirac at first did this, would introduce the negative-energy stuff. Unfortunately, this conceptual mistake has been so deep-rooted that we always issue the warning about the absence of "negative-energy solutions", for the clarity of the 21st-century minds.

For a lot of reasons, unspoken or to be spoken, the solutions of the Dirac equation are fitted (or accommodated) nicely into the 4-dimensional Minkowski space-time. The four-component nature of these solutions appears quite naturally, explaining the two different "spins" and the particle-antiparticle nature.

Yes; they touch upon the origins of these concepts. Any deep insights on this would offer the origin of "spin" and the origin of the particle-antiparticle nature. "Spin" is something demanded by the "transition" from the Einstein relation to Dirac's learization, although historically the Nobel physics was awarded in 1945 to Pauli for Pauli spin.

Since the certain solutions of the Dirac equation can be accommodated into the 4-dimensional Minkowski space-time, it implies that the boundary conditions are patched up nicely for atoms and nuclei, etc., and it may also imply that some topological constraints, so far unspoken, are in place. These questions are the 21st-century questions that our 21st-century scientists would face and eventually would provide the answers - the purpose of our efforts to update this textbook on "Relativistic Quantum Mechanics and Quantum Fields". Here we refer to "scientists" rather than purely "physicists", for obvious reasons.

What is really the quantum 4-dimensional Minkowski space-time? In fact, we do not know exactly the detailed answers. It seems, from the observed 3 K cosmic microwave background (CMB), that the force-fields gauge-group structure is born with this space-time so, the Lorentz group and the force-fields gauge group are borne together.

The spin of the electron emerges from the Dirac equation (for the electron) in our Universe with the 4-dimensional Minkowski space-time. It is no longer there if, instead of the 4-dimension, you consider 3-dimension, 5-dimension, or others - that is, spin should be born with the Dirac equation in the 4-dimensional Minkowski space-time. Only the real history made "the spin" a deep mystery.

Everything starts from the point-like description of the quantum 4-dimensional Minkowsk space-time, from complex scalar fields, spin-one force-fields gauge fields, and spin-1/2 Dirac fields; they are functions of the points (or point-like points) in the quantum 4-dimensional Minkowski space-time. It is a big surprise to know that there are no others, eventually we may be able to prove that (i.e., nothing more).

Appendix: Introduction on the Curved Space-Time

General relativity equates concepts in differential geometry (in mathematics) to those in quantum field theory (in physics), a miracle in the early 20th century. In this 21st century, this equivalence would be taken for granted while both of differential geometry and quantum field theory belong to branches of the fully-developed knowledge.

We adopt the textbook by Bernard F. Schutz, entitled "A first course in general relativity", to illustrate curved space-time for general relativity. Thus, in this appendix, the subscripts and the superscripts do have their own meanings — so differing from all the main texts (in which, for instance, the Lorentz indices appear only as subscripts).

Using the 2-dimensional manifold as example, we write

$$\begin{aligned} &\hat{x} = \vec{e}_x, \ \hat{y} = \hat{e}_y \\ &\hat{e}_r = \cos\theta \vec{e}_x + \sin\theta \vec{e}_y \\ &\hat{e}_\theta = -r\sin\theta \vec{e}_x + r\cos\theta \vec{e}_y \end{aligned}$$

These are the "basis vectors".

$$g(\vec{e}_\alpha, \vec{e}_\beta) = \delta_{\alpha\beta} \text{ for } (\vec{e}_x, \vec{e}_y)$$

$$\begin{aligned} g_{\alpha'\beta'} &= g(\vec{e}_{\alpha'}, \vec{e}_{\beta'}) = \vec{e}_{\alpha'} \cdot \vec{e}_{\beta'} \\ &\Rightarrow g_{rr} = 1, \ g_{\theta\theta} = r^2, \ g_{r\theta} = 0 \end{aligned}$$

metric tensors

$$\begin{aligned} ds^2 &= |dr\vec{e}_r + rd\theta\vec{e}_\theta|^2 \\ &= dr^2 + r^2 d\theta^2 \end{aligned}$$

which is used to describe the "infinitesimal distance".

$$\begin{aligned} \text{Vector} \quad & \vec{V} = V^\alpha \vec{e}_\alpha \\ \frac{\partial \vec{V}}{\partial x^\beta} &= \frac{\partial V^\alpha}{\partial x^\beta}\vec{e}_\alpha + V^\alpha \frac{\partial \vec{e}_\alpha}{\partial x^\beta} \\ & \frac{\partial \vec{e}_\alpha}{\partial x^\beta} \equiv \Gamma^\mu_{\alpha\beta}\vec{e}_\mu \text{ Christoffer symbols} \end{aligned}$$

$$\begin{aligned} \Rightarrow \frac{\partial \vec{V}}{\partial x^\beta} &= (\frac{\partial V^\alpha}{\partial x^\beta} + V^\mu \Gamma^\alpha_{\mu\beta})\vec{e}_\alpha \\ &\equiv (V^\alpha{}_{,\beta} + V^\mu \Gamma^\alpha_{\mu\beta})\vec{e}_\alpha \\ &\equiv V^\alpha{}_{;\beta}\vec{e}_\alpha \end{aligned}$$

The "covariant derivative" is defined by

$$(\vec{\nabla}\vec{V})^{\alpha}{}_{\beta} \equiv V^{\alpha}{}_{;\beta} \qquad \binom{1}{1} \text{ tensor}$$

The divergence of a vector is given by $= V^{\alpha}{}_{;\alpha}$.
The tensors of higher ranks are

$$\begin{aligned}
P_{\alpha;\beta} &= P_{\alpha,\beta} - P_{\mu}\Gamma^{\mu}{}_{\alpha\beta} \\
\nabla_{\beta}T_{\mu\nu} &= T_{\mu\nu,\beta} - T_{\alpha,\nu}\Gamma^{\alpha}{}_{\mu\beta} - T_{\mu\alpha}\Gamma^{\alpha}{}_{\nu\beta} \\
\nabla_{\beta}A^{\mu\nu} &= A^{\mu\nu}{}_{,\beta} + A^{\alpha\nu}\Gamma^{\mu}{}_{\alpha\beta} + A^{\mu\alpha}\Gamma^{\nu}{}_{\alpha\beta}
\end{aligned}$$

$$\begin{aligned}
&\Gamma^{\gamma}{}_{\beta\mu} = \frac{1}{2}g^{\alpha\gamma}(g_{\alpha\beta,\mu} + g_{\alpha\mu,\beta} - g_{\beta\mu,\alpha}) \\
&\text{or} \\
&\Gamma^{\alpha}{}_{\mu\nu} = \frac{1}{2}g^{\alpha\beta}(g_{\beta\mu,\nu} - g_{\mu\nu,\beta} + g_{\nu\beta,\mu}) \\
&\Rightarrow g_{\alpha\beta;\gamma} = 0, \qquad \text{in any "coordinate" system}
\end{aligned}$$

The "Riemann curvature tensor" is

$$\begin{aligned}
R^{\alpha}{}_{\beta\mu\nu} &\equiv \Gamma^{\alpha}{}_{\beta\nu,\mu} - \Gamma^{\alpha}{}_{\beta\mu,\nu} + \Gamma^{\alpha}{}_{\sigma\mu}\Gamma^{\sigma}{}_{\beta\nu} - \Gamma^{\alpha}{}_{\sigma\nu}\Gamma^{\sigma}{}_{\beta\mu} \\
&= \frac{1}{2}g^{\alpha\sigma}(g_{\sigma\nu,\beta\mu} - g_{\sigma\mu,\beta\nu} + g_{\beta\mu,\sigma\nu} - g_{\beta\nu,\sigma\mu})
\end{aligned}$$

$$\begin{aligned}
&\text{flat manifold} \Leftrightarrow R^{\alpha}{}_{\beta\mu\nu} = 0 \\
&R_{\alpha\beta\mu\nu} = -R_{\beta\alpha\mu\nu} = -R_{\alpha\beta\nu\mu} = R_{\mu\nu\alpha\beta} \\
&R_{\alpha\beta\mu\nu} + R_{\alpha\nu\beta\mu} + R_{\alpha\mu\nu\beta} = 0
\end{aligned}$$

The Bianchi identities are

$$R_{\alpha\beta\mu\nu;\lambda} + R_{\alpha\beta\nu\lambda;\mu} + R_{\alpha\beta\lambda\mu;\nu} = 0$$

Ricci tensor: $R_{\alpha\beta} \equiv R^{\mu}{}_{\alpha\mu\beta} = R_{\beta\alpha}$

Ricci scalar: $R \equiv R_{\alpha}{}^{\alpha}$

Einstein tensor: $G_{\alpha\beta} = R_{\alpha\beta} - \frac{1}{2}g_{\alpha\beta}R$
$G^{\alpha\beta}{}_{;\beta} = 0$

Einstein equation; $G^{\alpha\beta} = 8\pi G_N T^{\alpha\beta}$

The First Example: The Robertson-Walker metric to describe the early Universe.

$$
\begin{aligned}
ds^2 &= -dt^2 + R^2(t)\left\{\frac{dr^2}{1-kr^2} + r^2 d\theta^2 + r^2\sin^2\theta d\phi^2\right\} \\
\Rightarrow \quad & R_{00} = 3\frac{\ddot{R}}{R} \\
& R_{ij} = -g_{ij}\left\{\frac{\ddot{R}}{R} + 2\frac{\dot{R}^2}{R^2} + \frac{2k}{R^2}\right\}
\end{aligned}
$$

Perfect fluid

$$
T_{\mu\nu} = \begin{pmatrix} \rho & 0 & 0 & 0 \\ 0 & p & 0 & 0 \\ 0 & 0 & p & 0 \\ 0 & 0 & 0 & p \end{pmatrix}
$$

$$
R = -6\left\{\frac{\ddot{R}}{R} + \frac{\dot{R}^2}{R^2} + \frac{k}{R^2}\right\}
$$

$$
\begin{aligned}
G_{\mu\nu} &= -8\pi G_N T_{\mu\nu} \\
\Rightarrow \quad & \frac{\dot{R}^2}{R^2} + \frac{k}{R^2} = +\frac{8\pi}{3}G_N\rho \\
& 2\frac{\ddot{R}}{R} + \frac{\dot{R}^2}{R^2} + \frac{k}{R^2} = -8\pi G_N p
\end{aligned}
$$

The Second Example: Used to describe the stable spherical stellar systems.

$$
\begin{aligned}
ds^2 &= -e^{2\phi(r)}dt^2 + e^{2\Lambda(r)}dr^2 + r^2(d\theta^2 + \sin^2\theta d\phi^2) \\
\Rightarrow \quad & \Gamma^t{}_{tr} = -\phi',\ \Gamma^r{}_{tt} = -\phi'\exp(2\phi - 2\Lambda) \\
& \Gamma^r{}_{rr} = \Lambda',\ \Gamma^r{}_{\theta\theta} = -r\exp(-2\Lambda) \\
& \Gamma^r{}_{\phi\phi} = -r\sin^2\phi\exp(-2\Lambda) \\
& \Gamma^\theta{}_{r\theta} = \Gamma^\phi{}_{r\phi} = r^{-1},\ \Gamma^\theta{}_{\phi\phi} = -\sin\theta\cos\theta \\
& \Gamma^\phi{}_{\theta\phi} = \cot\theta \\
\Rightarrow \quad & R_{trtr} = [\phi'' - (\phi')^2 - \phi'\Lambda']\exp(2\phi), \\
& R_{trt\theta} = R_{trt\phi} = R_{trr\theta} = R_{trr\phi} = 0, \\
& R_{t\theta t\theta} = -r\phi'\exp(2\phi - 2\Lambda), \\
& R_{t\theta t\phi} = R_{t\theta r\theta} = R_{t\theta r\phi} = R_{t\theta\theta\phi} = 0, \\
& R_{t\phi t\phi} = -r\phi'\sin^2\theta\exp(2\phi - 2\Lambda), \\
& R_{t\phi r\theta} = R_{t\phi r\phi} = R_{t\phi\theta\phi} = 0,\ R_{r\theta r\phi} = R_{r\theta\theta\phi} = 0, \\
& R_{r\theta r\theta} = r\Lambda',\ R_{r\phi r\phi} = r\sin^2\theta\Lambda',\ R_{\theta\phi\theta\phi} = \frac{\sin^2\theta}{r^2}[1 - e^{-2\Lambda}], \\
& R_{r\phi\theta\phi} = 0
\end{aligned}
$$

$$\begin{aligned}
R_{rr} &\equiv R^{\alpha}{}_{r\alpha r} = R^{t}{}_{rtr} + R^{\theta}{}_{r\theta r} + R^{\theta}{}_{r\phi r} \\
&= -\left\{\phi'' - (\phi')^2 - \phi'\Lambda'\right\} + \frac{2}{r}\Lambda' \\
R_{tt} &= R^{r}{}_{trt} + R^{\theta}{}_{t\theta t} + R^{\phi}{}_{t\phi t} \\
&= \left\{\phi'' - (\phi')^2 - \phi'\Lambda'\right\} e^{2\phi-2\Lambda} - 2\frac{\phi'}{r}e^{2\phi-2\Lambda} \\
R_{\theta\theta} &= R^{t}{}_{\theta t\theta} + R^{r}{}_{\theta r\theta} + R^{\phi}{}_{\theta\phi\theta} \\
&= +r\phi' e^{-2\Lambda} + r\Lambda' e^{-2\Lambda} + (1 - e^{-2\Lambda}) \\
R_{\phi\phi} &= R^{t}{}_{\phi t\phi} + R^{r}{}_{\phi r\phi} + R^{\theta}{}_{\phi\theta\phi} \\
&= +r\phi' \sin^2\theta e^{-2\Lambda} + r\Lambda' \sin^2\theta e^{-2\Lambda} + \sin^2\theta(1 - e^{-2\Lambda})
\end{aligned}$$

$$R = -2\left\{\phi'' - (\phi')^2 - \phi'\Lambda'\right\} e^{-2\Lambda} + \frac{4}{r}(\phi' + \Lambda')e^{-2\Lambda} + \frac{2}{r^2}(1 - e^{-2\Lambda})$$

Here we have the sign discrepancy in $-\phi'^2$ (the 1st term).

$$\begin{aligned}
G_{00} &= R_{tt} - \frac{1}{2}g_{tt}R = \frac{2}{r}\Lambda' e^{2\phi-2\Lambda} + \frac{1}{r^2}e^{2\phi}(1 - e^{-2\Lambda}) \\
G_{rr} &= -\frac{2}{r}\phi' - \frac{1}{r^2}(e^{2\Lambda} - 1) \\
G_{\theta\theta} &= \left\{\phi'' - (\phi')^2 - \phi'\Lambda' - \frac{1}{r}(\phi' + \Lambda')\right\} r^2 e^{-2\Lambda} \\
G_{\phi\phi} &= \left\{\phi'' - (\phi')^2 - \phi'\Lambda' - \frac{1}{r}(\phi' + \Lambda')\right\} r^2 \sin^2\theta e^{-2\Lambda}
\end{aligned}$$

$$T_{\mu\nu} = \begin{pmatrix} \rho(r) & 0 & 0 & 0 \\ 0 & p(r) & 0 & 0 \\ 0 & 0 & p(r) & 0 \\ 0 & 0 & 0 & p(r) \end{pmatrix} \Rightarrow T^{\mu}{}_{\nu} = \begin{pmatrix} -\rho & 0 & 0 & 0 \\ 0 & p & 0 & 0 \\ 0 & 0 & p & 0 \\ 0 & 0 & 0 & p \end{pmatrix}$$

$$\Rightarrow \tfrac{2\Lambda'}{r} + \tfrac{1}{r^2}(e^{+2\Lambda} - 1) = 8\pi G_N \rho e^{-2\phi+2\Lambda} \tag{1}$$

$$-\tfrac{2\phi'}{r} + \tfrac{1}{r^2}(e^{2\Lambda} - 1) = -8\pi G_N p \tag{2}$$

$$\phi'' - (\phi')^2 - \phi'\Lambda' - \tfrac{1}{r}(\phi' + \Lambda') = 8\pi G_N p e^{2\Lambda} \tag{3}$$

We could use the black-hole solution as the special case for checking:

$$e^{2\Lambda} = e^{-2\phi} = \frac{1}{1 - \frac{2GM}{r}}$$

The summary on the stellar metric:

$$ds^2 = -e^{2\phi(r)}dt^2 + e^{2\Lambda(r)}dr^2 + r^2(d\theta^2 + \sin^2\theta d\varphi^2) \tag{4}$$

Einstein Equations:

$$\frac{2\Lambda'}{r} + \frac{1}{r^2}(e^{2\Lambda} - 1) = 8\pi G_N \rho e^{2\Lambda} \tag{5}$$
$$-\frac{2\phi'}{r} + \frac{1}{r^2}(e^{2\Lambda} - 1) = -8\pi G_N p e^{2\Lambda} \tag{6}$$
$$\phi'' + (\phi')^2 - \phi'\Lambda' + \frac{1}{r}(\phi' - \Lambda') = 8\pi G_N p e^{2\Lambda} \tag{7}$$

...

As it can be checked by setting

$$2\phi = -2\Lambda = \ell n(1 - \frac{2GM}{r}) \tag{8}$$

Thus, the Schwarzschild metric is indeed the solution of the Einstein equation.

We mentioned, at the beginning of the book (Chapter Zero), that we may cast the Schwarzschild metric in the form:

$$(ds)^2 = -(cdt^*)^2 + (dr^*)^2 + r^2 d^2\Omega \tag{9}$$

indicating that, inside the horizon, r^* and t^* become purely imaginary and outside $(-\infty, +\infty$ (outside the domain of "observation").

In the following we appendix the energy-momentum tensor for the ideal fluid:

$$\begin{aligned} T^{\mu\nu} &= (\rho + p)u^\mu u^\nu + pg^{\mu\nu} \\ &u^r = 0,\ u^\theta = 0,\ u^\phi = 0 \\ &-1 = \vec{u} \cdot \vec{u} \Rightarrow u^t = \frac{dt}{d\tau} = e^{-\phi} \\ &\Rightarrow T^{00} = \rho e^{-2\phi}, \quad T^{rr} = pe^{-2\Lambda} \\ &T^{\theta\theta} = pr^{-2}, \quad T^{\phi\phi} = pr^{-2}\sin^{-2\theta} \end{aligned} \tag{10}$$

Exercises: *Chapter 1*

1. Try to derive the T-pairing for the Dirac field and the electromagnetic field:

(i) The propagator for the Dirac field, Eq. (45).

(ii) The propagator for the electromagnetic field, Eq. (44). Try to discuss the loss of the Lorentz-covariant form.

This exercise re-appears in Chapter 8, indicating the importance of the underlying mathematics. The appearance of the $\pm i\epsilon$ fixes the undefined part of the T-pairing.

2. For the Appendix:

(i) Try to derive the metric for the stellar structure, Eq. (A.4-7). That is, there is only $r-$dependence in $g_{\mu\nu}$.

Chapter 2. The Dirac Theory for Free Electrons

2.1 Dirac's Relativistic Equation

Einstein's relation is of quadratic order in energy, thus a derivative of the second order in time. Dirac set out in seeking for an equation with derivatives of the first order in time, and hence also of the first order in x, y, and z.

To satisfy that the probability is non-negative, namely, $w \geq 0$, Dirac assumes to have n components $\psi_1, \psi_2, ..., n$ being undetermined at the outset, and takes

$$\begin{aligned} w &= (\psi_1^*, \psi_2^*, ...) \begin{pmatrix} \psi_1 \\ \psi_2 \\ \vdots \end{pmatrix} \\ &= \sum_{k=1}^{n} \psi_k^* \psi_k. \end{aligned} \tag{1}$$

The most general partial differential equation of the first order in t, x, y, z may be put in the form

$$-E\psi_k + \sum_j (\alpha_{kj} \cdot \mathbf{p})\psi_j + m_0 \sum_j \beta_{kj}\psi_j = 0,$$

or[1]

$$\frac{\partial \psi_k}{\partial t} + \sum_j (\alpha_{kj} \cdot \nabla)\psi_j + im_0 \sum_j \beta_{kj}\psi_j = 0, \tag{2}$$

where α_{kj}, β_{kj} are matrices

$$(\alpha_x)_{kj}, \qquad (\alpha_y)_{kj}, \qquad (\alpha_z)_{kj}, \qquad \beta_{kj} \tag{3}$$

which operate only on the subscript $\mathbf{j}$ of ψ_j, and do not operate on the momentum $\mathbf{p}$.

Let $\tilde{\alpha}$ be the transpose of α, i.e., $\tilde{\alpha}_{mn} = \alpha_{nm}$. The complex conjugate and transpose of Eq. (2) is

$$\frac{\partial \psi_k^*}{\partial t} + \sum_j (\nabla \psi_j^* \cdot \tilde{\alpha}_{jk}^*) - im_0 \sum_j \psi_j^* \tilde{\beta}_{jk}^* = 0. \tag{4}$$

From Eqs. (2) and (4), one obtains

$$\begin{aligned} &\sum_k (\psi_k^* \frac{\partial \psi_k}{\partial t} + \frac{\partial \psi_k^*}{\partial t}\psi_k) + \sum_k \sum_j \{\psi_k^*(\alpha_{kj} \cdot \nabla)\psi_j + (\tilde{\alpha}_{kj}^* \cdot \nabla \psi_j^*)\psi_k\} \\ &+ im_0 \sum_k \sum_j \{\psi_k^* \beta_{kj} \psi_j - \tilde{\beta}_{kj}^* \psi_j^* \psi_k\} = 0. \end{aligned} \tag{5}$$

---

[1]

$$\frac{1}{c}\frac{\partial \psi_k}{\partial t} + \sum_j (\alpha_{kj} \cdot \nabla)\psi_j + \frac{im_0 c}{\hbar} \sum_j \beta_{kj}\psi_j = 0. \tag{2'}$$

If α, β satisfy the conditions

$$(\alpha_x^*)_{jk} = (\alpha_x)_{kj}, \quad \text{etc.,} \qquad (\beta^*)_{jk} = \beta_{kj}, \tag{6}$$

it is possible to define w as in Eq. (1) and a 3-current

$$\mathbf{I} = \sum_{k,j} \psi_k^* \alpha_{kj} \psi_j \tag{7}$$

so that Eq. (5) becomes

$$\frac{\partial w}{\partial t} + \nabla \cdot \mathbf{I} = 0. \tag{8}$$

This satisfies the conditions (i), (ii).

The conditions Eq. (6) are also necessary conditions to obtain Eq. (8). They show that

$$\alpha_x, \quad \alpha_y, \quad \alpha_z, \quad \beta \quad \text{must be hermitian.} \tag{9}$$

Eq. (2) can be written in matrix form

$$\frac{\partial \psi}{\partial t} + (\alpha \cdot \nabla)\psi + im_0 \beta \psi = 0, \tag{10}$$

or[2, 3]

$$i\frac{\partial \psi}{\partial t} = H\psi, \tag{11}$$

$$H = (\alpha \cdot \mathbf{p}) + \beta m_0. \tag{12}$$

To determine the number of components of ψ_k (i.e., the α and β matrices), we employ the follow of argument: The solution ψ of Eq. (10) must also satisfy the Klein-Gordon equation.[4] Operating on Eq. (10) by

$$-\frac{\partial}{\partial t} + (\alpha \cdot \nabla) + im_0 \beta, \tag{13}$$

[2]

$$\frac{1}{c}\frac{\partial \psi}{\partial t} + (\alpha \cdot \nabla)\psi + \frac{im_0 c}{\hbar}\beta\psi = 0. \tag{10$'$}$$

[3]

$$H = c(\alpha \cdot \mathbf{p}) + \beta m_0 c^2. \tag{12$'$}$$

[4] *Note that the solution of the second order Klein-Gordon equation is not necessarily a solution of the first order equation (10).*

one gets[5]

$$-\frac{\partial^2\psi}{\partial t^2}+\sum_x^z\alpha_x^2\frac{\partial^2\psi}{\partial x^2}+\sum_{x>y}(\alpha_x\alpha_y+\alpha_y\alpha_x)\frac{\partial^2\psi}{\partial x\partial y}$$
$$+im_0\sum_x^z(\alpha_x\beta+\beta\alpha_x)\frac{\partial\psi}{\partial x}-m_0^2\beta^2\psi=0. \tag{14}$$

The necessary and sufficient condition for this to be identical with the Klein-Gordon equation Eq. (11), Ch. 1, is

$$\begin{aligned}\alpha_x\alpha_y+\alpha_y\alpha_x&=2\delta_{xy}, \quad x,y,z \quad \text{cyclically},\\ \alpha_x\beta+\beta\alpha_x&=0, \quad x=x,y,z;\\ \beta^2&=1,\end{aligned} \tag{15}$$

i.e., $\alpha_x, \alpha_y, \alpha_z$, anticommute with one another. Among the 10 relations in Eq. (15), we have

$$\alpha_x^2=\alpha_y^2=\alpha_z^2=1, \tag{16}$$

so that

$$\alpha_x^{-1}=\alpha_x, \quad \beta^{-1}=\beta. \tag{17}$$

To determine the order $n\times n$ of these matrices, it can first be shown that n cannot be odd; n cannot be 2; and the lowest n is $n=4$.[6]

5

$$-\frac{1}{c^2}\frac{\partial^2\psi}{\partial t^2}+\sum_x^z\alpha_x^2\frac{\partial^2\psi}{\partial x^2}+\sum_{x>y}(\alpha_x\alpha_y+\alpha_y\alpha_x)\frac{\partial^2\psi}{\partial x\partial y}$$
$$+\frac{im_0c}{\hbar}\sum_x^z(\alpha_x\beta+\beta\alpha_x)\frac{\partial\psi}{\partial x}-\frac{m_0^2c^2}{\hbar^2}\beta^2\psi=0. \tag{14'}$$

6 *Thus, take for example,*

$$\begin{aligned}\alpha_x\beta&=-\beta\alpha_x\\ &=-I\beta\alpha_x, \quad I=n\times n \quad \textit{unit matrix.}\end{aligned}$$

Take the determinant of both sides,

$$\begin{aligned}[det.\alpha_x][det.\beta]&=[det.(-1)][det.\beta][det.\alpha_x]\\ &=(-1)^n[det.\alpha_x][det.\beta].\end{aligned}$$

Since $det.\alpha_x\neq 0$ *and* $det.\beta\neq 0$ *[as from Eqs. (15) and (16)], we find that* n *must be even. Next, assume*

$$\beta^{-1}\alpha_x\beta=-\alpha_x \quad (\beta^{-1}=\beta)$$

Taking α, β to be 4×4 matrices, one may have infinitely many choices - "representations".

(i) One representation is to use the Pauli matrices (VIII-1), Vol. I,

$$\sigma_x = \begin{pmatrix} 0 & 1 \\ 1 & 0 \end{pmatrix}, \quad \sigma_y = \begin{pmatrix} 0 & -i \\ i & 0 \end{pmatrix}, \quad \sigma_z = \begin{pmatrix} 1 & 0 \\ 0 & -1 \end{pmatrix}, \tag{18}$$

and construct the 4×4 α, β matrices by

$$\begin{aligned} \alpha_1 = \alpha_x &= \begin{pmatrix} 0 & \sigma_x \\ \sigma_x & 0 \end{pmatrix}, \quad \alpha_2 = \alpha_y = \begin{pmatrix} 0 & \sigma_y \\ \sigma_y & 0 \end{pmatrix}, \\ \alpha_3 = \alpha_z &= \begin{pmatrix} 0 & \sigma_z \\ \sigma_z & 0 \end{pmatrix}, \quad \alpha_4 = \beta = \begin{pmatrix} I & 0 \\ 0 & -I \end{pmatrix}. \end{aligned} \tag{19}$$

where $I = \begin{pmatrix} 1 & 0 \\ 0 & 1 \end{pmatrix}$. Corresponding to the dimension 4×4 for α, β, there are 4 components in Ψ_k .

A related representation, as used by W. Pauli, T. D. Lee and others, are followed in the present book, is

$$\gamma_k = \begin{pmatrix} 0 & -i\sigma_k \\ i\sigma_k & 0 \end{pmatrix}, \quad k = 1, 2, 3, \ (x, y, z), \quad \gamma_4 = \begin{pmatrix} I & 0 \\ 0 & -I \end{pmatrix}. \tag{20a}$$

or, more explicitly,

$$\begin{aligned} \gamma_1 &= \begin{pmatrix} & & & -i \\ & & -i & \\ & i & & \\ i & & & \end{pmatrix}, \quad \gamma_2 = \begin{pmatrix} & & & -1 \\ & & 1 & \\ & 1 & & \\ -1 & & & \end{pmatrix}, \\ \gamma_3 &= \begin{pmatrix} & & -i & 0 \\ & & 0 & i \\ i & 0 & & \\ 0 & -i & & \end{pmatrix}, \quad \gamma_4 = \beta. \end{aligned} \tag{20b}$$

$$\gamma_1 = -i\beta\alpha_x, \quad \gamma_2 = -i\beta\alpha_y, \quad \gamma_3 = -i\beta\alpha_z, \quad \gamma_4 = \beta. \tag{20c}$$

---

and

$$Tr\ (\beta^{-1}\alpha_x\beta) = -Tr\ \alpha_x$$

But $Tr\ (\beta^{-1}\alpha_x\beta) = Tr\ \alpha_x$, *hence*

$$Tr\ \alpha_x = 0. \tag{17a}$$

Similarly

$$Tr\ \alpha_y = Tr\ \alpha_z = 0, \quad Tr\ \beta = 0.$$

It is easy to demonstrate from Eq. (15) that the four matrices $\alpha_x, \alpha_y, \alpha_z, \beta$ *are linearly independent. In the case of* 2×2 *matrices, there are only three linearly independent, traceless matrices, say,* $\sigma_x, \sigma_y, \sigma_z$. *Thus,* $n = 2$ *is ruled out for the choice of* $\{\alpha_x, \alpha_y, \alpha_z, \beta\}$.

(ii) Majorana representation:

The γ_μ^M in the Majorana representation are related to the Pauli γ_μ in Eq. (20) by a unitary transformation

$$\gamma_\mu^M = S\gamma_\mu S^\dagger \tag{21a}$$

where

$$S = \frac{1}{\sqrt{2}}(1 + i\gamma_2), \quad S^\dagger = S^{-1} = \frac{1}{\sqrt{2}}(1 - i\gamma_2),$$

so that, except for $\mu = 2$ (with $\gamma_2^M = \gamma_2$),

$$\gamma_\mu^M = i\gamma_2\gamma_\mu. \tag{21b}$$

Explicitly, [7]

$$\gamma_1^M = \begin{pmatrix} \sigma_z & 0 \\ 0 & \sigma_z \end{pmatrix}, \quad \gamma_2^M = \begin{pmatrix} 0 & -i\sigma_y \\ i\sigma_y & 0 \end{pmatrix},$$
$$\gamma_3^M = \begin{pmatrix} -\sigma_x & 0 \\ 0 & -\sigma_x \end{pmatrix}, \quad \gamma_4^M = \begin{pmatrix} 0 & -\sigma_y \\ -\sigma_y & 0 \end{pmatrix}. \tag{21c}$$

The γ_μ^M are all hermitian. $\gamma_1^M, \gamma_2^M, \gamma_3^M$ are all real and symmetric; γ_4^M is imaginary and antisymmetric.

Eq. (10) can be expressed in terms of the γ_μ's by multiplying (10), from the left, by $-i\beta$. One gets

$$(\gamma_\mu \frac{1}{i}\frac{\partial}{\partial x_\mu} - im_0)\Psi = 0 \tag{22}$$

or[8]

$$(\gamma_\mu \frac{\partial}{\partial x_\mu} + m_0)\Psi = 0, \tag{22a}$$

---

[7] If $\gamma_1, \gamma_2, \gamma_3$ are as in Eq. (20a) but one chooses

$$\gamma_4 = \begin{pmatrix} 0 & -I \\ -I & 0 \end{pmatrix},$$

then $\gamma_1^M, \gamma_2^M, \gamma_3^M$ are the same as in Eq. (21c) but

$$\gamma_4^M = \begin{pmatrix} -\sigma_y & 0 \\ 0 & \sigma_y \end{pmatrix}. \tag{21d}$$

[8]

$$(\gamma_\mu \frac{\partial}{\partial x_\mu} + \frac{m_0 c}{\hbar})\Psi = 0. \tag{22a$'$}$$

where the summation convention (over repeated indices from 1 to 4) is used. The relativistic equation for an electron (charge -e) in an electromagnetic field $(A, i\phi)$ can be obtained by replacing $p_k, \frac{1}{i}\frac{\partial}{\partial x_4}$ in Eq. (22) by[9]

$$\Pi_k = \frac{1}{i}\frac{\partial}{\partial x_k + eA_k}, \qquad \Pi_4 = \frac{1}{i}\frac{\partial}{\partial x_4} + ie\phi \tag{23}$$

leading to

$$(\gamma_\mu \Pi_\mu - im_0)\Psi = 0, \tag{24a}$$

or,[10]

$$\{i\frac{\partial}{\partial t} + e\phi - \alpha \cdot (\mathbf{p} + e\mathbf{A}) - \beta m_0\}\Psi = 0. \tag{24b}$$

Note that this is just another application of the principle of minimal substitution, Eq. (12), Ch. 1, which yields a gauge-invariant interaction in the context of gauge field theories.

2.2 Solution of Dirac's Equation for a Free Electron

On using the α, β representation of Eq. (19), the Dirac equation (10) becomes

$$\begin{aligned}
(\frac{1}{i}\frac{\partial}{\partial t} + m_0)\psi_1 + \frac{1}{i}(\frac{\partial}{\partial x} - i\frac{\partial}{\partial y})\psi_4 + \frac{1}{i}\frac{\partial}{\partial z}\psi_3 &= 0, \\
(\frac{1}{i}\frac{\partial}{\partial t} + m_0)\psi_2 + \frac{1}{i}(\frac{\partial}{\partial x} + i\frac{\partial}{\partial y})\psi_3 - \frac{1}{i}\frac{\partial}{\partial z}\psi_4 &= 0, \\
(\frac{1}{i}\frac{\partial}{\partial t} - m_0)\psi_3 + \frac{1}{i}(\frac{\partial}{\partial x} - i\frac{\partial}{\partial y})\psi_2 + \frac{1}{i}\frac{\partial}{\partial z}\psi_1 &= 0, \\
(\frac{1}{i}\frac{\partial}{\partial t} - m_0)\psi_4 + \frac{1}{i}(\frac{\partial}{\partial x} + i\frac{\partial}{\partial y})\psi_1 - \frac{1}{i}\frac{\partial}{\partial z}\psi_2 &= 0.
\end{aligned} \tag{25}$$

If one assumes plane waves for ψ_μ,

$$\psi_\mu = A_\mu \exp\{i(\mathbf{p} \cdot \mathbf{r} - Et)\}, \quad A_\mu = \text{const.}, \tag{26}$$

9

$$\Pi_k = \frac{\hbar}{i}\frac{\partial}{\partial x_k} + \frac{e}{c}A_k, \quad \Pi_4 = \frac{\hbar}{i}\frac{\partial}{\partial x_4} + i\frac{e}{c}\phi. \tag{23$'$}$$

10

$$\{\frac{\hbar}{i}\frac{\partial}{\partial t} - e\phi + c\alpha \cdot (\mathbf{p} + \frac{e}{c}\mathbf{A}) + \beta m_0 c^2\}\Psi = 0. \tag{24b$'$}$$

one gets for A_μ,[11]

$$\begin{aligned}
-(E-m_0)A_1 + \qquad +p_3A_3 \quad + (p_1 - ip_2)A_4 &= 0,\\
-(E-m_0)A_2 + (p_1+ip_2)A_3 - p_3A_4 \quad &= 0,\\
p_3A_1 \qquad + (p_1-ip_2)A_2 - (E+m_0)A_3 \quad &= 0,\\
(p_1+ip_2)A_1 - p_3A_2 \qquad - (E+m_0)A_4 &= 0
\end{aligned} \tag{27}$$

The condition for existence of nontrivial solution for A_μ is

$$[E^2 - m_0^2 - (p_1^2 + p_2^2 + p_3^2)]^2 = 0, \tag{28}$$

whose roots (each doubly degenerate) are

$$\left.\begin{matrix} E_+ \\ E_- \end{matrix}\right\} = \pm\sqrt{\mathbf{p}^2 + m_0^2}. \tag{28a}$$

Eq. (28) is the relativistic energy-momentum relation, and the positive and negative energy are also there in the classical theory. In classical theory, E_- are independent and one may ignore E_- as physically meaningless. But in the theory of Dirac, there are transitions between E_+ and E_- states and one may no longer ignore the negative energy E_- states. We shall come back to this in the following section.

On substituting E_+ into Eq. (27), one obtains[12]

$$\begin{aligned}
A_1^{(+)} &= \frac{1}{E_+ - m_0}\{p_3A_3^{(+)} + (p_1 - ip_2)A_4^{(+)}\},\\
A_2^{(+)} &= \frac{1}{E_+ - m_0}\{(p_1 + ip_2)A_3^{(+)} - p_3A_4^{(+)}\},
\end{aligned} \tag{29a}$$

or,

$$\begin{aligned}
A_3^{(+)} &= \frac{1}{E_+ + m_0}\{p_3A_1^{(+)} + (p_1 - ip_2)A_2^{(+)}\},\\
A_4^{(+)} &= \frac{1}{E_+ + m_0}\{(p_1 + ip_2)A_1^{(+)} - p_3A_2^{(+)}\}.
\end{aligned} \tag{29b}$$

If $E_+ - m_0 \ll m_0$, then

$$E_+ + m_0 \gg E_+ - m_0, \tag{30}$$

and

$$A_3^{(+)},\ A_4^{(+)} \ll A_1^{(+)},\ A_2^{(+)}. \tag{31}$$

---

[11]

$$-(E - m_0c^2)A_1 + cp_3A_3 + c(p_1 - ip_2)A_4 = 0, \text{ etc.} \tag{27'}$$

[12]

$$A_2^{(+)} = -\frac{c}{E_+ - m_0c^2}\{(p_1 + ip_2)A_3^{(+)} - p_3A_4^{(+)}\}, \text{etc.} \tag{29a'}$$

$A_1^{(+)}, A_2^{(+)}$ are called the big components; $A_3^{(+)}, A_4^{(+)}$ the small components.

On substituting E_- into Eq. (27), one has only to replace E_+ by E_- in Eqs. (29a)-(29b).[13]

$$A_1^{(-)},\, A_2^{(-)} \quad \ll \quad A_3^{(-)},\, A_4^{(-)}. \tag{32}$$

Let $\chi_\pm$ be the eigenfunction of σ_z ,

$$\sigma_z \chi_\pm = \begin{pmatrix} 1 & 0 \\ 0 & -1 \end{pmatrix} \chi_\pm = \pm\chi_\pm. \tag{37}$$

The eigenvalues of σ_z are ± 1 .

The normalized $\psi(x)$, including the spin function, are, from Eq. (33) and Eq. (36),

$$\psi = N \begin{pmatrix} \chi_\pm \\ \frac{(\sigma\cdot\mathbf{p})}{\lambda|E|+m_0}\chi_\pm \end{pmatrix} \frac{1}{(2\pi)^{\frac{3}{2}}} \exp\{i(\mathbf{p}\cdot\mathbf{r} - Et)\}. \tag{38}$$

2.3 The Early Puzzles in Terms of Negative-Energy States

One realizes that, when the Dirac equation was invented in 1928, the positron was not there thus, the interpretation of the seemingly negative-energy states became a necessary task. Nowadays we adopt the following interpretation, namely, in the Dirac equation, there are two positive-energy solutions for the spin-up and spin-down states while the other two

[13] *The results (29)-(32) can be expressed in a compact form. Let the column matrix ψ_μ in Eq. (26) be put in the form*

$$\psi_\mu(\mathbf{r}, t) = \begin{pmatrix} \psi_1(\mathbf{r}) \\ \psi_2(\mathbf{r}) \\ \psi_3(\mathbf{r}) \\ \psi_4(\mathbf{r}) \end{pmatrix} e^{-iEt} \equiv \begin{pmatrix} \Phi \\ \varphi \end{pmatrix} e^{-iEt}, \tag{33}$$

where

$$\Phi = \begin{pmatrix} \Phi_1 \\ \Phi_2 \end{pmatrix}, \quad \varphi = \begin{pmatrix} \varphi_1 \\ \varphi_2 \end{pmatrix}. \tag{33a}$$

Using the α, β of Eq. (19) in Eq. (10), one obtains

$$\begin{gathered} (E - m_0)\Phi - (\sigma\cdot\mathbf{p})\varphi = 0, \\ -(\sigma\cdot\mathbf{p})\Phi + (E + m_0)\varphi = 0, \end{gathered} \tag{34}$$

where σ are the Pauli matrices Eq. (18). To solve for Φ and φ, one may make use of the identity

$$(\sigma\cdot\mathbf{A})(\sigma\cdot\mathbf{B}) = (\mathbf{A}\cdot\mathbf{B}) + i(\sigma\cdot[\mathbf{A}\times\mathbf{B}]) \tag{35}$$

and obtain the two roots E_+, E_- in Eq. (28a) again, and[13]

$$\varphi = \frac{(\sigma\cdot\mathbf{p})}{\lambda\mid E\mid +m_0}\Phi, \quad \lambda = \pm \quad for \left\{ \begin{array}{c} E_+ \\ E_- \end{array} \right., \tag{36}$$

This is the same as Eqs. (29)-(32) above.

solutions are for the positive-energy spin-up and spin-down states of the positron (the sign of $ip \cdot x$ is switched for the antiparticle).

(1) Divorce of momentum from velocity

From Eqs. (11) and (12)[14]

$$\begin{aligned} i\frac{\partial \Psi}{\partial t} &= H\Psi \\ &= \{(\alpha \cdot \mathbf{p}) + \beta m_0\}\Psi, \end{aligned} \tag{39}$$

and the equations of motion in Heisenberg's picture, one gets[15]

$$\dot{x}_k = i(Hx_k - x_k H) = \alpha_k, \tag{40}$$

$$\dot{p}_k = i(Hp_k - p_k H) = 0. \tag{41}$$

The expectation value v_x of the velocity $\dot{x}$ is hence

$$v_x = \int \tilde{\Psi}^* \alpha_x \Psi d\tau. \tag{42}$$

If

$$\Psi(\mathbf{r}, t) = \psi(\mathbf{r})e^{-iEt}, \tag{43}$$

then Eq. (39) gives

$$E\psi - (\alpha \cdot \mathbf{p})\psi - \beta m_0 \psi = 0, \tag{44}$$

The complex conjugate and transpose of this is ($\tilde{\alpha}^* = \alpha$, etc)

$$E\tilde{\psi}^* - \mathbf{p}\tilde{\psi}^* \cdot \alpha - m_0 \tilde{\psi}^* \beta = 0. \tag{45}$$

Multiplying Eq. (44) from the left by $\tilde{\psi}^* \alpha_k$, and Eq. (45) from the right by $\alpha_k \psi$, integrating, and using Eq. (15), one gets

$$E \int \tilde{\psi}^* \alpha_k \psi d\tau = \int \tilde{\psi}^* p_k \psi d\tau. \tag{46}$$

14

$$\begin{aligned} i\hbar\frac{\partial \Psi}{\partial t} &= H\Psi \\ &= \{c(\alpha \cdot \mathbf{p}) + \beta m_0 c^2\}\Psi, \end{aligned} \tag{39$'$}$$

15

$$\dot{x}_k = \frac{i}{\hbar}(Hx_k - x_k H) = c\alpha_k. \tag{40$'$}$$

Hence, from Eq. (42), one obtains[16]

$$v_k = \frac{1}{E} < p_k > . \tag{47}$$

For E_- , the average $\mathbf{v}_-$ and the average $\mathbf{p}$ are in opposite directions! In classical dynamics one would associate a *negative* mass with an electron in a negative energy state! Such a particle has the strange property that when a force is applied in one direction, the acceleration is in the opposite direction.

This divorce between the "velocity" $\dot{x}_k$ and the momentum p_k can already be seen from Eqs. (40) and (41). The α_k's do not commute among themselves, nor with H. Hence the $\dot{x}_k$ are unlike any physical velocity in the classical sense. The p_k are constants of motion, as seen from Eq. (41).

(2) The Zitterbewegung of Schrödinger

The Heisenberg equations of motion are

$$\begin{aligned} \dot{\alpha}_k &= i(H\alpha_k - \alpha_k H), \\ \dot{\beta} &= i(H\beta - \beta H). \end{aligned} \tag{48}$$

From Eq. (12) and Eq. (15), one obtains

$$\begin{aligned} H\alpha_k + \alpha_k H &= 2p_k, \\ H\beta + \beta H &= 2m_0. \end{aligned} \tag{49}$$

Hence[17]

$$\begin{aligned} \dot{\alpha}_k &= 2i(p_k - \alpha_k H) = 2i(H\alpha_k - p_k), \\ \dot{\beta} &= 2i(m_0 - \beta H) = 2i(H\beta - m_0). \end{aligned} \tag{50}$$

From Eq. (41), $\dot{p}_k = 0$, and $\dot{H} = 0$, hence, on differentiating Eq. (50),

$$\ddot{\alpha}_k = -2i\dot{\alpha}_k H, \tag{51}$$

and on integration,

$$\dot{\alpha}_k = (\dot{\alpha}_k)_{t=0} e^{-2iHt}. \tag{52}$$

Substituting this into Eq. (50), one gets

$$\alpha_k = \frac{1}{2iH}(\dot{\alpha}_k)_{t=0} e^{-2iHt} + \frac{1}{H} p_k. \tag{53}$$

16

$$v_k = \frac{c^2}{E} < p_k > . \tag{47$'$}$$

17

$$\dot{\alpha}_k = \frac{2i}{\hbar}(cp_k - \alpha_k H) = \frac{2i}{\hbar}(H\alpha_k - cp_k). \tag{50$'$}$$

From Eq. (40), one gets[18]

$$\dot{x}_k = \frac{1}{2iH}(\dot{\alpha}_k)_{t=0}e^{-2iHt} + \frac{1}{H}p_k. \tag{54}$$

The velocity $\dot{x}_k$ has a classical part $\frac{1}{H}p_k = \frac{\partial E}{\partial p_k} = v_k$, and a high frequency $2H \geq 2m_0$ part found by Schrödinger. This Zitterbewegung (trembling motion) arises from the interference effect between the positive and the negative energy states, as we shall see below.

Let $\Psi(\mathbf{r}, t)$ and its Fourier transform $\Phi(p, t)$ be

$$\Psi(\mathbf{r}, t) = (2\pi)^{-3} \int \Phi(\mathbf{p}, t)e^{-i\mathbf{p}\cdot\mathbf{r}}d\mathbf{p}, \tag{55a}$$

$$\Phi(\mathbf{p}, t) = \int \Psi(\mathbf{r}, t)e^{i\mathbf{p}\cdot\mathbf{r}}d\mathbf{r}. \tag{55b}$$

The equation (39) in p-representation is then

$$i\frac{\partial\Phi}{\partial t} - (\alpha \cdot \mathbf{p})\Phi - m_0\beta\Phi = 0. \tag{56}$$

Let

$$\Phi = B_+e^{-i|E|t} + B_-e^{i|E|t}. \tag{57}$$

Then

$$\begin{aligned}(E_+ - (\alpha \cdot \mathbf{p}) - m_0\beta)B_+ &= 0,\\ (E_- - (\alpha \cdot \mathbf{p}) - m_0\beta)B_- &= 0.\end{aligned} \tag{58a}$$

Taking the complex conjugate and transpose, one gets

$$\begin{aligned}E_+\tilde{B}_+^* - \tilde{B}_+^*((\alpha \cdot \mathbf{p}) + m_0\beta) &= 0,\\ E_-\tilde{B}_-^* - \tilde{B}_-^*((\alpha \cdot \mathbf{p}) + m_0\beta) &= 0.\end{aligned} \tag{58b}$$

From Eqs. (58a)-(58b),[19]

$$E_+\tilde{B}_+^*B_-E_- = \tilde{B}_+^*\{(\alpha \cdot \mathbf{p}) + m_0\beta\}^2B_-$$

[18]

$$\dot{x}_k = \frac{\hbar c}{2iH}(\dot{\alpha}_k)_{t=0}e^{-2iHt/\hbar} + \frac{c^2}{H}p_k. \tag{54'}$$

[19] *B_+, B_- are column, $\tilde{B}_+^*, \tilde{B}_-^*$ are row matrices, and*

$$\tilde{B}_+^*B_- = \sum_\mu (B_+^*)_\mu(B_-)_\mu.$$

Using Eqs. (28a) and (15), one gets

$$-(p^2 + m_0^2)\tilde{B}_+^* B_- = (p^2 + m_0^2)\tilde{B}_+^* B_-$$

so that

$$\tilde{B}_+^* B_- = \sum (\tilde{B}_+^*)_\mu (B_-)_\mu = 0, \tag{59}$$

i.e., the positive and negative energy wave functions are orthogonal.

Let us calculate the expectation value v_x of the velocity $\dot{x}$ of an electron in the momentum representation. From Eqs. (40) and (42), one gets[20]

$$\begin{aligned} v_k &= \int \tilde{\Phi}^* \alpha_k \Phi dp \\ &= \int \tilde{B}_+^* \alpha_k B_+ dp + \int \tilde{B}_-^* \alpha_k B_- dp \\ &+ e^{2i|E|t} \int \tilde{B}_+^* \alpha_k B_- dp + e^{-2i|E|t} \int \tilde{B}_-^* \alpha_k B_+ dp \\ &= \frac{1}{E_+} < p_k > + \frac{1}{E_-} < p_k > + e^{2i|E|t} \int \tilde{B}_+^* \alpha_k B_- dp \\ &+ e^{-2i|E|t} \int \tilde{B}_-^* \alpha_k B_+ dp. \end{aligned} \tag{60}$$

This shows that the Zitterbewegung of Eq. (54) has its origin in the interference of E_+ and E_- states.

Before the discovery of the positron in 1932, the presence of the negative-energy states, the "divorce" between velocity and momentum, and the Zitterbewegung, were considered as troublesome features of the Dirac equation. However, quantization of the Dirac field resolves conceptual difficulties associated with these problems.

(3) The Klein paradox

Consider the motion of an electron travelling from left to right in one dimension in a potential $V(x)$ given by

$$V(x) = \begin{cases} 0, x < 0; \\ V = \text{constant}, 0 < x \end{cases}. \tag{61}$$

For $x < 0$, the wave function is

$$\psi_i = A \ \exp\{i(p_0 x - Et)\}. \tag{62a}$$

At $x = 0$, there are a reflected and a transmited wave

$$\psi_r = B \ \exp\{i(-p_0 x - Et)\}, \tag{62b}$$

$$\psi_t = C \ \exp\{i(px - Et)\} \tag{62c}$$

[20] *For the first two terms following the last equal sign, see Eq. (47).*

ψ_i, ψ_r, ψ_t and A , B , C are all column matrices. Dirac's equations are[21]

$$(E - \alpha p_0 - \beta m_0)A = 0, \tag{63a}$$

$$(E + \alpha p_0 - \beta m_0)B = 0, \tag{63b}$$

$$(E - V - \alpha p - \beta m_0)C = 0. \tag{63c}$$

The energy-momentum relations are

$$E^2 = p_0^2 + m_0^2, \quad \text{for } x < 0,$$

$$(E - V)^2 = p^2 + m_0^2, \quad \text{for } 0 < x. \tag{64}$$

At $x = 0$, the continuity of ψ requires that

$$A + B = C. \tag{65}$$

From Eq. (63) and this, one gets

$$(E - \beta m_0)(A + B) = \alpha p_0(A - B),$$

$$(E - \beta m_0)(A + B) = (V + \alpha p)(A + B).$$

Hence

$$(V + \alpha(p_0 + p))B = -(V - \alpha(p_0 - p))A.$$

Multiplying both sides by $V - \alpha(p_0 + p)$ and using Eq. (64), one gets

$$B = -\frac{2V(E - \alpha p_0)}{V^2 - (p_0 + p)^2}A. \tag{66}$$

Taking the adjoints of Eqs. (62a)–(62c), one obtains the complex conjugate and transpose of Eqs. (63a)–(63c),

$$\tilde{A}^*(E - \alpha p_0 - \beta m_0) = 0, \tag{67a}$$

$$\tilde{B}^*(E + \alpha p_0 - \beta m_0) = 0, \tag{67b}$$

$$\tilde{C}^*(E - V - \alpha p - \beta m_0) = 0. \tag{67c}$$

21

$$(E - c\alpha p_0 - \beta m_0 c^2)A = 0, \text{ etc.} \tag{63a$'$}$$

Multiplying Eq. (63a) by $\tilde{A}^*\alpha$ from the left, Eq. (67a) by αA from the right, one gets, on adding and using Eqs. (15) and (19),

$$\tilde{A}^*\alpha A = \frac{1}{E} p_0 \tilde{A}^* A, \tag{68}$$

which is once more Eq. (47).

In a procedure similar to that in obtaining Eq. (66), one gets

$$\tilde{B}^* = -\tilde{A}^* \frac{2V(E-\alpha p_0)}{V^2-(p_0+p)^2}. \tag{69}$$

From Eqs. (66) and (69), one obtains

$$\begin{aligned}
\tilde{B}^* B &= (\frac{2V}{V^2-(p_0+p)^2})^2 \tilde{A}^*(E-\alpha p_0)^2 A \\
&= (\frac{2V}{V^2-(p_0+p)^2})^2 \{(E^2+p_0^2)\tilde{A}^* A - 2Ep_0\tilde{A}^*\alpha A\} \\
&= (\frac{2V}{V^2-(p_0+p)^2})^2 (E^2-p_0^2)\tilde{A}^* A,
\end{aligned}$$

The reflection coefficient R is

$$R = \frac{\tilde{B}^* B}{\tilde{A}^* A} = (\frac{2Vm_0}{V^2-(p_0+p)^2})^2. \tag{70}$$

For a fixed E, when V increases from 0 to $E-m_0$, R increases from 0 to the value 1 , i.e., the transmitted electron has $p=0$ and the electron is totally reflected.

When V continues to increase from $E-m_0$ to the value $E+m_0$, it is seen from Eq. (64) that p is pure imaginary. Thus,

$$p = iq, \quad q = \text{real}, \quad \text{for } E-m_0 < V < E+m_0. \tag{71}$$

Eq. (62c) for the transmitted wave is now

$$\begin{aligned}
\psi_t &= D \quad \exp\{-(qx+iEt)\}, \\
\psi_t^* &= D^* \quad \exp\{-(qx-iEt)\}.
\end{aligned} \tag{72}$$

In Eq. (66), p is replaced by iq , and in Eq. (69) , p is replaced by $-iq$, so that the reflection coefficient R in Eq. (70) is now[22]

$$\begin{aligned}
R &= \frac{(2V)^2(E^2-p_0^2)}{[(V+p_0)^2+q^2][(V-p_0)^2+q^2]} \\
&= 1, \quad \text{for} \quad E-m_0 \le V \le E+m_0,
\end{aligned} \tag{73}$$

22

$$R = \frac{(2V)^2(E^2-c^2p_0^2)}{[(V+cp_0)^2+c^2q^2][(V-cp_0)^2+c^2q^2]}. \tag{73$'$}$$

since it can be verified from Eq. (64) that

$$(V \pm p_0)^2 + q^2 = 2V(E \pm p_0).$$

For $V > E + m_0$, Eq. (64) shows that $p^2 > 0$ so that R is again given by Eq. (70). But from Eq. (64),

$$E - V = \pm\sqrt{p^2 + m_0^2},$$

and for $V > E + m_0$, $E - V$ is negative. Hence we must take

$$E - V = -\sqrt{p^2 + m_0^2} < 0. \tag{74}$$

From the relation Eq. (47), one now has[23]

$$v = \frac{\partial E}{\partial p} = \frac{1}{E - V} p, \quad E - V < 0, \tag{75}$$

so that v and p are in *opposite* directions. But this is the property of an electron in a negative energy state found in Eq. (47).

Intuitively, we may visualize the situation as follows: For $x \leq 0$, we have $V = 0$ so that the positive-energy spectrum starts at $E = m_0$ while the negative-energy solutions appear for $E \leq -m_0$. For $x > 0$, the entire energy spectrum is lifted upward by an amount equal to the potential $V(> 0)$. A given incident electron wave of energy $E(> 0)$ is reflected totally by a potential barrier V with $E - m_0 \leq V \leq E + m_0$ since, for $x > 0$, E lies between (and outside) the positive and negative spectra. For $V \leq E - m_0$, transmission arises from transition into a positive energy solution in the $x > 0$ region. For $V > E + m_0$, transmission occurs because transition into a negative-energy solution in the $x > 0$ region becomes possible.

Thus when an electron of positive energy E impinges on a potential $V > E + m_0$, it penetrates into the barrier but makes a transition to a negative energy state.[24]

When V becomes infinitely high, one sees from Eq. (70) that

$$\lim_{V\to\infty} R = \frac{E - p_0}{E + p_0}. \tag{76}$$

(4) The "hole theory" of positron

The possibility of transitions of an electron from E_+ to E_- energy states, as shown by Klein's study, shows that the negative energy states cannot be regarded as completely

[23]

$$V = \frac{\partial E}{\partial p} = \frac{c^2}{E - V} p. \tag{75'}$$

[24] *It was shown by F. Sauter, Zeits. f. Phys. 69, 742; 73, 547 (1931), that if the gradient of V at $x = 0$ in Eq. (61) is not infinite but is smaller than $\frac{2m_0c^2}{\lambda}$, $\lambda = \frac{h}{m_0 c}$, the compton wavelength, the Klein transition to negative energy states will not occur. But the existence of this transition in principle is relevant here, since one can no longer simply ignore the E_- states as in classical theories.*

independent of the positive (ordinary) energy states. The question why all the electrons do not fall into the negative energy states is answered by the theory of Dirac (1930) that all these states are filled up in the sense of the Pauli principle so that a completely filled sea of E_- states constitutes the norm-the vacuum. When an electron in this sea is excited into an ordinary state (by γ-rays, say, of energy $> 2m_0c^2$), a hole is left in the *sea*, and the absence of a negative charge from the norm reveals itself as a positive charge $+e$. Dirac at first suggested that this positive charge is our proton; but it was soon pointed out by R. Oppenheimer (1930) and H. Weyl (1931) that the mass of this positive charge must be the same as that of our ordinary electron.

In 1932, C. D. Anderson discovered in a Wilson cloud chamber of cosmic rays a track which, from its direction of motion and its curvature in a magnetic field, can be definitely identified as that due to a positive charge $+e$ with the mass of an electron. This was immediately identified as the "hole" in the Dirac theory, and is named "positron".

Historically, before the discovery by Anderson, there already existed some experimental evidence for the production of positrons. It was found by C. Y. Chao at California Institute of Technology in 1930 in experiments on the absorption of hard γ-rays that there is some excess absorption not accounted for by the then known processes such as photoelectric and Compton effect. This excess absorption is later readily accounted for by the process.

$$h\nu \to e^+ + e^-, \qquad h\nu > 2m_0c^2 \tag{77a}$$

in which γ -rays, of energy greater than the sum of the rest energy of the electron and the positron, excite, in the intense field in the immediate neighborhood of an atomic nucleus, say, an electron from the sea of electrons in negative energy states into an ordinary (E_+) states, thereby producing a pair of $e^- + e^+$. This process is known as "pair production".

The reverse of the pair production process of Eq. (77a) also takes place

$$e^+ + e^- \to h\nu + h\nu \tag{77b}$$

in which e^+ and e^- annihilate each other, giving rise to two (occasionally three) quanta of γ -rays (for energy and momentum conservation.) Accurate measurements, including coincidence counter devices, definitely establish Eq. (77b).

These experimental verifications of Dirac's theory of positrons (or, anti-electron) are of course most satisfying; but at the same time they also call emphasis on the view that, since an infinite sea of electrons in negative energy states are always present, and especially in the presence of strong electric fields (such as near an atomic nucleus) the positive and negative energy states are not distinctly separated as in the case of free electrons, the Dirac theory cannot, from the rigorous point of view, be regarded as an exact theory of a single electron. In other words, a theory of the electron is intimately bound to a theory of fields.

(5) The first nontrivial solution of the Dirac equation - Wolfgang Kroll

In 1930, Wolfgang Kroll published his dissertation work on the Dirac equation - the solution of the electron under the external magnetic field in the y−direction.

The first component is in fact the nowadays third component, etc., in the Kroll's paper.[25] We should give the credit to Kroll in attempting to solve directly the Dirac equation.

[25] See, e.g., W-Y. Pauchy Hwang, The Universe, **3-3**, pp. 1-4 (2015); Wolfgang Kroll, Zeit. Physik **A66**, 69-108 (1930).

2.4 Electron Spin

We shall show that the Dirac relativistic equation already contains the electron spin.

In Eq. (19), we have defined the Dirac $\alpha_1, \alpha_2, \alpha_3, \alpha_4 = \beta$ matrices in terms of the Pauli matrices $\sigma_x, \sigma_y, \sigma_z$.

$$\alpha_1 = \begin{pmatrix} 0 & \sigma_x \\ \sigma_x & 0 \end{pmatrix}, \quad \alpha_2 = \begin{pmatrix} 0 & \sigma_y \\ \sigma_y & 0 \end{pmatrix},$$
$$\alpha_3 = \begin{pmatrix} 0 & \sigma_z \\ \sigma_z & 0 \end{pmatrix}, \quad \alpha_4 = \beta = \begin{pmatrix} I & 0 \\ 0 & -I \end{pmatrix}. \tag{78}$$

Now to follow the literature (Dirac), we define the 4×4 ρ_1, ρ_2, ρ_3 matrices

$$\rho_1 = \begin{pmatrix} 0 & I \\ I & 0 \end{pmatrix}, \quad \rho_2 = i\begin{pmatrix} 0 & -I \\ I & 0 \end{pmatrix}, \quad \rho_3 = \begin{pmatrix} I & 0 \\ 0 & -I \end{pmatrix} = \beta,$$

$$\rho_1^2 = \rho_2^2 = \rho_3^2 = \begin{pmatrix} I & 0 \\ 0 & I \end{pmatrix}. \tag{79}$$

and the 4×4 $\sigma_1, \sigma_2, \sigma_3$ matrices [not to be confused with the Pauli 2×2 matrices $\sigma_x, \sigma_y, \sigma_z$ in Eq. (18)],

$$\sigma_k = \rho_1 \alpha_k, \quad k = 1, 2, 3, \tag{80a}$$

or, written explicitly in the footnote.[26]

The Hamiltonian

$$H = (\alpha \cdot \mathbf{p}) + \beta m_0$$

in terms of the σ_k matrices Eq. (80b) is[27]

$$H = \rho_1(\sigma \cdot \mathbf{p}) + \alpha_4 m_0. \tag{84}$$

The angular momentum $\mathbf{L}(L_x, L_y, L_z)$

$$\mathbf{L} = \mathbf{r} \times \mathbf{p} \tag{85}$$

satisfy the relations

$$L_x p_x - p_x L_x = 0, \tag{86a}$$

$$L_x p_y - p_y L_x = i p_z, \tag{86b}$$

[26] *The $\sigma_1, \sigma_2, \sigma_3$ matrices satisfy the following relations:*

$$\sigma_1 = -i\sigma_2\sigma_3 = -i\alpha_2\alpha_3 = -i\gamma_2\gamma_3, \tag{80b}$$

$$\sigma_2 = -i\sigma_3\sigma_1 = -i\alpha_3\alpha_1 = -i\gamma_3\gamma_1, \tag{80c}$$

$$\sigma_3 = -i\sigma_1\sigma_2 = -i\alpha_1\alpha_2 = -i\gamma_1\gamma_2, \tag{80d}$$

$\sigma_1, \sigma_2, \sigma_3$ *are hermitian,*

$$\alpha_k = \rho_1\sigma_k, \quad k = 1, 2, 3, \tag{80e}$$

$$\alpha_4 = \rho_3 = \beta;$$

$$\sigma_i\sigma_j + \sigma_j\sigma_i = 2\delta_{ij}, \quad i, j = 1, 2, 3, \tag{81}$$

$$\sigma_i\sigma_j = i\sigma_k, \quad \textit{cyclic permutation of } i, j, k = 1, 2, 3.$$

$$\rho_j\sigma_k - \sigma_k\rho_j = 0, \quad k, j = 1, 2, 3 \tag{82}$$

$$\rho_1\alpha_k - \alpha_k\rho_1 = 0, \tag{83a}$$

$$\rho_2\alpha_k + \alpha_k\rho_2 = 0, \tag{83b}$$

$$\rho_3\alpha_k + \alpha_k\rho_3 = 0. \tag{83c}$$

[27]

$$H = c\rho_1(\sigma \cdot \mathbf{p}) + \alpha_4 m_0 c^2. \tag{84'}$$

$$L_x p_z - p_z L_x = -ip_y, \tag{86c}$$

and relations obtained by cyclic permutations of x, y, z.

From Eqs. (84) and (86b)-(86c),

$$\begin{aligned} L_x H - HL_x &= \rho_1[L_x(\sigma\cdot\mathbf{p}) - (\sigma\cdot\mathbf{p})L_x] \\ &= \rho_1[\sigma\cdot(L_x\mathbf{p} - \mathbf{p}L_x)] \\ &= i\rho_1(\sigma_2 p_3 - \sigma_3 p_2) \\ L_y H - HL_y &= i\rho_1(\sigma_3 p_1 - \sigma_1 p_3) \\ L_z H - HL_z &= i\rho_1(\sigma_1 p_2 - \sigma_2 p_1) \end{aligned}$$

or,[28]

$$\mathbf{L}H - H\mathbf{L} = i\rho_1 \quad [\sigma\times\mathbf{p}]. \tag{87}$$

Similarly,

$$\begin{aligned} \sigma_1 H - H\sigma_1 &= \sigma_1\rho_1(\sigma\cdot\mathbf{p}) - \rho_1(\sigma\cdot\mathbf{p})\sigma_1 + (\sigma_1\alpha_4 - \alpha_4\sigma_1)m_0 \\ &= -2i\rho_1(\sigma_2 p_3 - \sigma_3 p_2), \\ \sigma_2 H - H\sigma_2 &= -2i\rho_1(\sigma_3 p_1 - \sigma_1 p_3), \\ \sigma_3 H - H\sigma_3 &= -2i\rho_1(\sigma_1 p_2 - \sigma_2 p_1), \end{aligned}$$

or[29]

$$\sigma H - H\sigma = -2i\rho_1 \quad [\sigma\times\mathbf{p}]. \tag{88}$$

From Eqs. (87) and (88), one obtains

$$(\mathbf{L} + \frac{1}{2}\sigma)H - H(\mathbf{L} + \frac{1}{2}\sigma) = 0, \tag{89}$$

showing that

$$\mathbf{J} \equiv \mathbf{L} + \frac{1}{2}\sigma \tag{90}$$

is a constant of motion, while the orbital angular momentum $\mathbf{L}$ is not. The natural interpretation of $\mathbf{J}$ is that it is the total angular momentum of the electron, and this suggest that $\frac{1}{2}\sigma$ is an intrinsic angular momentum - the spin. If we introduce the components

$$S_1 = \frac{1}{2}\sigma_1, \quad S_2 = \frac{1}{2}\sigma_2, \quad S_3 = \frac{1}{2}\sigma_3, \tag{91}$$

28

$$\mathbf{L}H - H\mathbf{L} = i\hbar c\rho_1 \quad [\sigma\times\mathbf{p}]. \tag{87$'$}$$

29

$$\sigma H - H\sigma = -2ic\rho_1 \quad [\sigma\times\mathbf{p}]. \tag{88$'$}$$

then

$$\mathbf{J} = \mathbf{L} + \mathbf{S}. \tag{92}$$

From Eqs. (84) and (88), one can show that[30]

$$(L_z + \frac{1}{2}\sigma_3)H - H(L_z + \frac{1}{2}\sigma_3) = 0, \tag{93}$$

$$J^2(L_z + \frac{1}{2}\sigma_3) - (L_z + \frac{1}{2}\sigma_3)J^2 = 0, \tag{94}$$

and $H, J^2, (L_3 + \frac{1}{2}\sigma_3)$ have simultaneous eigenvectors.

In spherical polar coordinates (r, ϑ, φ)

$$J_3 = L_3 + S_3 = \frac{1}{i}\frac{\partial}{\partial\phi} + \frac{1}{2}\sigma_3.$$

$$\sigma_3\Psi = \begin{pmatrix} 1 & 0 & 0 & 0 \\ 0 & -1 & 0 & 0 \\ 0 & 0 & 1 & 0 \\ 0 & 0 & 0 & -1 \end{pmatrix} \begin{pmatrix} \psi_1 \\ \psi_2 \\ \psi_3 \\ \psi_4 \end{pmatrix} = \begin{pmatrix} c\psi_1 \\ -\psi_2 \\ \psi_3 \\ -\psi_4 \end{pmatrix}. \tag{95}$$

The equation

$$(L_3 + \frac{1}{2}\sigma_3)\Psi = m\Psi$$

becomes[31]

$$\begin{aligned} (\frac{1}{i}\frac{\partial}{\partial\varphi} + \frac{1}{2} - m_0)\begin{pmatrix} \psi_1 \\ \psi_3 \end{pmatrix} &= 0, \\ (\frac{1}{i}\frac{\partial}{\partial\varphi} - \frac{1}{2} - m_0)\begin{pmatrix} \psi_2 \\ \psi_4 \end{pmatrix} &= 0, \end{aligned} \tag{96}$$

so that eigenvalues m are given by

$$m \pm \frac{1}{2} = \pm\text{integer}. \tag{97}$$

30

$$(L_z + \frac{1}{2}\hbar\sigma_3)H - H(L_z + \frac{1}{2}\hbar\sigma_3) = 0, \text{ etc.} \tag{93$'$}$$

31

$$\hbar(\frac{1}{i}\frac{\partial}{\partial\varphi} + \frac{1}{2} - m_0)\begin{pmatrix} \psi_1 \\ \psi_3 \end{pmatrix} = 0, \text{ etc.} \tag{96$'$}$$

On defining

$$L_\pm = L_1 \pm iL_2, \qquad S_\pm = S_1 \pm iS_2, \tag{98}$$

then

$$\begin{aligned}\mathbf{J}^2 &= \mathbf{L}^2 + \mathbf{S}^2 + 2(\mathbf{L}\cdot\mathbf{S})\\ &= \mathbf{L}^2 + \mathbf{S}^2 + \mathbf{L}_+\mathbf{S}_- + L_-S_+ + 2L_zS_z.\end{aligned} \tag{99}$$

From Eqs. (80b) and (91), one finds[32]

$$S_+ = \begin{pmatrix} 0 & 1 & 0 & 0\\ 0 & 0 & 0 & 0\\ 0 & 0 & 0 & 1\\ 0 & 0 & 0 & 0\end{pmatrix}, \quad S_- = \begin{pmatrix} 0 & 0 & 0 & 0\\ 1 & 0 & 0 & 0\\ 0 & 0 & 0 & 0\\ 0 & 0 & 1 & 0\end{pmatrix}, \tag{100}$$

$$S^2 = \frac{1}{4}(\sigma_1^2 + \sigma_2^2 + \sigma_3^2) = \frac{3}{4}I, \tag{101}$$

I being a 4×4 unit matrix.

To obtain the eigenvalues ξ of J^2 , from Eqs. (99), (95), and (100), one obtains the 4 equations

$$(\mathbf{L}^2 + \frac{3}{4} + L_3 - \xi)\begin{pmatrix}\psi_1\\ \psi_3\end{pmatrix} + L_-\begin{pmatrix}\psi_2\\ \psi_4\end{pmatrix} = 0,$$

$$(\mathbf{L}^2 + \frac{3}{4} - L_3 - \xi)\begin{pmatrix}\psi_2\\ \psi_4\end{pmatrix} + L_+\begin{pmatrix}\psi_1\\ \psi_3\end{pmatrix} = 0, \tag{102}$$

which can be written[33]

$$\begin{aligned}(\mathbf{L}^2 + \frac{3}{4} + L_3 - \xi)\psi_1 + L_-\psi_2 = 0,\\ (\mathbf{L}^2 + \frac{3}{4} - L_3 - \xi)\psi_2 + L_+\psi_1 = 0,\end{aligned} \tag{103}$$

and two other equations in which ψ_3, ψ_4 replace the ψ_1, ψ_2 above respectively.

32

$$S^2 = \frac{1}{4}(\sigma_1^2 + \sigma_2^2 + \sigma_3^2)\hbar^2 = \frac{3}{4}\hbar^2 I. \tag{101$'$}$$

33

$$(\mathbf{L}^2 + \frac{3}{4}\hbar^2 + L_3\hbar - \xi)\psi_1 + L_-\hbar\psi_2 = 0. \tag{103$'$}$$

In spherical polar coordinates (r, ϑ, φ) , we have

$$L_{\pm} = \pm e^{\pm i\varphi}\{\frac{\partial}{\partial\theta} \pm i\cot\vartheta\frac{\partial}{\partial\varphi}\},$$
$$\mathbf{L}^2 = -\{\frac{1}{\sin\vartheta}\frac{\partial}{\partial\vartheta}(\sin\vartheta\frac{\partial}{\partial\vartheta}) + \frac{1}{\sin^2\vartheta}\frac{\partial^2}{\partial\varphi^2}\}, \quad (104)$$

and Eq. (103) are

$$\{-[\frac{1}{\sin\vartheta}\frac{\partial}{\partial\vartheta}(\sin\vartheta\frac{\partial}{\partial\vartheta}) - \frac{1}{\sin^2\vartheta}(m-\frac{1}{2})^2] + \frac{3}{4} + (m-\frac{1}{2}) - \xi\}\psi_1$$
$$- e^{-i\varphi}[\frac{\partial}{\partial\vartheta} + (m+\frac{1}{2})\cot\vartheta]\psi_2 = 0,$$
$$\{-[\frac{1}{\sin\vartheta}\frac{\partial}{\partial\vartheta}(\sin\vartheta\frac{\partial}{\partial\vartheta}) - \frac{1}{\sin^2\vartheta}(m+\frac{1}{2})^2] + \frac{3}{4} - (m+\frac{1}{2}) - \xi\}\psi_2$$
$$+ e^{i\varphi}[\frac{\partial}{\partial\vartheta} - (m-\frac{1}{2})\cot\vartheta]\psi_1 = 0. \quad (105)$$

The method of solving this pair of equations is similar to that of solving the Schrödinger equation. The results are[34]:

$$\xi = j(j+1), \quad (106a)$$

$$-j \le m \le j, \quad (106b)$$

$$m = \pm\frac{1}{2}, \pm\frac{3}{2}, \pm\frac{5}{2}, ..., \quad (106c)$$

$$j = \frac{1}{2}, \frac{3}{2}, \frac{5}{2}, ..., \quad (106d)$$

$$\psi_1 = \sqrt{\frac{j+1-m}{2j+2}}Y_{j+\frac{1}{2},m-\frac{1}{2}}(\vartheta,\varphi)f(r) + \sqrt{\frac{j+m}{2j}}Y_{j-\frac{1}{2},m-\frac{1}{2}}(\vartheta,\varphi)g(r), \quad (107a)$$

$$\psi_2 = -\sqrt{\frac{j+1+m}{2j+2}}Y_{j+\frac{1}{2},m+\frac{1}{2}}(\vartheta,\varphi)f(r) + \sqrt{\frac{j-m}{2j}}Y_{j-\frac{1}{2},m+\frac{1}{2}}(\vartheta,\varphi)g(r), \quad (107b)$$

$$Y_{\ell,m}(\vartheta,\varphi) = (-1)^m\sqrt{\frac{(2\ell+1)!(\ell-m)!}{4\pi(\ell+m)!}}P_\ell^m(\cos\vartheta)e^{im\varphi}. \quad (107c)$$

34

$$\xi = j(j+1)\hbar^2. \quad (106a')$$

The functions $f(r)$, $g(r)$ must be obtained from the Dirac equation (24b).

The functions ψ_3, ψ_4 of Eq. (102) are given by a pair of equations similar to Eq. (105) and have the same angular parts as Eqs. (107a)-(107b), but the radial parts are $F(r)$, $G(r)$ in place of the $f(r)$ and $g(r)$. The relationship between $f(r)$, $g(r)$ and $F(r)$, $G(r)$ is similar to that between the ψ_1, ψ_2 and ψ_3, ψ_4 in the case of the free electrons. If the state is a positive energy state, $F(r)$, $G(r)$ are the "small", and $f(r)$, $g(r)$ the "big" components, and vice versa for negative energy states. We shall see this when we treat the hydrogen atom problem in Chapter 5.

2.5 Foldy-Wouthuysen Representation

In Sect. 2, Eqs. (33)-(38), we have expressed the 4-component Dirac wave function $\Psi_\mu, \mu = 1, 2, 3, 4$, in two 2-component functions $\Phi_k, \varphi_k, k = 1, 2$.

Now, we shall introduce a unitary transformation operator[35]

$$U = \frac{\beta H + \mid E \mid}{\sqrt{2 \mid E \mid (\mid E \mid + m_0)}}, \tag{108}$$

and the transformation of

$$F = U\Psi. \tag{109}$$

Consequent upon this transformation, any operator Q is transformed into Q_F

$$Q_F = UQU^\dagger, \tag{110}$$

$$\Lambda_F = U\Lambda U^\dagger, \tag{111}$$

$$H_F = UHU^\dagger. \tag{112}$$

Here Λ (or Λ_F) is the energy projection operator. Calculations give the following results[36]

$$\begin{aligned} \mathbf{p}U - U\mathbf{p} = 0, \quad & \beta H \beta = 2m_0\beta - H, \\ H_F = \mid E \mid \beta, \quad & H_F^2 = E^2, \\ \Lambda_F = \beta. & \end{aligned} \tag{113}$$

35

$$U = \frac{\beta H + \mid E \mid}{\sqrt{2 \mid E \mid (\mid E \mid + m_0 c^2)}}. \tag{108$'$}$$

36

$$\mathbf{p}U - U\mathbf{p} = 0, \quad \beta H \beta = 2m_0 c^2 \beta - H. \tag{113$'$}$$

If one takes $E = E_+$,

$$m_s = \frac{1}{2}, \quad F = \begin{pmatrix} 1 \\ 0 \\ 0 \\ 0 \end{pmatrix} \frac{1}{(2\pi)^{\frac{3}{2}}} e^{ip_z z}, \tag{114a}$$

$$m_s = -\frac{1}{2}, \quad F = \begin{pmatrix} 0 \\ 1 \\ 0 \\ 0 \end{pmatrix} \frac{1}{(2\pi)^{\frac{3}{2}}} e^{ip_z z}, \tag{114b}$$

If one takes $E = E_-$,

$$m_s = \frac{1}{2}, \quad F = \begin{pmatrix} 0 \\ 0 \\ 0 \\ 1 \end{pmatrix} \frac{1}{(2\pi)^{\frac{3}{2}}} e^{ip_z z}, \tag{114c}$$

$$m_s = -\frac{1}{2}, \quad F = \begin{pmatrix} 0 \\ 0 \\ 1 \\ 0 \end{pmatrix} \frac{1}{(2\pi)^{\frac{3}{2}}} e^{ip_z z}. \tag{114d}$$

On introducing the operators

$$\frac{1}{2}(1+\beta), \quad \frac{1}{2}(1-\beta), \quad \text{and} \quad F = \begin{pmatrix} f_1 \\ f_2 \\ f_3 \\ f_4 \end{pmatrix} \tag{115}$$

then

$$\frac{1}{2}(1+\beta)F = \begin{pmatrix} f_1 \\ f_2 \\ 0 \\ 0 \end{pmatrix} \equiv F_+ \tag{116a}$$

$$\frac{1}{2}(1-\beta)F = \begin{pmatrix} 0 \\ 0 \\ f_3 \\ f_4 \end{pmatrix} \equiv F_- \tag{116b}$$

In the Foldy-Wouthuysen ($F_\pm$) representation, the positive energy states $\binom{f_1}{f_2}$ and the negative energy states $\binom{f_3}{f_4}$ do not mix. The 4-component Ψ_μ is now made up of two 2-component *spinors*

$$u = \begin{pmatrix} f_1 \\ f_2 \end{pmatrix}, \quad v = \begin{pmatrix} f_3 \\ f_4 \end{pmatrix} \tag{117}$$

The two components of u (and v) come from the spin angular momentum $m_s = \pm\frac{1}{2}$ values.

The separation of Ψ by U into unmixed positive and negative energy states, however, is possible only in the case of free electrons, for only in this case, the H and Λ do not mix the E_+ and E_- states. In the presence of external fields (as in the hydrogen atom, or in the Klein problem of Sect. 3. (3) above), the E_+ , E_- states are always mixed up and their separation cannot be achieved by the transformation U. In the jargon of the subject, one says that in an external field, there is a constant production of *virtual* electron-positron pairs.

Strictly speaking, in the pressence of external fields, it is not possible to regard the Dirac equation as a single-electron theory; the many-body feature is always present through the mixing with the negative energy states. Only in the non-relativistic limit is the one-particle theory valid; a fully relativistic treatment calls for the theory of quantized interacting electron-electromagnetic fields - namely, quantum electrodynamics.

References

Dirac, P. A. M., *Proc. Poy. Soc. London* **A 117**, 610; **118**, 351 (1928), gives the relativistic (linear) wave equations of four-component wave function.

Dirac, P. A. M., *ibid.* **A 126**, 360 (1930), hole theory of positron.

Oppenheimer, J. R., *Phys. Rev.* **35**, 939 (1930), hole theory of positron.

Pauli, W. Article in *Handb. der Physik*, Bd 24, 2nd ed. (1933), or Encycl. of Physics, Vol. 5/1, (1958).

Breit, G., *Proc. Nat Acad. Sci.*, (1928). The divorce between velocity and momentum in negative energy state.

Schrödinger, E., *Berlin Berichte*, 418 (1930); 63 (1931); Zeits. f. Physik **70**, 808 (1931). The Zitterbewegung, the even and odd operators. Also see Fock, V., *Zeits. f. Physik* **55**, 127 (1929); **68**, 527; **70**, 811 (1931).

Klein, O., *Zeits. f. Physik* **53**, 157 (1929), the "Klein paradox". See also Sauter, F., ibid. **69**, 742; **73**, 547 (1931).

Foldy, L. L. and Wouthuysen, S. A., *Phys. Rev.* **78**, 29 (1950); **87**, 688 (1952); see also B. Kursunoglu, *ibid.* **101**, 1419 (1956).

Exercises: *Chapter 2*

1. Clarification of the following aspects is useful for the materials covered in the present chapter:

(i) show that the four matrices $\{\alpha, \beta\}$ are linearly independent.

(ii) The u-spinor is specified by

$$u(\mathbf{p}, s) = N \begin{pmatrix} 1 \\ \frac{\sigma \cdot \mathbf{p}}{E+m_0} \end{pmatrix} \chi_s,$$

with N the normalization factor and χ_s the two-compenent Pauli spinor. Rewrite Eq. (38) in terms of the u-spinor.

On the other hand, the v-spinor is specified by

$$v(\mathbf{p}, s) = N \begin{pmatrix} -\frac{\sigma \cdot \mathbf{p}}{|E|+m_0} \\ 1 \end{pmatrix} \chi_{-s}.$$

Use the v-spinor to write out solutions to the free Dirac equation. (In a form analogous to Eq. (38).)

Obtain the condition which ensures the orthogonality between he u-spinors and v-spinors.

Find the expression for N if these spinors are normalized to unity.

(iii) On the problem of Klein paradox, prove Eq. (73). That is, the reflection coefficient R is unity for $E - m_0 \leq V \leq E + m_0$.

2. Two operators Λ^+ and Λ^- are defined by

$$\Lambda^+ = \frac{\gamma_\mu p_\mu + i m_0}{2 i m_0},$$

and

$$\Lambda^- = \frac{-\gamma_\mu p_\mu + i m_0}{2 i m_0}.$$

(i) Show that Λ^+ and Λ^- are projection operators, i.e.,

$$(\Lambda^\pm)^2 = \Lambda^\pm,$$

$$\Lambda^+ \Lambda^- = \Lambda^- \Lambda^+ = 0.$$

(ii) Show that

$$\sum_s u(\mathbf{p}, s)\bar{u}(\mathbf{p}, s) \propto \Lambda^+,$$

and

$$\sum_s v(\mathbf{p}, s)\bar{v}(\mathbf{p}, s) \propto \Lambda^-,$$

with $\bar{u} \equiv u^\dagger \gamma_4$ and $\bar{v} \equiv v^\dagger \gamma_4$.

3. Decide the condition under which one may obtain the simultaneous eigenvalues of the Hamiltonian and σ_3 of a free Dirac particle. Write down the solutions and their eigenvalues explicitly.

4. Show that the Dirac equation, Eq. (24b), is invariant under gauge transformations.

Chapter 3. γ_μ Matrices; Helicity; Charge Conjugation

3.1 Properties of the γ_μ Matrices

In Eqs. (20a) and (20b), Ch. 2, we have defined the γ_μ, $\mu = 1, ..., 4$, matrices which have the property

$$\gamma_\mu \gamma_\nu + \gamma_\nu \gamma_\mu = 2\delta_{\mu\nu}, \quad \mu, \nu = 1, ..., 4, \tag{1a}$$

$$\gamma_\mu^\dagger = \gamma_\mu. \tag{1b}$$

We shall in the following gather together the properties of these matrices in the form of theorems.

Theorem 1. The lowest order of 4 matrices satisfying the conditions Eqs. (1a) and (1b) is 4×4.

The proof is given in the footnote following Eq. (17), Ch. 2.

Let us form the following 16 4×4 matrices Γ_A, $A = 1, ..., 16$,

$$I \qquad \text{(unit matrix; } \Gamma_{16})$$

$$\gamma_1 \quad \gamma_2 \quad \gamma_3 \quad \gamma_4$$

$$i\gamma_1\gamma_2 \quad i\gamma_2\gamma_3 \quad i\gamma_3\gamma_4 \quad i\gamma_1\gamma_4 \quad i\gamma_2\gamma_4 \quad i\gamma_1\gamma_3$$

$$i\gamma_1\gamma_2\gamma_3 \quad i\gamma_2\gamma_3\gamma_4 \quad i\gamma_1\gamma_3\gamma_4 \quad i\gamma_1\gamma_2\gamma_4$$

$$\gamma_1\gamma_2\gamma_3\gamma_4 (\equiv \gamma_5) \tag{2}$$

Theorem 2. The Γ_A, $A = 1, ..., 16$, *are hermitian.*

The proof follows from Eqs. (1a) and (1b). For example

$$(i\gamma_1\gamma_2)^\dagger = -i\gamma_2^\dagger\gamma_1^\dagger = -i\gamma_2\gamma_1 = i\gamma_1\gamma_2, \quad \text{etc.}$$

Theorem 3.

$$\gamma_5\gamma_\mu + \gamma_\mu\gamma_5 = 0, \quad \mu = 1, ..., 4. \tag{3}$$

Theorm 4.

$$(\Gamma_A)^2 = I, \quad \text{i.e., the } \Gamma\text{'s are unitary.} \tag{4}$$

Theorem 5.

$$\Gamma_A \Gamma_B = \pm \Gamma_C, \quad \text{or} \quad \pm i\Gamma_C. \tag{5}$$

In view of Eqs. (1), any product $\Gamma_A \Gamma_B$ can be expressed as a product of at most 4 different $\gamma_1, \gamma_2, \gamma_3, \gamma_4$. But Eq. (2) contains all such products.
Theorem 6.

$$\Gamma_A \Gamma_B + \Gamma_B \Gamma_A = 0, \qquad \text{or} \qquad \Gamma_A \Gamma_B - \Gamma_B \Gamma_A = 0. \tag{6}$$

Theorem 7.

$$\text{If} \qquad \Gamma_A \Gamma_B = \Gamma_A \Gamma_C, \qquad \text{then} \qquad \Gamma_B = \Gamma_C. \tag{7}$$

If by hypothesis $\Gamma_A \Gamma_B = \Gamma_A \Gamma_C$, then

$$\Gamma_A \Gamma_A \Gamma_B = \Gamma_A \Gamma_A \Gamma_C.$$

$$\Gamma_B = \Gamma_C \quad \text{by Theorem 4.}$$

Theorem 8. Except for I, the trace of each Γ_A is zero.

Proof:

(i) From Eq. (1a),

$$\begin{aligned}
&\frac{1}{2}(\gamma_\mu \gamma_\mu + \gamma_\mu \gamma_\mu) = I \quad \text{(no summation over } \mu \text{ here);} \\
&\frac{1}{2}(\gamma_\nu \gamma_\mu \gamma_\mu + \gamma_\nu \gamma_\mu \gamma_\mu) = \gamma_\nu, \quad \nu \neq \mu \\
&\frac{1}{2}(\gamma_\nu \gamma_\mu \gamma_\mu - \gamma_\mu \gamma_\nu \gamma_\mu) = \gamma_\nu \quad \text{by Eq. (1a)} \\
&\frac{1}{2}(Tr(\gamma_\nu \gamma_\mu \gamma_\mu) - Tr(\gamma_\mu \gamma_\nu \gamma_\mu)) = Tr\gamma_\nu
\end{aligned}$$

But the lefthand side is zero by a theorem of the trace.

(ii)

$$\frac{1}{2}(\gamma_\mu \gamma_\nu + \gamma_\nu \gamma_\mu) = 0, \qquad \mu \neq \nu;$$

$$\frac{1}{2}(Tr\gamma_\mu \gamma_\nu + Tr\gamma_\nu \gamma_\mu) = 0,$$

or

$$Tr(\gamma_\mu \gamma_\nu) = 0.$$

(iii) Take $\gamma_1\gamma_2\gamma_3\gamma_4\gamma_\mu$ where μ is among 1, 2, 3, 4. By three permutations of pair of indices, the product can be brought to the order $\gamma_\mu\gamma_1\gamma_2\gamma_3\gamma_4$. By Eq. (1a),

$$\gamma_1\gamma_2\gamma_3\gamma_4\gamma_\mu + \gamma_\mu\gamma_1\gamma_2\gamma_3\gamma_4 = 0,$$

$$Tr(\gamma_1\gamma_2\gamma_3\gamma_4\gamma_\mu) = 0.$$

By letting $\gamma_\mu = \gamma_1, \gamma_2, \gamma_3, \gamma_4$ in succession, one gets

$$Tr(\gamma_\mu\gamma_\nu\gamma_\sigma) = 0 \quad \text{for } \mu \neq \nu \neq \sigma \neq \mu.$$

(iv) From

$$\gamma_1\gamma_2\gamma_3\gamma_4 + \gamma_1\gamma_2\gamma_3\gamma_4 = \gamma_1\gamma_2\gamma_3\gamma_4 - \gamma_4\gamma_1\gamma_2\gamma_3$$

$$2Tr(\gamma_1\gamma_2\gamma_3\gamma_4) = Tr(\gamma_1\gamma_2\gamma_3\gamma_4) - Tr(\gamma_4\gamma_1\gamma_2\gamma_3) = 0.$$

Theorem 9. The 16 Γ_A *in Eq. (2) are linearly independent.*

If they are not linearly independent, i.e., if there exists a relation

$$\sum_{A=1}^{16} C_A\Gamma_A = 0, \tag{8}$$

then multiplying this by an arbitrary Γ_B and using Eq. (4), one gets

$$\sum_{A=1}^{16}{}' C_A\Gamma_A\Gamma_B + C_B I = 0, \tag{9}$$

where $\sum'$ indicates the exclusion of the term C_B from the summation, so that in the sum, Γ_B is different from Γ_A. By Theorems 5 and 8, $Tr(\Gamma_A\Gamma_B) = 0$ in each term of the sum $\sum'$. But $TrI = n$. Hence $C_B = 0$. As Γ_B is arbitrary among the 16 Γ_A 's, it follows that all $C_A = 0$ in Eq. (8).

Theorem 10. An arbitrary 4×4 *matrix can be expressed as a linear combination of the 16* Γ_A *'s,*

$$X = \sum_{A=1}^{16} C_A\Gamma_A, \tag{10}$$

a 4×4 matrix has 16 (complex) numbers, and the 16 numbers C_A serve to represent them. The C_A are given by

$$C_A = \frac{1}{4}Tr(X\Gamma_A). \tag{10a}$$

Theorem 11. If an arbitrary 4×4 *matrix* F *commutes with* γ_μ, $\mu = 1, ..., 4$,

$$\gamma_\mu - \gamma_\mu F = 0, \quad \mu = 1, 2, 3, 4,$$

then

$$F\Gamma_A - \Gamma_A F = 0 \quad \text{for all } \Gamma_A. \tag{11}$$

Theorem 12. Except for $\Gamma_{16} \equiv I$, *for every* Γ_A , *there is at least one* Γ_B *such that*

$$\Gamma_B \Gamma_A \Gamma_B = -\Gamma_A. \tag{12}$$

The following table gives the Γ_B for each Γ_A satisfying Eq. (12). The numbering starts with $I = \Gamma_{16}$, $\gamma_1 = \Gamma_1$, $\gamma_2 = \Gamma_2$, etc.

A	2	3	4	5	6	7	8
B	3, 4, 5	4, 5, 2	3, 5, 2	2, 3, 4	2, 3	3, 4	4, 5
	6, 9, 11	6, 7, 10	7, 8, 11	8, 9, 10	7, 9, 10, 11	6, 8, 10, 11	7, 9, 10, 11
	13	11	15	12			

A	9	10	11	12	13	14	15	16
B	5, 2	5, 3	2, 4	5	2	3	4	2, 3, 4, 5
	6, 8, 10, 11	6, 7, 8, 9	6, 7	8, 9, 10	6, 9, 11	6, 7, 10	7, 8, 11	12, 13
	8, 9			13, 14, 15	14, 15, 12	15, 12, 13	12, 13, 14	14, 15

(12a)

Theorem 13. If F *commutes with all* Γ_A, F *must be a multiple of the unit matrix* I.

Proof: We write F in the form Eq. (10)

$$\begin{aligned} F &= \sum_{A=1}^{16} C_A \Gamma_A \\ &= C_B \Gamma_B + \sum_A{}' C_A \Gamma_A, \end{aligned} \tag{13a}$$

where Γ_B is any one of the 16 Γ_A except I, and $\sum'$ indicates the exclusion of A = B. Let Γ_μ be one of the matrices that satisfy Eq. (12) (see (12a)),

$$\Gamma_\mu \Gamma_B \Gamma_\mu = -\Gamma_B, \quad \text{(no summation over } \mu \text{ here)}$$

They by Eq. (6),

$$\Gamma_\mu F \Gamma_\mu = -C_B \Gamma_B + \sum_A{}' (\pm) C_A \Gamma_A. \tag{13b}$$

By hypothesis, $\Gamma_\mu F \Gamma_\mu = \Gamma_\mu \Gamma_\mu F = F$. Multiplying Eqs. (13a) and (13b) by Γ_B and taking the trace, one finds

$$C_B = -C_B, \qquad C_B = 0.$$

Since Γ_B is arbitrary (except I), hence

$$F = cI.$$

Theorem 14. Any two sets of γ_μ *satisfying*

$$\gamma_\mu \gamma_\nu + \gamma_\nu \gamma_\mu = 2\delta_{\mu\nu}, \qquad \gamma'_\mu \gamma'_\nu + \gamma'_\nu \gamma'_\mu = 2\delta_{\mu\nu}, \tag{14a}$$

differ from each other only by a similarity transformation

Proof: It is obvious that Eq. (14) is a sufficient condition for γ'_μ to satisfy Eq. (14a) if $\gamma_\mu{}'s$ satisfy it.

To show the necessity part, let the Γ_A and the Γ'_A formed from the γ_μ, γ'_μ respectively be similarly ordered as in Eq. (2). By Theorem 5,

$$\Gamma_A \Gamma_B = \epsilon_{AB} \Gamma_C, \qquad \Gamma'_A \Gamma'_B = \epsilon_{AB} \Gamma'_C$$

where

$$\epsilon = \pm 1, \pm i$$

Let F be an arbitrary nonzero 4×4 matrix, and define S by

$$S^{-1} = \sum_{B=1}^{16} \Gamma'_B F \Gamma_B.$$

Then

$$\begin{aligned} \Gamma'_A S^{-1} \Gamma_A &= \sum_{B=1} \Gamma'_A \Gamma'_B F \Gamma_B \Gamma_A \\ &= \sum_c \epsilon_{AB} \Gamma'_C F \frac{1}{\epsilon_{AB}} \Gamma_C = S^{-1}, \end{aligned}$$

$$\Gamma'_A S^{-1} = S^{-1} \Gamma_A,$$

$$\Gamma'_A = S^{-1} \Gamma_A S,$$

which includes Eq. (14).

Theorem 15. If in Theorem 14 the two sets of γ_μ, γ'_μ*, in addition to satisfying Eq. (14a), are hermitian, then they differ by a unitary transformation,*

$$\gamma'_\mu = U^\dagger \gamma_\mu U. \tag{15}$$

Theorem 16. The transformation matrix S in Eq. (14) is unique, up to a complex multiplicative constant.

Let S_1 and S_2 be two matrices satisfying Eq. (14).

$$\gamma'_\mu = S_1^{-1}\gamma_\mu S_1, \quad \gamma'_\mu = S_2^{-1}\gamma_\mu S_2.$$

Hence

$$\begin{aligned}\gamma'_\mu &= S_1^{-1}(S_2\gamma'_\mu S_2^{-1})S_1\\ &= (S_1^{-1}S_2)\gamma'_\mu(S_2^{-1}S_1)\\ &= (S_1^{-1}S_2)\gamma'_\mu(S_1^{-1}S_2)^{-1},\end{aligned}$$

$$\gamma'_\mu(S_1^{-1}S_2) = (S_1^{-1}S_2)\gamma'_\mu, \qquad \mu = 1,2,3,4.$$

By Theorem 11, $(S_1^{-1}S_2)$ commutes with all Γ_A.

By Theorem 13,

$$\begin{aligned}S_1^{-1}S_2 &= \text{const. } I.\\ S_2 &= \text{const. } S_1.\end{aligned}$$

Theorem 17. The Dirac equations[1]

$$\begin{aligned}&\frac{\partial\Psi}{\partial t} + [(\alpha\cdot\nabla) + im_0\beta]\Psi = 0,\\ &\frac{\partial\tilde{\Psi}^*}{\partial t} + (\nabla\tilde{\Psi}^*\cdot\alpha) - im_0\tilde{\Psi}^*\beta = 0,\end{aligned} \tag{16}$$

are invariant under the following unitary transformation

$$\begin{aligned}&\alpha'_\mu = U^\dagger\alpha_\mu U, \qquad \beta' = U^\dagger\beta U,\\ &\Psi' = U^\dagger\Psi, \qquad \tilde{\Psi}'^* = \tilde{\Psi}^* U.\end{aligned} \tag{17}$$

The proof is obvious.

3.2 Helicity and Neutrinos

Let $\mathbf{n}$ be a unit vector along the momentum p,

$$\mathbf{n}(\xi,\eta,\zeta) \tag{18}$$

[1]

$$\frac{\partial\Psi}{\partial t} + [c(\alpha\cdot\nabla) + i\frac{m_0c^2}{\hbar}\beta]\Psi = 0. \tag{16'}$$

where ξ, η, ζ are the direction consines of $\mathbf{n}$. Let $\sigma_x, \sigma_y, \sigma_z$ be the 2×2 Pauli matrices. Then the component of σ along $\mathbf{p}$ are

$$\begin{aligned}\sigma_n &= (\sigma\cdot\mathbf{n})\\ &= \sigma_x\xi + \sigma_y\eta + \sigma_z\zeta,\\ &= \begin{pmatrix}\zeta & \xi - i\eta\\ \xi + i\eta & -\zeta\end{pmatrix},\end{aligned}\tag{19a}$$

and

$$\sigma_n \text{ is hermitian,}\tag{19b}$$

$$(\sigma\cdot\mathbf{n})^2 = I.\tag{19c}$$

(1) Helicity, eigenvalues and eigenfunctions
The helicity operator (a 4×4 matrix) is defined as

$$\sigma_n = \begin{pmatrix}(\sigma\cdot\mathbf{n}) & 0\\ 0 & (\sigma\cdot\mathbf{n})\end{pmatrix},\tag{20}$$

$$(\sigma_n)^2 = \begin{pmatrix} I & 0\\ 0 & I\end{pmatrix}.\tag{21}$$

The eigenvalues and eigenfunctions of σ_n in Eq. (19a)

$$\sigma_n\begin{pmatrix}\psi_1\\ \psi_2\end{pmatrix} = h\begin{pmatrix}\psi_1\\ \psi_2\end{pmatrix}\tag{22}$$

are

$$h = \begin{cases} 1, & \frac{\psi_1}{\psi_2} = \frac{1+\zeta}{\xi+i\eta}\left(=\frac{\xi-i\eta}{1-\zeta}\right),\\ -1, & \frac{\psi_1}{\psi_2} = \frac{-\xi+i\eta}{1+\zeta}\left(=\frac{-1+\zeta}{\xi+i\eta}\right),\end{cases}\quad .\tag{23}$$

For the helicity σ_n in Eq. (20),

$$\sigma_n\begin{pmatrix}u_1\\ u_2\\ u_3\\ u_4\end{pmatrix} = h\begin{pmatrix}u_1\\ u_2\\ u_3\\ u_4\end{pmatrix},\tag{24}$$

this equation decomposes into two equations

$$\sigma_n\begin{pmatrix}u_1\\ u_2\end{pmatrix} - h\begin{pmatrix}u_1\\ u_2\end{pmatrix} = 0,\quad \sigma_n\begin{pmatrix}u_3\\ u_4\end{pmatrix} - h\begin{pmatrix}u_3\\ u_4\end{pmatrix} = 0,\tag{25}$$

which are both identical with Eq. (22), so that

$$h = \pm 1 \text{ as in Eq. (23).}\tag{26}$$

Let the subscript u and ℓ denote "upper" and "lower" , and

$$\phi_u = \begin{pmatrix} u_1 \\ u_2 \end{pmatrix}, \quad \phi_\ell = \begin{pmatrix} u_3 \\ u_4 \end{pmatrix}, \tag{27}$$

each of which is a 2-component spinor. Eq. (25) can be written

$$\sigma_n \phi_u = h\phi_u, \qquad \sigma_n \phi_\ell = h\phi_\ell \tag{28}$$

The normalized ϕ_u, ϕ_ℓ are, from Eq. (23),

$$\begin{aligned} h = 1, \quad & \phi_{u+} = \phi_{\ell+} = \sqrt{\frac{1}{2(1+\zeta)}} \begin{pmatrix} 1+\zeta \\ \xi + i\eta \end{pmatrix}, \\ h = -1, \quad & \phi_{u-} = \phi_{\ell-} = \sqrt{\frac{1}{2(1+\zeta)}} \begin{pmatrix} -\xi + i\eta \\ 1+\zeta \end{pmatrix}, \end{aligned} \tag{29}$$

so that

$$\tilde{\phi}^*_{u+}\phi_{u+} = \tilde{\phi}^*_{u-}\phi_{u-} = \tilde{\phi}^*_{\ell+}\phi_{\ell+} = \tilde{\phi}^*_{\ell-}\phi_{\ell-},$$

$$\tilde{\phi}^*_{u+}\phi_{u-} = \tilde{\phi}^*_{\ell+}\phi_{\ell-} = 0.$$

We define the nomenclature:

$$h = \begin{cases} 1, & \text{positive helicity, right-handed,} \\ -1, & \text{negative helicity, left-handed.} \end{cases} \tag{30}$$

For a free electron, it is seen from Eq. (88), Ch. 2, that the Hamiltonian commutes with σ_n,

$$H\sigma_n - \sigma_n H = 0, \tag{31}$$

so that H and the helicity have simultaneous eigenfunctions. Dirac's equation is[2]

$$((\alpha \cdot \mathbf{p}) + m_0\beta)\Psi = E\Psi. \tag{32}$$

In terms of the spinors ϕ_u, ϕ_ℓ, we have

$$\Psi = \begin{pmatrix} \phi_u \\ \phi_\ell \end{pmatrix} = \begin{pmatrix} a_u \\ a_\ell \end{pmatrix} \exp(ip_\mu x_\mu), \tag{33}$$

where

$$p_\mu x_\mu = \mathbf{p} \cdot \mathbf{r} - Et. \tag{33a}$$

---

[2]

$$(c(\alpha \cdot \mathbf{p}) + m_0 c^2)\Psi = E\Psi. \tag{32'}$$

Using Eq. (19), Ch. 2, we obtain from Eq. (32)

$$(E - m_0)a_u - (\sigma \cdot \mathbf{p})a_\ell = 0,$$
$$-(\sigma \cdot \mathbf{p})a_u + (E + m_0)a_\ell = 0, \tag{34}$$

where σ are the 2×2 Pauli matrices, and $\mathbf{p}$ is the momentum but no longer an operator. The eigenvalues of E are

$$E = \pm\sqrt{\mathbf{p}^2 + m_0^2} = E_\pm, \tag{35}$$

and[3]

$$a_\ell = \frac{(\sigma \cdot \mathbf{p})}{E + m_0} a_u \quad (\text{or}, \ \frac{E - m_0}{(\sigma \cdot \mathbf{p})} a_u). \tag{36}$$

For

$$E = \left\{ \begin{array}{c} E_+ \\ E_- \end{array} \right\}, \quad a_u = \left\{ \begin{array}{c} \text{large} \\ \text{small} \end{array} \right\} \quad \text{and} \quad a_\ell = \left\{ \begin{array}{c} \text{small} \\ \text{large} \end{array} \right\} \quad \text{components.} \tag{37}$$

This is what has been obtained in

$$h = \left\{ \begin{array}{c} +1 \\ -1 \end{array} \right\}, \qquad \frac{(\sigma \cdot \mathbf{p})}{|\,\mathbf{p}\,|}\Psi = \pm\Psi. \tag{38}$$

Let

$$a = \frac{|\,\mathbf{p}\,|}{|\,E\,| + m_0}, \qquad N^2 = \frac{|\,E\,| + m_0}{4\,|\,E\,|\,(1+\zeta)}. \tag{39}$$

The simultaneous eigenvalues and normalized eigenfunctions of H and σ_n are as follows,

$$E_+, \quad h = 1, \quad \Psi = N \begin{pmatrix} 1+\zeta \\ \xi + i\eta \\ a(1+\zeta) \\ a(\xi + i\eta) \end{pmatrix} \exp(ip_\mu x_\mu), \tag{40a}$$

$$E_+, \quad h = -1, \quad \Psi = N \begin{pmatrix} -\xi + i\eta \\ 1+\zeta \\ -a(-\xi + i\eta) \\ -a(1+\zeta) \end{pmatrix} \exp(ip_\mu x_\mu), \tag{40b}$$

$$E_-, \quad h = 1, \quad \Psi = N \begin{pmatrix} a(1+\zeta) \\ a(\xi + i\eta) \\ 1+\zeta \\ \xi + i\eta \end{pmatrix} \exp(-ip_\mu x_\mu), \tag{40c}$$

---

[3]

$$a_\ell = \frac{c(\sigma \cdot \mathbf{p})}{E + m_0 c^2} a_u, \quad \text{etc.} \tag{36$'$}$$

$$E_-, \quad h = -1, \quad \Psi = N \begin{pmatrix} -a(-\xi + i\eta) \\ -a(1+\zeta) \\ -\xi + i\eta \\ 1+\zeta \end{pmatrix} \exp(-ip_\mu x_\mu). \tag{40d}$$

On account of $h = \pm 1$, the E_+ and E_- state are each doubly degenerate.

(2) Neutrino, helicity and chirality

The Dirac equations Eq. (34) will simplify when the mass m_0 of the particle approaches zero. As in Eq. (18), let $\mathbf{n}$ be a unit vector along $\mathbf{p}$, and here

$$\mathbf{n} = \frac{1}{E}\mathbf{p}, \tag{41}$$

so that

$$\begin{aligned} &\mathbf{n} \uparrow \; \uparrow \mathbf{p} \quad \text{for} \quad E_+, \\ &\mathbf{n} \uparrow \; \downarrow \mathbf{p} \quad \text{for} \quad E_-. \end{aligned} \tag{42}$$

Eq. (34) now become

$$\begin{aligned} \phi_u &= (\sigma \cdot \mathbf{n})\phi_\ell, \\ \phi_\ell &= (\sigma \cdot \mathbf{n})\phi_u, \\ \sigma &= \text{Pauli matrices.} \end{aligned} \tag{43}$$

Eq. (33) becomes

$$\Psi = \begin{pmatrix} \phi_u \\ \phi_\ell \end{pmatrix} = \begin{pmatrix} \phi_u \\ (\sigma \cdot \mathbf{n})\phi_u \end{pmatrix} \tag{44a}$$

and

$$(\sigma \cdot \mathbf{n})\Psi = \begin{pmatrix} (\sigma \cdot \mathbf{n})\phi_u \\ (\sigma \cdot \mathbf{n})^2\phi_u \end{pmatrix} = \begin{pmatrix} \phi_\ell \\ \phi_u \end{pmatrix} \tag{44b}$$

so that $(\sigma \cdot \mathbf{n})$ interchanges ϕ_u and ϕ_ℓ.

Consider $\gamma_5 = \gamma_1\gamma_2\gamma_3\gamma_4$ in Eq. (2).

$$\gamma_5 = -\begin{pmatrix} 0 & 0 & 1 & 0 \\ 0 & 0 & 0 & 1 \\ 1 & 0 & 0 & 0 \\ 0 & 1 & 0 & 0 \end{pmatrix} = -\begin{pmatrix} 0 & I \\ I & 0 \end{pmatrix} \tag{45a}$$

so that

$$\gamma_5 \begin{pmatrix} \phi_u \\ \phi_\ell \end{pmatrix} = -\begin{pmatrix} \phi_\ell \\ \phi_u \end{pmatrix}. \tag{45b}$$

Hence

$$-\gamma_5 = (\sigma \cdot \mathbf{n}). \tag{46}$$

If we define two 2 component spinors α, β by

$$\alpha = \frac{1}{2}(\phi_u + \phi_\ell) = \frac{1}{2}(1 + \sigma \cdot \mathbf{n})\phi_u$$
$$\beta = \frac{1}{2}(\phi_u - \phi_\ell) = \frac{1}{2}(1 - \sigma \cdot \mathbf{n})\phi_u \tag{47}$$

then

$$(\sigma \cdot \mathbf{n})\alpha = \alpha, \qquad (\sigma \cdot \mathbf{n})\beta = -\beta, \tag{48}$$

or

$$(\sigma \cdot \mathbf{n})\begin{pmatrix}\alpha \\ \beta\end{pmatrix} = \begin{pmatrix}\alpha \\ -\beta\end{pmatrix}$$

i.e., α, β are the 2-component eigenfunctions of the helicity operator $(\sigma \cdot \mathbf{n})$ corresponding to the helicity $h = 1, -1$ respectively.

The above case of parallel and antiparallel σ and $\mathbf{p}$ comes about only for $m_0 = 0$, i.e., the particle travelling with the speed of light. For $m_0 \neq 0$, it is always possible to make a Lorentz transformation to a frame in which the particle is at rest, and the above relations do not hold.

The above theory, in which two-component wave functions for a particle of zero rest mass was in fact considered by H. Weyl (1929) before the neutrino was proposed by Pauli in 1931, has the following feature. Let P be the parity, or inversion, operator

$$P: \quad \mathbf{r} \to -\mathbf{r}. \tag{49}$$

The momentum $\mathbf{p}$ is a polar vector whereas the angular momentum σ is a pseudo (or, axial) vector,

$$P\,\mathbf{p} = -\mathbf{p}\,P, \quad P\sigma = \sigma\,P,$$

$$P(\sigma \cdot \mathbf{n}) = -(\sigma \cdot \mathbf{n})P. \tag{50}$$

Applying P on the equation Eq. (48), one gets

$$-(\sigma \cdot \mathbf{n})(P\alpha) = (P\alpha)$$
$$-(\sigma \cdot \mathbf{n})(P\beta) = -(P\beta) \tag{51}$$

which differ from Eq. (48) by a negative sign, i.e., are not invariant under the inversion operation. It is for this reason Pauli rejected this theory (in his 1933 article in the Handbuch der Physik, 2nd edition).

This basis for the objection based on the violation of the "conservation of parity" was removed by the theory of T. D. Lee and C. N. Yang on the non-conservation of parity in weak interactions,[4] and its subsequent confirmation by the experiments of C. S. Wu et al.

[4] *Weak interactions are those that govern such processes as*

and Lederman. The 2-component theory of neutrinos is revived by A. Salam, L. Landau, and Lee and Yang (1957).

The concept of helicity is related to the concept of neutrino and anti-neutrino. First of all, the electron e^- and the positron e^+ are called the 'particle' and the 'anti-particle' by convention. Once we start with this 'definition', the neutrino associated with e^- is an anti-neutrino denoted by $\bar{\nu}_e$, whereas the neutrino associated with e^+ is a neutrino, denoted by ν_e.[5] These two neutrinos are actually *different*. Experiments show that, with $\mathbf{s} = \frac{\sigma}{2}$,

$$\text{the neutrino } \nu_e \text{ has negative helicity, } \mathbf{s} \uparrow\downarrow \mathbf{p},$$

$$\text{the anti-neutrino } \bar{\nu}_e \text{ has positive helicity, } \mathbf{s} \uparrow\uparrow \mathbf{p}. \tag{55}$$

We shall introuce another related concept "chirality" here.

Let us transform the γ_μ in Eqs. (20a) and (20b), Ch. 2, by a unitary transformation

$$\gamma'_\mu = U\gamma_\mu U^{-1}, \tag{57}$$

$$U = \frac{1}{\sqrt{2}}\begin{pmatrix} 1 & 1 \\ -1 & 1 \end{pmatrix}, \quad U^\dagger = U^{-1} = \frac{1}{\sqrt{2}}\begin{pmatrix} 1 & -1 \\ 1 & 1 \end{pmatrix}, \tag{58}$$

(i) β -decays,

$$n \to p + e^- + \bar{\nu}_e,$$

$$^{11}_{6}C \to {}^{11}_{5}B + e^+ + \nu_e, \tag{52}$$

(ii) $\pi - \mu$ decays,

$$\pi^+ \to \mu^+ + \nu_\mu,$$

$$\pi^- \to \mu^- + \bar{\nu}_\mu, \tag{53}$$

(iii) $\mu - e$ decay

$$\mu^- \to e^- + \bar{\nu}_e + \nu_\mu \tag{54}$$

[5] *This naming of ν_e and $\bar{\nu}_e$ is not arbitrary. We introduce the concept of a quantum number called the "lepton number" L, such that a "particle" has $L = 1$, and an "anti-particle" $L = -1$. From the systematics of all elementary particle reactions, it has been found that the "selection rule"*

$$\triangle L = 0 \tag{56}$$

holds. That e^+ has $L = -1$ follows from the pair-production process

$$h\nu \to e^+ + e^-.$$

It is to be noted that in Eq. (53), the neutrinos associated with the muons μ^-, μ^+ are denoted by $\bar{\nu}_\mu, \nu_\mu$. That they are distinct from the neutrinos associated with the e^-, e^+ is experimentally established by Lederman (1962).

The notations in reactions (52), (53) and (54) all confirm the helicity relation (55) and lepton number conservation law (56).

(See Theorem 14 and 15 of Sect. 1). It is found that

$$\gamma'_k = \gamma_k, \quad k = 1, 2, 3,$$
$$\gamma'_4 = \begin{pmatrix} 0 & -1 \\ -1 & 0 \end{pmatrix},$$
$$\gamma'_5 = \begin{pmatrix} -1 & 0 \\ 0 & 1 \end{pmatrix}. \tag{58a}$$

Thus

$$\gamma'_5 \begin{pmatrix} \alpha \\ \beta \end{pmatrix} = \begin{pmatrix} -\alpha \\ \beta \end{pmatrix}. \tag{59}$$

The operator γ'_5 is called the chirality operator.[6] Comparison with the helicity $(\sigma \cdot \mathbf{n})$ in Eq. (48) shows that γ'_5 and $\sigma_n = (\sigma \cdot \mathbf{n})$ have opposite eigenvalues, i.e., γ'_5 and $-\sigma_n$ have the same property on the spinors $\binom{\alpha}{\beta}$.

3.3 Charge Conjugation

(1) Charge conjugate state Ψ_c

The Dirac equation for an electron (charge $-e$) in an electromagnetic field $(A, i\phi)$ is Eq. (24a),[7] Ch. 2.[8]

$$(\gamma_\mu(p_\mu + eA_\mu) - im_0)\Psi = 0, \tag{60}$$

$$p_k = \frac{1}{i}\frac{\partial}{\partial x_k}, \quad k = 1, 2, 3, \qquad p_4 = \frac{1}{i}\frac{\partial}{\partial x_4}, \qquad x_4 = it$$

$$A_k \text{ real}, \quad k = 1, 2, 3, \qquad A_4 = i\phi = \text{imaginary}. \tag{61}$$

Taking the complex conjugate of Eq. (60) (not transpose), one gets

$$\{-\gamma^*_k(p_k - eA_k) + \gamma^*_4(p_4 - eA_4) + im_0\}\Psi^* = 0, \tag{62}$$

[6] *For a given Dirac spinor ψ as in Eq. (44a), the projected states $\psi_L = \frac{1}{2}(1+\gamma_5)\psi$ and $\psi_R = \frac{1}{2}(1-\gamma_5)\psi$ are referred to as "left-handed" and "right-handed" components of ψ , respectively. (Here γ_5 is given by Eq. (2).) Eq. (46) indicates that H and $\frac{1}{2}(1 \pm \gamma_5)$ have simultaneous eigenfunctions. (See Eqs. (31) and (40a)-(40d).) The components of definite chirality (or helicity) are basic entries in the construction of the Glashow-Salam-Weinberg $SU(2)_L \times U(1)$ electroweak theory. (See Chapter 13.)*

[7]

$$(\gamma_\mu(p_\mu + \frac{e}{c}A_\mu) - im_0c)\Psi = 0. \tag{60'}$$

[8] *In the present section, we use the following notations: $\Psi_\mu, \Psi^\dagger_\mu, \Psi_c$ are column matrices, $\tilde{\Psi}_\mu, \tilde{\Psi}^*_\mu, \tilde{\Psi}_c, \Psi^\dagger = \tilde{\Psi}^*, \bar{\Psi} = \Psi^\dagger\gamma_4$ are row matrices.*
Different notations are used in the literature. Thus in Pauli's article in Handbuch der Physik , $i\tilde{\Psi}^*\gamma_4$ is denoted by $\Psi^\dagger$; in Davydov's Quantum Mechanics, $\tilde{\Psi}^*\gamma_4$ is denoted by $\bar{\Psi}$, as here.

where p_k, p_4, A_k, A_4 have the same meaning as in Eq. (61).

For a particle of the same rest mass m_0 and charge $+e$, the Dirac equation is

$$(\gamma_\mu(p_\mu - eA_\mu) - im_0)\Psi_c = 0. \tag{63}$$

Ψ_c is defined as the *charge conjugate state* wave function. Note that Ψ_c is *not* the positron state wave function. See the following.

To obtain the relation between Ψ and Ψ_c, let C be a 4×4 matrix, and multiply Eq. (62) by C from the left, and get (using the summation convention for k)

$$C\{\gamma_k^*(p_k - eA_k) - \gamma_4^*(p_4 - eA_4) - im_0\}C^{-1}C\Psi^* = 0. \tag{64}$$

Comparison of Eqs. (63) and (64) shows that

$$\Psi_c \quad \text{and} \quad (\pm)C\Psi^* \tag{65}$$

satisfy the same equation if C satisfies the relations

$$C\gamma_k^*C^{-1} = \gamma_k, \; k = 1, 2, 3, \quad -C\gamma_4^*C^{-1} = \gamma_4. \tag{66}$$

These conditions do not suffice for the determination of the matrix C. On taking the transpose of Eq. (62), multiplying by γ_4 from the right and referring to Eqs. (20a)-(20b), Ch. 2, with

$$\tilde{\gamma}_\mu = \gamma_\mu^*, \qquad \gamma_4^* = \gamma_4,$$

$$\gamma_k\gamma_\mu = -\gamma_\mu\gamma_k, \qquad \gamma_k\gamma_\mu^* = -\gamma_\mu^*\gamma_k, \tag{67}$$

$$(\mu \neq k) \tag{1}$$

one gets

$$\tilde{\Psi}^*\gamma_4\{\gamma_k(p_k - eA_k) + \gamma_4(p_4 - eA_4) + im_0\} = 0. \tag{68}$$

Taking the transpose of Eq. (63), multiplying by $-\gamma_4C^{-1}$ from the right, one gets

$$\tilde{\Psi}_c\gamma_4\tilde{\gamma}_kC^{-1}(p_k - eA_k) - \tilde{\Psi}_c\gamma_4\tilde{\gamma}_4C^{-1}(p_4 - eA_4) + im_0\tilde{\Psi}_c\gamma_4C^{-1} = 0. \tag{69}$$

Comparison of Eqs. (68) and (69) (noting $\tilde{\gamma}_kC^{-1} = \gamma_k^*C^{-1} = C^{-1}\gamma_k$) shows that

$$\tilde{\Psi}^*\gamma_4 \quad \text{and} \quad (\pm)\tilde{\Psi}_c\gamma_4C^{-1} \tag{70}$$

satisfy the same equation. From Eqs. (65) and (70), one gets (since $\gamma_4^* = \gamma_4$)

$$\gamma_4C = \pm C\gamma_4. \tag{71}$$

To be consistent with Eq. (66)

$$\gamma_4C = -C\gamma_4,$$

we may take

$$\tilde{C} = C = C^{-1}, \tag{72}$$

and in particular,

$$C = \gamma_2 = \begin{pmatrix} 0 & 0 & 0 & -1 \\ 0 & 0 & 1 & 0 \\ 0 & 1 & 0 & 0 \\ -1 & 0 & 0 & 0 \end{pmatrix}. \tag{73}$$

We finally take, from Eq. (65)

$$\Psi_c = C\Psi^*, \tag{74}$$

and from Eq. (70)

$$\tilde{\Psi}^*\gamma_4 = -\tilde{\Psi}_c\gamma_4 C^{-1},$$

which is identical with Eq. (74).[9]

Let us now apply Eq. (74) to the wave function Ψ of an electron (charge $-e$) in a positive energy E_+ state, helicity $\sigma_n = 1, p =\mid \mathbf{p} \mid$, namely Ψ in Eq. (40a).

$$\Psi = N \begin{pmatrix} 1+\zeta \\ \xi + i\eta \\ a(1+\zeta) \\ a(\xi + i\eta) \end{pmatrix} \exp(ip_\mu x_\mu), \quad (a, N \text{ in Eq. (39)}). \tag{75}$$

From Eqs. (74) and (72),

$$\Psi_c = C\Psi^* = N \begin{pmatrix} -a(\xi - i\eta) \\ a(1+\zeta) \\ \xi - i\eta \\ -(1+\zeta) \end{pmatrix} \exp(-ip_\mu x_\mu). \tag{76}$$

Comparing this with Ψ in Eq. (40d), one sees that Ψ_c is exactly the wave function of an electron e^- in a negative energy E_- state, helicity $\sigma_n = -1$, $p = -\mid \mathbf{p} \mid$ (except for a phase factor $e^{i\pi} = -1$).

To sum up the above tedious calculations, it is found that the wave function Ψ_c of e^+ is the same as that of e^- in a negative energy E_- state.

The following figure shows the above result

E_+	....	$p =\mid \mathbf{p} \mid$,	$\sigma_n = 1$,	$(\sigma \uparrow\uparrow \mathbf{p})$,		
m_0c^2	————————					
0	————————					
$-\,m_0c^2$	————————					
E_-	————————	$p = -\mid \mathbf{p} \mid$,	$\sigma_n = -1$,	$(\sigma \uparrow\downarrow \mathbf{p})$,	$\Psi_c = \gamma_2\Psi^*$.	(77)

9

$$\gamma_4 C^{-1} = \gamma_4\gamma_2 = -\gamma_2\gamma_4.$$

Thus. $\tilde{\Psi}^* = \tilde{\Psi}_c\gamma_2 = \tilde{\Psi}_c\tilde{\gamma}_2$, *so that* $\Psi^* = \gamma_2\Psi_c$ *or* $\Psi_c = \gamma_2\Psi^*$.

It is important to note that Ψ_c of e^+, although it is the same as that of e^- in a E_- state, does *not* represent a positron. A positron is represented by a 'hole' in the sea of negative energy states. Thus the 'hole' has the following characteristics: charge $+e$, energe E_+, momentum p. The 'hole', or positron, is called an anti-electron.

Coming back to Eq. (74) , one sees that Ψ_c is the charge-conjugate state of Ψ. But from Eq. (74), one obtains [using Eqs. (72)-(73)]

$$\Psi = C\Psi_c^*, \tag{78}$$

so that Ψ is also the charge-conjugate state of Ψ_c. But Ψ_c has just been seen to be the same as the function of e^- in a negative energy $E_- < 0$ state. Hence Eq. (78) shows that the charge conjugate state of e^- in a E_- state is the state of e^- in a E_+ state Ψ.

(2) Charge conjugate current

From Eqs. (1), (5), and (7) in Ch. 2., we write[10]

$$I_k = \tilde{\Psi}^*\alpha_k\Psi, \quad k = 1,2,3, \quad I_4 = i\tilde{\Psi}^*\Psi, \tag{79}$$

so that the equation of continuity Eq. (8), Ch. 2., now takes the form

$$\frac{\partial I_\mu}{\partial x_\mu = 0.} \tag{80}$$

In the γ_μ -representation Eq. (20a)

$$\begin{aligned} \alpha_k &= i\beta\gamma_k = i\gamma_4\gamma_k, & \alpha_4 &= \beta = \gamma_4, \\ I_k &= i\tilde{\Psi}^*\gamma_4\gamma_k\Psi, & I_4 &= i\tilde{\Psi}^*\gamma_4\gamma_4\Psi. \end{aligned} \tag{81}$$

We define the 4-electric current

$$J_\mu = -eI_\mu = -ie\tilde{\Psi}^*\gamma_4\gamma_\mu\Psi. \tag{82}$$

It will be shown in the following chapter that the I_μ defined in Eq. (81) transforms as a 4-vector under Lorentz transformations.

For the charge conjugate current, let us consider

$$\tilde{\Psi}_c^*\gamma_4\gamma_\mu\Psi_c. \tag{83}$$

From Eq. (74), we have, with $C = \gamma_2$, $\tilde{\gamma}_2^* = \gamma_2$,

$$\begin{aligned} \tilde{\Psi}_c^*\gamma_4\gamma_\mu\Psi_c &= \tilde{\Psi}\gamma_2\gamma_4\gamma_\mu\gamma_2\Psi^*, \\ \tilde{\Psi}_c^*\gamma_4\gamma_k\Psi_c &= \tilde{\Psi}\gamma_2\gamma_4\gamma_k\gamma_2\Psi^* = -\tilde{\Psi}\gamma_4\gamma_k^*\Psi^*, \quad \text{by Eq. (66)}, \\ \tilde{\Psi}_c^*\gamma_4\gamma_4\Psi_c &= \tilde{\Psi}\Psi^*. \end{aligned}$$

10

$$I_k = c\tilde{\Psi}^*\alpha_k\Psi, \quad I_4 = ic\tilde{\Psi}^*\Psi. \tag{79'}$$

Hence, if we define for the charge conjugate current

$$J_k^c = ie\{\tilde{\Psi}_c^*\gamma_4\gamma_k\Psi_c\}, \tag{84}$$

then

$$J_\mu^c = ie\tilde{\Psi}^*\gamma_4\gamma_\mu\Psi. \tag{84a}$$

Thus

$$J_\mu^c = -J_\mu. \tag{85}$$

3.4 Dirac Equation in Majorana Representation

In Chapter 2, Sec. 1, (ii), we have defined Majorana's representation of the Dirac matrices

$$\begin{aligned}
\gamma_1^M &= \begin{pmatrix} \sigma_z & 0 \\ 0 & \sigma_z \end{pmatrix}, \quad \gamma_2^M = \begin{pmatrix} 0 & -i\sigma_y \\ i\sigma_y & 0 \end{pmatrix}, \\
\gamma_3^M &= \begin{pmatrix} -\sigma_x & 0 \\ 0 & \sigma_x \end{pmatrix}, \quad \gamma_4^M = \begin{pmatrix} 0 & -\sigma_y \\ -\sigma_y & 0 \end{pmatrix}.
\end{aligned} \tag{86}$$

where $\sigma_x, \sigma_y, \sigma_z$ are the 2×2 Pauli matrices. [See Eq. (21) in Ch. 2.] In terms of the γ_μ^M matrices, the operator in Dirac's equation for a free electron[11]

$$\begin{aligned}
&(\gamma_\mu^M \frac{\partial}{\partial x_\mu} + m_0)\Psi = 0, \\
&\gamma_\mu^M \frac{\partial}{\partial x_\mu} + m_0 = (\gamma_\mu^M \frac{\partial}{\partial x_\mu} + m_0)^*,
\end{aligned} \tag{87}$$

is an operator in real numbers. If we take as in Eq. (26) of Ch. 2,

$$\Psi_\mu = A_\mu \quad \exp(i(\mathbf{p}\cdot\mathbf{r})), \quad A_\mu = \text{const.}, \tag{88}$$

as one solution for $p =\mid \mathbf{p} \mid$, $E = \sqrt{p^2 + m_0^2}$, then

$$\Psi_\mu^* = A_\mu^* \quad \exp(-i(\mathbf{p}\cdot\mathbf{r})), \quad A_\mu^* = \text{const.}, \tag{89}$$

is also a solution for $p = -\mid \mathbf{p} \mid$, $E = -\sqrt{p^2 + m_0^2}$. As in Eqs. (40a)-(40d), for E_+ and $p =\mid \mathbf{p} \mid$, the double degeneracy is distinguished by the helicity $h = \pm 1$, and similarly for E_- and $p = -\mid \mathbf{p} \mid$.

The helicity operator

$$\sigma_n = (\sigma \cdot \mathbf{n}), \quad \mathbf{n} = \frac{\mathbf{p}}{\mid \mathbf{p} \mid}. \tag{90}$$

---

[11]

$$(\gamma_\mu^M \frac{\partial}{\partial x_\mu} + \frac{m_0 c}{\hbar})\Psi = 0. \tag{87'}$$

has been defined in Eq. (20) in terms of $(\sigma \cdot \mathbf{n})$ where $\sigma_x, \sigma_y, \sigma_z$ are the Pauli 2×2 matrices. But σ_n can also be expressed in terms of the 4×4 $\gamma_1, \gamma_2, \gamma_3$ matrices of Eq. (80e) of Ch. 2,

$$\sigma_j = -i\gamma_k\gamma_\ell, \quad \text{cyclic permutation of j, k, } \ell = 1,\ 2,\ 3. \tag{91}$$

In Majorana represention, we have

$$\sigma_j^M = -i\gamma_k^M\gamma_\ell^M. \tag{92}$$

As the γ_k^M, $k = 1, 2, 3$, are all real and symmetric, the σ_j^M are all pure imaginary

$$(\sigma_j^M)^* = -\sigma_j^M. \tag{93}$$

Hence

$$(\sigma_n^M)^* = -\sigma_n^M. \tag{93a}$$

From Eqs. (24), (25), and (30), we know the eigenvalues of σ_n to be ± 1.

Let $\Psi_\mu^{(1)}, \Psi_\mu^{(2)}, \Psi_\mu^{(3)}, \Psi_\mu^{(4)}$ be solutions of the Dirac equation for the following states

$$\left\{ \begin{array}{c} \Psi_\mu^{(1)} \\ \Psi_\mu^{(2)} \end{array} \right\} \quad \text{are solutions for} \quad p = \mid \mathbf{p} \mid, \quad E_+, \quad \sigma_n^M = \left\{ \begin{array}{c} 1 \\ -1 \end{array} \right\}, \tag{94a}$$

$$\left\{ \begin{array}{c} \Psi_\mu^{(3)} \\ \Psi_\mu^{(4)} \end{array} \right\} \quad \text{are solutions for} \quad p = - \mid \mathbf{p} \mid, \quad E_-, \quad \sigma_n^M = \left\{ \begin{array}{c} -1 \\ 1 \end{array} \right\}. \tag{94b}$$

The eigenvalue equation for σ_n are

$$\begin{aligned} \sigma_n^M \Psi_\mu^{(1)} &= \Psi_\mu^{(1)}, \\ \sigma_n^M \Psi_\mu^{(2)} &= -\Psi_\mu^{(2)}, \\ \sigma_n^M \Psi_\mu^{*(1)} &= -\Psi_\mu^{*(1)}, \qquad \Psi^{*(1)} = \Psi^{(3)}; \\ \sigma_n^M \Psi_\mu^{*(2)} &= \Psi_\mu^{*(2)}, \qquad \Psi^{*(2)} = \Psi^{(4)}. \end{aligned} \tag{95}$$

Thus in the Majorana representation, Ψ_μ are separated into two pairs

$$\begin{pmatrix} \Psi^{(1)} \\ \Psi^{(2)} \end{pmatrix} \quad \text{and} \quad \begin{pmatrix} \Psi^{(3)} \\ \Psi^{(4)} \end{pmatrix} = \begin{pmatrix} \Psi^{*(1)} \\ \Psi^{*(2)} \end{pmatrix}.$$

Thus, if ψ_M is a solution of energy $E_\pm$ and Majorana spin $\sigma_n^M (= \pm 1)$, then ψ_M^* is also a Majorana spinor of energy $E_\mp$ and Majorana spin $-\sigma_n^M$. This aspect turns out to be of some importance in constructing theories for neutrinos.

References

Pauli, W., *Article in Handbuch der Physik Bd.* 24 (1933), or *Encycl. of Physics*, Vol. 5/1, (1958).

Good, R. H., *Reviews of Mod. Phys.* **27**, 187 (1955).

Eisele, J.A., *Modern Quantum Mechanics with Applications to Elementary Particle Physics* (Pergamon Press, New York, 1969).

Schiff, L.I., *Quantum Mechanics*, 3rd Edition (McGraw-Hill, New York, 1968).

Davydov, A. S., *Quantum Mechanics*, 2nd Edition (Pergamon Press, Oxford, 1965).

Exercises: *Chapter 3*

1. Theorem 1 states that the lowest order of 4 matrices satisfying the conditions (1a) and (1b) is 4×4. It is also of critical importance to ask whether the structure of the Dirac equation is *unique*. To this end, it is useful to examine Theorems 2-17 to see if Theorems 15-17 can be derived without any reference to the dimension of the γ matrices.

2. Demonstrate that the charge conjugation operator defined in Section 3.3 is unique up to a phase.

3. Assume that Ψ^M is a solution to the free Dirac equation in the Majorana representation. Obtain a solution to the free Dirac equation in the Pauli-Dirac representation, which can be expressed in terms of Ψ^M. Discuss the property of such solution under charge conjugation. You may now turn around the argument by showing a way to obtain the solution to the Dirac equation in the Majorana representation from that in the Pauli-Dirac representation.

Chapter 4. Transformations of the Dirac Equation

The properties of the Dirac relativistic equations of a charge particle under various transformtions such as gauge transformation of the electromagnetic fields, Lorentz transformations, space inversion will be the subject of our discussion in this chapter. The charge conjugation treated in the preceding chapter is another transformation.

4.1 Unitary Transformations

Theorem: Under the unitary transformation

$$\alpha'_\mu = S^{-1}\alpha_\mu S, \quad \beta' = S^{-1}\beta S, \tag{1}$$

$$\Psi' = S^{-1}\Psi, \quad \Psi'^{\dagger} = \Psi^{\dagger} S, \tag{2}$$

the form of the Dirac equation is invariant.

We take the form Eq. (16) of Ch. 3,[1]

$$\frac{\partial \Psi'}{\partial t} + (\alpha' \cdot \nabla)\Psi' + im_0\beta'\Psi' = 0, \tag{3}$$

$$\frac{\partial \Psi'^{\dagger}}{\partial t} + (\nabla\Psi'^{\dagger} \cdot \alpha') - im_0\Psi'^{\dagger}\beta' = 0. \tag{4}$$

The theorem follows from the four equations (1)-(4).

4.2 Gauge Transformations

Theorem: Under the transformation of the 4-potential[2]

$$A' = A + \nabla\chi, \quad \phi' = \phi - \frac{\partial \chi}{\partial t},$$
$$\partial_\mu\partial_\mu\chi = 0 \text{ with } \chi \text{ a scalar function}, \tag{5}$$

and the simultaneous transformation[3]

$$\Psi' = \Psi \exp(-ie\chi), \tag{6}$$

---

[1]

$$\frac{\partial \Psi'}{\partial t} + c(\alpha' \cdot \nabla)\Psi' + \frac{im_0c^2}{\hbar}\beta'\Psi' = 0. \tag{3'}$$

[2]

$$\phi' = \phi - \frac{1}{c}\frac{\partial \chi}{\partial t}. \tag{5'}$$

[3]

$$\Psi' = \Psi \exp(-\frac{ie}{\hbar c}\chi). \tag{6'}$$

the Dirac equation remains unchanged in form.[4]

We take the form Eqs. (24a) and (23), Ch. 2,

$$\{\gamma_\mu(\frac{1}{i}\frac{\partial}{\partial x_\mu} + eA'_\mu) - im_0\}\Psi' = 0. \tag{7}$$

The theorem follows from Eqs. (5), (6), and (7).

4.3 Lorentz Transformations

The Lorentz transformation

$$\begin{aligned} x' &= \gamma(x - \beta t), \quad y' = y, \quad z' = z, \quad t' = \gamma(t - \beta x), \\ x &= \gamma(x' + \beta t'), \quad y = y', \quad z = z', \quad t = \gamma(t' + \beta x'), \end{aligned} \tag{8a}$$

$$\beta = \frac{v}{c}, \qquad \gamma = \frac{1}{\sqrt{1-\beta^2}}, \tag{8b}$$

will be put in the form[5,6]

$$\begin{aligned} x'_\mu &= a_{\mu\nu} x_\nu, \quad x_4 = it, \\ x_\nu &= x'_\mu a_{\mu\nu}, \quad x'_4 = it', \end{aligned} \tag{8c}$$

$$a_{\mu\rho} a_{\nu\rho} = \delta_{\mu\nu}, \tag{8d}$$

$$\det |\, a_{\mu\nu} \,| = 1, \tag{8e}$$

or

$$\begin{pmatrix} x'_1 \\ x'_2 \\ x'_3 \\ x'_4 \end{pmatrix} = \begin{pmatrix} \gamma & 0 & 0 & i\gamma\beta \\ 0 & 1 & 0 & 0 \\ 0 & 0 & 1 & 0 \\ -i\gamma\beta & 0 & 0 & \gamma \end{pmatrix} \begin{pmatrix} x_1 \\ x_2 \\ x_3 \\ x_4 \end{pmatrix}, \tag{8f}$$

[4] *Eqs. (5) and (6) are referred to as "gauge transformation of the second kind". The invariance of the equation under such transformation is called "gauge invariance"'. It is now known that gauge field theories, i.e., field theories which respect gauge invariance, are of fundamental importance. However, it remains mysterious why theories with extra unphysical gauge degrees of freedom are so successful. Perhaps there is a much deeper meaning in connection with the gauge degree of freedom but any such meaning is yet to be found.*

[5] *Eq. (8a) states that the Jacobian of the transformation Eq. (8a), or Eqs. (8f) and (8g) is +1 . This condition restricts the transformations to proper Lorentz transformations. For space inversion and time reversal, the Jacobian is -1 . See Sect. 4 below.*

[6]

$$x'_\mu = a_{\mu\nu} x_\nu, \quad x_4 = ict. \tag{8c$'$}$$

$$(x_1, x_2, x_3, x_4) = (x_1', x_2', x_3', x_4')\begin{pmatrix} \gamma & 0 & 0 & i\gamma\beta \\ 0 & 1 & 0 & 0 \\ 0 & 0 & 1 & 0 \\ -i\gamma\beta & 0 & 0 & \gamma \end{pmatrix}. \tag{8g}$$

In the following, we shall use the notation[7]

$$\Psi^\dagger = i\tilde{\Psi}^*\gamma_4, \quad \text{or} \quad \bar{\Psi} = \tilde{\Psi}^*\gamma_4. \tag{9}$$

Dirac's equations are Eqs. (22a) and (24a), Ch. 2,

$$(\gamma_\mu \frac{\partial}{\partial x_\mu} + m_0)\Psi = 0, \tag{10}$$

$$(\gamma_\mu \Pi_\mu - im_0)\Psi = 0, \tag{11}$$

where[8]

$$\Pi_k = \frac{1}{i}\frac{\partial}{\partial x_k} + eA_k, \quad \Pi_4 = \frac{1}{i}\frac{\partial}{i\partial t} + ie\phi, \tag{12}$$

and

$$\Pi_\mu \Pi_\nu - \Pi_\nu \Pi_\mu = \frac{e}{i} F_{\mu\nu} \tag{13}$$

$$F_{\mu\nu} = \frac{\partial A_\nu}{\partial x_\mu} - \frac{\partial A_\mu}{\partial x_\nu} = \begin{pmatrix} 0 & H_z & -H_y & -iE_x \\ -H_z & 0 & H_x & -iE_y \\ H_y & -H_x & 0 & -iE_z \\ iE_x & iE_y & iE_z & 0 \end{pmatrix} = -F_{\nu\mu}. \tag{14}$$

In terms of the $\Psi^\dagger$ in Eq. (9), we obtain from Eqs. (10) and (11)

$$\frac{\partial \Psi^\dagger}{\partial x_\mu}\gamma_\mu - m_0\Psi^\dagger = 0, \tag{10a}$$

$$\Pi_\mu^* \Psi^\dagger \gamma_\mu - im_0\Psi^\dagger = 0. \tag{11a}$$

We define[9]

$$s_\mu = \Psi^\dagger \gamma_\mu \Psi \tag{15}$$

[7] *The notation $\Psi^\dagger$ here is that of Pauli in the Handbuch der Physik, Bd 24/1, (1933). $\Psi^\dagger, \tilde{\Psi}^*$ are row matrices. In Pauli's article, Ψ^* is a row matrix, so that Eq. (9) is $\Psi^\dagger = i\Psi^*\gamma_4$.*

[8]

$$\Pi_k = \frac{\hbar}{i}\frac{\partial}{\partial x_k} + \frac{e}{c}A_k, \quad \Pi_4 = \frac{\hbar}{i}\frac{\partial}{ic\partial t} + i\frac{e}{c}\phi. \tag{12'}$$

[9] *The proof that s_μ so defined is a 4-vector under Lorentz transformation will be given in the following. The J_μ in Eq. (82), Ch. 3, is in fact s_μ, except for a constant factor*

$$J_\mu = -es_\mu.$$

and the 4-divergence

$$\frac{\partial s_\mu}{\partial x_\mu} = 0. \tag{16}$$

On making the Lorentz transformation Eq. (8c)

$$x'_\mu = a_{\mu\nu} x_\nu, \quad \text{or} \quad x_\nu = x'_\mu a_{\mu\nu}$$

and

$$\Psi' = S\Psi, \quad \Psi^{\dagger'} = \Psi^\dagger S^{-1}, \tag{17}$$

then Eqs. (10) and (10a) become[10]

$$\gamma_\nu \frac{\partial}{\partial x'_\nu} \Psi' + m_0 \Psi' = 0, \tag{18}$$

$$\frac{\partial \Psi^{\dagger'}}{\partial x'_\nu} \gamma_\nu - m_0 \Psi^{\dagger'} = 0, \tag{19}$$

where

$$\gamma_\nu = (S^{-1} \gamma_\mu S) a_{\mu\nu}, \tag{20}$$

which can be written

$$S^{-1} \gamma_\mu S = a_{\mu\nu} \gamma_\nu. \tag{20a}$$

From Eqs. (9) and (17), we have

$$\Psi^{\dagger'} = i\tilde{\Psi}^* \gamma_4 S^{-1}. \tag{21}$$

We also have

$$\Psi^{\dagger'} = i\tilde{\Psi}^{*'} \gamma_4.$$

But from Eq. (17),

$$\tilde{\Psi}^{*'} = \tilde{\Psi}^* S^\dagger, \quad (S^\dagger = \tilde{S}^*).$$

Hence

$$\Psi^{\dagger'} = i\tilde{\Psi}^* S^\dagger \gamma_4. \tag{22}$$

---

[10]

$$(\gamma_\nu \frac{\partial}{\partial x'_\nu} + \frac{m_0 c}{\hbar}) \Psi' = 0, \quad \text{etc.} \tag{18'}$$

From Eqs. (21) and (22) , we have the condition

$$S^{\dagger}\gamma_4 S = \gamma_4. \tag{23}$$

To insure the Lorentz invariance of the Dirac equations (10), (10a), (18), and (19), the conditions are Eqs. (20a) and (23a)

$$S^{-1}\gamma_\mu S = a_{\mu\nu}\gamma_\nu, \tag{20a}$$

$$S^{\dagger}\gamma_4 S = \gamma_4 \ \text{(for proper Lorentz transformation).} \tag{23a}$$

We have to show that an S exists that satisfies these conditions.

Before we do that, we shall find the transformation properties of the 16 $\Gamma_A{}'s$ in Eq. (2), Ch. 3.

(i) I

$$\Psi^{\dagger'}\Psi' = \Psi^{\dagger}S^{-1}S\Psi = \Psi^{\dagger}\Psi, \quad \text{an invariant.}$$

(ii) $\gamma_1, \gamma_2, \gamma_3, \gamma_4$

$$\Psi^{\dagger'}\gamma_\mu\Psi' = \Psi^{\dagger}S^{-1}\gamma_\mu S\Psi = a_{\mu\nu}\Psi^{\dagger}\gamma_\nu\Psi, \quad \text{a 4-vector.}$$

(iii) $\gamma_1\gamma_2,\ \gamma_2\gamma_3,\ \gamma_3\gamma_4,\ \gamma_1\gamma_4,\ \gamma_2\gamma_4,\ \gamma_1\gamma_3$

$$\begin{aligned} M'_{\mu\nu} &= \Psi^{\dagger'}\gamma_\mu\gamma_\nu\Psi' = \Psi^{+}S^{-1}\gamma_\mu SS^{-1}\gamma_\nu S\Psi \\ &= a_{\mu\rho}a_{\nu\sigma}M_{\rho\sigma}, \quad \text{an antisymmetric tensor (or, a 6-vector).} \end{aligned} \tag{24}$$

(iv) $\gamma_1\gamma_2\gamma_3,\ \gamma_2\gamma_3\gamma_4,\ \gamma_1\gamma_3\gamma_4,\ \gamma_1\gamma_2\gamma_4$

$$\begin{aligned} M'_{\lambda\mu\nu} &= \Psi^{\dagger'}S^{-1}\gamma_\lambda SS^{-1}\gamma_\mu SS^{-1}\gamma_\nu S\Psi \\ &= a_{\lambda\rho}a_{\mu\sigma}a_{\nu\tau}M_{\rho\sigma\tau}, \quad \text{an antisymmetric tensor of the third rank.} \end{aligned}$$

(v) $\gamma_1\gamma_2\gamma_3\gamma_4 \equiv \gamma_5$

$$\Psi^{\dagger'}\gamma_1\gamma_2\gamma_3\gamma_4\Psi' = \Psi^{\dagger}\gamma_1\gamma_2\gamma_3\gamma_4\Psi,$$

or

$$\Psi^{\dagger'}\gamma_5\Psi' = \Psi^{\dagger}\gamma_5\Psi.$$

These relations hold for proper Lorentz transformations (i.e., not involving space inversion or reflection, or time reversal).

4.4 Space Inversion, Charge Conjugation, and Time Reversal

(1) Space inversion:

$$P: \quad \mathbf{r} \to -\mathbf{r}. \tag{25}$$

The transformation equations are

$$x'_\mu = a_{\mu\nu} x_\nu,$$

where

$$a_{\mu\nu} = \begin{pmatrix} -1 & 0 & 0 & 0 \\ 0 & -1 & 0 & 0 \\ 0 & 0 & -1 & 0 \\ 0 & 0 & 0 & 1 \end{pmatrix}. \tag{26}$$

The conditions Eqs. (20a) and (23) are then

$$\begin{aligned} S_p\gamma_k + \gamma_k S_p &= 0, \quad k = 1,2,3; \\ S_p\gamma_4 - \gamma_4 S_p &= 0, \\ S_p^\dagger \gamma_4 - \gamma_4 S_p^{-1} &= 0. \end{aligned} \tag{27}$$

These relations can be satisfied by choosing

$$S_p = \lambda\gamma_4, \quad \lambda = \text{constant}. \tag{28}$$

As two successive space inversions are equivalent to an identity transformation, one has, for $S_p(S_p\psi) = +\psi$,

$$(S_p)^2 = \lambda^2\gamma_4^2 = \lambda^2 = 1. \tag{29a}$$

But an inversion is equivalent to a reflection on a plane followed by a rotation through π about an axis perpendicular to that plane; two successive inversions are equivalent to a rotation through an angle 2π . It will be shown later that the S_r for rotation about an axis $j = 1,2,3$ (for x, y, z respectively) through an angle w is [11]

$$S_{rj} = \cos\frac{w}{2} + i\sigma_j \sin\frac{w}{2}.$$

and for $w = 2\pi$,

$$S_{2\pi} = -1.$$

Hence one also has

$$(S_p)^2 = S_{2\pi} = -1, \tag{29b}$$

[11] *See Eq. (58) below. For σ_j, $j = 1,2,3$, see Eq. (80e), Ch. 2.*

which corresponds to $S_p(S_p\psi) = -\psi$.
From Eqs. (29a) and (29b) , one has 4 solutions [12]

$$\lambda = i,\ -i,\ 1,\ -1. \tag{30}$$

From Eq. (17), $\Psi' = S\Psi$ and Eq. (29), one has

$$\Psi' = \lambda\gamma_4\Psi \quad \text{or} \quad \psi'(\mathbf{r},t) = \lambda\gamma_4\psi(\mathbf{r},t), \tag{31}$$

or, on taking the complex conjugate,

$$\Psi^{*'} = \lambda^*\gamma_4^*\Psi^*, \tag{31a}$$

or

$$\tilde{\Psi}^{*'} = \lambda^*\tilde{\Psi}^*\gamma_4, \quad \Psi^{\dagger'} = \lambda^*\Psi^\dagger\gamma_4.$$

The properties of the 16 Γ'_A in Eq. (2) of Ch. 3 under inversion are as follows,

$$\Psi^{\dagger'}\Psi' = \Psi^\dagger\Psi,$$

$$\begin{cases} \Psi^{\dagger'}\gamma_k\Psi' = -\Psi^\dagger\gamma_k\Psi, & k = 1,2,3; \\ \Psi^{\dagger'}\gamma_4\Psi' = \Psi^\dagger\gamma_4\Psi, & \text{a vector.} \end{cases}$$

$$\begin{cases} \Psi^{\dagger'}\gamma_k\gamma_j\Psi' = \Psi^\dagger\gamma_k\gamma_j\Psi, \\ \Psi^{\dagger'}\gamma_k\gamma_4\Psi' = -\Psi^\dagger\gamma_k\gamma_4\Psi; \end{cases} \tag{32}$$

$$\begin{aligned} \Psi^{\dagger'}\gamma_1\gamma_2\gamma_3\Psi' &= -\Psi^\dagger\gamma_1\gamma_2\gamma_3\Psi, \\ \Psi^{\dagger'}\gamma_k\gamma_j\gamma_4\Psi' &= \Psi^\dagger\gamma_k\gamma_j\gamma_4\Psi \quad \text{an axial vector,} \\ \Psi^{\dagger'}\gamma_1\gamma_2\gamma_3\gamma_4\Psi' &= -\Psi^\dagger\gamma_1\gamma_2\gamma_3\gamma_4\Psi, \quad \text{a pseudo-scalar.} \end{aligned}$$

In Eqs. (44b) and (46), Ch. 3, it is seen that $-\gamma_5$ and the helicity operator $(\sigma\cdot\mathbf{n})$ have the same property in their operation on the two 2-component spinors ϕ_u, ϕ_l , namely, interchange of ϕ_u and ϕ_l . From the last relation above

$$\Psi^{\dagger'}\gamma_5\Psi' = -\Psi^\dagger\gamma_5\Psi$$

it is seen that both $\Psi^\dagger\gamma_5\Psi$ and $(\sigma\cdot\mathbf{n})$ change sign upon space inversion.

From Eqs. (73), (74), and (78), it is shown that

$$\Psi_c = \gamma_2\Psi^*, \quad \Psi^* = \gamma_2\Psi_c. \tag{33}$$

Upon space inversion

$$\Psi'_c = \gamma_2\Psi^{*'}, \tag{1}$$

[12] *Cf. C. N. Yang and J. Tiomno, Phys. Rev.* **79**, *495 (1950).*

one has from Eq. (31a)

$$\begin{aligned}\Psi_c' &= \lambda^* \gamma_2 \gamma_4^* \Psi^* \\ &= -\lambda^* \gamma_4 \gamma_2 \Psi^* \quad \text{by Eq. (66), Ch. 3} \\ &= -\lambda^* \gamma_4 \Psi_c. \end{aligned} \tag{34}$$

On comparing with Eq. (31),

$$\Psi' = \lambda \gamma_4 \Psi,$$

it is seen that

$$\begin{aligned} &\text{(i) if} \quad \lambda = \pm i, \quad \Psi_c \quad \text{and} \quad \Psi \quad \text{transform in the same way,} \\ &\text{(ii) if} \quad \lambda = \pm 1, \quad \Psi_c \quad \text{and} \quad \Psi \quad \text{transform differently.} \end{aligned} \tag{35}$$

But

$$\begin{aligned}\tilde{\Psi}_c' \Psi' &= -\lambda \lambda^* \tilde{\Psi}_c \gamma_4 \gamma_4 \Psi \\ &= -\tilde{\Psi}_c \Psi, \end{aligned} \tag{36}$$

showing that the intrinsic parity of Ψ (of e^-) and the intrinsic parity of Ψ_c (of the charge conjugate e^+) are always opposite to each other.[13]

The relationship between charge conjugation and parity operation is as follows.

In Eqs. (48) and (55), Ch. 3, it has been seen that for

$$\begin{aligned} \text{anti-neutrino,} \quad h &= 1, \quad (\sigma \cdot \mathbf{n})\alpha = \alpha, \\ \text{neutrino,} \quad h &= -1, \quad (\sigma \cdot \mathbf{n})\beta = -\beta, \end{aligned} \tag{37a}$$

or,

$$(\sigma \cdot \mathbf{n}) \begin{pmatrix} \alpha \\ \beta \end{pmatrix} = \begin{pmatrix} \alpha \\ -\beta \end{pmatrix}.$$

Now, upon parity (space inversion) operation P, Eq. (25), shows

$$\begin{aligned} P(\sigma \cdot \mathbf{n}) &= -(\sigma \cdot \mathbf{n})P, \\ (\sigma \cdot \mathbf{n})(P\alpha) &= -(P\alpha), \\ (\sigma \cdot \mathbf{n})(P\beta) &= (P\beta). \end{aligned}$$

Comparison between these relations and Eq. (37a) shows that the transformation property of (P α) (α an anti-neutrino state) is the same as β , a neutrino state, and (P β) has the transformation property of α , i.e., a parity operation converts a neutrino into an anti-neutrino, and vice versa. This is expressed by the following expression

$$\begin{aligned} P\alpha &\propto \beta, \\ P\beta &\propto \alpha. \end{aligned} \tag{37b}$$

[13] *By intrinsic parity, it is meant the parity apart from that pertaining to the orbital motion of a particle.*

Consider next the charge conjugation Eqs. (74) and (73), Ch. 3.

$$\Psi_c = \gamma_2 \Psi^*$$

$$\gamma_2 = \begin{pmatrix} 0 & \begin{pmatrix} 0 & -1 \\ 1 & 0 \end{pmatrix} \\ \begin{pmatrix} 0 & 1 \\ -1 & 0 \end{pmatrix} & 0 \end{pmatrix}.$$

Carrying out in succession the P and C operations and using $\alpha^* = \alpha$ and $\beta^* = \beta$, we obtain

$$\begin{aligned} CP\begin{pmatrix} \alpha \\ \beta \end{pmatrix} &= CP(\sigma\cdot\mathbf{n})\begin{pmatrix} \alpha \\ -\beta \end{pmatrix} = -C(\sigma\cdot\mathbf{n})P\begin{pmatrix} \alpha \\ -\beta \end{pmatrix} \\ &= -C(\sigma\cdot\mathbf{n})\begin{pmatrix} \beta \\ -\alpha \end{pmatrix} = C\begin{pmatrix} \beta \\ \alpha \end{pmatrix} \\ &= \gamma_2\begin{pmatrix} \beta \\ \alpha \end{pmatrix} = \begin{pmatrix} 0 & -1 \\ 1 & 0 \end{pmatrix}\begin{Bmatrix} \alpha \\ -\beta \end{Bmatrix}. \end{aligned} \tag{37c}$$

In words, this states that the combined operation of parity P and charge conjugation C on a $\left\{\begin{matrix} \textit{antineutrino} \\ \textit{neutrino} \end{matrix}\right\}$ leads to a $\left\{\begin{matrix} \textit{antineutrino} \\ \textit{neutrino} \end{matrix}\right\}$; i.e., the nature of $\left\{\begin{matrix} \textit{antineutrino} \\ \textit{neutrino} \end{matrix}\right\}$ is conserved under the combined CP operation.

(2) Time reversal:

$$T : t \to -t. \tag{38}$$

Consider the Dirac equation for a particle of charge -e,

$$\{\gamma\cdot(\mathbf{p} + e\mathbf{A}) + \gamma_4(-\frac{\partial}{\partial t} + ie\phi) - im_0\}\psi = 0. \tag{39a}$$

under the time reversal transformation, $t \to t' = -t$, we have $\mathbf{A}(-t) = -\mathbf{A}(t)$ and $\phi(-t) = \phi(t)$ so that

$$\{\gamma\cdot(\mathbf{p} - e\mathbf{A}(t)) + \gamma_4(\frac{\partial}{\partial t} + ie\phi) - im_0\}\psi(-t) = 0. \tag{39b}$$

Introduce

$$T\psi(-t) = \psi'(t). \tag{40}$$

We find, from Eq. (39b),

$$\begin{aligned} &T(-i\gamma)T^{-1}\cdot\nabla\psi'(t) - T\gamma T^{-1}\cdot e\mathbf{A}(t)\psi'(t) \\ &+ T\gamma_4 T^{-1}\frac{\partial}{\partial t'}\psi'(t) + Ti\gamma_4 T^{-1}e\phi(t)\psi'(t) \\ &- im_0\psi'(t) = 0. \end{aligned} \tag{41}$$

In order that Eq. (41) reduces to Eq. (39a), we obtain

$$
\begin{aligned}
Ti\gamma T^{-1} &= -i\vec{\gamma},\\
T\gamma T^{-1} &= \gamma,\\
T\gamma_4 T^{-1} &= \gamma_4,\\
Ti\gamma_4 T^{-1} &= -i\gamma_4.
\end{aligned} \tag{42}
$$

In the Pauli-Dirac representation of γ-matrices, we find

$$T = T^0 K, \tag{43}$$

with K the complex conjugation operator and $T_0 = \lambda_T i\gamma_1\gamma_3$. (λ_T is a phase factor.) Or, we have

$$T\psi(\mathbf{x}, t) = \psi'(\mathbf{x}, -t) = \lambda_T i\gamma_1\gamma_3\psi^*(\mathbf{x}, t). \tag{44}$$

4.5 The Transformation Matrix S

In Sect. 3, it is shown that to establish the Lorentz invariance of the form of the Dirac equations, one has only to show the existence of a transformation S that satisfies the conditions Eqs. (20) and (23a). In Sect. 4, we have the S for the special transformation of space inversion, or parity, operation P,

$$S_p = \lambda\gamma_4, \quad \lambda = i,\ -i,\ 1,\ -1. \tag{33}$$

In the present section we shall obtain the S for other special Lorentz transformations.
(1) Infinitesimal Lorentz transformations
Let the

$$a_{\mu\nu} = \delta_{\mu\nu} + \epsilon_{\mu\nu}, \tag{45}$$

where

$$\epsilon_{\mu\nu} = -\epsilon_{\nu\mu}, \quad |\ \epsilon_{\mu\nu}\ | \ll 1. \tag{46}$$

The orthonormal condition Eq. (8d) is then

$$
\begin{aligned}
a_{\mu\nu}a_{\sigma\nu} &= (\delta_{\mu\nu} + \epsilon_{\mu\nu})(\delta_{\sigma\nu} + \epsilon_{\sigma\nu})\\
&= \delta_{\mu\sigma} + \epsilon_{\sigma\mu} + \epsilon_{\mu\sigma} + O(\epsilon^2)\\
&= \delta_{\mu\sigma}.
\end{aligned} \tag{47}
$$

Let $T^{\mu\nu}$ be a tensor, and $T^{\nu\mu}$ a tensor equal to $-T^{\mu\nu}$ and define a 4×4 matrix[14]

$$S = I + \frac{1}{2}\epsilon_{\mu\nu}T^{\mu\nu}, \tag{48a}$$

[14] *Note that μ, ν are not row-column indices of the tensor T. $T^{\mu\nu}$ is the name of the tensor itself. See Eq. (51) below.*

where I is a unit matrix, and

$$S^{-1} = I - \frac{1}{2}\epsilon_{\mu\nu}T^{\mu\nu}. \tag{48b}$$

The condition (20a) is then

$$(I - \frac{1}{2}\epsilon_{\mu\nu}T^{\mu\nu})\gamma_\rho(I + \frac{1}{2}\epsilon_{\alpha\beta}T^{\alpha\beta}) = (\delta_{\rho\sigma} + \epsilon_{\rho\sigma})\gamma_\sigma,$$

or

$$\gamma_\rho + \frac{1}{2}\epsilon_{\mu\nu}(\gamma_\rho T^{\mu\nu} - T^{\mu\nu}\gamma_\rho) = \gamma_\rho + \epsilon_{\rho\sigma}\gamma_\sigma$$

$$\frac{1}{2}\epsilon_{\mu\nu}(\gamma_\rho T^{\mu\nu} - T^{\mu\nu}\gamma_\rho) = \frac{1}{2}\epsilon_{\mu\nu}(\delta_{\mu\rho}\gamma_\nu - \delta_{\nu\rho}\gamma_\mu).$$

For this to hold for arbitrary $\epsilon_{\mu\nu}$, one must have

$$\gamma_\rho T^{\mu\nu} - T^{\mu\nu}\gamma_\rho = \delta_{\mu\rho}\gamma_\nu - \delta_{\nu\rho}\gamma_\mu. \tag{49}$$

The condition (23a) is then

$$\begin{aligned} S^\dagger &= \gamma_4 S^{-1}\gamma_4 \\ &= I - \frac{1}{2}\epsilon_{\mu\nu}\gamma_4 T^{\mu\nu}\gamma_4. \end{aligned} \tag{50}$$

From Eq. (48a),

$$S^\dagger = I + \frac{1}{2}\epsilon^*_{\mu\nu}T^{\dagger\mu\nu}. \tag{51}$$

Hence

$$\epsilon^*_{\mu\nu}T^{\dagger\mu\nu} = -\epsilon_{\mu\nu}\gamma_4 T^{\mu\nu}\gamma_4. \tag{52}$$

From Eq. (8f), one sees the generalized $a_{\mu\nu}$ to have the properties

$$\begin{aligned} \epsilon_{ji} &= -\epsilon_{ij} = \text{real}, \quad i,j = 1,2,3 \\ \epsilon_{4j} &= -\epsilon_{j4} = \text{imaginary}, \quad j = 1,2,3. \end{aligned} \tag{53}$$

From these and Eq. (52), one has

$$\begin{aligned} T^{\dagger ij} &= -\gamma_4 T^{ij}\gamma_4, \\ T^{\dagger 4j} &= \gamma_4 T^{4j}\gamma_4. \end{aligned} \tag{54}$$

It is seen that

$$T^{\mu\nu} = \frac{1}{2}\gamma_\mu\gamma_\nu \tag{55}$$

satisfies Eqs. (49) and (54). Thus

(i) Substituting Eq. (55) into (49), one gets

$$\frac{1}{2}(\gamma_\rho\gamma_\mu\gamma_\nu - \gamma_\mu\gamma_\nu\gamma_\rho) = \begin{cases} 0 & \text{for } \rho \neq \mu, \rho \neq \nu, \\ \gamma_\nu & \text{for } \rho = \mu, \rho \neq \nu, \\ -\gamma_\mu & \text{for } \rho \neq \mu, \rho = \nu, \end{cases}$$

$$\delta_{\mu\rho}\gamma_\nu - \delta_{\nu\rho}\gamma_\mu = \begin{cases} 0 & \text{for } \rho \neq \mu, \rho \neq \nu, \\ \gamma_\nu & \text{for } \rho = \mu, \rho \neq \nu, \\ -\gamma_\mu & \text{for } \rho \neq \mu, \rho = \nu. \end{cases} \qquad \text{q.e.d.}$$

(ii)

$$T^{\dagger ij} = \frac{1}{2}\gamma_j^\dagger\gamma_i^\dagger = \frac{1}{2}\gamma_j\gamma_i,$$

$$-\gamma_4 T^{ij}\gamma_4 = -\frac{1}{2}\gamma_4\gamma_i\gamma_j\gamma_4 = -\frac{1}{2}\gamma_i\gamma_j = \frac{1}{2}\gamma_j\gamma_i,$$

$$\frac{1}{2}\gamma_4\gamma_4\gamma_j\gamma_4 = \frac{1}{2}\gamma_j\gamma_4 = T^{j4}. \qquad \text{q.e.d.}$$

Hence

$$S = I + \frac{1}{4}\epsilon_{\mu\nu}\gamma_\mu\gamma_\nu, \tag{56}$$

insures the Lorentz invariance under infinitesimal Lorentz transformations.
(2) Finite rotations in 3-dimensional space

(i) Rotation about z-axis through an angle w

$$a_{\mu\nu} = \begin{pmatrix} cosw & sinw & 0 & 0 \\ -sinw & cosw & 0 & 0 \\ 0 & 0 & 1 & 0 \\ 0 & 0 & 0 & 1 \end{pmatrix} \tag{57}$$

For infinitesimal rotation $\triangle w$,

$$\epsilon_{12} = -\epsilon_{21} = \triangle w, \quad \text{all other } \epsilon_{\mu\nu} = 0.$$

$S_z(\triangle w)$ from Eq. (56) is

$$S_z(\triangle w) = I + \frac{1}{2}\triangle w\gamma_1\gamma_2$$

For finite rotation $w = \lim_{n\to\infty} n\triangle w$,

$$\begin{aligned}
S_z(w) &= \lim_{n\to\infty} \Pi S(\triangle w) = \lim_{n\to\infty} (1 + \frac{1}{2}\triangle w\gamma_1\gamma_2)^n \\
&= \lim_{n\to\infty} (1 + \frac{w}{2n}\gamma_1\gamma_2)^n \\
&= e^{\frac{1}{2}w\gamma_1\gamma_2} \\
&= \{1 - \frac{1}{2!}(\frac{w}{2})^2 + \frac{1}{4!}(\frac{w}{2})^4 - ...\} + \gamma_1\gamma_2\{(\frac{w}{2}) - \frac{1}{3!}(\frac{w}{2})^3 + ...\} \\
&= Icos\frac{w}{2} + \gamma_1\gamma_2 sin\frac{w}{2} \qquad (58) \\
&= Icos\frac{w}{2} + i\begin{pmatrix} 1 & 0 & 0 & 0 \\ 0 & -1 & 0 & 0 \\ 0 & 0 & 1 & 0 \\ 0 & 0 & 0 & -1 \end{pmatrix} sin\frac{w}{2} \\
&= \begin{pmatrix} e^{i\frac{w}{2}} & 0 & 0 & 0 \\ 0 & e^{-i\frac{w}{2}} & 0 & 0 \\ 0 & 0 & e^{i\frac{w}{2}} & 0 \\ 0 & 0 & 0 & e^{-i\frac{w}{2}} \end{pmatrix}. \qquad (58a)
\end{aligned}$$

(ii) Rotation about the x-axis through an angle w

$$a_{\mu\nu} = \begin{pmatrix} 1 & 0 & 0 & 0 \\ 0 & \cos w & \sin w & 0 \\ 0 & -\sin w & \cos w & 0 \\ 0 & 0 & 0 & 1 \end{pmatrix} \qquad (59)$$

$$\begin{aligned}
S_x(w) &= cos\frac{w}{2} + \gamma_2\gamma_3 sin\frac{w}{2} \qquad (60) \\
&= \begin{pmatrix} \cos\frac{w}{2} & i\sin\frac{w}{2} & 0 & 0 \\ i\sin\frac{w}{2} & \cos\frac{w}{2} & 0 & 0 \\ 0 & 0 & \cos\frac{w}{2} & i\sin\frac{w}{2} \\ 0 & 0 & i\sin\frac{w}{2} & \cos\frac{w}{2} \end{pmatrix}. \qquad (60a)
\end{aligned}$$

(iii) Rotation about the y-axis through an angle w

$$a_{\mu\nu} = \begin{pmatrix} \cos w & 0 & -\sin w & 0 \\ 0 & 1 & 0 & 0 \\ \sin w & 0 & \cos w & 0 \\ 0 & 0 & 0 & 1 \end{pmatrix}, \qquad (61)$$

$$\begin{aligned}
S_y(w) &= \cos\frac{w}{2} + \gamma_3\gamma_1 \sin\frac{w}{2} \qquad (62) \\
&= \begin{pmatrix} \cos\frac{w}{2} & \sin\frac{w}{2} & 0 & 0 \\ -\sin\frac{w}{2} & \cos\frac{w}{2} & 0 & 0 \\ 0 & 0 & \cos\frac{w}{2} & \sin\frac{w}{2} \\ 0 & 0 & -\sin\frac{w}{2} & \cos\frac{w}{2} \end{pmatrix}. \qquad (62a)
\end{aligned}$$

(iv) A general rotation

Let $OX'Y'Z'$ be a rectangular coordinate system fixed in space, and $OXYZ$ be a rotating system.

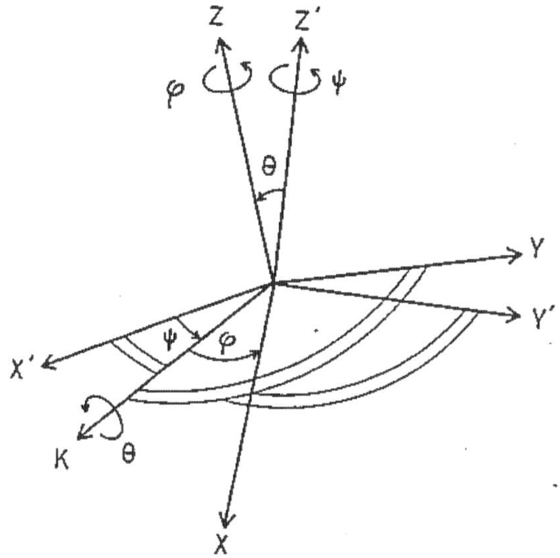

Initially, the two systems of axes coincide at the position indicated as $OX'Y'Z'$. First rotate the $OXYZ$ system about OZ' through an angle ψ, thereby bringing OX from OX' to OK. Next rotate the $OXYZ$ system about OK through an angle ϑ, thereby bringing the OZ axis from OZ' to OZ in the figure. Finally rotate the $OXYZ$ system about the OZ axis through an angle φ, thereby bringing the OX axis at OK to OX in the figure. The angles ψ, ϑ, φ are the Euler angles.

Now rotations form a non-abelian group, so that the order of the rotations is relevant. If the three rotations above are performed in the order described, then it can be shown that the product

$$S(\varphi)S(\vartheta)S(\psi)$$

of the transformation matrices is given by

$$S(\varphi, \vartheta, \psi) = \begin{pmatrix} \alpha & \gamma & 0 & 0 \\ \beta & \delta & 0 & 0 \\ 0 & 0 & \alpha & \gamma \\ 0 & 0 & \beta & \delta \end{pmatrix} \qquad (63)$$

where

$$\alpha = cos\frac{\vartheta}{2}exp(-i\frac{\psi+\varphi}{2}), \quad \beta = -isin\frac{\vartheta}{2}exp(i\frac{\psi-\varphi}{2}),$$

$$\gamma = -isin\frac{v}{2}exp(-i\frac{\psi-\varphi}{2}), \quad \delta = cos\frac{v}{2}exp(i\frac{\psi+\varphi}{2}), \tag{64}$$

are called the Cayley-Klein parameters. They satisfy the relation

$$\alpha\delta - \beta\gamma = 1. \tag{65}$$

The adjoint is

$$S^\dagger = \begin{pmatrix} \delta & -\gamma & 0 & 0 \\ -\beta & \alpha & 0 & 0 \\ 0 & 0 & \delta & -\gamma \\ 0 & 0 & -\beta & \alpha \end{pmatrix} \tag{66}$$

and it is seen that

$$SS^\dagger = S^\dagger S = 1 \tag{67}$$

so that S is unitary.

It is to be emphasized that in all the rotation matrices in Eqs. (58), (60), (62) and (63), the half-angles $\frac{w}{2}, \frac{\vartheta}{2}, \frac{\psi}{2}, \frac{\varphi}{2}$ appear.[15] The $\begin{pmatrix} \alpha & \gamma \\ \beta & \delta \end{pmatrix}$ is the transformation matrix for spinors – 2-component quantities. The appearance of spinors, as already referred to in Eq. (117) of Ch. 2 and Eqs. (27), (47) and (59) of Ch. 3, is characteristic of the Dirac equations.

(3) Special Lorentz transformation Eq. (8a)

For the special case of two inertial systems in uniform relative motion along their x-axes, the Lorentz transformation is Eq. (8a). We define an angle ϑ by

$$cosh\,\vartheta = \gamma, \quad sinh\,\vartheta = \beta\gamma,$$

so that

$$\beta = \frac{v}{c} < 1, \quad \gamma = \frac{1}{\sqrt{1-\beta^2}}, \tag{69}$$

so that

$$a_{\mu\nu} = \begin{pmatrix} cosh\,\vartheta & 0 & 0 & isinh\,\vartheta \\ 0 & 1 & 0 & 0 \\ 0 & 0 & 1 & 0 \\ -isinh\,\vartheta & 0 & 0 & cosh\,\vartheta \end{pmatrix} \tag{70}$$

[15] *The expressions (60), (62) and (58) for the $S_x(w), S_y(w), S_z(w)$ can be expressed in terms of the $\sigma_1, \sigma_2, \sigma_3$ matrices by means of the relations (80e) of Ch. 2, namely*

$$S_x(w) = exp(i\frac{w}{2}\sigma_1),$$
$$S_y(w) = exp(i\frac{w}{2}\sigma_2),$$
$$S_z(w) = exp(i\frac{w}{2}\sigma_3). \tag{68}$$

By a generalization of the results Eqs. (57), (58), and (58a), one obtains

$$S(\vartheta) = Icosh\frac{\vartheta}{2} + i\gamma_1\gamma_4 sinh\frac{\vartheta}{2} \tag{71}$$

$$= \begin{pmatrix} \cos h\frac{\vartheta}{2} & 0 & 0 & -\sin h\frac{\vartheta}{2} \\ 0 & \cos h\frac{\vartheta}{2} & -\sin h\frac{\vartheta}{2} & 0 \\ 0 & -\sin h\frac{\vartheta}{2} & \cos h\frac{\vartheta}{2} & 0 \\ -\sin h\frac{\vartheta}{2} & 0 & 0 & \cos h\frac{\vartheta}{2} \end{pmatrix} \tag{71a}$$

For the inverse transformation, $S^{-1}(\vartheta)$ can be obtained from $S(\vartheta)$ by replacing the velocity v by -v (i.e., β by $-\beta$, and ϑ by $-\vartheta$), and one sees from Eq. (71a) that

$$S^{-1}(\vartheta) \neq S^{\dagger}(\vartheta), \tag{72}$$

i.e., $S(\vartheta)$ for the special Lorentz transformation is not unitary.

The transformation of ψ under the special Lorentz transformation (8a) or (70) is, from Eq. (17) ,

$$\Psi' = S\Psi, \tag{73}$$

and with S given by Eq. (71a), one can write this transformation equation in the form

$$\begin{pmatrix} \Psi'_1 \\ \Psi'_2 \\ \Psi'_3 \\ \Psi'_4 \end{pmatrix} = \begin{pmatrix} c\Psi_1 - s\Psi_4 \\ c\Psi_2 - s\Psi_3 \\ -s\Psi_2 + c\Psi_3 \\ -s\Psi_1 + c\Psi_4 \end{pmatrix}, \quad c = \cos h\frac{\vartheta}{2}, \quad s = \sin h\frac{\vartheta}{2}, \tag{74}$$

which can be split up into a pair of equations

$$\begin{pmatrix} \Psi'_1 \\ \Psi'_4 \end{pmatrix} = \begin{pmatrix} c & -s \\ -s & c \end{pmatrix} \begin{pmatrix} \Psi_1 \\ \Psi_4 \end{pmatrix},$$
$$\begin{pmatrix} \Psi'_2 \\ \Psi'_3 \end{pmatrix} = \begin{pmatrix} c & -s \\ -s & c \end{pmatrix} \begin{pmatrix} \Psi_2 \\ \Psi_3 \end{pmatrix}. \tag{75}$$

The $\binom{\Psi_1}{\Psi_4}, \binom{\Psi_2}{\Psi_3}$ are two pairs of 2-components spinors (of rank 1); they transform according to Eq. (75), or [16]

$$phi'_1 = b_{11}\phi_1 + b_{12}\phi_2,$$
$$\phi'_2 = b_{21}\phi_1 + b_{22}\phi_2, \tag{76a}$$

with

$$b_{11}b_{22} - b_{12}b_{21} = 1. \tag{76b}$$

[16] *The transformation (71a), [or in the form (75)] and the transformation (63) involve the half-angles of the* $a_{\mu\nu}$ *, are not the transformation laws of vectors or tensors. In the theory of groups, tensors do not constitute all the representations of the Lorentz group; the transformations of the* ψ *of the Dirac equation form another representation of the Lorentz group — namely the spinors, characterized by Eqs. (76a) and (76b) for spinors of the first rank.*

References

Pauli, W., Article in *Handbuch der Physik*, Bd. 24 (1933), or in *Encycl. of Physics*, Vol. 5/1, Julius Springer (1958).

Bjorken, J. D. and Drell, S. D., *Relativistic Quantum Mechanics* (McGraw-Hill, N. Y. 1964).

Exercises: *Chapter 4*

1. In treating the Lorentz transformation of the Dirac equation in the present chapter, we have taken the view that the γ_μ's are constants (i.e. not subject to transformation) while the Dirac wave function $\psi(x)$ is transformed into $\psi'(x')$ (which is $S\psi(x)$). In fact, it is equivalent to assume that $\psi(x)$ is not transformed while γ_μ transforms like a Lorentz 4-vector,

$$\gamma'_\mu = a_{\mu\nu}\gamma_\nu.$$

(i) Prove that the two views are indeed equivalent.

(ii) Use the second view to discuss C, P, and T.

2. Choose a specific representation for γ-matrices such that the $T^{\mu\nu}$ in Eq. (55) assumes the following form:

$$T^{i4} = \frac{i}{2}\begin{pmatrix} \sigma_i & 0 \\ 0 & -\sigma_i \end{pmatrix}$$

where σ_i $(i = 1, 2, 3)$ are three 2×2 matrices.

3. **(i)** Show that the effect of two successive rotations on the Dirac wave function depends on the order of the two operations.

(ii) Try to generalize (i) to two arbitrary Lorentz transformations. Decide the commuting properties for these operators.

(iii) Work out the effect of the combined CPT operation on the Dirac wave function.

Chapter 5. The Dirac Electron in an Electromagnetic Field

For convenience, let us collect together the Dirac equations and the notations of symbols here[1]:

$$x_\mu\,(x_1, x_2, x_3, x_4 = it),$$

$$A_\mu\,(A_1, A_2, A_3, A_4 = i\phi),$$

$$\text{charge of electron} = -e,$$

$$\begin{aligned} \Pi_k &= \frac{1}{i}\frac{\partial}{\partial x_k} + eA_k, \quad k = 1, 2, 3, \\ \Pi_4 &= \frac{1}{i}\frac{\partial}{i\partial t} + ie\phi. \end{aligned} \tag{1}$$

$$\gamma_k = \begin{pmatrix} 0 & -i\sigma_k \\ i\sigma_k & 0 \end{pmatrix}, \quad k = x, y, z, \quad \sigma_k \ \text{ Pauli } 2\times 2 \text{ matrices;}$$

$$\gamma_4 = \begin{pmatrix} I & 0 \\ 0 & -I \end{pmatrix}.$$

$$\gamma_1 = i\begin{pmatrix} 0 & 0 & 0 & -1 \\ 0 & 0 & -1 & 0 \\ 0 & 1 & 0 & 0 \\ 1 & 0 & 0 & 0 \end{pmatrix}, \quad \gamma_2 = \begin{pmatrix} 0 & 0 & 0 & -1 \\ 0 & 0 & 1 & 0 \\ 0 & 1 & 0 & 0 \\ -1 & 0 & 0 & 0 \end{pmatrix},$$

$$\gamma_3 = i\begin{pmatrix} 0 & 0 & -1 & 0 \\ 0 & 0 & 0 & 1 \\ 1 & 0 & 0 & 0 \\ 0 & -1 & 0 & 0 \end{pmatrix}, \quad \gamma_4 = \begin{pmatrix} 1 & 0 & 0 & 0 \\ 0 & 1 & 0 & 0 \\ 0 & 0 & -1 & 0 \\ 0 & 0 & 0 & -1 \end{pmatrix}. \tag{2}$$

$$\Psi^\dagger = i\tilde{\Psi}^*\gamma_4, \text{ a row matrix,}$$

$$\Psi, \text{ a column matrix.} \tag{3}$$

---

[1]

$$x_4 = ict, \quad \Pi_k = \frac{\hbar}{i}\frac{\partial}{\partial x_k} + \frac{e}{c}A_k, \quad \Pi_4 = \frac{\hbar}{i}\frac{1}{ic}\frac{\partial}{\partial t} + i\frac{e}{c}\phi \tag{1'}$$

The Dirac equations are[2]

$$\begin{aligned} &\gamma_\mu \Pi_\mu \Psi - i m_0 \Psi = 0, \\ &\Pi_\mu^* \Psi^\dagger \gamma_\mu - i m_0 \Psi^\dagger = 0. \end{aligned} \tag{4}$$

These equations are covariant under the gauge transformation[3]

$$A' = A + \nabla\chi, \quad \phi' = \phi - \frac{\partial\chi}{\partial t}, \qquad \chi = \text{a scalar function}, \tag{5}$$

$$\Psi' = \Psi exp(-ie\chi). \tag{6}$$

On defining

$$\Pi_\pm = \Pi_1 \pm i\Pi_2, \tag{7}$$

one can write Eq. (4) for Ψ in the explicit form[4]

$$\begin{aligned} &(\frac{1}{i}\frac{\partial}{\partial t} - e\phi + m_0)\Psi_1 + \Pi_-\Psi_4 + \Pi_3\Psi_3 = 0, \\ &(\frac{1}{i}\frac{\partial}{\partial t} - e\phi + m_0)\Psi_2 + \Pi_+\Psi_3 - \Pi_3\Psi_4 = 0, \\ &(\frac{1}{i}\frac{\partial}{\partial t} - e\phi - m_0)\Psi_3 + \Pi_-\Psi_2 + \Pi_3\Psi_1 = 0, \\ &(\frac{1}{i}\frac{\partial}{\partial t} - e\phi - m_0)\Psi_4 + \Pi_+\Psi_1 - \Pi_3\Psi_2 = 0, \end{aligned} \tag{8}$$

which are the extension of the equations for a free electron [Eq. (25), Ch. 2] to the electron in an $(\mathbf{A}, i\phi)$ field.

2

$$\gamma_\mu \Pi_\mu \Psi - i m_0 c \Psi = 0. \tag{4'}$$

3

$$\phi' = \phi - \frac{1}{c}\frac{\partial\chi}{\partial t}. \tag{5'}$$

$$\Psi' = \Psi exp(-\frac{ie}{\hbar c}\chi). \tag{6'}$$

4

$$(\frac{\hbar}{i}\frac{1}{c}\frac{\partial}{\partial t} - \frac{e}{c}\phi + m_0 c)\Psi_1 + \Pi_-\Psi_4 + \Pi_3\Psi_3 = 0. \tag{8'}$$

5.1 Dirac Equations in Second-Order Form

One can apply the operator

$$(\gamma_\mu \Pi_\mu + im_0) = (\gamma_4 \Pi_4 + (\gamma \cdot \Pi) + im_0) \tag{9}$$

on Eq. (4), obtaining[5]

$$\{\Pi_4^2 + (\gamma \cdot \Pi)^2 + m_0^2 + \sum_k^3 (\gamma_k \gamma_4 \Pi_k \Pi_4 + \gamma_4 \gamma_k \Pi_4 \Pi_k)\}\Psi = 0. \tag{10}$$

On defining Π_0

$$\Pi_0 = -i\Pi_4 = (-\frac{1}{i}\frac{\partial}{\partial t} + e\phi), \tag{12}$$

Eq. (10) takes the form[6]

$$[-\Pi_0^2 + (\sigma \cdot \Pi)^2 + m_0^2 - \rho_1\{(\sigma \cdot \Pi)\Pi_0 - \Pi_0(\sigma \cdot \Pi)\}\Psi = 0. \tag{13}$$

---

[5] *This equation, on using Eqs. (13) and (14) of Ch. 4, can be put in the form*

$$\{\Pi_\mu \Pi_\mu + m_0^2 + \frac{e}{2i}\gamma_\mu \gamma_\nu F_{\mu\nu}\}\Psi = 0. \tag{10a}$$

From Eqs. (20c) and (80f) of Ch. 2, one has

$$\gamma_k = -i\beta\rho_1\sigma_k, \quad k = 1, 2, 3,$$

$$\begin{aligned}
(\gamma \cdot \Pi)^2 &= -(\beta\rho_1\sigma_j\Pi_j)(\beta\rho_1\sigma_k\Pi_k) \\
&= \beta\sigma_j\rho_1\rho_1\beta\sigma_k\Pi_j\Pi_k \quad \textit{(from Eq. (82), Ch. 2 and } \beta\rho_1 + \rho_1\beta = 0) \\
&= \beta\sigma_j\beta\sigma_k\Pi_j\Pi_k = \sigma_j\sigma_k\Pi_j\Pi_k \quad \textit{(from } \sigma_j\beta - \beta\sigma_j = 0) \\
&= (\sigma \cdot \Pi)^2. \\
\gamma_k\gamma_4 &= -i\beta\alpha_k\beta \quad \textit{(using Eq. (20c), Ch. 2)} \\
&= i\alpha_k = i\rho_1\sigma_k. \quad \textit{(using Eq. (80f), Ch. 2)}
\end{aligned} \tag{11}$$

[6]

$$[-\Pi_0^2 + (\sigma \cdot \Pi)^2 + m_0^2c^2 - \rho_1\{(\sigma \cdot \Pi)\Pi_0 - \Pi_0(\sigma \cdot \Pi)\}\Psi = 0. \tag{13'}$$

This equation can be rewritten as follows[7,8]

$$\{-(-\frac{1}{i}\frac{\partial}{\partial t}+e\phi)^2+(\frac{1}{i}\nabla+eA)^2+m_0^2+e(\sigma\cdot\mathbf{H})-ie\rho_1(\sigma\cdot\mathbf{E})\}\Psi=0. \tag{14}$$

On comparing Eq. (14) with the Klein-Gordon equation Eq. (14), Ch. 1 (in which e is now the charge of an electron -e), it is seen that the Dirac equation Eq. (14) has two extra terms[9]

$$\frac{e}{2m_0}(\sigma\cdot\mathbf{H})\Psi-i\frac{e}{2m_0}(\sigma\cdot\mathbf{E})\Psi. \tag{18}$$

The first term here with the operator

$$\mu_B(\sigma\cdot\mathbf{H})=2\mu_B(\mathbf{S}\cdot\mathbf{H}), \tag{19}$$

where $\mathbf{S}=\frac{1}{2}\sigma$ is the spin angular momentum operator. Eq. (91), Ch. 2, can immediately be interpreted as the interaction of the spin magnetic moment $\mu_B\mathbf{S}$ with the external magnetic field with the gyromagnetic ratio $g=2$.

---

[7] *If X is an arbitrary vector (operator) that commutes with σ , then on using Eq. (81), Ch. 2,*

$$\sigma_i\sigma_j+\sigma_j\sigma_i=2\delta_{ij},$$

one has

$$(\sigma\cdot B)^2=B^2+i\sigma_1(X_2X_3-X_3X_2)+i\sigma_2(X_3X_1-X_1X_3)+i\sigma_3(X_1X_2-X_2X_1). \tag{15}$$

Hence

$$\begin{aligned}(\sigma\cdot\Pi)^2&=\sum_{k=1}^{3}(\frac{1}{i}\frac{\partial}{\partial x_k}+eA_k)^2+e(\sigma\cdot curlA)\\&=\Pi^2+e(\sigma\cdot\mathbf{H}),\quad \mathbf{H}=\textit{magnetic field},\end{aligned} \tag{16}$$

$$\begin{aligned}(\sigma\cdot\Pi)\Pi_0-\Pi_0(\sigma\cdot\Pi)&=\frac{e}{i}(\sigma\cdot(\nabla\phi+\frac{\partial\mathbf{A}}{\partial t}))\\&=-\frac{e}{i}(\sigma\cdot\mathbf{E}),\qquad \mathbf{E}=\textit{electric field}.\end{aligned} \tag{17}$$

Or, without using $\hbar=c=l$,

$$(\sigma\cdot\Pi)^2=\Pi^2+\frac{e\hbar}{c}(\sigma\cdot\mathbf{H}), \tag{16'}$$

$$....=-\frac{e\hbar}{ic}(\sigma\cdot\mathbf{E}). \tag{17'}$$

[8]

$$\{-(-\frac{\hbar}{i}\frac{\partial}{c\partial t}+\frac{e}{c}\phi)^2+(\frac{\hbar}{i}\nabla+\frac{e}{c}A)^2+m_0^2c^2+\frac{e\hbar}{c}(\sigma\cdot\mathbf{H})-i\frac{e\hbar}{c}\rho_1(\sigma\cdot\mathbf{E})\}\Psi=0. \tag{14'}$$

[9]

$$\frac{e\hbar}{2m_0c}(\sigma\cdot\mathbf{H})\Psi-i\frac{e\hbar}{2m_0c}(\sigma\cdot\mathbf{E})\Psi \tag{18'}$$

The second term in Eq. (18) is a pure imaginary and seems to have no physical meaning. But we shall see below that in the case of the hydrogenic atom it leads to the so-called "S-state correction".[10]

The above comparison between the Dirac equation Eq. (14) and the Klein-Gordon equation Eq. (14) of Ch. 1, shows that the latter applies to spinless particles, such as the pions $\pi^{\pm}, \pi^0$.

5.2 Dirac Equation: Approximate, Iterated Form

Let E_+ be the energy of an electron in a positive-energy state, and

$$w = E_+ - m_0, \tag{20a}$$

and consider in this section the situation

$$\mid w \mid \ll m_0. \tag{20b}$$

Let Ψ_k in Eq. (8) be

$$\Psi_k = \Phi_k \exp(-im_0 t), \tag{21}$$

so that[11]

$$\begin{aligned}
(\frac{1}{i}\frac{\partial}{\partial t} - e\phi + m_0)\Psi_k &= exp(-im_0 t)(\frac{1}{i}\frac{\partial}{\partial t} - e\phi)\Phi_k, \\
(\frac{1}{i}\frac{\partial}{\partial t} - e\phi - m_0)\Psi_k &= exp(-im_0 t)(\frac{1}{i}\frac{\partial}{\partial t} - e\phi - 2m_0)\Phi_k.
\end{aligned} \tag{22}$$

It is seen from Eq. (8) that Ψ_1, Ψ_2 are the "big" components and Ψ_3, Ψ_4 the "small" components. Let us introduce the notations

$$\Phi \equiv \begin{pmatrix} \Phi_1 \\ \Phi_2 \end{pmatrix}, \qquad \varphi \equiv \begin{pmatrix} \varphi_1 \\ \varphi_2 \end{pmatrix} \equiv \begin{pmatrix} \Phi_3 \\ \Phi_4 \end{pmatrix}, \tag{23}$$

$$\mid \varphi \mid \ll \mid \Phi \mid. \tag{24}$$

By means of the 2×2 Pauli matrices

$$\sigma_x = \begin{pmatrix} 0 & 1 \\ 1 & 0 \end{pmatrix}, \quad \sigma_y = \begin{pmatrix} 0 & -i \\ i & 0 \end{pmatrix}, \quad \sigma_z = \begin{pmatrix} 1 & 0 \\ 0 & -1 \end{pmatrix}, \tag{25}$$

one can express Eq. (8) in the form

$$(\frac{1}{i}\frac{\partial}{\partial t} - e\phi)\begin{pmatrix} \Phi_1 \\ \Phi_2 \end{pmatrix} + (\sigma \cdot \Pi)\begin{pmatrix} \varphi_1 \\ \varphi_2 \end{pmatrix} = 0 \tag{26a}$$

---

[10] *See Eq. (39), the last term containing $\frac{d\phi}{dr}$, and Eq. (57) in the following.*

[11]

$$(\frac{\hbar}{i}\frac{\partial}{\partial t} - \frac{e}{c}\phi + m_0 c)\Psi_k = exp(-\frac{i}{\hbar}m_0 c^2 t)(\frac{\hbar}{i}\frac{\partial}{\partial t} - \frac{e}{c}\phi)\Phi_k, \quad etc. \tag{22'}$$

$$(\frac{1}{i}\frac{\partial}{\partial t} - e\phi - 2m_0)\begin{pmatrix}\varphi_1\\ \varphi_2\end{pmatrix} + (\sigma \cdot \Pi)\begin{pmatrix}\Phi_1\\ \Phi_2\end{pmatrix} = 0, \tag{26b}$$

or[12]

$$(-\frac{1}{i}\frac{\partial}{\partial t} + e\phi)\Phi - (\sigma \cdot \Pi)\varphi = 0, \tag{27a}$$

$$(-\frac{1}{i}\frac{\partial}{\partial t} + e\phi + 2m_0)\varphi - (\sigma \cdot \Pi)\Phi = 0. \tag{27b}$$

These equations are exact. The last equation[13]

$$\{1 + \frac{1}{2m_0}(-\frac{1}{i}\frac{\partial}{\partial t} + e\phi)\}\varphi = \frac{(\sigma \cdot \Pi)}{2m_0}\Phi, \tag{28}$$

shows that the second term on the left is of order $(\frac{v}{c})^2$.

Now in the non-relativistic (Schrödinger) approximation, one has

$$(-\frac{1}{i}\frac{\partial}{\partial t} + e\phi)\Phi = \frac{\Pi^2}{2m_0}\Phi, \tag{29}$$

or on using Eq. (20a)

$$(w + e\phi)\Phi = \frac{\Pi^2}{2m_0}\Phi, \tag{30a}$$

and the next approximation for φ in Eq. (28) is[14]

$$\varphi = \frac{1}{1 + \frac{\Pi^2}{4m_0^2}} \cdot \frac{(\sigma \cdot \Pi)}{2m_0}\Phi. \tag{30}$$

Let us define an operator F by

$$F \equiv [1 + \frac{\Pi^2}{4m_0^2}]^{-1} = [1 + \frac{w + e\phi}{2m_0}]^{-1}. \tag{31}$$

12

$$(-\frac{\hbar}{ic}\frac{\partial}{\partial t} + \frac{e}{c}\phi)\Phi - (\sigma \cdot \Pi)\varphi = 0, \quad etc. \tag{27a$'$}$$

13

$$\{1 + \frac{1}{2m_0c^2}(-\frac{\hbar}{i}\frac{\partial}{\partial t} + e\phi)\}\varphi = \frac{(\sigma \cdot \Pi)}{2m_0c}\Phi. \tag{28$'$}$$

14

$$\varphi = \frac{1}{1 + \frac{\Pi^2}{4m_0^2c^2}} \cdot \frac{(\sigma \cdot \Pi)}{2m_0c}\Phi. \tag{30$'$}$$

The equation Eq. (27a) can be approximated by

$$\left(-\frac{1}{i}\frac{\partial}{\partial t}+e\phi-\frac{1}{2m_0}(\sigma\cdot\Pi)F(\sigma\cdot\Pi)\right)\Phi=0. \tag{32}$$

Now

$$(\sigma\cdot\Pi)F(\sigma\cdot\Pi)=F(\sigma\cdot\Pi)^2+(\sigma\cdot\Pi F)(\sigma\cdot\Pi). \tag{33}$$

From Eq. (30a)

$$\frac{\Pi^2}{4m_0^2}=\frac{w+e\phi}{2m_0}, \tag{34}$$

and for $w+e\phi\ll 2m_0$, as from Eq. (20b), one expands F in Eq. (31) and obtains, using Eq. (16),[15]

$$\begin{aligned}F(\sigma\cdot\Pi)^2&=\left(1-\frac{\Pi^2}{4m_0^2}\right)(\sigma\cdot\Pi)^2\\&=\Pi^2-\frac{\Pi^4}{4m_0^2}+e(\sigma\cdot H)-\frac{\Pi^2e}{4m_0^2}(\sigma\cdot H).\end{aligned} \tag{35}$$

The calculation of the last term in Eq. (33) is long, but to the order $(\frac{v}{c})^2$, one may neglect the vector potential A and replace Π_k by p_k , i.e.,

$$(\sigma\cdot\Pi F)(\sigma\cdot\Pi)\to(\sigma\cdot\mathbf{p}F)(\sigma\cdot\mathbf{p}).$$

If the scalar potential ϕ is central field, then

$$(\sigma\cdot\mathbf{p}F)=\left(\sigma\cdot\frac{1}{i}\frac{d}{dr}F\right)=\frac{1}{ir}\frac{dF}{dr}(\sigma\cdot\mathbf{r}),$$

$$(\sigma\cdot\mathbf{p}F)(\sigma\cdot\mathbf{p})=\frac{dF}{dr}\left(-\frac{\partial}{\partial r}+\frac{1}{r}(\mathbf{L}\cdot\sigma)\right), \tag{36}$$

where $\mathbf{L}=[\mathbf{r}\times\mathbf{p}]$ is the angular momentum, and

$$\frac{dF}{dr}=-\left[1+\frac{w+e\phi}{2m_0}\right]^{-2}\cdot\frac{e}{2m_0}\frac{d\phi}{dr} \tag{37}$$

$$\cong-\frac{e}{2m_0}\frac{d\phi}{dr}\qquad\text{[From Eq. (34)]}. \tag{38}$$

15

$$\begin{aligned}F(\sigma\cdot\Pi)^2&=\left(1-\frac{\Pi^2}{4m_0^2c^2}\right)(\sigma\cdot\Pi)^2\\&=\Pi^2-\frac{\Pi^4}{4m_0^2c^2}+\frac{e\hbar}{c}(\sigma\cdot\mathbf{H})-\frac{\Pi^2e\hbar}{4m_0^2c^2}(\sigma\cdot\mathbf{H}).\end{aligned} \tag{35'}$$

Substituting Eqs. (35), (36), and (37) into (32), one obtains[16]

$$\{-\frac{1}{i}\frac{\partial}{\partial t}+e\phi-\frac{\Pi^2}{2m_0}(1-\frac{\Pi^2}{4m_0^2})-\frac{e}{2m_0}(1-\frac{\Pi^2}{4m_0^2})(\sigma\cdot\mathbf{H})$$
$$+[1+\frac{w+e\phi}{2m_0}]^{-2}\frac{e}{4m_0^2}\frac{d\phi}{dr}[\frac{1}{r}(\mathbf{L}\cdot\sigma)-\frac{\partial}{\partial r}]\}\Phi=0. \tag{39}$$

The first three terms are the Schrödinger non-relativistic theory Eq. (29), the term in Π^4 is the Sommerfeld correction (treated in Chapter 6, (VI-22) Vol. I); the term $\frac{e}{2m_0}(\sigma\cdot\mathbf{H})$ has been referred to in Eq. (19) already, (the term $\frac{\Pi^2}{4m_0^2}\mu_B(\sigma\cdot\mathbf{H})$ is a higher order correction to it, and can be neglected); the last two terms can be traced to the last term in Eq. (14); they are the spin-orbit interactions and will be examined in detail in the following section.

Equation (39) has been obtained in approximate theories by Pauli and Darwin, before the Dirac theory from which Eq. (39) can be derived.

5.3 Hydrogenic Atoms in Dirac's Theory — Approximate Solution

For hydrogenic atoms,[17]

$$\phi(r)=\frac{Ze}{4\pi r},\quad \frac{d\phi}{dr}=-\frac{Ze}{4\pi r^2};$$
$$A=0,\quad \text{i.e.,}\quad H=0;$$
$$S=\frac{1}{2}\sigma,$$
$$J=L+S=L+\frac{1}{2}\sigma,$$
$$2(L\cdot S)=J^2-L^2-S^2. \tag{40}$$

Let

$$\Phi(r,t)=\Psi(r)e^{-iwt}. \tag{41}$$

---

[16]

$$\{-\frac{\hbar}{i}\frac{\partial}{\partial t}+e\phi-\frac{\Pi^2}{2m_0}(1-\frac{\Pi^2}{4m_0^2c^2})-\frac{e\hbar}{2m_0c}(1-\frac{\Pi^2}{4m_0^2c^2})(\sigma\cdot\mathbf{H})$$
$$+[1+\frac{w+e\phi}{2m_0c^2}]^{-2}\frac{e\hbar^2}{4m_0^2c^2}\frac{d\phi}{dr}[\frac{1}{r}(\mathbf{L}\cdot\sigma)-\frac{\partial}{\partial r}]\}\Phi=0. \tag{39'}$$

[17]

$$\phi(r)=\frac{Ze}{r},\ etc. \tag{40'}$$

Eq. (39) leads to the stationary value problem[18]

$$\{\frac{p^2}{2m_0} - \frac{Ze^2}{4\pi r} - \frac{1}{2m_0}(w + \frac{Ze^2}{4\pi r})^2 + \mu_B^2[1 + \frac{1}{2m_0}(w + \frac{Ze^2}{4\pi r})]^{-2}\frac{Z}{4\pi r^2}[\frac{2}{r}(L\cdot S) - \frac{\partial}{\partial r}]\}\Psi(r) = w\Psi(r). \quad (42)$$

The lefthand side may be regarded as an approximate Hamiltonian $\bar{H}$, and from Eq. (42), it is seen that [19]

$$\bar{H}J^2 - J^2\bar{H} = 0,$$
$$\bar{H}S^2 - S^2\bar{H} = 0,$$
$$\bar{H}L^2 - L^2\bar{H} = 0. \quad (43)$$

The eigenvalues of $\bar{H}, J^2, S^2, L^2$ are

$$w, \quad j(j+1), \quad s(s+1), \quad \ell(\ell+1), \quad (s = \frac{1}{2}). \quad (44)$$

From Eq. (40), the eigenvalues of $2(L\cdot S)$ are given by

$$2(L\cdot S)\Psi = \{j(j+1) - \ell(\ell+1) - s(s+1)\}\Psi. \quad (45)$$

The Schrödinger approximation as obtained from Eq. (39) is

$$\{-\frac{1}{2m_0}(\frac{d^2}{dr^2} + \frac{2}{r}\frac{d}{dr} - \frac{\ell(\ell+1)}{r^2}) - e\phi\}R_0(r) = wR_0(r). \quad (46)$$

$R_0(r)$ is the Schrödinger wave function of Chapter 3, (III-139), Vol. I. The other terms may be treated by the perturbation method. Let

$$H_r = -\frac{1}{2m_0}(w + \frac{Ze^2}{4\pi r})^2, \quad (47)$$

$$H_{s.o.} = \mu_B^2[1 + \frac{w + \frac{Ze^2}{4\pi r}}{2m_0}]^{-2}\frac{2Z}{4\pi r^3}(L\cdot S), \quad (48a)$$

[18]

$$\{\frac{p^2}{2m_0} - \frac{Ze^2}{r} - \frac{1}{2m_0c^2}(w + \frac{Ze^2}{r})^2 + \mu_B^2[1 + \frac{1}{2m_0c^2}(w + \frac{Ze^2}{r})]^{-2}\frac{Z}{r^2}[\frac{2}{r}(L\cdot S) - \frac{\partial}{\partial r}]\}\Psi(r) = w\Psi(r). \quad (42')$$

[19] *Note that* $\mathbf{L}^2$ *does not commute with the exact Hamiltonian* H

$$i\frac{\partial}{\partial t}\Phi = H\Phi,$$

$$H = (\alpha\cdot\Pi) - e\phi + \beta m_0.$$

$$H_s = -\mu_B^2[1 + \frac{w + \frac{Ze^2}{4\pi r}}{2m_0}]^{-2}\frac{Z}{4\pi r^2}\frac{\partial}{\partial r}. \tag{48b}$$

The term H_r is the Sommerfeld relativistic correction and has been calculated in Chapter 6, (VI-22), Vol. I. The result is[20]

$$\langle n,\ell|\, H_r\, |n,\ell\rangle = -\frac{Z^4\alpha^2}{n^4}(\frac{n}{\ell+\frac{1}{2}} - \frac{3}{4}), \qquad \alpha = \frac{e^2}{4\pi} \simeq \frac{1}{137}. \tag{49}$$

The term $H_{s.o.}$ is simplified for the approximation[21]

$$1 + \frac{w + e\phi}{2m_0} \simeq 1, \tag{50}$$

and is then the spin-orbit interaction already treated in Chapter 8, Sect. 1, (2), (VIII-11, 25a, b, 28), Vol. I. [22] For $\ell \neq 0$,

$$\langle n,\ell|\, H_{s.o.}\, |n,\ell\rangle = \frac{Z^4\alpha^2}{n^3}\frac{1}{\ell(\ell+\frac{1}{2})(\ell+1)} \cdot \begin{cases} \frac{\ell}{2}, & j = \ell + \frac{1}{2}, \quad \ell = 1, 2, ... \\ -\frac{\ell+1}{2}, & j = \ell - \frac{1}{2}, \quad \ell = 1, 2, \end{cases} \tag{51}$$

The term H_s in Eq. (48b) can also be calculated for the approximation Eq. (50). Then

$$\begin{aligned}\langle n,\ell|\, H_s\, |n,\ell\rangle &= -Z\mu_B^2 \int_0^\infty R_{n,\ell}\frac{1}{r^2}\frac{d}{dr}R_{n,\ell}r^2 dr \\ &= \frac{1}{2}Z\mu_B^2|\, R_{n,\ell}(0)\,|^2, \end{aligned} \tag{55}$$

---

[20]

$$\langle n,\ell|\, H_r\, |n,\ell\rangle = -\frac{Z^4\alpha^2}{n^4}(\frac{n}{\ell+\frac{1}{2}} - \frac{3}{4}), \qquad \alpha = \frac{e^2}{\hbar c}. \tag{49'}$$

[21]

$$1 + \frac{w + e\phi}{2m_0c^2} \simeq . \tag{50'}$$

[22] *The matrix element of $H_{s.o.}$ consists of two factors, namely,*

$$\langle n,\ell|\, \frac{Z}{r^3}\, |n,\ell\rangle, \quad and \quad \langle j,m|\, \ell\cdot s\, |j,m\rangle. \tag{52}$$

For S state, $\ell = 0$, one has the indeterminate result $\frac{0}{0}$, and one has to examine the situation more carefully. For $\ell = 0$, the matrix element $\langle n,0|\, \frac{Z}{r^3}\, |n,0\rangle$

$$\propto \int_0^\infty R_{n,0}\frac{1}{r^3}R_{n,0}\,r^2 dr \quad \xrightarrow{(r\to 0)} \quad \int_0^\infty \frac{1}{r}dr, \tag{53}$$

is divergent at $r = 0$. But at $r = 0$, one must use the expression (48a), and then one finds the matrix element $\langle n,0|\, \frac{Z}{r^3}\, |n,0\rangle$ to be finite. Hence for S state,

$$\langle n,0|\, H_{s.o.}\, |n,0\rangle = 0. \tag{54}$$

which does not vanish only for S state ($\ell = 0$) , and[23]

$$\begin{aligned}\langle n,0|\, H_s \,|n,0\rangle &= \frac{1}{2} Z\mu_B^2 \frac{4Z^3}{n^3 a^3}, \quad a_0 = \frac{1}{m_0 e^2}, \quad \text{Bohr radius;} \\ &= \frac{Z^4\alpha^2}{n^3} (in \quad \frac{m_0 e^4}{2} = \frac{e^2}{2a}). \end{aligned} \tag{56}$$

which is called the S-state correction. This result may be combined together with Eq. (51) (which is valid for $\ell \neq 0$) , so that now one may write

$$\begin{aligned}&\langle n,\ell|\, H_{s.o.} + H_s \,|n,\ell\rangle \\ &= \frac{Z^4\alpha^2}{n^3} \frac{1}{\ell(\ell+\frac{1}{2})(\ell+1)} \begin{cases} \frac{\ell}{2}, & j = \ell + \frac{1}{2}, \ell = 0,1,.. \\ -\frac{\ell+1}{2}, & j = \ell - \frac{1}{2}, \ell = 1,2,.. \end{cases}\end{aligned} \tag{57}$$

The total contribution from $H_r, H_{s.o.}, H_s$ in Eqs. (47), (48a), and (48b) is then

$$\langle n,\ell|\, H_r + H_{s.o.} + H_s \,|n,\ell\rangle = -\frac{Z^4\alpha^2}{n^4}(\frac{n}{j+\frac{1}{2}} - \frac{3}{4}). \tag{58}$$

This energy is degenerate for the two states

$$(n, \ell = j - \frac{1}{2}) \quad \text{and} \quad (n, \ell' = j + \frac{1}{2}), \tag{59}$$

for example

$$\begin{aligned} &{}^2S_{\frac{1}{2}} \quad \text{and} \quad {}^2P_{\frac{1}{2}}, \\ &{}^2P_{\frac{3}{2}} \quad \text{and} \quad {}^2D_{\frac{3}{2}}, \quad \text{etc.} \end{aligned} \tag{59a}$$

Let us find the wave functions of the approximate Hamiltonian $\bar{H}$ in Eq. (42). In Eq. (43), we have seen that J^2, S^2, L^2 commute with $\bar{H}$. In addition to these,

$$J_z = L_z + S_z \tag{60}$$

also commutes with $\bar{H}$

$$\bar{H}J_z - J_z\bar{H} = 0, \tag{61}$$

and J_z has simultaneous eigenvalues m with J^2, S^2, L^2,

$$-J \leq m \leq J. \tag{62}$$

[23]

$$\begin{aligned} a_0 &= \frac{\hbar^2}{m_0 e^2}, \quad \alpha = \frac{e^2}{\hbar c}, \\ &= \frac{Z^4\alpha^2}{n^3} (\text{in units of} \quad \frac{m_0 e^4}{2\hbar^2} = \frac{e^2}{2a_0}). \end{aligned} \tag{56'}$$

But L_z and S_z separately do not commute with H so that m_ℓ, m_s are not "exact" quantum numbers.

A state is now defined by the quantum numbers

$$n,\ j,\ \ell,\ s = \frac{1}{2} \quad \text{and} \quad m. \tag{63}$$

A state having n, ℓ, j, m is a combination

$$\Psi = R_{n,\ell}(r)[aY_{\ell,m_\ell} + bY_{\ell,m'_\ell}], \tag{64}$$

where

$$m_\ell + \frac{1}{2} = m'_\ell - \frac{1}{2} = m. \tag{64a}$$

The ratio $\frac{a}{b}$ is determined by the requirement that Ψ of Eq. (64) be also an eigenfunction of $(L \cdot S)$. The theory has already been given in Chapter 8, Sect. 1, (3) (VIII-31a, 31b), Vol. I. Thus for $j = \ell + \frac{1}{2}$, $\quad (\ell \cdot s) = \frac{\ell}{2}$,

$$\Phi_{n,\ell,j,m} = \begin{pmatrix} \Phi_1 \\ \Phi_2 \end{pmatrix} = \frac{1}{\sqrt{2\ell+1}} R_{n,\ell}(r) \begin{pmatrix} \sqrt{\ell+m+\frac{1}{2}} & Y_{\ell,m-\frac{1}{2}} \\ -\sqrt{\ell-m+\frac{1}{2}} & Y_{\ell,m+\frac{1}{2}} \end{pmatrix}, \tag{65a}$$

$j = \ell - \frac{1}{2}$, $\quad (\ell \cdot s) = -\frac{\ell+1}{2}$,

$$\Phi_{n,\ell,j,m} = \begin{pmatrix} \Phi_1 \\ \Phi_2 \end{pmatrix} = \frac{1}{\sqrt{2\ell+1}} R_{n,\ell}(r) \begin{pmatrix} \sqrt{\ell-m+\frac{1}{2}} & Y_{\ell,m-\frac{1}{2}} \\ \sqrt{\ell+m+\frac{1}{2}} & Y_{\ell,m+\frac{1}{2}} \end{pmatrix}. \tag{65b}$$

5.4 Hydrogenic Atoms in Dirac's Theory — Exact Solution

For hydrogenic atoms, we have[24]

$$A = 0, \qquad \phi(r) = \frac{Ze}{4\pi r},$$

$$\begin{aligned} p_\pm &\equiv \frac{1}{i}\left(\frac{\partial}{\partial x} \pm i\frac{\partial}{\partial y}\right) \\ &= \frac{1}{i}e^{\pm i\varphi}\left[sin\vartheta\frac{\partial}{\partial r} + cos\vartheta\frac{1}{r}\frac{\partial}{\partial\vartheta} \pm \frac{i}{rsin\vartheta}\frac{\partial}{\partial\varphi}\right], \\ p_3 &= \frac{1}{i}\frac{\partial}{\partial z} \\ &= \frac{1}{i}\left[\cos\vartheta\frac{\partial}{\partial r} - \sin\vartheta\frac{\partial}{r\partial\vartheta}\right]. \end{aligned} \tag{66}$$

[24] *In ordinary units, $\phi(r) = \frac{Ze}{r}$ in (67), (68), (73), (74), (77), (78), (80) and m_0 is m_0c^2 in (68).*

$$E = w + m_0 = w + E_0, \qquad w < 0 \quad \text{for bound states.}$$

The exact equations are, from Eq. (8),

$$\begin{aligned}
-(E + \frac{Ze^2}{4\pi r} - E_0)\Phi_1 + p_-\Phi_4 + p_3\Phi_3 &= 0,\\
-(E + \frac{Ze^2}{4\pi r} - E_0)\Phi_2 + p_+\Phi_3 - p_3\Phi_4 &= 0,\\
-(E + \frac{Ze^2}{4\pi r} + E_0)\Phi_3 + p_-\Phi_2 + p_3\Phi_1 &= 0,\\
-(E + \frac{Ze^2}{4\pi r} + E_0)\Phi_4 + p_+\Phi_1 - p_3\Phi_2 &= 0.
\end{aligned} \tag{67}$$

For hydrogenic atoms in their stationary states,

$$w + \frac{Ze^2}{4\pi r} \ll w + \frac{Ze^2}{4\pi r} + 2m_0. \tag{68}$$

Thus Φ_1, Φ_2 are the big components and Φ_3, Φ_4 the small components (for positive energy $E_+ > 0$ states).

The functions $\binom{\Phi_1}{\Phi_2}$ and $\binom{\Phi_3}{\Phi_4}$ are coupled in Eq. (67) so that the Schrödinger radial wave functions $R_{n,\ell}(r)$ are not solutions. From Eq. (65a), we may write

$$\begin{cases} j = \ell + \frac{1}{2}, & \Phi_1(r) = g(r)\sqrt{\frac{\ell+m+\frac{1}{2}}{2\ell+1}}Y_{\ell,m-\frac{1}{2}}(\vartheta,\varphi),\\ m & \Phi_2(r) = -g(r)\sqrt{\frac{\ell-m+\frac{1}{2}}{2\ell+1}}Y_{\ell,m+\frac{1}{2}}(\vartheta,\varphi) \end{cases} \tag{69}$$

Substituting these into the last two equations in (67), one gets [25]

$$(E+\frac{Ze^2}{4\pi r}+E_0)\Phi_3=\frac{1}{i}\sqrt{\frac{\ell-m+\frac{3}{2}}{2\ell+3}}(\frac{dg}{dr}-\ell\frac{g}{r})Y_{\ell+1,m-\frac{1}{2}}(\vartheta,\varphi),$$

$$(E+\frac{Ze^2}{4\pi r}+E_0)\Phi_4=\frac{1}{i}\sqrt{\frac{\ell+m+\frac{3}{2}}{2\ell+3}}(\frac{dg}{dr}-\ell\frac{g}{r})Y_{\ell+1,m+\frac{1}{2}}(\vartheta,\varphi). \quad (70)$$

If we introduce a function f(r) by letting

$$\Phi_3=-i\sqrt{\frac{\ell-m+\frac{3}{2}}{2\ell+3}}f(r)Y_{\ell+1,m-\frac{1}{2}},$$

$$\Phi_4=-i\sqrt{\frac{\ell+m+\frac{3}{2}}{2\ell+3}}f(r)Y_{\ell+1,m+\frac{1}{2}}, \quad (72)$$

then Eq. (70) becomes

$$(E+\frac{Ze^2}{4\pi r}+E_0)f(r)=\frac{dg}{dr}-\ell\frac{g}{r}. \quad (73)$$

Substituting this into the first two equations of (67) yields the same equation

$$(E-\frac{Ze^2}{4\pi r}-E_0)g(r)=-\frac{df}{dr}-(\ell+2)\frac{f}{r}. \quad (74)$$

[25] *Use is made of Eq. (66) and the following relations*

$$\frac{\partial}{\partial z}(gY_{\ell,m})=\sqrt{\frac{(\ell+m+1)(\ell-m+1)}{(2\ell+1)(2\ell+3)}}Y_{\ell+1,m}(\frac{dg}{dr}-\ell\frac{g}{r})$$
$$+\sqrt{\frac{(\ell+m)(\ell-m)}{(2\ell-1)(2\ell+1)}}Y_{\ell-1,m}(\frac{dg}{dr}+(\ell+1)\frac{g}{g}),$$

$$(\frac{\partial}{\partial x}+i\frac{\partial}{\partial y})(gY_{\ell,m})=\sqrt{\frac{(\ell+m+1)(\ell+m+2)}{(2\ell+1)(2\ell+3)}}Y_{\ell+1,m+1}(\frac{dg}{dr}-\ell\frac{g}{r})$$
$$-\sqrt{\frac{(\ell-m-1)(\ell-m)}{(2\ell-1)(2\ell+1)}}Y_{\ell-1,m+1}(\frac{dg}{dr}+(\ell+1)\frac{g}{r}), \quad (71)$$

$$(\frac{\partial}{\partial x}-i\frac{\partial}{\partial y})(gY_{\ell,m})=-\sqrt{\frac{(\ell-m+1)(\ell-m+2)}{(2\ell+1)(2\ell+3)}}Y_{\ell+1,m-1}(\frac{dg}{dr}-\ell\frac{g}{r})$$
$$+\sqrt{\frac{(\ell+m-1)(\ell+m)}{(2\ell-1)(2\ell+1)}}Y_{\ell-1,m-1}(\frac{dg}{dr}+(\ell+1)\frac{g}{r}).$$

Similarly, from Eq. (65b), let

$$\begin{cases} j = \ell - \frac{1}{2} & \Phi_1 = g(r)\sqrt{\frac{\ell-m+\frac{1}{2}}{2\ell+1}}Y_{\ell,m-\frac{1}{2}}(\vartheta,\varphi), \\ m & \Phi_2 = g(r)\sqrt{\frac{\ell+m+\frac{1}{2}}{2\ell+1}}Y_{\ell,m+\frac{1}{2}}(\vartheta,\varphi), \end{cases} \tag{75}$$

and by a procedure similar to Eq. (72),

$$\Phi_3 = -i\sqrt{\frac{\ell+m-\frac{1}{2}}{2\ell-1}}f(r)Y_{\ell-1,m-\frac{1}{2}},$$

$$\Phi_4 = i\sqrt{\frac{\ell-m-\frac{1}{2}}{2\ell-1}}f(r)Y_{\ell-1,m+\frac{1}{2}}, \tag{76}$$

and corresponding to Eqs. (73) and (74)

$$(E + \frac{Ze^2}{4\pi r} + E_0)f(r) = \frac{dg}{dr} + (\ell+1)\frac{g}{r}, \tag{77}$$

$$(E + \frac{Ze^2}{4\pi r} - E_0)g(r) = -\frac{df}{dr} + (\ell-1)\frac{f}{r}. \tag{78}$$

The 4 equation (73), (74), (77) and (78) can be expressed in the form of a pair of equations by the following notation[26]:

$$j = \begin{cases} \ell + \frac{1}{2} : \\ \ell - \frac{1}{2} : \end{cases} \qquad \xi = \begin{cases} -(j+\frac{1}{2}) = -(\ell+1) \\ (j+\frac{1}{2}) = \ell \end{cases} \tag{79}$$

$$\begin{aligned} &\frac{df}{dr} + (1-\xi)\frac{f}{r} + (E + \frac{Ze^2}{4\pi r} - E_0)g(r) = 0, \\ &\frac{dg}{dr} + (1+\xi)\frac{g}{r} - (E + \frac{Ze^2}{4\pi r} + E_0)f(r) = 0. \end{aligned} \tag{80}$$

Introducing

$$G(r) \equiv rg(r), \qquad F(r) \equiv rf(r), \tag{81}$$

one has from Eq. (80),[27]

$$\begin{aligned} &\frac{dF}{dr} - \xi\frac{F}{r} = [m_0(1-\epsilon) - \frac{\beta}{r}]G, \\ &\frac{dG}{dr} + \xi\frac{G}{r} = [m_0(1+\epsilon) + \frac{\beta}{r}]F. \end{aligned} \tag{82}$$

---

[26] *For $j = \ell - \frac{1}{2}$, ℓ must not be 0; otherwise $\Phi_1, \Phi_2, \Phi_3, \Phi_4$ all vanish, as seen from Eqs. (75) and (76).*

[27]

$$\begin{aligned} &\frac{dF}{dr} - \xi\frac{F}{r} = [\frac{m_0 c}{\hbar}(1-\epsilon) - \frac{\beta}{r}]G, \\ &\frac{dG}{dr} + \xi\frac{G}{r} = [\frac{m_0 c}{\hbar}(1+\epsilon) + \frac{\beta}{r}]F. \end{aligned} \tag{82$'$}$$

$$\epsilon = \frac{E}{m_0 c^2} < 1, \qquad \alpha = \frac{e^2}{\hbar c}(\simeq \frac{1}{137.037}), \qquad \beta = Z\alpha. \tag{83$'$}$$

$$\epsilon \equiv \frac{E}{m_0} < 1, \qquad \alpha = \frac{e^2}{4\pi}(\simeq \frac{1}{137.037}), \qquad \beta \equiv Z\alpha. \tag{83}$$

On introducing a dimensionless length ρ

$$\rho = 2m_0\sqrt{1-\epsilon^2}\, r, \tag{84}$$

and a pair of functions $u(\rho)$, $v(\rho)$,

$$\begin{aligned} F(\rho) &= \sqrt{1-\epsilon}\, e^{-\frac{1}{2}\rho}\rho^{\gamma}(u-v), \\ G(\rho) &= \sqrt{1+\epsilon}\, e^{-\frac{1}{2}\rho}\rho^{\gamma}(u+v), \end{aligned} \tag{85}$$

one gets for $u(\rho)$, $v(\rho)$

$$\begin{aligned} &\frac{du}{d\rho} - [1 - \frac{1}{\rho}(\gamma + \frac{\beta\epsilon}{\sqrt{1-\epsilon^2}})]u + \frac{1}{\rho}(\xi + \frac{\beta}{\sqrt{1-\epsilon^2}})v = 0, \\ &\frac{dv}{d\rho} + \frac{1}{\rho}(\gamma - \frac{\beta\epsilon}{\sqrt{1-\epsilon^2}})v + \frac{1}{\rho}(\xi - \frac{\beta}{\sqrt{1-\epsilon^2}})u = 0. \end{aligned} \tag{86}$$

Let

$$u = \sum_{s=0} c_s\rho^s, \qquad v = \sum_{s=0} d_s\rho^s. \tag{87}$$

Then one gets

$$-c_{s-1} + (s + \gamma + \frac{\beta\epsilon}{\sqrt{1-\epsilon^2}})c_s + (\xi + \frac{\beta}{\sqrt{1-\epsilon^2}})d_s = 0,$$

$$(\xi - \frac{\beta}{\sqrt{1-\epsilon^2}})c_s + (s + \gamma - \frac{\beta\epsilon}{\sqrt{1-\epsilon^2}})d_s = 0. \tag{88}$$

The indicial equation is (for $s = 0$)

$$\begin{vmatrix} \gamma + \frac{\beta\epsilon}{\sqrt{1-\epsilon^2}} & \xi + \frac{\beta}{\sqrt{1-\epsilon^2}} \\ \xi - \frac{\beta}{\sqrt{1-\epsilon^2}} & \gamma - \frac{\beta\epsilon}{\sqrt{1-\epsilon^2}} \end{vmatrix} = 0, \tag{89}$$

$$\begin{aligned} \gamma &= \sqrt{\xi^2 - Z^2\alpha^2} \\ &= \text{real} \quad \text{for}\, \xi^2 > (\frac{Z}{137})^2. \end{aligned} \tag{90}$$

To obtain the eigenvalues of Eq. (82) or (86), on eliminating d_s between the two equations in (88), one gets [using Eq. (90a)],

$$c_s = \frac{s + \gamma - \frac{\beta\epsilon}{\sqrt{1-\epsilon^2}}}{(s+\gamma)^2 - \gamma^2}\, c_{s-1}. \tag{91}$$

For the series in $u(\rho), v(\rho)$ to terminate at a certain power, i.e.,

$$c'_n = 0, \qquad n' = \text{a positive integer},$$

one has

$$n' + \gamma - \frac{\beta\epsilon}{\sqrt{1-\epsilon^2}} = 0. \tag{92}$$

From this expression, it is seen that only for[28]

$$\epsilon = \frac{E}{m_0} = \frac{w + m_0}{m_0} < 1,$$

i.e.,

$$w < 0, \tag{93}$$

will n' be real. The condition that n' = positive integer determines the eigenvalues ϵ, or w.

From Eq. (79)

$$\xi^2 = (j + \frac{1}{2})^2.$$

Let us introudce a positive integer n

$$n = n' + |\,\xi\,| = n' + j + \frac{1}{2}. \tag{94}$$

$|\,\xi\,| = \ell + 1$, and n' can be identified with the radial quantum number n_r in Sommerfeld's theory (I-53), Vol. I. Thus n is the principal quantum number.

From Eq. (92), one obtains[29]

$$\frac{w + m_0}{m_0} = [1 + \frac{Z^2\alpha^2}{[n - j - \frac{1}{2} + \sqrt{(j+\frac{1}{2})^2 - Z^2\alpha^2}]^2}]^{-\frac{1}{2}} \tag{95}$$

which is the exact eigenvalue formula from Dirac's theory.

For $Z^2\alpha^2 \ll 1$, one obtains, on expanding [30]

$$w = -\frac{Z^2}{n^2} - \frac{Z^4\alpha^2}{n^4}\left\{ \begin{array}{l} \frac{n}{\ell+1} - \frac{3}{4} \\ \frac{n}{\ell} - \frac{3}{4} \end{array} \right. + \ldots \quad \text{for} \quad j = \left\{ \begin{array}{l} \ell + \frac{1}{2} \\ \ell - \frac{1}{2} \end{array} \right. ,$$

$$\text{in units of } \frac{e^2}{2a_0}. \tag{95a}$$

28

$$\epsilon = \frac{E}{m_0c^2} = \frac{w + m_0c^2}{m_0c^2} < 1. \tag{93$'$}$$

29

$$\frac{w + m_0c^2}{m_0c^2} = [1 + \frac{Z^2\alpha^2}{[n - j - \frac{1}{2} + \sqrt{(j+\frac{1}{2})^2 - Z^2\alpha^2}]^2}]^{-\frac{1}{2}}. \tag{95$'$}$$

30 *If multiplied by $R\hbar c$, R = Rydberg constant, the above result is in c.g.s. units.*

To obtain the wave functions, one has from the recursion relation Eq. (91) together with (92),

$$\begin{aligned} c_s &= -\frac{n'-s}{s(2\gamma+s)}c_{s-1} \\ &= (-1)^s \frac{(n'-1)...(n'-s)}{s!(2\gamma+1)...(2\gamma+s)}c_0. \end{aligned} \tag{96}$$

Thus $u(\rho)$ is the confluent hypergeometric series

$$u(\rho) = c_0\, F(-n'+1,\, 2\gamma+1;\, \rho), \tag{97}$$

which has been defined in Eq. (III-147), Vol. I.

From Eqs. (88) and (92), one has

$$d_s = -\frac{-\xi+\beta/\sqrt{1-\epsilon^2}}{n'-s}c_s, \tag{98}$$

so that from the second equation in Eq. (88) again,

$$\frac{d_s}{c_s} = \frac{d_0}{c_0}\frac{n'}{n'-s},$$

$$d_s = (-1)^s \frac{n'(n'-1)...(n'-s+1)}{s!(2\gamma+1)...(2\gamma+s)}d_0, \quad \text{from Eq. (96)}.$$

Thus $v(\rho)$ is

$$v(\rho) = d_0\, F(-n',\, 2\gamma+1;\, \rho). \tag{99}$$

From Eq. (98),

$$\frac{d_0}{c_0} = -\frac{-\xi+\frac{Z\alpha}{\sqrt{1-\epsilon^2}}}{n'}.$$

We shall take

$$c_0 = -C\frac{n'}{\sqrt{-\xi+Z\alpha/\sqrt{1-\epsilon^2}}},$$

$$d_0 = C\sqrt{-\xi+\frac{Z\alpha}{\sqrt{1-\epsilon^2}}}, \tag{100}$$

where the constant C is to be determined from the normalization condition for the Φ_1, Φ_2, Φ_3, of Eq. (67)

$$\int \{|\,\Phi_1\,|^2 + |\,\Phi_2\,|^2 + |\,\Phi_3\,|^2 + |\,\Phi_4\,|^2\}dr = 1. \tag{101}$$

After a lengthy calculation, one obtains

$$\begin{aligned}
f(r) &= -\sqrt{1-\epsilon}W(r)\{n'F(-n'+1, 2\gamma+1; \frac{2Z}{Na_0}r) \\
&\quad + (N-\xi)F(-n', 2\gamma+1; \frac{2Z}{Na_0}r)\}, \\
g(r) &= -\sqrt{1+\epsilon}W(r)\{-n'F(-n'+1, 2\gamma+1; \frac{2Z}{Na_0}r) \\
&\quad + (N-\xi)F(-n', 2\gamma+1; \frac{2Z}{Na_0}r)\}, \\
N^2 &= n^2 - 2n'(\xi - \sqrt{\xi^2 - Z^2\alpha^2}), \\
W(r) &= \sqrt{\frac{\Gamma(2\gamma+n'+1)}{\Gamma(2\gamma+1)\sqrt{n'!}}}\frac{1}{\sqrt{4N(N-\xi)}}(\frac{2Z}{Na_0})^{\gamma+\frac{1}{2}}e^{-Zr/Na_0}.
\end{aligned} \tag{102}$$

5.5 Dirac Equation and Many-Body Features

In Chapter 2, Sect. 3, it has been pointed out that the negative energy states in the Dirac theory (in fact in any relativistic theory) bring forth a consequence of basic importance, namely, the Dirac equation cannot strictly be regarded as theory of *one* particle, but is bound in a fundamental way with an infinite sea of particles in the negative energy states. That the negative energy states cannot be simply ignored as in the classical theory is first shown by Klein's study [Chap. 2, Sect. 3, (3)] and is then experimentally shown by the discovery of the positron and of the process of pair production. [Chap. 2, Sect. 3, (4)].

Then, in the mid and later 1940's, the development of quantum electrodynamics (the theory of the quantized coupled electron and electromagnetic field) and the experimental discovery of the Lamb shift and the "anomaly" of the gyromagnetic ratio g of the electron by Kusch points without doubt to the many-body or field nature of the relativistic theory of the electron.

This many-body nature of the Dirac equation can be brought out in a very elementary way without using the method of quantum electrodynamics. We see in Section 4, Eq. (95), that for a (hypothetical) heavy enough atom

$$Z > \frac{(j+\frac{1}{2})}{\alpha} \tag{103}$$

i.e.,

$$Z > 137, \tag{103}$$

imaginary numbers appear in the energy formula. Let us trace this a little way back to the equation (82) for the radial wave functions, namely,

$$\begin{aligned}
\frac{dF}{dr} - \xi\frac{F}{r} &= [\frac{1}{\lambda}(1-\epsilon) - \frac{Z\alpha}{r}]G, \\
\frac{dG}{dr} + \xi\frac{G}{r} &= [\frac{1}{\lambda}(1+\epsilon) + \frac{Z\alpha}{r}]F,
\end{aligned} \tag{104}$$

where

$$\lambda = \frac{\hbar}{m_0 c} = \frac{1}{2\pi} \times \text{Compton wave length of the electron.}$$

Introducing ρ as in Eq. (84), eliminating the small component F(r) between the two equations and writing

$$G(\rho) = e^{-\rho/2} W(\rho), \tag{105}$$

one obtains

$$\begin{aligned}(\frac{1}{2}\sqrt{\frac{1+\epsilon}{1-\epsilon}} + \frac{Z\alpha}{\rho})(\frac{d^2W}{d\rho^2} - \frac{dW}{d\rho} + \frac{Z\alpha\epsilon}{\sqrt{1-\epsilon^2}}\frac{W}{\rho} + \frac{Z^2\alpha^2 - \xi^2}{\rho^2}W) \\ + \frac{Z\alpha}{\rho^2}(\frac{dW}{d\rho} - W) - \frac{1}{2}\sqrt{\frac{1+\epsilon}{1-\epsilon}}\frac{\xi}{\rho^2}W(\rho) = 0.\end{aligned} \tag{106}$$

Setting

$$W(\rho) = \rho^\gamma \sum_{s=0} a_s \rho^s, \tag{107}$$

one gets the indicial equation

$$\gamma^2 - \xi^2 + Z^2\alpha^2 = 0. \tag{108}$$

If

$$Z > \frac{\xi}{\alpha} = 137\xi, \tag{109}$$

γ is imaginary, and it is easy to see from the recursion formula for the a_s's that the series does not teminate into a polynomial but approaches

$$W(\rho) \simeq e^\rho, \tag{110}$$

so that $G(\rho)$ approaches $e^{\frac{1}{2}\rho}$ and is not quadratically integrable. Thus no stable states exist for $Z > 137$, in contradistinction to the non-relativistic Schrödinger theory.

The interpretation of this result is as follows. As Z becomes large, the electric field near the nucleus is intense and mixes up the positive and negative energy states so that the many-body properties become increasingly important, rendering the one-particle concept invalid. In this connection, it is interesting to see that the fine structure constant α , which is the coupling strength parameter in quantum field theory, plays the role of a critical value for the breaking down of the one-particle theory.

5.6 "Extra Moments" to the Dirac Equation?

In Eq. (10a), it has been seen that the Dirac equation can be put in the form[31] (summation convention)

$$[\Pi_\mu^2 + m_0^2 + \frac{e}{2i}\gamma_\mu\gamma_\nu F_{\mu\nu}]\Psi = 0. \tag{111}$$

In Eq. (14), it is seen that

$$\frac{e}{2i}\gamma_\mu\gamma_\nu F_{\mu\nu} = e(\sigma \cdot H) - ie\rho_1(\sigma \cdot E), \tag{112}$$

representing a moment σ.

But the Dirac equation Eq. (4)

$$(\gamma_\mu\Pi_\mu - im_0)\Psi = 0$$

can be extended by two extra terms without destroying the Lorentz covariance,[32]

$$\begin{aligned}(\gamma_\mu\Pi_\mu - im_0)\Psi = g_1(\frac{e}{4m_0})\gamma_\mu\gamma_\nu F_{\mu\nu}\Psi \\ - g_2 e(\frac{1}{m_0})^2\gamma_\mu\partial_\nu\partial_\nu A_\mu\Psi,\end{aligned} \tag{113}$$

where g_1, g_2 are two dimensionless constants; $\gamma_\mu\gamma_\nu F_{\mu\nu}$ is a scalar under Lorentz transformations;

$$\partial_\nu\partial_\nu A_\mu = -4\pi j_\mu, \tag{114}$$

is a 4-vector, so that $\gamma_\mu\partial_\nu\partial_\nu A_\mu$ is also a scalar. The g_2 term in Eq. (113) may be regarded as representing a replacement of the A_μ in Eq. (4) by

$$A_\mu + g_2(\frac{1}{m_0})^2\partial_\nu\partial_\nu A_\mu. \tag{115}$$

We shall in the following consider only the additional term g_1 (i.e., disregarding the g_2 term).

Denote by

$$G = g_1(\frac{e}{4m_0})\gamma_\mu\gamma_\nu F_{\mu\nu} \tag{116}$$

31

$$[\Pi_\mu^2 + m_0^2c^2 + \frac{\hbar e}{2ic}\gamma_\mu\gamma_\nu F_{\mu\nu}]\Psi = 0. \tag{111'}$$

32

$$\begin{aligned}(\gamma_\mu\Pi_\mu - im_0c)\Psi = g_1(\frac{e\hbar}{4m_0c^2})\gamma_\mu\gamma_\nu F_{\mu\nu}\Psi \\ - g_2\frac{e}{c}(\frac{\hbar}{m_0c})^2\gamma_\mu\partial_\nu\partial_\nu A_\mu\Psi.\end{aligned} \tag{113'}$$

and carry out a calculation parallel to that from Eq. (9) to (14), one obtains the following equation[33]

$$[\Pi^2 + m_0^2 + (1+g_1)\frac{e}{2i}\gamma_\mu\gamma_\nu F_{\mu\nu}] = [G^2 + \gamma_\sigma\Pi_\sigma G - G\gamma_\sigma\Pi_\sigma]. \tag{117}$$

The extra magnetic moment arising from the term G (and g_1) is suggested by Pauli. So far there seems to have been no experimental evidence for such an extra moment in the case of an electron.[34,35]

---

[33]

$$[\Pi^2 + m_0^2c^2 + (1+g_1)\frac{e\hbar}{2ic}\gamma_\mu\gamma_\nu F_{\mu\nu}] = [G^2 + \gamma_\sigma\Pi_\sigma G - G\gamma_\sigma\Pi_\sigma]. \tag{117'}$$

[34] *There seems to be a story in connection with this Pauli moment. In 1947, G. Breit (at Yale University) and I. I. Rabi (Columbia University) discussed the possibility of detecting, by means of the atomic beam facilities at Columbia, any extra magnetic moment — deviation from the value $g = 2$ of Dirac's theory. In the careful measurements of the value of g by P. Kusch (1947) and the analysis by H. M. Foley, the value*

$$g = 2(1 + \frac{\alpha}{2\pi} + ...), \tag{118}$$

was found, which is in excellent agreement with the theoretical value just predicted by J. Schwinger (at Harvard University) on his covariant treatment of quantum electrodynamics. The Lamb shift (1947) and this "g-2" discovery are most important as direct supports for the then new developments of quantum electrodynamics.

[35] *Another modification of the Dirac equation has been examined, by G. Feinberg, Phys. Rev. 112, 1637 (1958); E. F. Salpeter, ibid. 112, 1642 (1958), namely*

$$(\gamma_\mu\Pi_\mu - im_0)\Psi = \xi\frac{e}{4m_0}\gamma_5\gamma_\mu\gamma_\nu F_{\mu\nu}\Psi. \tag{113a}$$

The extra term containing the pseudo-scalar γ_5 corresponds to a parity non-conserving electric dipole moment. ξ is a numerical parameter. This term leads to terms of the form

$$\begin{aligned} &\xi\mu_B(\sigma\cdot\mathbf{E}_{in}), \qquad \xi\mu_B(\sigma\cdot\mathbf{E}_{ext}), \\ &-\xi\frac{\mu_B}{2m_0}(\sigma\cdot[\mathbf{H}\times\mathbf{p}]), \\ &i\xi\frac{\mu_B}{2m_0}(\mathbf{p}\cdot\mathbf{H}). \end{aligned}$$

$\mathbf{E}_{in}$ is the internal electric field due to the nucleus, and $\mathbf{E}_{ext}$ the external field. Of the various terms, only the $\xi\mu_B(\sigma\cdot\mathbf{E}_{in})$, where $\mathbf{E}_{in} = \frac{Ze}{4\pi r^2}$, is important, and for the ground state is ${}^2S_{\frac{1}{2}}$ of hydrogen,

$$\begin{aligned} E(1s\,{}^2S_{\frac{1}{2}}) &= -\frac{1}{2}\xi^2\alpha^2\,(Rydberg), \qquad \alpha = \frac{e^2}{4\pi} \simeq \frac{1}{137} \\ &= -8.7\times10^4\xi^2 MC/sec. \\ &= -2.9\xi^2 cm^{-1}, \end{aligned}$$

which is a downward shift (opposite to the Lamb shift). If the optical method for the Lyman α-line can detect at best a shift of $\simeq 0.01cm^{-1}$, it is not sensitive enough for an upper limit for ξ . Salpeter shows that for the $2s\,{}^2S_{\frac{1}{2}}$ state, a considerably lower value can be concluded for the upper limit of $\xi(\xi < 0.004)$.

Exercises: *Chapter 5*

1. Solve the free Dirac equation in a spherical cavity of radius R subject to the linear boundary condition proposed in the MIT bag model:

$$(\gamma_j r_j / r)\psi(\mathbf{r}) = \psi(\mathbf{r}) \qquad \text{at } r = R.$$

Show that the eigenvalue ω is given by

$$\omega = \frac{1}{R}\sqrt{x^2 + (mR)^2},$$

with x determined by

$$tanx = \frac{x}{1 - mR - \sqrt{x^2 + (mR)^2}}.$$

For $m = 0$, one finds $x = 2.0428$ as the lowest eigenvalue. Find the next eigenvalue if you are familiar with the numerical method. Otherwise, find the forms for the first few excited states.

2. Solve the problem of a Dirac particle moving in a confining potential:

$$(1 + \gamma_4)\frac{kr^2}{4} + (a + b\gamma_4).$$

(i) Fix the parameters such that the ground-state solution will be of the form:

$$\psi(\mathbf{r}, s) = \begin{pmatrix} u(r) \\ i\sigma \cdot \mathbf{r} v(r) \end{pmatrix} \chi_s,$$

with χ_s a two-component Pauli spinor and $v(r)/u(r) =$ constant.

(ii) Fix the parameters such that the mean square radius is $0.517\ fm^2$ for the ground state (a value for the nucleon in the MIT bag).

3. **(i)** Solve the problem of an electron moving in a constant magnetic field in the z direction.

(ii) Solve the problem of a neutron as a neutral Dirac particle moving in a constant magnetic field. (Note that the neutron has an anomalous magnetic moment of $-1.913\ n.m.$)

(iii) Compare the above two cases. Could you confine neutrons in a suitably designed magnetic field?

Part B. Introduction to Quantum Field Theory

Chapter 6. Classical Fields

6.1 Introduction

In classical physics, "particles" and "fields" are two different concepts; they deal with different phenomena; they are characterized by discreteness and continuity respectively. A system may contain a vast number of particles (such as the molecules in a gas), but as long as they are denumerable, the basic theory is classical dynamics. But if the variables of a system are not denumerable, such as the electric field intensity or the velocity field of a fluid, the basic theory is the so-called field theory.

The fundamental theory in classical dynamics is the Newtonian dynamics, which can be expressed in different mathematical forms, such as the Lagrangian form, the Hamiltonian form, and the variational form.

The phenomena and the laws of fields are usually expressed in partial differential equations - such as those of fluid dynamics and of electromagnetic fields. But these equations are also expressed in the form of Lagrange and Hamilton equations, and these latter equations can in turn be derived from a Lagrangian density function via a variational principle.

In classical physics, electromagnetic fields are regarded as continuous functions - non-denumerably infinite numbers of variables. If all electromagnetic phenomena, including the propagation of light, are continuous, then the descriptions by the Maxwell equations, or by means of variational equations, are equivalent.

But since the quantum theory of radiation of Einstein in 1905, direct experiments such as the photo-electric effect and the Compton effect, give strong evidence for the particle properties of electromagnetic radiation. This leads us to seek a mathematical theory to describe the quantum properties of continuous fields — analogous to the quantization of atomic and molecular phenomena in quantum mechanics. To extend quantum mechanics to the quantized properties of fields, the theory is called the theory of quantized fields.

There are various kinds of fields. One kind is the "classical fields", such as the electromagnetic field. The concept of the "quantum" of radiation originates with Einstein's theory of the photon in 1905; but the formal mathematical method of "quantization" of particle dynamics begins with Bohr's theory of the hydrogen atom in 1913, in the form of the quantum conditions, as generalized by Sommerfeld and Wilson,

$$\oint p_k dq_k = n_k h. \tag{1}$$

This "old quantum theory" has amazingly successful results for the hydrogen atom, but then has met with very basic difficulties in other systems. Then quantum mechanics was born, and the quantization condition can be formulated in the form of the commutation relation

$$pq - qp = \frac{\hbar}{i}. \tag{2}$$

On this relation, quantum mechanics in the form of the matrix theory and the Schrödinger theory has been established. Quantum mechanics represents both the particle and the wave properties of what in classical physics have been described as "particles" and "waves", mutually exclusive properties now joined by the Einstein-de Broglie relations

$$E = h\nu, \qquad p = \frac{h}{\lambda}. \tag{3}$$

Quantum mechanics has successfully treated all known phenomena in atomic, molecular, solid state, and nuclear physics.

Now we wish to go from particles back to the older problem — the quantum theory of the electromagnetic field, to extend the method of quantization for particle dynamics (coordinates and momenta, angular momenta and energies) to the continuous fields.

The method consists of the following program:

(1) Express the laws of electromagnetic field, namely, Maxwell's field equations, in terms of the 4-potential $A_1, A_2, A_3, A_4 = i\phi$, in the form of the Hamiltonian equations in classical dynamics,

(2) define the field variables (such as the 4-potential) and their canonical conjugates,

(3) quantize the field variables by relations similar to the relation (2).

This program, when applied to the so-called "free field" (such as electromagnetic field free from the charge density ρ and current density $\mathbf{j}$), can be readily carried out, and in fact the quantized field is represented by the photons.

The electromagnetic field having a 4-potential is not the simplest classical field. If we regard the Klein-Gordon equation[1]

$$(\nabla^2 - \frac{\partial^2}{\partial t^2})\Psi = (\frac{1}{\lambda_c})^2\Psi. \qquad \lambda_c = \frac{1}{m}, \tag{4}$$

as the equation of a classical scalar Ψ field, the quantization process is even simpler. We shall see that the quantized field of Eq. (4) is appropriate for the description of the π mesons (pions, π^0, π^+, π^-).

But there are other kinds of fields than the above-mentioned classical fields. We must return to the quantization of particle systems.

In the last four chapters, we have treated the Dirac relativistic wave equation of the electron. It has the following successes: it is Lorentz covariant; it contains the electron spin with the gyromagnetic ratio $g = 2$; it predicts the anti-particle positron which is experimentally discovered; it gives the correct fine structure of the hydrogenic atom levels.

As already mentioned in the introductory chapter (Ch. 0), there are "difficulties" associated with the Dirac equation. The first one has to do with the question of extending the theory for a single electron to a system of many electrons, since it is not clear if the negative-energy spectra associated with different electrons can be treated in a consistent manner. Another "difficulty" is of an even more basic nature, namely, a theory started out to represent a single electron ends up being inseparably bound with a many-body effect on account of the infinite sea of electrons in negative energy states. Using Dirac's "hole" picture, we see that, in the presence of a strong electric field, an electron does not exist as a lone

[1] *λ_c is the Compton wave length. In ordinary units, it is $\hbar/(mc)$.*

electron but is a system containing infinitely many (and not constant in number) electrons and positrons. Thus, strictly speaking, the single-electron theory is valid only in the sense of a limiting case of a free electron and must be replaced by a theory of "many-electrons".

Just as the representation of "photons" is the electromagnetic field, the appropriate representation of a system of an infinitely large and variable number of "particles" is a "field". Thus, we see that, starting from the particle point of view, one seeks a theory of particles that satisfies the relativity principle and finds that the appropriate theory is a many-body theory, and the appropriate representation of a many-body system is "field".

Therefore, the problem is then to formulate a many-electron field. One may start from the Dirac equation for a free electron:

$$(\gamma_\mu \frac{1}{i}\frac{\partial}{\partial x_\mu} - im_0)\Psi = 0, \tag{5}$$

and regard it as the equation of a classical field Ψ - in this case a 4-component field. Then treat Ψ_j as the field variables (similarly to the 4-potential A_μ of the electromagnetic field), and their canonical conjugates. Then introduce the quantization conition in the form of the commutation relations (2), and as expected, one obtains the electron as the particles of the quantized field. Such a quantization procedure is called "second quantization". The quantized Ψ possesses both the wave and the particle properties (just as the quantized electromagnetic field possesses the photon and the wave property).

Unfortunately, the story did not end here. The quantum theory of a pure electron field presents no difficulties but, for the system of the electron coupled to the electromagnetic field, a relativistic quantum field theory, known as quantum electrodynamics or QED, presents serious deep-rooted difficulties, namely the persistent presence of infinities in the results of calculations of physical quantities.

The theory of quantized electromagnetic fields begins with the work of Dirac in 1927, followed immediately by the work of Jordan and Wigner, and by Fermi in 1930. A breakthrough came in the mid 1940's with the work of Tomonaga in Japan and Schwinger, Feynman, and Dyson in the United States. Although infinities still remain, the theory succeeds in "subtracting" them away in a consistent way so that finite results can be obtained, which have been found to be in excellent agreement with the observed Lamb shifts and the "g anomaly". The decade from the mid 1940's to the mid 1950's is a period of fervent studies of quantum electrodynamics, both in further calculations on this "renormalization" theory and in attempts to rid the theory of the infinities (not just to isolate and bury them, so to speak).

During the last two decades, physicists have arrived at a successful description of elementary particles and the nature of their interactions and their unification – the so-called "Standard Model" which consists of quantum chromodynamics (*QCD*) for describing strong interactions among quarks and gluons and the Glashow-Salam-Weinberg (GSW) theory for a unified description of electromagnetic and weak interactions. Both QCD and the GSW electroweak theory are constructed using QED as a prototype theory. Till now, however, a satisfactory answer is yet to be found toward the question why QED and more generally the Standard Model, in which infinities appear but get subtracted away in a certain way, have been so successful.

In this part of the book, we thus wish to introduce in a pedagogic manner basic elements associated with QED, ranging from free classical fields and their quantization, to calculations of scattering matrix, or S-matrix, elements, and to concepts of "regularization" and "renormalization". A primary objective of our presentation is to set up the framework in order to introduce in the third part of this book basic ingredients associated with the Standard Model. Another objective is to stimulate further thoughts, or developments, concerning the treatment (or "renormalization") of infinities in QED.

6.2 Classical Field Equation

In classical dynamics, the state of a system of N particles is defined by $3N$ pairs of generalized coordinates q_k and their conjugate momenta p_k.

In classical field theories, a field is described by a continuous function of real variables $\phi(\mathbf{r}, t)$, and n fields by n such functions $\phi_\alpha(\mathbf{r}, t), \alpha = 1, 2, 3, ..., n$. Each field is described by partial differential equations of independent variables $\mathbf{r}$, t, called the field equations. Thus in classical electromagnetic field theory, ther are four fields for $(A_1, A_2, A_3, A_4 = i\phi)$, forming a vector field with four components.

If the Klein-Gordon equation is regarded as an equation for a classical field Ψ, then the field is a scalar field, i.e., a Ψ having only one component.

The equations of motion of particles in classical dynamics can be obtained from a variational principle. We may extend this view to a field in the following manner.

A field is regarded as a dynamical system having a non-denumerably infinite number of degrees of freedom, whose generalized coordinates and conjugate momenta are the components $\varphi_\alpha(\mathbf{r}, t)$ and $\dot{\varphi}_\alpha = \frac{\partial \phi_\alpha}{\partial t}$. Here the $\mathbf{r}$ are *not* the coordinates, but are just independent parameters. Let us imagine a volume V in the 3-dimentional space $\mathbf{r}$ be divided in small cells, $\triangle V^{(s)}$ at $(\mathbf{r}_s, t)$, and let $\triangle V^{(s)}$ be smaller and smaller. At a point x_s at time t, $\varphi_\alpha^{(s)}$ is almost constant within $\triangle V^{(s)}$. We shall choose as "coordinates" for the field φ the average values of φ_α at various x_s

$$\varphi_\alpha^{(s)}(x_s, t), \qquad \alpha = 1, 2, 3, ... \quad s = 1, 2, 3, ... \tag{6}$$

and s is denumerable. Now let $\triangle V^{(s)}$ approach zero in the limit, and the degree of freedom of the field becomes non-denumerably infinite. In this sense, we can extend the variational method of obtaining the Lagrange and Hamiltonian equations in classical dynamics to obtain the field equations.

Let us introduce the notations

$$\partial_\mu = \frac{\partial}{\partial x_\mu}, \qquad \mu = 1, 2, 3, 4$$

$$d^3x = dx_1 dx_2 dx_3$$

$$d^4x = dx_1 dx_2 dx_3 dt \quad \text{(scalar, invariant)},$$

and define the Lagrangian density $\mathcal{L}$ and Lagrangian L

$$\mathcal{L} = \mathcal{L}(\varphi_\alpha, \partial_\mu \varphi_\alpha), \tag{7}$$

$$L = \int_{V_3} \mathcal{L}(\varphi_\alpha, \partial_\mu \varphi_\alpha) d^3x, \tag{8}$$

$$\text{Action function} = S = \int_{t_1}^{t_2} L dt = \int_{V_4} \mathcal{L}(\varphi_\alpha, \partial_\mu \varphi_\alpha) d^4x, \tag{9}$$

where V_4 is a 4-dimensional volume $V_4 = V_3 \times (t_2 - t_1).d^4x$ is scalar (invariant) under Lorentz transformations. If $\mathcal{L}$ is a scalar, then S is also an invariant. (Conversely, if S is an invariant, then $\mathcal{L}$ is an invariant; but L is not an invariant.) Let $\delta\varphi_\alpha$ be variations such that

$$\delta\varphi_\alpha = \text{arbitrary within} \quad V_4, \tag{10}$$

$$\delta\varphi_\alpha = 0 \quad \text{on the boundary surface} \quad S \quad \text{of} \quad V_4,$$

and formulate the variational principle

$$\delta S = \delta \int_{V_4} \mathcal{L} d^4x = 0. \tag{11}$$

From

$$\delta\mathcal{L} = \frac{\partial \mathcal{L}}{\partial \varphi_\alpha} \delta\varphi_\alpha + \frac{\partial \mathcal{L}}{\partial(\partial_\mu \varphi_\alpha)} \delta(\partial_\mu \varphi_\alpha)$$

and on integrating by parts,[2]

$$\delta S = \int_{V_4} \{\frac{\partial \mathcal{L}}{\partial \varphi_\alpha} - \partial_\mu (\frac{\partial \mathcal{L}}{\partial(\partial_\mu \varphi_\alpha)})\} \delta\varphi_\alpha d^4x \tag{12}$$

from which one obtains the Euler equation

$$\frac{\partial \mathcal{L}}{\partial \varphi_\alpha} - \partial_\mu (\frac{\partial \mathcal{L}}{\partial(\partial_\mu \varphi_\alpha)}) = 0, \tag{13}$$

or

$$\frac{\partial \mathcal{L}}{\partial \varphi_\alpha} - \partial_k \frac{\partial \mathcal{L}}{\partial(\partial_k \varphi_\alpha)} - \partial_t \frac{\partial \mathcal{L}}{\partial(\partial_t \varphi_\alpha)} = 0, \tag{13a}$$

[2] *The integral*

$$\int_{V_4} \partial_\mu (\frac{\partial \mathcal{L}}{\partial(\partial_\mu \varphi_\alpha)} \delta\varphi_\alpha) d^4x = \int_S \frac{\partial \mathcal{L}}{\partial(\partial_\mu \varphi_\alpha)} \delta\varphi_\alpha dS = 0,$$

where the integral of a 4-divergence can be transformed into a surface integral over the boundary $\sum$ *of* V_4, *and* $\delta\varphi_\alpha = 0$ *on* $\sum$ *according to Eq. (10).*

which is a second order differential equation if $\mathcal{L}$ is a function of the first order partial derivatives $\partial_\mu \varphi_\alpha$.
We introduce the $\varphi_\alpha^{(s)}$ of Eq. (6) and replace the integral L in Eq. (8) by

$$L = \lim_{\triangle V^{(s)} \to 0} \sum_s \mathcal{L}(\varphi_\alpha^{(s)}, \partial_\mu \varphi_\alpha^{(s)}) \triangle V^{(s)} \tag{14}$$

and take $\varphi_\alpha^{(s)}$ as the coordinates (like the q_k for particles), and define their conjugate variables p_α and conjugate momentum densities $\pi_\alpha^{(s)}$ by

$$p_\alpha = \frac{\delta \mathcal{L}}{\delta \dot{\varphi}_\alpha^{(s)}} \triangle V^{(s)} \tag{15}$$

$$\pi_\alpha^{(s)} = \frac{p_\alpha}{\triangle V^{(s)}} = \frac{\partial \mathcal{L}}{\partial \dot{\varphi}_\alpha^{(s)}} \tag{16}$$

As in classical dynamics, we define the Hamiltonian density $\mathcal{H}$ and the Hamiltonian H[3]

$$\mathcal{H}(\varphi_\alpha, \pi_\alpha, \partial_k \varphi_\alpha) = \pi_\alpha^{(s)} \dot{\varphi}_\alpha^{(s)} - \mathcal{L} \tag{17}$$

$$\begin{aligned} H &= \lim_{\triangle V^{(s)} \to 0} \sum_s \mathcal{H} \triangle V^{(s)} \\ &= \int_{V_3} \mathcal{H} d^3 x. \end{aligned} \tag{18}$$

On carrying out the variation δH of the integral (18) and the integration by parts, one obtains

$$\delta H = \int_{V_3} \{ [\frac{\partial \mathcal{H}}{\partial \varphi_\alpha} - \partial_k \frac{\partial \mathcal{H}}{\partial(\partial_k \varphi_\alpha)}] \delta \varphi_\alpha + \frac{\partial \mathcal{H}}{\partial \pi_\alpha} \delta \pi_\alpha \} d^3 x, \tag{19}$$

$$\frac{\delta H}{\delta \varphi_\alpha} = \frac{\partial \mathcal{H}}{\partial \varphi_\alpha} - \partial_k \frac{\partial \mathcal{H}}{\partial(\partial_k \varphi_\alpha)}, \tag{20}$$

$$\frac{\delta H}{\delta \pi_\alpha} = \frac{\partial \mathcal{H}}{\partial \pi_\alpha}. \tag{21}$$

From Eq. (17), one obtains

$$\frac{\partial \mathcal{H}}{\partial \varphi_\alpha} - \partial_k \frac{\partial \mathcal{H}}{\partial(\partial_k \varphi_\alpha)} = -\frac{\partial \mathcal{L}}{\partial \varphi_\alpha} + \partial_k \frac{\partial \mathcal{L}}{\partial(\partial_k \varphi_\alpha)}, \tag{22}$$

[3] *The $\mathcal{H}$ defined by the Legendre transformation Eq. (17) is a function of $\partial_k \varphi_\alpha^{(s)}$, $k = 1, 2, 3$, but not of $\partial_t \varphi_\alpha^{(s)} = \dot{\varphi}_\alpha^{(s)}$.*

$$\frac{\partial \mathcal{H}}{\partial \pi_\alpha} = \dot{\varphi}_\alpha. \tag{23}$$

From Eqs. (21), (16), (13), (22), and (20), one obtains

$$\dot{\varphi}_\alpha = \frac{\delta H}{\delta \pi_\alpha}, \qquad \dot{\pi}_\alpha = -\frac{\delta H}{\delta \varphi_\alpha}, \tag{24}$$

which are the field equations in canonical form.

From Eqs. (18), (17), (16), and (13), it can be shown that

$$\frac{dH}{dt} = -\int_{V_3} \frac{\partial \mathcal{L}}{\partial t} d^3x \tag{25}$$

so that if $\frac{\partial \mathcal{L}}{\partial t} = 0$, then $\frac{dH}{dt} = 0$, i.e., H is a constant of motion.

If $\mathcal{L}$ is replaced by

$$\mathcal{L}' = \mathcal{L} + \partial_\mu g_\mu(\varphi_\alpha), \tag{26}$$

where g_μ is an arbitrary 4-vector function of φ_α so that $\partial_\mu g_\mu$ is a 4-divergence, then the Euler-Lagrange equation obtained from

$$\delta \int \mathcal{L}' d^4x = 0, \tag{27}$$

is the same as Eq. (13).[4]

6.3 Noether's Theorem

The Lagrangian density $\mathcal{L}$ as specified by Eq. (7) contains all the information concerning dynamical invariants such as energy-momentum 4-vector and angular momentum tensor, just like in the case of classical dynamics for a system of particles where constants of motion are contained in the Lagrangian. This aspect is summarized by the well-known Noether's theorem.

Noether's Theorem: To every continuous transformation of field functions and simultaneously coordinates which depends on n_0 continuous parameters and which ensures that the variation of the action is zero, there corresponds n_0 dynamical invariants which are combinations of field functions and their derivatives that are conserved in time.

Proof:

We define the continuous transformation in question as follows:

$$x_\mu \to x'_\mu = x_\mu + \delta x_\mu = x_\mu + X_{\mu\alpha}\delta w_\alpha, \tag{28a}$$

$$\phi_i(x) \to \phi'_i(x') = \phi_i(x) + \delta\phi_i(x) = \phi_i(x) + \Psi_{i\alpha}\delta w_\alpha, \tag{28b}$$

where δw_α ($\alpha = 1, 2, ..., n_0$) are n_0 infinitesimal parameters, $X_{\mu\alpha}$ specify the variation in coordinates, and $\Psi_{i\alpha}$ characterize the change in field functions $\phi_i(x)$.

[4] *$g_\mu(\varphi_\alpha)$ have only to satisfy the condition that $g_\mu = 0$ on the boundary surface $\sum$ of V_4.*

The theorem states that, if the corresponding variation in the action S is zero,

$$\delta S = \delta \int \mathcal{L}(x) d^4x \equiv \int \mathcal{L}'(x') d^4x' - \int \mathcal{L}(x) d^4x = 0, \tag{29}$$

then we must be able to find n_0 dynamical quantities $\Theta_{\mu\alpha}$ such that

$$\partial_\mu \Theta_{\alpha\mu} = 0 \quad (\alpha = 1, 2, ..., n_0). \tag{30}$$

The procedure to find the expression for $\Theta_{\alpha\mu}$ is described below: Define the variation in the form of the field function,

$$\bar{\delta}\phi_i(x) \equiv \phi'_i(x) - \phi_i(x), \tag{31}$$

or,

$$\begin{aligned} \bar{\delta}\phi_i(x) &= \delta\phi_i(x) - (\partial_\mu \phi_i)\delta x_\mu \\ &= (\Psi_{i\alpha} - (\partial_\mu \phi_i) X_{\mu\alpha})\delta w_\alpha. \end{aligned} \tag{32}$$

We have

$$\mathcal{L}'(x') \equiv \mathcal{L}(\phi'_i(x'), \partial'_\mu \phi'_i(x')) = \mathcal{L}(x) + \delta\mathcal{L}(x), \tag{33a}$$

with

$$\delta\mathcal{L} = \frac{\partial \mathcal{L}}{\partial \phi_i}\delta\phi_i + \frac{\partial \mathcal{L}}{\partial(\partial_\mu \phi_i)}\delta(\partial_\mu \phi_i) = \bar{\delta}\mathcal{L}(x) + \frac{d\mathcal{L}}{dx_\mu}\delta x_\mu, \tag{33b}$$

$$\bar{\delta}\mathcal{L}(x) = \frac{\partial \mathcal{L}}{\partial \phi_i}\bar{\delta}\phi_i + \frac{\partial \mathcal{L}}{\partial(\partial_\mu \phi_i)}\bar{\delta}(\partial_\mu \phi_i). \tag{33c}$$

Eq. (29) becomes

$$\delta S = \int (\bar{\delta}\mathcal{L}(x) + \frac{d\mathcal{L}}{dx_\mu}\delta x_\mu) d^4x + \int \mathcal{L}(x) d^4x' - \int \mathcal{L}(x) d^4x. \tag{34}$$

Noting that

$$d^4x' = dx'_1 dx'_2 dx'_3 dt' = \frac{\partial(x'_1, x'_2, x'_3, t')}{\partial(x_1, x_2, x_3, t)} d^4x \cong (1 + \frac{\partial \delta x_\mu}{\partial x_\mu}) d^4x, \tag{35}$$

we obtain

$$\delta S = \int \{\bar{\delta}\mathcal{L}(x) + \frac{d}{dx_\mu}(\mathcal{L}(x)\delta x_\mu)\} d^4x. \tag{36}$$

Using Eq. (13), we find from Eq. (33c)

$$\bar{\delta}\mathcal{L}(x) = \partial_\mu \{\frac{\partial \mathcal{L}}{\partial(\partial_\mu \phi_i)}\bar{\delta}\phi_i\}, \tag{37}$$

so that Eq. (36) becomes

$$\begin{aligned}\delta S &= \int \frac{d}{dx_\mu}\{\frac{\partial\mathcal{L}}{\partial(\partial_\mu\phi_i)}\bar{\delta}\phi_i + \mathcal{L}(x)\delta x_\mu\}d^4x \\ &\equiv -\int(\partial_\mu\Theta_{\alpha\mu}(x))d^4x\delta w_\alpha,\end{aligned} \tag{38}$$

with

$$\Theta_{\alpha\mu} = -\frac{\partial\mathcal{L}}{\partial(\partial_\mu\phi_i)}(\Psi_{i\alpha} - (\partial_\nu\phi_i)X_{\nu\alpha}) - \mathcal{L}(x)X_{\mu\alpha}. \tag{39}$$

Since δw_α is arbitrary, we obtain from Eq. (29)

$$\partial_\mu\Theta_{\alpha\mu}(x) = 0. \tag{30$'$}$$

This completes the proof of Noether's theorem.

(1) The Energy-Momentum Tensor $T_{\mu\nu}$. Consider Noether's theorem in the case of translation:

$$x_\mu \to x'_\mu = x_\mu + \delta a_\mu = x_\mu + \delta_{\mu\nu}\delta w_\nu, \qquad \phi_i(x) \to \phi'_i(x') = \phi_i(x). \tag{40}$$

We obtain the energy-momentum tensor,[5,6,7,8]

$$\begin{aligned} T_{\mu\nu} &= \partial_\mu \varphi_\alpha \frac{\partial \mathcal{L}}{\partial(\partial_\nu \varphi_\alpha)} - \delta_{\mu\nu} \mathcal{L} \\ &\neq T_{\nu\mu}. \end{aligned} \tag{41}$$

$$T_{jk}, \quad j, k = 1, 2, 3, \quad \text{is the stress tensor,} \tag{42}$$

$$\begin{aligned} &\frac{1}{i} T_{j4} = \partial_j \varphi_\alpha \frac{\partial \mathcal{L}}{\partial \dot{\varphi}_\alpha} = \pi_\alpha \partial_j \varphi_\alpha = \mathcal{P}_j, \\ &\quad \mathcal{P}_j, j = 1, 2, 3, \text{ form the momentum vector,} \end{aligned} \tag{43}$$

$$\begin{aligned} &i T_{4j} = \dot{\varphi}_\alpha \frac{\partial \mathcal{L}}{\partial(\partial_j \varphi_\alpha)} = \mathcal{S}_j, \\ &\mathcal{S}_j, \quad j = 1, 2, 3, \text{ from energy flux density vector.} \end{aligned} \tag{44}$$

$$T_{44} = \dot{\varphi}_\alpha \frac{\partial \mathcal{L}}{\partial \dot{\varphi}_\alpha} - \mathcal{L} = \mathcal{H}, \qquad \text{the energy density.} \tag{45}$$

5

$$\frac{1}{ic} T_{j4} = \partial_j \varphi_\alpha \frac{\partial \mathcal{L}}{\partial \dot{\varphi}_\alpha} = \pi_\alpha \partial_j \varphi_\alpha = \mathcal{P}_j. \tag{43$'$}$$

6

$$ic T_{4j} = \dot{\varphi}_\alpha \frac{\partial \mathcal{L}}{\partial(\partial_j \varphi_\alpha)} = \mathcal{S}_j. \tag{44$'$}$$

7

$$c T_{44} = \dot{\varphi}_\alpha \frac{\partial \mathcal{L}}{\partial \dot{\varphi}_\alpha} - \mathcal{L} = \mathcal{H}. \tag{45$'$}$$

8 *$T_{\mu\nu}$ defined in Eq. (41) is not symmetric in general. In Eqs. (26) and (27), it is known that adding a 4-divergence $\partial_\mu g_\mu$ to $\mathcal{L}$ leaves the field equations unchanged. But if one adds to $\mathcal{L}$ a $\partial_\mu g_\mu$ satisfying the relations*

$$\partial_\nu \varphi_\alpha \frac{\partial}{\partial(\partial_\mu \varphi_\alpha)} - \partial_\mu \varphi_\alpha \frac{\partial}{\partial(\partial_\nu \varphi_\alpha)} \}(\mathcal{L} + \partial_\beta g_\beta) = 0, \tag{46}$$

and

$$\partial_\beta \left(\frac{\partial g_\beta}{\partial \varphi_\alpha} - \partial_\mu \frac{\partial g_\beta}{\partial(\partial_\mu \varphi_\alpha)} \right) = 0, \tag{47}$$

then one can make

$$T_{\mu\nu} = T_{\nu\mu}. \tag{48}$$

If $\mathcal{L}$ does not explicitly depend on x_μ, one has from Eq. (41)

$$\partial_\nu T_{\mu\nu} = -(\frac{\partial \mathcal{L}}{\partial \varphi_\alpha} - \partial_\nu \frac{\partial \mathcal{L}}{\partial(\partial_\nu \varphi_\alpha)})\partial_\mu \varphi_\alpha. \tag{49}$$

Hence we have, as a special case of Noether's theorem,

Theorem: If $\mathcal{L}$ satisfies the Lagrange equation Eq. (13), then

$$\partial_\nu T_{\mu\nu} = 0. \tag{50}$$

Theorem: If Eq. (50) holds, then $\mathcal{L}$ satisfies the Lagrange equation Eq. (13).

The physical meaning of Eq. (50) is seen as follows. On integrating over a volume V_3, one has

$$\begin{aligned} 0 &= \int_{V_3} \partial_\nu T_{\mu\nu} d^3x \\ &= \int_{V_3} (\partial_j T_{\mu j} + \partial_4 T_{\mu 4}) d^3x \\ &= \int_S (T_{\mu j} dS_j) + \frac{d}{dx_4} \int_{V_3} T_{\mu 4} d^3x, \quad \mu = 1, 2, 3, 4. \end{aligned} \tag{50a}$$

The first term is a surface integral and vanishes if $T_{\mu j} = 0$ on the surface S of V_3. Then the above equation gives, from Eqs. (43) and (45),

$$\frac{d}{dt} \int_{V_3} \mathcal{P}_j d^3x = 0, \quad j = 1, 2, 3, \tag{51}$$

$$\frac{d}{dt} \int_{V_3} \mathcal{H} d^3x = 0, \tag{52}$$

which are the momentum and energy conservation relations.

Again from Eq. (50), one has[9]

$$\frac{\partial}{\partial x_j} T_{4j} + \frac{1}{i} \frac{\partial}{\partial t} T_{44} = 0, \tag{53}$$

$$\frac{\partial}{\partial x_j} T_{kj} + \frac{1}{i} \frac{\partial}{\partial t} T_{k4} = 0, \tag{54}$$

9

$$\frac{\partial}{\partial x_j} T_{4j} + \frac{1}{ic} \frac{\partial}{\partial t} T_{44} = 0. \tag{53$'$}$$

$$\frac{\partial}{\partial x_j} T_{kj} + \frac{1}{ic} \frac{\partial}{\partial t} T_{k4} = 0. \tag{54$'$}$$

which are, on using Eqs. (44), (43), and (45),

$$div\mathcal{S} + \frac{\partial}{\partial t}\mathcal{H} = 0, \tag{53a}$$

$$\frac{\partial}{\partial x_j}T_{kj} + \frac{\partial}{\partial t}\mathcal{P}_k = 0. \tag{54a}$$

Eq. (53a) is the continuity equation (energy conservation). If the field is an electromagnetic field, then $\mathcal{S}$ is the Poynting vector.

The $T_{\mu\nu}$ tensor is[10]

$$T_{\mu\nu} = \begin{pmatrix} T_{11} & T_{12} & T_{13} & i\mathcal{P}_1 \\ T_{21} & T_{22} & T_{23} & i\mathcal{P}_2 \\ .T_{31} & T_{32} & T_{33} & i\mathcal{P}_3 \\ \frac{1}{i}\mathcal{S}_1 & \frac{1}{i}\mathcal{S}_2 & \frac{1}{i}\mathcal{S}_3 & \mathcal{H} \end{pmatrix}. \tag{55}$$

If $\mathcal{L}$ is an explicit function of x_μ , then Eq. (50) does not hold, and energy and momentum are not conserved, the field interacting with external sources.

(2) The Angular-Momentum Tensor.[11]

The angular momentum density $\mathcal{M}_{jk}$ is defined by[12]

$$\begin{aligned} \mathcal{M}_{jk} &= x_j\mathcal{P}_k - x_k\mathcal{P}_j \\ &= \frac{1}{i}(x_jT_{k4} - x_kT_{j4}), \end{aligned} \tag{56}$$

---

[10]

$$T_{\mu\nu} = \begin{pmatrix} T_{11} & T_{12} & T_{13} & ic\mathcal{P}_1 \\ T_{21} & T_{22} & T_{23} & ic\mathcal{P}_2 \\ .T_{31} & T_{32} & T_{33} & ic\mathcal{P}_3 \\ \frac{1}{ic}\mathcal{S}_1 & \frac{1}{ic}\mathcal{S}_2 & \frac{1}{ic}\mathcal{S}_3 & \mathcal{H} \end{pmatrix}. \tag{55'}$$

[11] *Rotations in space and Lorentz transformations (boosts) yield*

$$x_\mu \to x'_\mu = x_\mu + x_\nu\delta w_{\mu\nu} = x_\mu + X_{\mu\alpha\beta}\delta w_{\alpha\beta}, \qquad (\alpha < \beta) \tag{63}$$

with $\delta w_{\nu\mu} = -\delta w_{\mu\nu}$ *and* $X_{\mu\alpha\beta} = x_\beta\delta_{\mu\alpha} - x_\alpha\delta_{\mu\beta}$. *The corresponding change in the field function is, with* $\alpha < \beta$,

$$\phi_i(x) \to \phi'_i(x') = \phi_i(x) + \delta\phi_i(x) = \phi_i(x) + A_{ij\alpha\beta}\phi_j(x)\delta w_{\alpha\beta}. \tag{64}$$

Using Noether's theorem, we obtain the angular momentum tensor,

$$M_{\alpha\beta\mu} = (x_\beta T_{\alpha\mu} - x_\alpha T_{\beta\mu}) - \frac{\partial\mathcal{L}}{\partial(\partial_\mu\phi_i)}A_{ij\alpha\beta}\phi_j(x). \tag{65}$$

For a scalar field, $A_{ij\alpha\beta} = 0$ *so that Eq. (58) and Eq. (65) are related. For a vector field, we have*

$$A_{ij\alpha\beta} = \delta_{i\alpha}\delta_{j\beta} - \delta_{i\beta}\delta_{j\alpha}, \tag{66}$$

so that Eq. (58) should be modified. So is the case for the Dirac field.

[12] *Note that, in ordinary units, i stands for ic in Eqs. (56), (58), and (60).*

and the total angular momentum of the field in volume V_3 is

$$M_{jk} = \int_{V_3} \mathcal{M}_{jk} d^3x. \tag{57}$$

Let us define a $4 \times 4 \times 4$ tensor $m_{\mu\nu\rho}$ by

$$m_{\mu\nu\rho} = \frac{1}{i}(x_\mu T_{\nu\rho} - x_\nu T_{\mu\rho}), \tag{58}$$

so that

$$m_{jk4} = \mathcal{M}_{jk}. \tag{59}$$

Then

$$\begin{aligned} \partial_\rho m_{\mu\nu\rho} &= \frac{1}{i}(T_{\nu\mu} + x_\mu \partial_\rho T_{\nu\rho} - T_{\mu\nu} - x_\nu \partial_\rho T_{\mu\rho}) \\ &= \frac{1}{i}(T_{\nu\mu} - T_{\mu\nu}) \quad \text{if Eq. (50) holds,} \qquad (60) \\ &= 0 \qquad\qquad \text{if Eq. (48) holds.} \qquad (61) \end{aligned}$$

Integrating this over V_3 , one has

$$\begin{aligned} 0 &= \int_{V_3} \partial_j m_{\mu\nu j} d^3x + \frac{1}{i}\int_{V_3} \frac{\partial}{\partial t} m_{\mu\nu 4} d^3x \\ &= \int_S m_{\mu\nu j} dS + \frac{1}{i}\frac{d}{dt}\int_{V_3} \mathcal{M}_{\mu\nu} d^3x, \end{aligned}$$

where we have written, as extension of Eq. (59),

$$\mathcal{M}_{\mu\nu} \equiv m_{\mu\nu 4}.$$

The surface integral vanishes if $m_{\mu\nu j} = 0$ on S. Thus

$$\frac{d}{dt} M_{\mu\nu} = \frac{d}{dt}\int_{V_3} \mathcal{M}_{\mu\nu} d^3x = 0, \tag{62}$$

which includes Eq. (57). Thus if $T_{\mu\nu} = T_{\nu\mu}$, one has the conservation of the generalized angular momentum $M_{\mu\nu}$.

6.4 The Klein-Gordon Field in Lagrangian Form

(1) The simplest free field is a scalar, real field satisfying the Klein-Gordon equation

$$(\partial_\alpha \partial_\alpha - \mu^2)\phi(x_\nu) = 0,$$

$$\mu = \frac{m_0 c}{\hbar} = \frac{2\pi}{\lambda_c}. \tag{67}$$

A Lagrangian density leading to this equation is[13,14]

$$\mathcal{L}(\phi, \partial_\nu \phi) = -\frac{1}{2}(\partial_\nu \phi \partial_\nu \phi + \mu^2 \phi^2). \tag{68}$$

$\mathcal{L}$ is Lorentz invariant for a scalar or a pseudo-scalar ϕ, since it is quadratic in ϕ.

The Hamiltonian density is, from Eq. (17) is[15]

$$\begin{aligned}\mathcal{H}(\phi, \pi, \partial_k \phi) &= \pi^2 + \frac{1}{2}(\partial_\nu \phi \partial_\nu \phi + \mu^2 \phi^2) \\ &= \frac{1}{2}\{\pi^2 + \mu^2 \phi^2 + (\nabla \phi)^2\}.\end{aligned} \tag{70}$$

The canonical energy-momentum tensor is, from Eq. (28),[16]

$$T_{\mu\nu} = -\partial_\mu \phi \partial_\nu \phi - \delta_{\mu\nu} \mathcal{L} = T_{\nu\mu}. \tag{71}$$

In analogy with Eqs. (61) and (62), one obtains the angular momentum conservation equation

$$\frac{dM_{\mu\nu}}{dt} = \frac{1}{i}\frac{d}{dt}\int_{V_3}(x_\mu T_{\nu 4} - x_\nu T_{\mu 4}) d^3x = 0. \tag{72}$$

From Eqs. (69)-(71), it is seen that

$$T_{44} = \mathcal{H}. \tag{73}$$

From Eqs. (43), (44), (53a), and (54a), one obtains

$$\mathcal{S}_j = iT_{4j} = -\dot{\phi}\partial_j \phi = -\pi \partial_j \phi,$$

---

[13] *At any given instant,*

$$\frac{\partial \mathcal{L}}{\partial \phi} = -\mu^2 \phi,$$

$$\frac{\partial \mathcal{L}}{\partial(\partial_k \phi)} = -\partial_k \phi, \quad k = 1, 2, 3, \quad \frac{\partial \mathcal{L}}{\partial \dot{\phi}} = \pi \quad \textit{from Eq. (16).} \tag{69}$$

From Eq. (13), one obtains Eq. (67).

[14]

$$\mathcal{L}(\phi, \partial_\nu \phi) = -\frac{c^2}{2}(\partial_\nu \phi \partial_\nu \phi + \mu^2 \phi^2). \tag{68$'$}$$

[15]

$$\begin{aligned}\mathcal{H}(\phi, \pi, \partial_k \phi) &= \pi^2 + \frac{c^2}{2}(\partial_\nu \phi \partial_\nu \phi + \mu^2 \phi^2) \\ &= \frac{1}{2}\{\pi^2 + c^2\mu^2 \phi^2 + c^2(\nabla \phi)^2\}.\end{aligned} \tag{70$'$}$$

[16]

$$T_{\mu\nu} = -c^2 \partial_\mu \phi \partial_\nu \phi - \delta_{\mu\nu} \mathcal{L} = T_{\nu\mu}. \tag{71$'$}$$

$$\mathcal{P}_j = \frac{1}{i}T_{j4} = \dot{\phi}\partial_j\phi = \pi\partial_j\phi,$$
$$= \frac{1}{2}(\pi\partial_j\phi + (\partial_j\phi)\pi), (\mathcal{P} \quad \text{hermitian}), \tag{74}$$

$$\partial_t\mathcal{H} + \nabla\cdot\mathcal{S} = 0,$$

$$\partial_t\mathcal{P}_k + \partial_j T_{kj} = 0. \tag{75}$$

(2) Another Klein-Gordon field is a complex scalar field $\Psi(x_\nu)$ which may be regarded as made up of two independent real scalar fields $\psi_1(x_\nu), \psi_2(x_\nu)$:

$$\Psi(x_\nu) = \frac{1}{\sqrt{2}}(\psi_1(x_\nu) + i\psi_2(x_\nu)),$$

$$\Psi^*(x_\nu) = \frac{1}{\sqrt{2}}(\psi_1(x_\nu) - i\psi_2(x_\nu)), \tag{76}$$

so that Ψ and Ψ^* are also independent. $\psi_1(x), \psi_2(x)$ are hermitian, while Ψ, Ψ^* are not hermitian operators.

The Lagrangian density $\mathcal{L}(\psi_1, \psi_2, \partial_\nu\psi_1, \partial_\nu\psi_2)$ is[17]

$$\mathcal{L} = -(\partial_\nu\Psi\partial_\nu\Psi^* + \mu^2\Psi\Psi^*)$$
$$= -\frac{1}{2}(\partial_\nu\psi_j\partial_\nu\psi_j + \mu^2\psi_j\psi_j), \quad \text{(summed over } j = 1, 2). \tag{77}$$

The Lagrange equations[18]

$$\frac{\partial\mathcal{L}}{\partial\Psi} - \partial_\nu\frac{\partial\mathcal{L}}{\partial(\partial_\nu\Psi)} = 0, \qquad (\Psi = \Psi, \Psi^*), \tag{78}$$

17

$$\mathcal{L} = -c^2(\partial_\nu\Psi\partial_\nu\Psi^* + \mu^2\Psi\Psi^*)$$
$$= -\frac{c^2}{2}(\partial_\nu\psi_j\partial_\nu\psi_j + \mu^2\psi_j\psi_j). \tag{77$'$}$$

18 *The Lagrange equations*

$$\frac{\partial\mathcal{L}}{\partial\psi_j} - \partial_\nu\frac{\partial\mathcal{L}}{\partial(\partial_\nu\psi_j)} = 0, \qquad j = 1, 2; \tag{80a}$$

give

$$(\partial_\alpha\partial_\alpha - \mu^2)\begin{pmatrix}\psi_1\\ \psi_2\end{pmatrix} = 0,$$

$$\Pi_j = \frac{\partial\mathcal{L}}{\partial\dot{\psi}_j} = \dot{\psi}_j(x_\nu),$$

$$\Pi = \frac{1}{\sqrt{2}}(\Pi_1 - i\Pi_2), \qquad \Pi^* = \frac{1}{\sqrt{2}}(\Pi_1 + i\Pi_2), \tag{80b}$$

Π_1, Π_2 *are hermitian operators.*

give the Klein-Gordon equations

$$(\partial_\alpha \partial_\alpha - \mu^2) \begin{pmatrix} \Psi \\ \Psi^* \end{pmatrix} = 0, \qquad \mu = \frac{m_0 c}{\hbar}.$$

The conjugate momenta $\Pi(x_\nu), \Pi^*(x_\nu)$ are

$$\Pi = \frac{\partial \mathcal{L}}{\partial \dot{\Psi}} = \dot{\Psi}^*(x_\nu),$$
$$\Pi^* = \frac{\partial \mathcal{L}}{\partial \dot{\Psi}^*} = \dot{\Psi}(x_\nu). \tag{79}$$

Π, Π^* are non-hermitian operators.
The Hamiltonian density $\mathcal{H}$ and Hamiltonian H are

$$\begin{aligned} \mathcal{H} &= \Pi \dot{\Psi} + \Pi^* \dot{\Psi}^* - \mathcal{L} \\ &= \Pi \Pi^* + (\nabla \Psi \cdot \nabla \Psi^* + \mu^2 \Psi \Psi^*), \\ H &= \int_{V_3} \mathcal{H} d^3 x. \end{aligned} \tag{81}$$

The energy-momentum tensor $T_{\mu\nu}$ is[19]

$$\begin{aligned} T_{\mu\nu} &= -(\partial_\mu \Psi \partial_\nu \Psi^* + \partial_\mu \Psi^* \partial_\nu \Psi) - \delta_{\mu\nu} \mathcal{L} \\ &= T_{\nu\mu}. \end{aligned} \tag{82}$$

It is seen from the definition (77) that $\mathcal{L}$ is a real scalar (invariant under proper Lorentz transformations). $\mathcal{L}$ is invariant under the following gauge transformation (of the first kind):

$$\Psi' = e^{i\alpha} \Psi, \qquad \Psi'^* = e^{-i\alpha} \Psi^*, \tag{83}$$

where α is an arbitrary real constant,

$$\delta \mathcal{L} = 0. \tag{84}$$

For infinitesimal transformations

$$\Psi(x_\nu) \to \Psi'(x_\nu) = (1 + i\alpha) \Psi(x_\nu),$$

$$\Psi^*(x_\nu) \to \Psi'^*(x_\nu) = (i - i\alpha) \Psi^*(x_\nu),$$

i.e.,

$$\delta \Psi(x_\nu) = i\alpha \Psi, \qquad \delta \Psi^* = -i\alpha \Psi^*, \tag{85}$$

19

$$T_{\mu\nu} = -c^2 (\partial_\mu \Psi \partial_\nu \Psi^* + \partial_\mu \Psi^* \partial_\nu \Psi) - \delta_{\mu\nu} \mathcal{L}. \tag{82'}$$

$$\delta\mathcal{L} = (\frac{\partial\mathcal{L}}{\partial\Psi} - \partial_\nu\frac{\partial\mathcal{L}}{\partial(\partial_\nu\Psi)})\delta\Psi + (\frac{\partial\mathcal{L}}{\partial\Psi^*} - \partial_\nu\frac{\partial\mathcal{L}}{\partial(\partial_\nu\Psi^*)})\delta\Psi^*$$
$$+ \partial_\nu(\frac{\partial\mathcal{L}}{\partial(\partial_\nu\Psi)}\delta\Psi + \frac{\partial\mathcal{L}}{\partial(\partial_\nu\Psi^*)}\delta\Psi^*) = 0.$$

From the Lagrange equations (78) and (85), one has[20]

$$\partial_\nu(\frac{\partial\mathcal{L}}{\partial(\partial_\nu\Psi)}\Psi - \frac{\partial\mathcal{L}}{\partial(\partial_\nu\Psi^*)}\Psi^*) = 0,$$

or

$$\partial_\nu((\partial_\nu\Psi^*)\Psi - (\partial_\nu\Psi)\Psi^*) = 0. \tag{86}$$

e may define a 4-current density[21]

$$j_\nu = ie\{(\partial_\nu\Psi^*)\Psi - (\partial_\nu\Psi)\Psi^*\}, \tag{87}$$

so that (86) becomes an equation of continuity

$$\partial_\nu j_\nu = 0, \tag{86a}$$

or[22]

$$j_k = ie((\partial_k\Psi^*)\Psi - (\partial_k\Psi)\Psi^*),$$
$$j_4 = e(\Psi\dot{\Psi}^* - \Psi^*\dot{\Psi}) = i\rho$$
$$\nabla\cdot\mathbf{j} + \frac{\partial\rho}{\partial t} = 0. \tag{88}$$

[20] *Derivation of a conserved current $j_\mu(x)$ is just an application of Noether's theorem. Under the gauge transformation of the first kind, we have*

$$x_\mu \to x'_\mu = x_\mu \quad or \quad X_{\mu\alpha} = 0, \tag{90a}$$

and, from Eq. (85),

$$\Psi_{j\alpha} = i\psi \quad for \quad j = \psi$$
$$- i\psi^* \quad for \quad j = \psi^*. \tag{90b}$$

Thus, the Noether's current is

$$\theta_\mu = -\frac{\partial\mathcal{L}}{\partial(\partial_\mu\psi)}\cdot i\psi - \frac{\partial\mathcal{L}}{\partial(\partial_\mu\psi^*)}(-i\psi^*)$$
$$= i\{(\partial_\mu\psi^*)\psi - \psi^*(\partial_\mu\psi)\}, \tag{90}$$

which is proportional to the current given by Eq. (87).

[21]

$$j_\nu = i\frac{e}{\hbar}c^2\{(\partial_\nu\Psi^*)\Psi - (\partial_\nu\Psi)\Psi^*\}. \tag{87'}$$

[22]

$$j_k = i\frac{ec^2}{\hbar}((\partial_k\Psi^*)\Psi - (\partial_k\Psi)\Psi^*), \qquad j_4 = \frac{ec}{\hbar}(\Psi\dot{\Psi}^* - \Psi^*\dot{\Psi}) = ic\rho. \tag{88'}$$

The total charge of the field is

$$Q = \int_{V_3} \rho d^3x = -ie \int_{V_3} (\Psi \dot{\Psi}^* - \Psi^* \dot{\Psi}) d^3x. \tag{89}$$

This we shall see is the electric charge operator.

6.5 The Electromagnetic Field in Lagrangian Form

The fields φ_α in the preceding sections are the 4-potentials

$$\varphi_\alpha = A_1, A_2, A_3, A_4 = i\phi.$$

The electric and magnetic field are

$$\mathbf{E} = -\nabla\phi - \frac{\partial}{\partial t}\mathbf{A}, \qquad \mathbf{B} = \nabla \times \mathbf{A}. \tag{92}$$

Ampère's law and Coulomb's laws are, for free fields,

$$\nabla \times \mathbf{B} = \frac{\partial}{\partial t}\mathbf{E}, \qquad \nabla \cdot \mathbf{E} = 0. \tag{93}$$

The Lorentz relation (Lorentz gauge) is

$$\partial_\mu A_\mu = 0. \tag{94}$$

The skew-symmetric (antisymmetric) electromagnetic field tensor $F_{\mu\nu}$ is defined by

$$\begin{aligned} F_{\mu\nu} &\equiv \partial_\mu A_\nu - \partial_\nu A_\mu = -F_{\nu\mu} \\ &= \begin{pmatrix} 0 & B_3 & -B_2 & -iE_1 \\ & 0 & B_1 & -iE_2 \\ & & 0 & -iE_3 \\ & & & 0 \end{pmatrix}. \end{aligned} \tag{95}$$

The Lagrangian density $\mathcal{L}$ is chosen to be

$$\begin{aligned} \mathcal{L} &= -\frac{1}{4}[F_{\mu\nu}F_{\mu\nu} + 2(\partial_\mu A_\mu)^2] \\ &= \frac{1}{2}(\mathbf{E}\cdot\mathbf{E} - \mathbf{B}\cdot\mathbf{B}) - \frac{1}{2}(\nabla\cdot\mathbf{A} + \frac{\partial\phi}{\partial t})^2. \end{aligned} \tag{96}$$

Under the gauge transformation

$$A_\mu \to A_\mu + \partial_\mu \chi, \qquad \chi = \text{a scalar function}, \tag{97}$$

with

$$\partial_\mu \partial_\mu \chi = 0, \tag{98}$$

both the Lorentz relation (94) and the Lagrangian density $\mathcal{L}$ are invariant.

In any arbitrary Lorentz frame, it is always possible to choose a $\Lambda(x_\mu)$ satisfying (98) to make

$$A_4 = i\phi + \frac{\partial}{\partial x_4}\Lambda = 0, \tag{99}$$

for a source-free field, so that the Lorentz relation (94) reduces to

$$\partial_k A_k = 0 \qquad (i.e., \quad \nabla \cdot \mathbf{A} = 0). \tag{100}$$

The Euler equation (13) now gives the field equation for A_μ,

$$\partial_\mu \partial_\mu A_\nu = 0, \tag{101}$$

or

$$\partial_\mu^2 A_\nu = 0, \qquad \nu = 1, 2, 3, 4. \tag{101a}$$

The canonical conjugates π_μ of the A_μ are, from Eqs. (16), (96a), and (95),

$$\pi_\mu = \frac{\partial \mathcal{L}}{\partial \dot{A}_\mu} = i(F_{4\mu} + \partial_\nu A_\nu \delta_{\mu 4}), \tag{102}$$

$$\pi_k = -E_k, \qquad\qquad \pi_4 = i\partial_\nu A_\nu. \tag{102a}$$

If one introduces the Lorentz relation (94), then

$$\pi_4 = 0. \tag{103}$$

The Hamiltonian density $\mathcal{H}$ is

$$\begin{aligned} \mathcal{H} &= \pi_\mu \dot{A}_\mu - \mathcal{L} \\ &= \pi_\mu \dot{A}_\mu + \frac{1}{4}[F_{\mu\nu}F_{\mu\nu} + 2(\partial_\alpha A_\alpha)^2]. \end{aligned} \tag{104}$$

From Eqs. (92) and (102),

$$\begin{aligned} \dot{A}_k &= -E_k - \partial_k \phi = \pi_k + i\partial_k A_4, \\ \dot{A}_4 &= \pi_4 - i\partial_k A_k, \end{aligned} \tag{105}$$

$$\begin{aligned} \mathcal{H} &= \pi_\mu \pi_\mu + i(\pi_k \partial_k A_4 - \pi_4 \partial_k A_k) + \frac{1}{4}[F_{\mu\nu}F_{\mu\nu} + 2(\partial_\mu A_\mu)^2] \\ &= \frac{1}{2}(\mathbf{E}\cdot\mathbf{E} + \mathbf{B}\cdot\mathbf{B}) + i(\pi_k \partial_k A_4 - \pi_4 \partial_k A_k) + \frac{1}{2}\pi_4^2. \end{aligned} \tag{104a}$$

As

$$\begin{aligned} i\int_{V_3} \Pi_k \partial_k A_4 d^3x &= \int_{V_3} (\mathbf{E}\cdot\nabla\phi) d^3x \\ &= -\int \phi(\nabla\cdot\mathbf{E}) d^3x \\ &= 0, \qquad\qquad \text{(See Eq. (93))}, \end{aligned}$$

and on using Lorentz gauge (94), $\pi_4 = 0$ (Eq. (103)) and $\partial_k A_k = 0$ from Eq. (100), one has for the Hamiltonian H

$$H = \int_{V_3} \mathcal{H} d^3x$$
$$= \int_{V_3} \frac{1}{2}(\mathbf{E}\cdot\mathbf{E} + \mathbf{B}\cdot\mathbf{B}) d^3x. \tag{106}$$

The electromagnetic field tensor $T_{\mu\nu}$ is, from Eqs. (41), (95), (96a), and (94)

$$T_{\mu\nu} = (\partial_\mu A_\alpha)F_{\alpha\nu} - \delta_{\mu\nu}\mathcal{L}$$
$$= F_{\mu\alpha}F_{\alpha\nu} + (\partial_\alpha A_\mu)F_{\alpha\nu} - \delta_{\mu\nu}\mathcal{L}. \tag{107}$$

It can be shown that the second term on the right contributes nothing to the momentum or energy conservation relations in Eqs. (50)-(52).[23] Hence one has

$$T_{\mu\nu} = F_{\mu\alpha}F_{\alpha\nu} - \delta_{\mu\nu}\mathcal{L} \tag{108}$$

or

$$T_{\mu\nu} = \begin{pmatrix} E_1E_1 + B_1B_1 - \mathcal{H} & E_1E_2 + B_1B_2 & E_1E_3 + B_1B_3 & i\mathcal{P}_1 \\ E_2E_1 + B_2B_1 & E_2E_2 + B_2B_2 - \mathcal{H} & E_2E_3 + B_2B_3 & i\mathcal{P}_2 \\ E_3E_1 + B_3B_1 & E_3E_2 + B_3B_2 & E_3E_3 + B_3B_3 - \mathcal{H} & i\mathcal{P}_3 \\ \frac{1}{i}\mathcal{S}_1 & \frac{1}{i}\mathcal{S}_2 & \frac{1}{i}\mathcal{S}_3 & \mathcal{H} \end{pmatrix}, \tag{109}$$

where

$$\mathcal{H} = \frac{\infty}{\in}(\mathbf{E}\cdot\mathbf{E} + \mathbf{B}\cdot\mathbf{B}), \tag{110}$$

$$\mathcal{S} = [\mathbf{E}\times\mathbf{B}], \qquad \text{the Poynting vector (energy flux density),}$$

$$\mathcal{P} = -\mathcal{S}, \qquad \text{momentum density.}$$

---

[23] *Taking the 4-divergence and integrating over the volume V_3 of the field, and by repeated application of the relation*

$$\partial_\mu F_{\nu\mu} = \partial_\mu\partial_\nu A_\mu - \partial_\mu^2 A_\nu = \partial_\nu\partial_\mu A_\mu = 0,$$

one obtains

$$\int_{V_3} \partial_\nu(\partial_\alpha A_\mu F_{\alpha\nu})d^3x = \int_{V_3} \partial_\nu(\partial_\alpha A_\mu)F_{\alpha\nu}d^3x$$
$$= \int_{V_3} \partial_k(\partial_\alpha A_\mu)F_{\alpha k}d^3x + \int_{V_3} \partial_4(\partial_\alpha A_\mu)F_{\alpha 4}d^3x$$
$$= -\int_{V_3} \partial_\alpha A_\mu \partial_k F_{\alpha k}d^3x - \int_{V_3} \partial_\alpha A_\mu \partial_4 F_{\alpha 4}d^3x + \int_{V_3} \partial_4((\partial_\alpha A_\mu)F_{\alpha 4})d^3x$$
$$= -\int \partial_\alpha A_\mu \partial_\nu F_{\alpha\nu}d^3x + \partial_4 \int \partial_\alpha(A_\mu F_{\alpha 4})d^3x - \partial_4 \int A_\mu \partial_\alpha F_{\alpha 4}d^3x$$
$$= \partial_4 \int \partial_k(A_\mu F_{k4})d^3x = -\partial_4 \int_s A_\mu F_{4k}\cdot dS = 0.$$

6.6 The Dirac Field in Lagrangian Form

We shall regard the Dirac relativistic equations of the electron, not as the quantum mechanical (wave) equation of a particle, but as a "classical" field equation — the spinor field $\Psi(x)$ (see Chapter 2, 3). We shall quantize the Dirac free field, i.e., the field given by the equations (22a), Ch. 2.

To obtain the Dirac equations as the equations of a classical field, one starts with the Lagrangian density[24,25]

$$\mathcal{L} = -\frac{1}{2}\{\Psi^\dagger\gamma_4[\gamma_\mu\partial_\mu + m_0]\Psi + [(\partial_\mu\Psi)^\dagger\gamma_\mu + m_0\Psi^\dagger]\gamma_4\Psi\}. \tag{111}$$

The Euler equations (13)

$$\frac{\partial\mathcal{L}}{\partial\Psi^\dagger_\alpha} - \partial_\mu\frac{\partial\mathcal{L}}{\partial(\partial_\mu\Psi^\dagger_\alpha)} = 0,$$
$$\frac{\partial\mathcal{L}}{\partial\Psi_\alpha} - \partial_\mu\frac{\partial\mathcal{L}}{\partial(\partial_\mu\Psi_\alpha)} = 0, \qquad \alpha = 1,2,3,4,$$

give the equations[26]

$$(\gamma_\mu\partial_\mu + m_0)\Psi = 0, \tag{112}$$

$$(\partial_\mu\Psi)^\dagger\gamma_\mu + m_0\Psi^\dagger = 0. \tag{113}$$

Eq. (113) is, since $\partial_4^* = -\partial_4$,

$$\partial_j\Psi^\dagger\gamma_j - \partial_4\Psi^\dagger\gamma_4 + m_0\Psi^\dagger = 0. \tag{113a}$$

It is seen from Eqs. (112) and (113) that for Ψ, $\Psi^\dagger$ satisfying the Dirac equations.

The tensor component T_{44} [Eq. (45)] is[27]

$$T_{44} = \frac{\partial\mathcal{L}}{\partial\dot{\Psi}}\dot{\Psi} + \dot{\Psi}^\dagger\frac{\partial\mathcal{L}}{\partial\dot{\Psi}^\dagger} - \mathcal{L}, \tag{114}$$

[24] *Here* $\Psi^\dagger$ *is a row matrix* Ψ_1^*, Ψ_2^*, Ψ_3^*, Ψ_4^*, *i.e.*, $\Psi^\dagger = \tilde{\Psi}^*$.

[25]

$$\mathcal{L} = -\frac{\hbar c}{2}\{\Psi^\dagger\gamma_4[\gamma_\mu\partial_\mu + \frac{m_0c}{\hbar}]\Psi + [(\partial_\mu\Psi)^\dagger\gamma_\mu + \frac{m_0c}{\hbar}\Psi^\dagger]\gamma_4\Psi\}. \tag{111'}$$

[26]

$$(\gamma_\mu\partial_\mu + \frac{m_0c}{\hbar})\Psi = 0. \tag{112'}$$

[27]

$$cT_{44} = \frac{\partial\mathcal{L}}{\partial\dot{\Psi}}\dot{\Psi} + \dot{\Psi}^\dagger\frac{\partial\mathcal{L}}{\partial\dot{\Psi}^\dagger} - \mathcal{L}. \tag{114'}$$

and the Hamiltonian density $\mathcal{H}$ is (using $\mathcal{L} = 0$)[28]

$$\mathcal{H} = \frac{i}{2}(\Psi^\dagger \dot{\Psi} - \dot{\Psi}^\dagger \Psi). \tag{115}$$

The energy H is

$$H = \int_{V_3} \mathcal{H} d^3x. \tag{116}$$

The electric charge 4-current J_μ is, from Eq. (82), Ch. 3,[29,30]

$$\begin{aligned} J_\mu &= -ie\Psi^\dagger \gamma_4 \gamma_\mu \Psi, \\ J_4 &= i\rho, \qquad\qquad \rho = e\Psi^\dagger \Psi, \end{aligned} \tag{117}$$

and the total charge Q is

$$Q = \int_{V_3} \rho d^3x. \tag{118}$$

Under the transformation specified by Eq. (64) (which is to be compared with Eqs. (8c), (45), and (46) in Ch. 4), we find, with $\mu < \nu$,

$$\psi(x) \to \psi'(x') = (1 + \frac{1}{2}\delta w_{\mu\nu}\gamma_\mu \gamma_\nu)\psi(x), \tag{119}$$

which yields

$$A_{\psi\psi\mu\nu} = \frac{1}{2}\gamma_\mu\gamma_\nu, \qquad A_{\psi\bar{\psi}\mu\nu} = 0. \tag{120a}$$

Analogously, we have

$$A_{\bar{\psi}\bar{\psi}\mu\nu} = -\frac{1}{2}\gamma_\mu\gamma_\nu, \qquad A_{\bar{\psi}\psi\mu\nu} = 0. \tag{120b}$$

Thus, we obtain from Eq. (65) the angular momentum tensor

$$M_{\mu\alpha\beta} = (x_\beta T_{\mu\alpha} - x_\alpha T_{\mu\beta}) + \frac{1}{4}\bar{\psi}\{\gamma_\mu\gamma_\alpha\gamma_\beta + \gamma_\alpha\gamma_\beta\gamma_\mu\}\psi. \tag{121}$$

This completes our treatment of the Dirac field in Lagrangian form.

---

[28]

$$\mathcal{H} = \frac{i\hbar}{2}(\Psi^\dagger \dot{\Psi} - \dot{\Psi}^\dagger \Psi). \tag{115'}$$

[29] *J_μ can also be obtained from Noether's theorem.*

[30]

$$J_\mu = -ice\Psi^\dagger \gamma_4 \gamma_\mu \Psi, \qquad J_4 = ic\rho. \tag{117'}$$

Appendix

Electromagnetic Fields

For convenient reference, we collect in this appendix together some formulas relevant for the electromagnetic field. As in §.6.5., we adopt the m.k.s.a. unit system (instead of $\hbar = c = 1$).

Faraday's law

$$\nabla \times \mathbf{E} = -\frac{\partial \mathbf{B}}{\partial t} \tag{A1}$$

Amperè's law

$$\nabla \times \mathbf{H} = \mathbf{j} + \frac{\partial \mathbf{D}}{\partial t}, \quad j = \rho \mathbf{v}, \tag{A2}$$

Coulomb's law

$$\nabla \cdot \mathbf{B} = 0, \tag{A3}$$

Coulomb's law

$$\nabla \cdot \mathbf{D} = \rho, \tag{A4}$$

$$\mathbf{B} = \mu_0 \mathbf{H}\, (\mu_0 = 4\pi \times 10^{-17}\, \text{henry/m}),$$

$$\mathbf{D} = \epsilon_0 \mathbf{E}\, (\epsilon_0 = \frac{1}{36\pi} \times 10^{-9}\, \text{farad/m}), \tag{A5}$$

$$\frac{1}{\sqrt{\mu_0 \epsilon_0}} = c = 3 \times 10^8 \text{m/sec}.$$

Of the four equations (A1) - (A4), only two are independent. From the divergence of both sides of (A1), if $\nabla \cdot \mathbf{B} = 0$ at any one instant, $\nabla \cdot \mathbf{B} = 0$ at all times. This gives (A3). From the divergence of (A2), the equation of continuity

$$\nabla \cdot \mathbf{j} + \frac{\partial \rho}{\partial t} = 0, \tag{A6}$$

leads to (A4).

On introducing the 4-potential $A_\mu = (A, \frac{i}{c}\phi)$

$$\mathbf{B} = \nabla \times \mathbf{A}, \tag{A7}$$

$$\mathbf{E} = -\nabla\phi - \frac{\partial \mathbf{A}}{\partial t}, \tag{A8}$$

and the Lorentz relation

$$\nabla \cdot \mathbf{A} + \frac{1}{c^2}\frac{\partial \phi}{\partial t} = 0, \tag{A9}$$

one gets from (A1) - (A4)

$$\partial_\mu \partial_\mu \mathbf{A} = -\mu_0 \mathbf{j}, \tag{A10}$$

$$\partial_\mu \partial_\mu \phi = -\frac{1}{\epsilon_0}\rho. \tag{A11}$$

If one makes the gauge transformation

$$\mathbf{A}' = \mathbf{A} + \nabla\chi, \qquad \phi' = \phi - \frac{\partial \chi}{\partial t}, \qquad \chi = \text{a scalar function}, \tag{A12}$$

with the condition

$$\partial_\mu \partial_\mu \chi = 0, \tag{A13}$$

the **B, E** fields, the equation (A10), (A11) and the Lorentz relation (A9) are all unchanged.

In tensor notations, the above equations take the following forms. Define

$$x_\mu = (x_1, x_2, x_3, x_4 = ict), \tag{A14}$$

$$j_\mu = (\rho v_1, \rho v_2, \rho v_3, ic\rho), \tag{A15}$$

$$F_{\mu\nu} = \partial_\mu A_\nu - \partial_\nu A_\mu = -F_{\nu\mu}, \tag{A16}$$

then

$$(A1) + (A3): \qquad \partial_k F_{\ell m} + \partial_\ell F_{mk} + \partial_m F_{k\ell} = 0, \tag{A17}$$

$$(A2) + (A4): \qquad \partial_\nu F_{\rho\nu} = \mu_0 j_\rho, \tag{A18}$$

$$(A9): \qquad \partial_\nu A_\nu = 0, \tag{A19}$$

$$(A10) + (A11): \qquad \partial_\mu \partial_\mu A_\nu = -\mu_0 j_\nu, \tag{A20}$$

$$(A12): \qquad A'_\nu = A_\nu + \partial_\nu \chi. \tag{A21}$$

If a 4-vector f_μ is defined by

$$f_\mu = F_{\mu\nu} j_\nu, \tag{A22}$$

then

$$f_k = \rho(E_k + [\mathbf{v} \times \mathbf{B}]_k), \qquad k = 1, 2, 3,$$

$$f_4 = \frac{i}{c}\mathbf{E} \cdot \mathbf{j} = \frac{i}{c}\rho\mathbf{E} \cdot \mathbf{v}, \tag{A23}$$

and

$$(A21) + (A18): \qquad f_\rho = \frac{1}{\mu_0} F_{\rho\nu} \partial_\lambda F_{\nu\lambda}. \tag{A24}$$

In empty space ($\rho = 0$, $\mathbf{j} = 0$), it is possible to choose a function Λ to make A_4 in (A21) zero, i.e.,

$$\phi = -icA_4 = 0. \tag{A25}$$

With this choice, the Lorentz relation (A19) becomes

$$\nabla \cdot \mathbf{A} = 0. \tag{A26}$$

This relation is not Lorentz invariant, but in any Lorentz frame, it is always possible to choose a gauge (Λ) such that $\phi = 0$.[31]

For $\rho = 0$, one may choose $\nabla \cdot \mathbf{A} = 0$, so that of the three components A_1, A_2, A_3, only two are linearly independent.

On expressing $A_j(x), x = x_\mu$, in Fourier integral representation, one has

$$A_j(x) = \frac{1}{(2\pi)^{3/2}} \int a_j(k) e^{ik_\nu x_\nu} d^3k, \tag{A27}$$

$$\begin{aligned} k_\nu &= (k_j, k_4) = (\mathbf{k}, k_4), \\ \hbar k_4 &= \frac{iE}{c}, \qquad\qquad E = \hbar\omega, \\ k_\nu^2 &= k_\nu k_\nu = \mathbf{k}^2 - \frac{w^2}{c^2} = 0. \end{aligned} \tag{A28}$$

This last relation expresses the fact that the quantum (photon) of electromagnetic field has zero rest mass.[32]

From (A26), (A27), one has

$$\nabla \cdot \mathbf{A} = \partial_j A_j = \frac{1}{(2\pi)^{3/2}} \int k_j a_j(k) e^{ik_\nu x_\nu} d^3k = 0, \tag{A29}$$

[31] *Only when $\rho = 0$ may we make $\phi = 0$; for $\rho \neq 0$, making $\phi = 0$ contradicts (A11).*

[32] *That the photon has zero rest mass is also reflected in the field equations (A10, 11) or (A17, 18) containing only $\partial_\nu A_\nu$, but not the A_ν themselves. The Klein-Gordon and the Dirac equations both contain the mass m_0 of the particle, when the equations are regarded as field equations of the Ψ.*

so that

$$k_j a_j = (k \cdot a) = 0, \tag{A29a}$$

i.e., the vector potential A is perpendicular to the direction of propagation k of the electromagnetic wave.

From (A8), it is seen that for monochromatic electromagnetic wave, **E** is parallel to **A**. Hence in vacuum ($\rho = 0$), electromagnetic waves are *transverse* waves. But this transverse property is not Lorentz invariant, coming as it does from $\nabla \cdot \mathbf{A} = 0$ which is not Lorentz invariant.

The solutions of equations (A11), (A10) are, on account of their time-reversal invariance, the "retarded" and the "advanced" potentials[33]

$$\left.\begin{array}{l}\phi_{ret}(\mathbf{r}, t)\\ \phi_{adv}(\mathbf{r}, t)\end{array}\right\} = \frac{1}{4\pi\epsilon_0}\int d\mathbf{r}'\frac{\rho(\mathbf{r}', t')}{|\,\mathbf{r}-\mathbf{r}'\,|}, \qquad t' = \left\{\begin{array}{l} t - \frac{|\mathbf{r}-\mathbf{r}'|}{c}\\ t + \frac{|\mathbf{r}-\mathbf{r}'|}{c}\end{array}\right., \tag{A30}$$

$$\left.\begin{array}{l}\mathbf{A}_{ret}(\mathbf{r}, t)\\ \mathbf{A}_{adv}(\mathbf{r}, t)\end{array}\right\} = \frac{\mu_0}{4\pi}\int d\mathbf{r}'\frac{\mathbf{j}(\mathbf{r}', t')}{|\,\mathbf{r}-\mathbf{r}'\,|}, \qquad t' = \left\{\begin{array}{l} t - \frac{|\mathbf{r}-\mathbf{r}'|}{c}\\ t + \frac{|\mathbf{r}-\mathbf{r}'|}{c}\end{array}\right.. \tag{A31}$$

In the Appendix of Chapter **8**, we shall see that these integrals may be expressed in terms of Green's functions which are Lorentz invariant and which appear in the quantization conditions of the A_ν fields.

---

[33] *The general solutions of (A11), (A10) are obtained by adding to (A30), (A31) arbitrary solutions of the respective homogeneous equations*

$$\partial_\mu \partial_\mu \phi(r, t) = 0, \qquad \partial_\mu \partial_\mu A(r, t) = 0.$$

Exercises: *Chapter 6*

1. **(i)** Assume $\mathcal{L}(x) = \mathcal{L}(u^i(x), \partial_\mu u^i(x), \partial_\mu \partial_\nu u^i(x))$. Derive the generalized Euler-Lagrar equation, i.e. the analogue of Eq. (3).

(ii) Consider a vector field $B_\mu(x)$,

$$\mathcal{L}(x) = -\frac{1}{4}B_{\mu\nu}B_{\mu\nu} - \frac{1}{2}m^2 B_\mu B_\mu + \varepsilon(\partial_\lambda B_{\mu\nu})B_\lambda B_{\mu\nu}$$

with $B_{\mu\nu} \equiv \partial_\mu B_\nu - \partial_\nu B_\mu$ and ε some constant. Use (a) to derive field equations.

(iii) Add a total derivative to $\mathcal{L}$(x):
$\delta\mathcal{L}(x) = -\frac{\varepsilon}{2}\partial_\lambda\{B_{\mu\nu}B_\lambda B_{\mu\nu}\}$.
Then $\mathcal{L}'(x) = \mathcal{L}(x) + \delta\mathcal{L}(x)$ no longer depends on $\partial_\mu\partial_\nu u^i(x)$. Use Eq. (3) to derive the field equations. Discuss whether these equations are equivalent to those in (b).

(iv) Now impose the constraint $\partial_\mu B_\mu = 0$ and discuss its role.

2. *(Examples on Noether Theorem)*

(i) As the first exercise, consider a rotation about the z-axis:

$$x_1' = x_1 - \varepsilon x_2,$$
$$x_2' = x_2 + \varepsilon x_1,$$
$$x_3' = x_3,$$

with ε an infinitesimal. Obtain the expression for the conserved quantity in the case of the Dirac electron field.

(ii) Consider the gauge transformation of the first kind:

$$\phi_i(x) \to e^{i\alpha}\phi_i(x).$$

Obtain the resultant conserved quantity, respectively, for a complex scalar field, the electromagnetic field, and the Dirac electron field.

Chapter 7. Many-Body Systems

Many-body systems can be divided into the system of Fermions and the system of Bosons, depending on whether the exchange of two identical particles is antisymmetric or symmetric. This phenomenon should be recognized as a basic manifestation of the quantum principle.

Lately, cosmic background (CB) neutrinos (of three flavors, and antineutrinos), because of their nonzero masses, would follow, in the 1-1 correspondence, the heavy visual ordinary-matter objects, such as the Earth, the Venus, the Sun, and the stars. Thus, in the early Universe, the CB ν's would form the neutrino halos, capable of accounting for the 25% dark matter of our Universe (against the 5% visual ordinary matter).

In the Standard Model[1], the CB neutrinos are the *only* natural long-lived dark matter, and nothing else. We should realize that the Standard-Model description of our Universe is only one that is logical and rationale, including the 25% dark-matter world.

The many-body systems, such as the system of fermions, in fact play a fundamental role in our early Universe, thus in Cosmology.

To begin with, by indistinguishable particles, we mean that under any permutation P of the particles, the operator of any physical quantity Q, such as the Hamiltonian or the electric moment, of the system and their expectation values $< Q >$ are invariant,

$$PQ = QP,$$

$$< Q > = < \Psi \mid P^{-1}QP \mid \Psi > = < Q >_P . \tag{1}$$

A question that arises is the effect of permutations of the indistinguishable particles on the state (or, the wave function of the state) of a system.

7.1 Permutations:

(1) Permutation group

Consider a system of n indistinguishable particles. There are $n!$ permutations P of these n particles. The simplest permutation operation is the transposition of a pair of particles j and k . As an example, the transposition of particle B, A in a system of four particles A, B, C, D may be represented by the scheme[2]

$$T_{AB} = \begin{pmatrix} A & B & C & D \\ B & A & C & D \end{pmatrix}. \tag{2}$$

A permutation consisting of two transpositions T_{AB} and T_{AC} in succession is

$$T_{AC}T_{AB} = \begin{pmatrix} A & B & C & D \\ B & C & A & D \end{pmatrix} = P_{ABC}, \tag{3}$$

[1] *W-Y. Pauchy Hwang, arXiv:1304.4705v2 [hep-ph] 25 August 25 2015; "The Standard Model".*

[2] *The permutations (transpositions) are read vertically downward.*

which may be denoted by P_{ABC} in which A, B, C is cyclically permuted. But the transpositions T_{AB} and T_{AC} taken in the reversed order lead to a different permutation

$$T_{AB}T_{AC} = \begin{pmatrix} A & B & C & D \\ C & A & B & D \end{pmatrix} = P_{ACB}. \tag{4}$$

It is seen that

$$P_{ABC}P_{ACB} = P_{ACB}P_{ABC} = 1. \tag{5}$$

i.e., P_{ABC} and P_{ACB} are the inverse of each other, and this can be generalized and expressed by saying that to every permutation P, there is an inverse P^{-1},

$$P^{-1}P = PP^{-1} = 1. \tag{5a}$$

It is readily seen that the $n!$ permutations P_j have the following properties

(i) Two successive permutations P_i, P_j give rise to another P_k,

$$P_iP_j = P_k. \tag{6}$$

(ii) The P_i has the associative property,

$$P_i(P_jP_k) = (P_iP_j)P_k. \tag{7}$$

(iii) Every P_i has an inverse satisfying Eq. (5a).

(iv) There is the identity permutation which leaves the system unchanged.

These properties define the P_i as a *group* - called the permutation group, or, the symmetric group.[3]

In the example of four particles above, it is seen that T_{AB}, $T_{AC}T_{AB}$ are elements of the group (of 4! elements) and from Eqs. (3) and (4), it is seen that

$$P_iP_j \neq P_jP_i. \tag{8}$$

The group is non-commutative, and is said to be non-abelian.

Any permutation P_j can be regarded as made up of a number ν_j of transpositions of pairs. We shall call the P_j's "even" and "odd" by an ϵ_j such that

$$\epsilon_j = \begin{cases} 1, & \text{for } \nu_j = \text{even integer;} \\ -1, & \text{for } \nu_j = \text{odd integer.} \end{cases} \tag{9}$$

It is clear that P_j and P_j^{-1} have the same evenness or oddness, and that the ϵ of the product P_k of two permutations P_iP_j is

$$\epsilon_k = \epsilon_i\epsilon_j. \tag{9a}$$

[3] *The name "symmetric group" must not be confused with symmetry group.*

(2) Permutation operator

Consider a state $| a\rangle$ of a system of n particles. Its representative in the coordinate representation is the wave function

$$\psi(x_1, x_2, ..., x_n).$$

Upon a permutation P of the particles, this becomes

$$P\psi(x_1, x_2, ..., x_n) = \psi(x_{\alpha_1}, x_{\alpha_2}, ..., x_{\alpha_n}).$$

Let us introduce an operator U_p

$$\psi(x_\alpha) = U_P \psi(x). \tag{10}$$

The U_{P_j} obviously have the group properties (i) - (iv) above, for example,

$$U_{P_i} U_{P_j} = U_{P_k}, \qquad U_{P^{-1}} = U_P^{-1}. \tag{11}$$

U_P is also unitary,[4]

$$U_P^\dagger = U_P^{-1}. \tag{12}$$

The operator U_P commutes with any observable (hermitian operator Q),[5]

$$QU_P - U_P Q = 0 \tag{12a}$$

Let U_T be an operator for the transposition Eq. (2),

$$U_T \psi(x..., x_j, .., x_k, ..) = \psi(..., x_k, .., x_j, ..). \tag{13}$$

It is obvious

$$U_T U_T = 1$$

so that the eigenvalues of U_T are ± 1, and corresponding to these eigenvalues ± 1, the function ψ is $\left\{ \begin{array}{c} symmetric \\ antisymmetric \end{array} \right\}$,

$$\begin{aligned} U_T \psi_s &= \psi_s \\ U_T \psi_a &= -\psi_a. \end{aligned} \tag{14}$$

[4] *This follows from the invariance of the integral of any two functions* $\psi_1(x)$, $\psi_2(x)$,

$$\begin{aligned} (\psi_1(x), \psi_2(x)) &= (\psi_1(x_\alpha), \psi_2(x_\alpha)) \\ &= (U_P \psi_1(x), U_P \psi_2(x)) \\ &= (U_P^\dagger U_P \psi_1(x), \psi_2(x)), \qquad etc. \end{aligned}$$

[5] *This follows from Eq. (1),*

$$\begin{aligned} (\psi(x), Q\psi(x)) &- (U_P \psi(x), QU_P \psi(x)) \\ &= (\psi(x), U_P^\dagger QU_P \psi(x)), \qquad etc. \end{aligned}$$

If T^{-1} and T are defined as Eq. (5a), then

$$U_T = U_{T^{-1}} = (U_T)^{-1}. \tag{15}$$

From Eqs. (9), (11), and (15), it is seen that

$$\begin{aligned} U_P\psi_s &= \psi_s, \\ U_P\psi_a &= \epsilon_P\psi_a, \end{aligned} \tag{16}$$

The question is, in Nature, what is the symmetry property of a system of indistinguishable particles.

The answer is the empirical law of nature:
For particles with half-odd-integral spin, called fermions, such as electrons, protons, neutrons, muons, neutrinos, the wave function is always the antisymmetric ψ_a kind in Eq. (14).
For particles with integral spins, such as photons, pions, α-particles, and particles built up from even number of half-odd-integral spins, called bosons, the wave function is always of the ψ_s kind.

7.2 Symmetric and Antisymmetric Wave Functions for Fermions and Bosons

Consider a system of n indistinguishable, noninteracting particles. Let K stand for the complete set of quantum numbers of a single particle.[6]

For bosons, the symmetric wave function $\Psi^{(s)}$ is

$$\Psi^{(s)}_{K_1,K_2,..,K_n} = \frac{1}{\sqrt{n!}}\sum_P U_P\psi_{K_1}(x_1)\psi_{K_2}(x_2)...\psi_{K_n}(x_n) \tag{17}$$

where the sum is over all $n!$ permutations of $x_1, x_2, ..., x_n$, and U_P is the permutation operator Eq. (10).

For fermions, the wave function is antisymmetric, and for non-interacting particles can be put in a determinant form

$$\begin{aligned} \Psi^{(a)}_{K_1,K_2,..,K_n} &= \frac{1}{\sqrt{n!}}\begin{vmatrix} \psi_{K_1}(x_1) & \psi_{K_1}(x_2) & ... & \psi_{K_1}(x_n) \\ .. & .. & & .. \\ .. & .. & & .. \\ .. & .. & & .. \\ \psi_{K_n}(x_1) & \psi_{K_n}(x_2) & ... & \psi_{K_n}(x_n) \end{vmatrix} \\ &= \frac{1}{\sqrt{n!}}\sum_P \epsilon_P U_P\psi_{K_1}(x_1)\psi_{K_2}(x_2)...\psi_{K_n}(x_n) \end{aligned} \tag{18}$$

where the summation is over the $n!$ permutations of $x_1, ..., x_n$, and $\epsilon_P = \pm 1$ according as the permutation is $\left\{ \begin{matrix} even \\ odd \end{matrix} \right\}$ number of transpositions of pairs of x_i and x_j.

[6] *This means the totality of quantum numbers for the common eigenstates of a complete set of commuting observables for a single particle, in the sense of Ch. 5, Sect. 4(1), Vol. I ("Quantum Mechanics"). For example, for an electron in a central field, K stands for (n, ℓ, m_ℓ, m_s), or (n, ℓ, j, m) .*

From Eq. (18), it follows that

$$\Psi^{(a)}_{K_1,K_2,\ldots K_n} = 0, \qquad \text{when two sets} \quad K_j = K_i. \tag{19}$$

This is the Pauli exclusion principle.

For convenience, we introduce the following hermitian projection operators[7]

$$S = \frac{1}{n!}\sum_P U_P, \tag{20}$$

$$A = \frac{1}{n!}\sum_P \epsilon_P U_P, \tag{21}$$

$$C = 1 - S - A, \tag{22}$$

where the sum is over all $n!$ permutations (of the symmetric group).
From Eqs. (23a) and (23b),

$$\begin{aligned} U_P(S\Psi) &= S\Psi, \\ U_P(A\Psi) &= \epsilon_P(A\Psi). \end{aligned} \tag{25}$$

Comparison with Eq. (16) shows that

$$\begin{aligned} &S\Psi \qquad \text{is a symmetric function} \\ &A\Psi \qquad \text{is an antisymmetric function} \end{aligned} \tag{26}$$

---

[7] *The hermitian character is seen as follows:*

$$S^\dagger = \frac{1}{n!}\sum_P U_P^\dagger = \frac{1}{n!}\sum U_{P^{-1}} = S, \tag{20a}$$

$$A^\dagger = \frac{1}{n!}\sum_P \epsilon_P U_P^\dagger = \frac{1}{n!}\sum_P \epsilon_{P^{-1}} U_{P^{-1}} = A, \tag{21a}$$

$$C^\dagger = 1 - S^\dagger - A^\dagger = 1 - S - A = C. \tag{22a}$$

That they are projection operators is seen as follows

$$U_{P_i}S = \frac{1}{n!}\sum U_{P_i}U_P = \frac{1}{n!}\sum U_{P_j} = S, \tag{20b}$$

$$\begin{aligned} U_{P_i}A &= \frac{1}{n!}\sum U_{P_i}\epsilon_P U_P = \frac{1}{n!}\epsilon_{P_i}\sum \epsilon_{P_i}\epsilon_P U_{P_i}U_P \\ &= \frac{1}{n!}\epsilon_{P_i}\sum \epsilon_{P_j}U_{P_j} = \epsilon_{P_i}A \quad \text{[Eqs. (9a), (11)]} \end{aligned} \tag{21b}$$

From Eqs. (20b) and (21b), one obtains

$$SU_{P_i} = S = U_{P_i}S, \tag{23a}$$

$$AU_{P_i} = \epsilon_{P_i}A = U_{P_i}A, \tag{23b}$$

$$CU_{P_i} = U_{P_i}C, \tag{23c}$$

and

$$\begin{aligned} S^2 &= \frac{1}{n!}\sum U_P S = S, \qquad \text{[Eq. (23a)]}, \\ A^2 &= \frac{1}{n!}\sum \epsilon_P^2 A = A. \end{aligned} \tag{24}$$

and reference to Eq. (14) shows that

$$\begin{aligned} U_T(S\Psi) &= (S\Psi), \\ U_T(A\Psi) &= -(A\Psi). \end{aligned} \tag{27}$$

From the orthogonality of $\Psi^{(s)}$ and $\Psi^{(a)}$ function, one has

$$(A\Psi, S\Psi) = 0,$$

$$A^\dagger S = AS = 0, \qquad S^\dagger A = SA = 0, \tag{28}$$

and from Eqs. (24) and (28), one obtains

$$C^2 = C, \tag{29}$$

so the C is also a projection operator.

From Eqs. (20), (21), and (22), one sees that an arbitrary ψ may be projected into three parts

$$\Psi = S\Psi + A\Psi + C\Psi. \tag{30}$$

Consider

$$S + A = \frac{1}{n!}\sum_P (1 + \epsilon_P) U_P.$$

Let U_{P_e} be the permutation operator corresponding to permutations made up of even number of transpositions so that $\epsilon_{P_e} = 1$. Then by Eqs. (23a) and (23b),

$$\begin{aligned} U_{P_e}(S + A) &= \tfrac{2}{n!}\textstyle\sum_{P_e} U_{P_e} = S + A, \\ U_{P_e}\{(S + A)\Psi\} &= \{(S + A)\Psi\}, \end{aligned} \tag{31}$$

i.e., $S + A$ picks out those states which do not change sign upon even number of transpositions T_{ij}. The rest, C, are states that do not have this symmetry.

As an example, for two particles, n = 2, and

$$U_P = U_T;$$

$$S = \frac{1}{2}(1 + U_T), \qquad A = \frac{1}{2}(1 - U_T);$$

$$C = 0, \tag{32}$$

so that an arbitrary state can always be written

$$\begin{aligned} \Psi &= S\Psi + A\Psi \\ &= \Psi^{(s)} + \Psi^{(a)}. \end{aligned} \tag{33}$$

For a system of n particles, the wave function in Eq. (10) is the representative in the coordinate representation of a ket $\mid K_1, ..., K_n\rangle$

$$\Psi_{K_1, K_2, ..., K_n}(x_1, ..., x_n) = \langle x_1, ..., x_n \mid K_1, K_2, ..., K_n\rangle. \tag{34}$$

Now we introduce another representation

$$\mid n_1, n_2, n_3, ...\rangle, \tag{35}$$

where n_j is the number of particles which have the set of quantum number K_j, and the total number of particles is

$$\sum_j n_j = n. \tag{36}$$

The representation (35) is called the "occupation number representation", and its representatives in the coordinate representation are

$$\Psi_{n_1,n_2,...}(x) = \langle x_1, x_2, ..., x_n \mid n_1, n_2, ...\rangle. \tag{37}$$

By means of the projection operators S and A and Eq. (26), we can obtain the wave function for bosons and fermions

$$\Psi^B_{n_1,n_2,...}(x_1, ..., x_n) = c_n\langle x_1, ..., x_n \mid S \mid n_1, n_2, ...\rangle, \tag{38}$$

$$\Psi^F_{n_1,n_2,...}(x_1, ..., x_n) = c_n\langle x_1, ..., x_n| \; A \; |n_1, n_2, ...\rangle, \tag{39}$$

where c_n is the normalization constant.

If all the $n_1, n_2, ... = 1$, these two wave functions are just $\Psi^{(s)}, \Psi^{(a)}$, respectively, of Eqs. (17) and (18),

$$\langle x \mid S \mid 1, 1, 1, ...\rangle = \frac{1}{\sqrt{n!}}\Psi^{(s)}(x) \tag{40}$$

$$\langle x \mid A \mid 1, 1, 1, ...\rangle = \frac{1}{\sqrt{n!}}\Psi^{(a)}(x) \tag{41}$$

In $\mid n_1, n_2, ...\rangle$ in Eq. (35), there may be infinite numbers of sets of quantum number K_j, so that there are infinitely many states $\mid n_1, n_2, ...\rangle$. [8] We assume that the $\mid n_1, n_2, ...\rangle$ are orthogonal and form a complete set,

$$\langle n_1, n_2, ... \mid n'_1, n'_2, ...\rangle = \delta_{n_1,n'_1}\delta_{n_2,n'_2}..., \tag{42}$$

$$\sum_{(n)} \mid n_1, n_2, ...\rangle\langle n_1, n_2, ... \mid = 1, \tag{43}$$

where the summation is over all $n_1, n_2, n_3, ...$ subject to Eq. (36).

We shall now calculate the c_n in Eqs. (38) and (39). We remember that in the case of $\Psi^B_{n_1,n_2,...}(x)$, the $n_1, n_2, n_3, ...$ do not have to be 1, and the normalization constant $c_n = n!$ in Eq. (40) is not for the general case. Let

$$\Psi_{K_1,K_2,...}(x_1, x_2, x_3, ...) \tag{44}$$

[8] *For example, for a particle in a central field,* $K = (n, \ell, m, m_s)$, *and* $n = 1, 2, 3, ...$ *infinite. Since the total number of particles n is fixed, there are many* $n'_j s$ *which are zero.*

be the general wave function in which the particles are in various (single-particle) states K_1, K_2, no longer restricted to $n_j = 1$ for all j, and let Ψ^B of Eq. (38) be

$$\begin{aligned} \Psi^B(x) &= c_n S\Psi_{K_1K_2...}(x) \\ &= c_n \frac{1}{n!}\sum_P U_P \Psi_{K_1K_2...}(x_1, x_2, ...). \end{aligned} \tag{45}$$

The permutation operator U_P acts on the $x_1, x_2, x_3, ...$, and the sum is over all $n!$ permutations. An equivalent view is to leave the $x_1, x_2, ...$ alone and permute the $K_1, K_2, ...$ by the inverse permutation U_{P-1}, i.e.,

$$U_P \Psi_{K_1,K_2,...}(x_1, x_2, ...) = \Psi_{U_{P-1}(K_1,K_2,...)}(x_1, x_2, ...) \tag{46}$$

As n_j particles have the set of quantum numbers K_j, K_j appears n_j times in $\Psi_{K_1K_2...(x_1,x_2,...)}$ so that $n_1!n_2!... = \Pi_j n_j!$ permutations have no effect on $\Psi_{K_1K_2...}(x_1, x_2, ...)$.

From Eq. (45),

$$\begin{aligned} 1 = (\Psi^B, \Psi^B) &= | c_n |^2 (S\Psi_{K_1K_2...}, S\Psi_{K_1K_2...}) \\ &= | c_n |^2 (\Psi_{K_1K_2...}, S\Psi_{K_1K_2...}) \quad \text{by Eq. (20a)} \\ &= | c_n |^2 \frac{1}{n!} \sum (\Psi_{K_1K_2...}, \Psi_{U_{P-1}(K_1K_2...)}) \quad \text{by Eq. (46)} \end{aligned}$$

one obtains, for bosons,

$$| c_n |^2 = \frac{n!}{\prod_j n_j!}, \tag{47}$$

where

$$\sum_{j=1} n_j = n, \qquad \text{by Eq. (36).}$$

For fermions, one has, in place of Eq. (45),

$$\Psi^F(x) = c_n A\Psi_{K_1K_2...}(x). \tag{48}$$

But now

$$n_j = 1, \quad \text{or} \quad 0,$$

and

$$| c_n |^2 = n!, \tag{49}$$

again with

$$\sum_{n=1} n_j = n, \qquad \text{by Eq. (36).}$$

7.3 Fock Representation: Creation and Annihilation Operators

As exposed in Chapter 6, Sect. 1, we approach the problem of the quantum theory of fields by regarding a classical field as a dynamical system of nondenumerably infinite number of degrees of freedom and applying the quantization condition to the field variables. In brief, a quantized field is represented by a system of nondenumerably infinite number of particles. The most convenient mathematical method is to represent a field by such a system of harmonic oscillators. The quantum mechanics of harmonic oscillators is well-known, and the Fock representation is particularly suitable for application to the problem of quantized fields. This Fock representation has been introduced in Appendix B, Ch. 3, Vol. I. We shall summarize the results there for one oscillator, and extend them to a system of oscillators.

(1) Harmonic oscillator

Without repeating the details, we shall simply collect the results of Appendix B, Ch. 3, Vol. I ("*Quantum Mechanics*"), below,

$$H = \hbar\omega \frac{1}{2}(\xi^2 - \frac{d^2}{d\xi^2}), \tag{B1}$$

$$b^\dagger = \frac{1}{\sqrt{2}}(\xi - \frac{d}{d\xi}), \qquad b = \frac{1}{\sqrt{2}}(\xi + \frac{d}{d\xi}). \tag{B2}$$

$$bb^\dagger - b^\dagger b = 1, \tag{B4}$$

$$b^\dagger b \mid \lambda\rangle = \lambda \mid \lambda\rangle, \tag{B7}$$

$$b^\dagger b(b \mid \lambda\rangle) = (\lambda - 1)(b \mid \lambda\rangle), \tag{B9}$$

$$b^\dagger b(b^n \mid \lambda\rangle) = (\lambda - n)(b^n \mid \lambda\rangle). \tag{B11}$$

Concept of the lowest state:

$$b \mid \lambda_{min}\rangle = 0, \tag{B12}$$

i.e.,

$$b^\dagger b \mid \lambda_{min}\rangle = 0. \tag{B13}$$

From Eq. (B7), one has

$$\lambda_{min} = 0.$$

We introduce the notation $\mid 0\rangle$ for the lowest state $\mid \lambda_{min}\rangle$

$$\mid 0\rangle \equiv \mid \lambda_{min}\rangle. \tag{50}$$

$$bb^\dagger \mid \lambda\rangle = (\lambda + 1) \mid \lambda\rangle. \tag{B17}$$

$$b^\dagger b(b^\dagger \mid \lambda\rangle) = (\lambda + 1)(b^\dagger \mid \lambda\rangle). \tag{B10}$$

$$H = (b^\dagger b + \frac{1}{2})\hbar\omega, \tag{B3}$$

$$H \mid \lambda\rangle = (\lambda + \frac{1}{2}) \mid \lambda\rangle\hbar\omega, \tag{B3, 7}$$

$$H(b^\dagger \mid \lambda\rangle) = (\lambda + \frac{3}{2})(b^\dagger \mid \lambda\rangle)\hbar\omega. \tag{[(B3)+(B10)]}$$

This last relations shows that $b^\dagger \mid \lambda\rangle$ is the eigenstate of H for the eigenvalue $\lambda + 1$, i.e., $b^\dagger \mid \lambda\rangle$ changes $\mid \lambda\rangle$ into $\mid \lambda + 1\rangle$. When normalized,

$$b^\dagger \mid \lambda\rangle = \sqrt{\lambda + 1} \mid \lambda + 1\rangle. \tag{B18}$$

Similarly

$$b \mid \lambda\rangle = \sqrt{\lambda} \mid \lambda - 1\rangle. \tag{B23}$$

The corresponding eigenvalues and eigenstates of H are:

$$\frac{H}{\hbar\omega} = \frac{1}{2}, \quad \frac{3}{2}, \quad \frac{5}{2}, \quad ...(n + \frac{1}{2})... \tag{B14}$$

$$\mid \lambda\rangle = \mid 0\rangle, \quad b^\dagger \mid 0\rangle, \quad \frac{1}{\sqrt{2}}(b^\dagger)^2 \mid 0\rangle, ... \frac{1}{\sqrt{n!}}(b^\dagger)^n \mid 0\rangle... \tag{B19}$$

The operators $b^\dagger, b^-$ are adjoint to each other, but are not self-adjoint (i.e., not hermitian). Therefore they are not "observables" as the $p.q$. Now the operator $b^\dagger$ does not commute with any operator $A(q,p)$, i.e., $b^\dagger$ by itself constitutes "a complete set of commutable operators" in the sense of Ch. 5, Sec. 4, Vol. I. Hence one may take the $b^\dagger$ representation, and the set of kets (B19) above constitutes a complete set, in terms of which an arbitrary ket $\mid P\rangle$ can be expressed

$$\mid P\rangle = \sum_{k=0} a_k (b^\dagger)^k \mid 0\rangle, \tag{51}$$

i.e., $\mid P\rangle$ is expressed in terms of the eigenkets of H.

In the $b^\dagger$ - representation, $b^\dagger$ is a multiplicative factor[9] but b has to be obtained from (B4) above.

By repeated application of (B-4), one obtains

$$b(b^\dagger)^n - (b^\dagger)^n b = n(b^\dagger)^{n-1}. \tag{52}$$

If one writes

$$f(b^\dagger) = \sum_{k=0} a_k (b^\dagger)^k, \tag{53}$$

then from Eq. (51),

$$b \mid P\rangle = bf(b^\dagger) \mid 0\rangle, \tag{54}$$

and from Eq. (52),

$$(bf - fb) \mid 0\rangle = \sum_{n=1} a_n n (b^\dagger)^{n-1} \mid 0\rangle. \tag{55}$$

From (B12) and (50), $b \mid 0\rangle = 0$, hence

$$\begin{aligned} bf(b^\dagger) \mid 0\rangle &= \sum_{n=1} a_n n (b^\dagger)^{n-1} \mid 0\rangle \\ &= \frac{\partial f(b^\dagger)}{\partial b^\dagger} \mid 0\rangle. \end{aligned} \tag{56a}$$

[9] *Just as in the Schrödinger, or coordinate, representation, q or x is a multiplicative number, while p is determined by the commutation relation.*

This gives the operator equation, for $\mid 0\rangle$,

$$bf(b^\dagger) = \frac{\partial f(b^\dagger)}{\partial b^\dagger} \tag{56b}$$

which is consistent with the relation (B4).

Assume $\mid 0\rangle$ to have been normalized,

$$\langle 0 \mid 0\rangle = 1.$$

From Eq. (56),

$$\begin{aligned} \langle 0 \mid b^n (b^\dagger)^n \mid 0\rangle &= \langle 0 \mid b^{n-1} b (b^\dagger)^n \mid 0\rangle \\ &= n\langle 0 \mid b^{n-1}(b^\dagger)^{n-1} \mid 0\rangle \\ &= n! \end{aligned} \tag{57}$$

By repeated application of (B18), (B23), one finds

$$\langle 0 \mid b^m (b^\dagger)^n \mid 0\rangle = 0 \qquad \text{for} \qquad m \neq n. \tag{58}$$

Let

$$\mid Q\rangle = \sum_{n=0} c_n (b^+)^n \mid 0\rangle. \tag{59a}$$

Then

$$\langle Q \mid = \langle 0 \mid \sum_{n=0} c_n^* (b)^n \tag{59b}$$

and from Eq. (51),

$$\begin{aligned} \langle Q \mid P\rangle &= \sum_{n,m} c_n^* a_m \langle 0 \mid (b)^n (b^+)^m \mid 0\rangle \\ &= \sum_n c_n^* a_n n! \qquad \text{by Eqs. (57)-(58)} \end{aligned} \tag{60}$$

and

$$1 = \langle P \mid P\rangle = \sum_n \mid a_n \mid^2 n! \tag{61}$$

This leads to the interpretation of $\mid a_n \mid^2 n!$ as the probability that $\mid P >$ is in the n th state of the harmonic oscillator, i.e., the probability of the eigenvalue $(n + \frac{1}{2})\hbar\omega$.

(2) A system of harmonic oscillators

The results of the preceding subsection can readily be extended to a system of independent oscillators so that the Hamiltonian H is

$$H = \sum_k H_K,$$

$$H_K = \frac{1}{2}(\xi_k^2 - \frac{d^2}{d\xi_k^2})\hbar\omega_k, \tag{62}$$

$$H_j H_k - H_k H_j = 0,$$

$$b_k^\dagger = \frac{1}{\sqrt{2}}(\xi_k - \frac{d}{d\xi_k}), \qquad b_k = \frac{1}{\sqrt{2}}(\xi_k + \frac{d}{d\xi_k}),$$

$$b_k b_j^\dagger - b_j^\dagger b_k = \delta_{kj}, \qquad \text{for all} \quad k, j, \tag{63}$$

$$b_k \mid 0 >= 0, \qquad \text{for all} \quad k,$$

and as operators on $(b_k^\dagger) \mid 0\rangle$,

$$b_k = \frac{\partial}{\partial b_k^\dagger}. \tag{64}$$

The $\mid 0\rangle$ here is the lowest state of all H_k , and

$$\langle 0 \mid 0\rangle = 1,$$

and similarly to (B19),

$$(b_1^\dagger)^{n_1}(b_2^\dagger)^{n_2}(b_3^\dagger)^{n_3}... \mid 0\rangle \tag{65}$$

represents the state in which the jth oscillator is in the ${n_j}^{th}$ state, so that the eigenvalue corresponding to Eq. (65) is

$$\sum_j (n_j + \frac{1}{2})\hbar\omega_j. \tag{66}$$

If

$$\mid P\rangle = \sum_{n_1, n_2, ..} a_{n_1, n_2, ..}(b_1^\dagger)^{n_1}(b_2^\dagger)^{n_2}... \mid 0\rangle,$$

then

$$\langle P \mid P\rangle = \sum_{n_1, n_2, ..} \mid a_{n_1 n_2 ..} \mid^2 n_1! n_2! n_3! ...$$

i.e.,

$$\mid a_{n_1 n_2 ...} \mid^2 n_1! n_2! ... \tag{67}$$

is the probability that $\mid P >$ is a state in which the jth oscillator is in the ${n_j}^{th}$ state, the kth oscillator in the ${n_k}^{th}$ state, etc.

When the number of oscillators N is infinite, the system has an infinite number of degrees of freedom. This system is to represent a "field".

(3) Creation and annihilation operators

We have introduced the occupation-number representation in which a state is represented by the number n_j of particles in the single-particle state $\mid k_j\rangle$, where k_j stands for the totality of quantum numbers specifying the one-particle state,

$$\mid n_1, n_2, n_3, ...\rangle. \tag{68}$$

The coordinate representative of this ket is,

$$\Psi_{n_1 n_2 ...}(x_1, x_2, ...x_n) = \langle x_1, x_2, ...x_n \mid n_1, n_2, n_3, ...\rangle. \tag{69}$$

(i) System of bosons

The state of a system is changed by a change in the occupation numbers n_j. The annihilation and creation operator, a and $a^\dagger$ respectively, are defined by

$$a_j \mid n_1, n_2, ..., n_j, ...\rangle = \sqrt{n_j} \mid n_1, n_2, ..., n_j - 1, ...\rangle, \tag{70}$$

$$a_j^\dagger \mid n_1, n_2, ..., n_j, ...\rangle = \sqrt{n_j + 1} \mid n_1, n_2, ..., n_j + 1, ...\rangle, \tag{71}$$

a_j decreasing the number of n_j in state $\mid k_j\rangle$ by 1, $a_j^\dagger$ increasing n_j by 1.

In the x-representative $\Psi_{n_1 n_2 ...}(x)$, we have

$$\begin{aligned}
&(a_j^\dagger \Psi_{n_1 n_2 ..., n_j - 1, ...}, \Psi_{n_1 n_2 ..., n_j ...}) \\
&\quad = \sqrt{n_j}(\Psi_{n_1 n_2 ... n_j ...}, \Psi_{n_1 n_2 ... n_j ...}) \\
&\quad = \sqrt{n_j},
\end{aligned}$$

$$\begin{aligned}
&(\Psi_{n_1 n_2 ... n_j - 1 ...}, a_j \Psi_{n_1 n_2 ... n_j ...}) \\
&\quad = \sqrt{n_j}(\Psi_{n_1 n_2 ... n_j - 1 ...}, \Psi_{n_1 n_2 ... n_j - 1 ...}) \\
&\quad = \sqrt{n_j}.
\end{aligned}$$

From these two equations, it is seen that

$$a_j \quad \text{and} \quad a_j^\dagger \qquad \text{are adjoint of each other.} \tag{72}$$

From the definitions (70) and (71), one obtains the following operator relations:

$$[a_i, a_j]_- = 0, \qquad [a_i^\dagger, a_j^\dagger]_- = 0, \tag{73a}$$

$$[a_i, a_j^\dagger]_- = \delta_{ij}, \tag{73b}$$

where

$$[A, B]_- = AB - BA. \tag{74}$$

The vacuum state of a system of harmonic oscillators is

$$\mid 0 >=\mid 0, 0, 0, ..., 0, ... > . \tag{75}$$

We introduce the "occupation number operator" N_j

$$N_j = a_j^\dagger a_j. \tag{76}$$

From Eq. (72), one has

$$(\Psi_{n_1 n_2 ...}, a_j^\dagger a_j \Psi_{n_1 n_2 ...}) = (a_j \Psi_{n_1 n_2 ...}, a_j \Psi_{n_1 n_2 ...}),$$

and

$$(a_j^\dagger a_j \Psi_{n_1 n_2 ...}, \Psi_{n_1 n_2 ...}) = (a_j \Psi_{n_1 n_2 ...}, a_j \Psi_{n_1 n_2 ...}).$$

Hence

$$(a_j^\dagger a_j)^\dagger = (a_j^\dagger a_j),$$

i.e.,

$$N_j = (a_j^\dagger a_j) \quad \text{is hermitian.} \tag{77}$$

From Eq. (76), one has

$$\begin{aligned} N_j \mid n_1, n_2, ..., n_j, ...\rangle &= a_j^\dagger \sqrt{n_j} \mid n_1, n_2, ..., n_j - 1, ...\rangle \\ &= n_j \mid n_1, n_2, ..., n_j, ...\rangle. \end{aligned} \tag{78}$$

i.e., the eigenvalues of N_j are n_j.

The $N_j, a_j^\dagger, a_j$ have the following relations

$$\begin{aligned} N_j a_j^\dagger \mid ..., n_j, ...\rangle &= (n_j + 1) a_j^\dagger \mid ..., n_j, ...\rangle, \\ a_j^\dagger N_j \mid ..., n_j, ...\rangle &= n_j a_j^\dagger \mid ..., n_j, ...\rangle, \\ N_j a_j \mid ..., n_j, ...\rangle &= (n_j - 1) a_j \mid ..., n_j, ...\rangle, \\ a_j N_j \mid ..., n_j, ...\rangle &= n_j a_j \mid ..., n_j, ...\rangle. \end{aligned} \tag{79}$$

i.e., the operator relations

$$[N_i, a_j^\dagger]_- = a_j^\dagger \delta_{ij}, \tag{80a}$$

$$[N_i, a_j]_- = -a_j \delta_{ij}. \tag{80b}$$

By repeated application of the above relations, one obtains

$$N_j (a_j^\dagger)^k \mid ...n_j...\rangle = (n_j + k)(a_j^\dagger)^k \mid ...n_j...\rangle, \tag{81}$$

$$N_j (a_j)^k \mid ...n_j...\rangle = (n_j - k)(a_j)^k \mid ...n_j...\rangle, \tag{82}$$

where k is an arbitrary integer $\langle n_j$. For $k = n_j$, the last relation gives

$$\begin{aligned} N_j (a_j)^{n_j} \mid ...n_j...\rangle &= N_j \mid ..., 0, ...\rangle \\ &= 0 \end{aligned}$$

If

$$N_j \mid ..., 0, ...\rangle = 0 \qquad \text{for all j,}$$

then

$$\mid 0, 0, ..., 0, ... >\equiv \mid 0\rangle \quad \text{is the vacuum state.} \tag{83}$$

From Eq. (71), one has

$$\mid n_1, n_2, ..., n_j, ...\rangle = \prod \frac{1}{\sqrt{n_j!}} (a_j^\dagger)^{n_j} \mid 0\rangle. \tag{84}$$

If all oscillators are in the same state $\mid K\rangle$ state, then

$$\begin{aligned} \mid n_1, n_2, ...\rangle &\equiv \mid n\rangle \\ &= \frac{1}{\sqrt{n!}} (a^\dagger)^n \mid 0\rangle. \end{aligned} \tag{84a}$$

From Eqs. (70) and (71), one finds the non-vanishing matrix elements of $a_j^\dagger, a_j$

$$\begin{aligned} \langle n_j+1 \mid a_j^\dagger \mid n_j\rangle &= \langle n_j \mid a_j \mid n_j+1\rangle \\ &= \sqrt{n_j+1}. \end{aligned} \tag{85}$$

(ii) System of fermions

For fermions, the Pauli exclusion principle postulates that the occupation number n_j be only either 0 or 1, and that the state function be antisymmetric with respect to the transposition of any two indistinguishable particles.

These two requirements can be expressed by the following form for the state ket

$$\mid k_1, k_2, ..., k_i, ..., k_j, ...\rangle = a_1^\dagger a_2^\dagger ... a_i^\dagger ... a_j^\dagger ... \mid 0\rangle$$

so that upon the transposition of i and j,

$$\begin{aligned} \mid k_1, k_2, ..., k_j, ...k_i, ...\rangle &= a_1^\dagger a_2^\dagger ... a_j^\dagger ... a_i^\dagger ... \mid 0\rangle \\ &= - \mid k_1, k_2, ..., k_i, ..., k_j, ...\rangle. \end{aligned} \tag{86}$$

This condition, together with $n_j = 0$ or 1, leads to the following relations

$$\begin{aligned} (a_i)^2 \mid \Psi\rangle &= (a_i^\dagger)^2 \mid \Psi\rangle = 0, \\ a_i a_j \mid \Psi\rangle &= -a_j a_i \mid \Psi\rangle, \\ a_i^\dagger a_j^\dagger \mid \Psi\rangle &= -a_j^\dagger a_i^\dagger \mid \Psi\rangle, \\ a_i a_j^\dagger \mid \Psi\rangle &= -a_j^\dagger a_i \mid \Psi\rangle, \qquad i \neq j, \\ (a_i a_i^\dagger + a_i^\dagger a_i) \mid \Psi\rangle &= \mid \Psi\rangle. \end{aligned}$$

On defining the anticommutation operator (of Jordan and Wigner, 1927)

$$[A, B]_+ = AB + BA, \tag{87}$$

the above conditions can be expressed in the form

$$[a_i, a_j]_+ = 0, \qquad [a_i^\dagger, a_j^\dagger]_+ = 0, \tag{88a}$$

$$[a_i, a_j^\dagger]_+ = \delta_{ij}. \tag{88b}$$

For fermions, the relations (Eqs. 70, 71) are no longer valid. Instead, one now has

$$a_j \mid n_1, n_2, ..., n_j, ...\rangle = (-1)^{f_j} n_j \mid n_1, n_2, ..., n_j - 1, ...\rangle \tag{89}$$

$$a_j^\dagger \mid n_1, n_2, ..., n_j, ...\rangle = (-1)^{f_j} (1 - n_j) \mid n_1, n_2, ..., n_j + 1, ...\rangle \tag{90}$$

where

$$f_j = \sum_{i=1}^{j-1} n_i. \tag{91}$$

When in Eq. (89) $n_j = 0$ initially, $a_j \mid n_1, ..., n_j, ...\rangle = 0$ does not exist and this is expressed by the vanishing of the righthand side. Similarly, when in Eq. (90) $n_j = 1$ initially, $a_j^\dagger \mid ..., n_j = 1, ...\rangle$ does not exist.

7.4 Neutrino Halos in our Universe

According to the Standard Model, only the neutrinos (used in this article to represent neutrinos of all species, including all antineutrinos) are long-lived dark-matter particles, so that the dark-matter part of a visual macroscopic object, such as the Earth, the Venus, the Sun, and stars, has to be a neutrino halo.[10] The neutrino halo of a visual ordinary-matter object should be weighted five times the visual ordinary-matter host, according to what we understand on the dark matter.

But neutrino halos are Fermi-Dirac gases with incompressible sizes owing to Pauli's exclusive principle. The Fermi-Dirac sphere of a neutrino halo of the visual ordinary-matter object is something brand new in the studies of our Universe. According to (approximate) Newton's universal gravitational law, each neutrino halo would follow the (mother, but smaller) visual ordinary-matter object, one on one with five times in weight, till the aggregates of about 10^{60} smallest units of matter, each.

If we treat all neutrino species as the same, we have, in the limit of the very low temperature and very high densities,

$$\begin{aligned} U &= \tfrac{3}{5} N\epsilon_F[1 + \tfrac{5}{12}\pi^2(\tfrac{kT}{\epsilon_F})^2 + ...]; \\ P &= \tfrac{2}{3}\tfrac{U}{V} = \tfrac{2}{5}\tfrac{\epsilon_F}{v}[1 + \tfrac{5\pi^2}{12}(\tfrac{kT}{\epsilon_F})^2 + ...]; \\ \epsilon_F &= \tfrac{\hbar^2}{2m}(\tfrac{6\pi^2}{gv})^{2/3}; \qquad v \equiv \tfrac{V}{N}. \end{aligned} \tag{92}$$

In this limit, it is characterized by a large Fermi energy ϵ_F, of very small kinetic energy. That is, the very small mass is accompanied by a very large Fermi energy ϵ_F.

We could use these formulae to the neutrino gas in the case of the neutrino halo for the "visual" collapsing star, to see if the neutrino halo would stop the collapse of the star.

First of all, we note from the factor $\hbar^2/(2m)$ that it is an effect arising from the quantum principle and that it is also due to the non-zero mass in this limit. Secondly, it may be difficult to determine v, the volume per particle, in the same limit. But the white dwarf gives an example. It gives, for a dwarf star of a solar mass ($10^{33}\, gm$),

$$\epsilon_F \approx \frac{\hbar^2}{2m_e}\frac{1}{v^{2/3}} \approx 20\, MeV. \tag{93}$$

Using this formula for the neutrino halo with $m_\nu = 0.058\, eV$ (which is the largest observed neutrino mass),

$$\epsilon_F^\nu \approx \frac{\hbar^2}{2m_\nu}\frac{1}{v^{2/3}} \approx 200\, TeV. \tag{94}$$

This estimate comes from the one-solar-mass white dwarf. To apply it to the neutrino halo of the Earth, we need to do a couple of re-scalings (to bring it to a much smaller neutrino halo etc.) Thus, $600\, GeV$, as required by the AMS experiment (i.e., twice of the positron energy in the maximum), sounds a reasonable guess. The reported observation by the *IceCube* collaboration is the behavior of the upper-limit events and the location of the

[10] *W-Y. Pauchy Hwang, The Universe,* **4-2**, *1 (2015).*

maximum is unknown. Of course, it is an important topic to evaluate the maximum of the neutrino halo of the Earth; that is, the Fermi momentum k_F^ν is the maximum.

For the Earth, the Fermi momentum $k_F^\nu \approx 600\,GeV$ of the neutrino halo could be regarded as the measured value of k_F^ν by the AMS experiments, judging from the gross body of the positron data. In fact, we could not think of an alternative measurement of this Fermi momentum k_F^ν.

For the AMS experiments, our interpretation of the Fermi surface of the Fermi-Dirac sphere of the neutrino halo of the Earth offers a very good outlet in this game. Using the dark-matter particle without knowing what the dark-matter is, and thus our Universe is treated as a completely unknown entity, should be, in our opinion, avoided altogether.

We could generalize the uncertainty relation as follows:

$$(\epsilon_F v^{2/3}) \cdot m \geq \hbar^2, \tag{95}$$

the "higher-order" uncertainty relation, in the area rather than in the length. It also gives some dynamic meaning to the term "mass". In addition, these formulae indicate that the quantum principle is in operation.

Actually, there are three species of neutrinos, plus their antineutrinos. When they are stacking up, there are different configurations. It is unlikely that the final configurations would be completely random. But to figure out these, it might take us a few centuries so, we may classify it as "one of the questions of all centuries".

We know that neutrinos oscillate among themselves. In stacking up to form a Fermi-Dirac sphere, they have to be the mass eigen-states so that, in oscillating with the energy, they don't affect their neighbors [in this case, they are stacking up to form the lowest-energy state(s)]. Basically, the oscillations happen with the eigen-energies for the mass eigen-states, rather than for the flavor states. The flavor states are *not eigen-states of any sort*, this poses some other question to think about.

To simplify the situations, the choice of the ground-state configurations is similar to the problem of the "spontaneous symmetry breaking (SSB) with complicated multiple configurations". The happenings of the various reactions, such as $\nu(Solar) + \bar{\nu}(CB; k_F^\nu) \rightarrow \nu + \bar{\nu}$ (invisible), $\nu(Solar) + \bar{\nu}(CB; k_F^\nu) \rightarrow e^- + e^+$, and others, in view of their slow rates, may be treated as "perturbations". Thus, the problem seems to be traceable.

The situation is even more complicated than these. Our estimate of the Fermi energy k_F^ν follows closely the case of the white dwarf, but both cases in fact have k_F much bigger than the rest mass. This is probably not justified even if it may be argued as a simple labeling system. By and large, the Fermi-Dirac sphere is a momentum-space concept and in addition it involves the Dirac space. Its further meanings remain to be discovered.

It is more than interesting to understand the nontrivial structure of our Universe, judging from that the highly nontrivial Fermi-Dirac sphere of the neutrino halo of the Earth (or, of the Venus), should be raveled some day to come. This fundamental issue, unknown to the Kerson's time, regarding neutrino halos and their Fermi-Dirac spheres, should eventually become clear.

7.5 Black Holes Do Not Exist in our Universe

Our Universe is dictated by two laws — Einstein's relativity principle and the quantum principle. At the beginning of the 21st Century, we realized that there exist the *smallest units of matter*, such as electrons, neutrinos, quarks, and others, which are described by the Standard Model, as first named by S. Weinberg.

In our Universe, there are Cosmic Microwave Background (CMB) and Cosmic Background Neutrinos (CBν's). Here "neutrinos" stand for neutrinos of all three flavors and their antineutrinos. CMB photons, in view of their massless feature, are almost uniformly distributed, at present nearly 3 K with one part in 10^5 in fluctuations. Neutrinos have a tiny mass, with $0.058\, eV$ the largest, so clustering into neutrino halos.

The world represented by the Standard Model is in fact the world of ours. There is nothing surprising to us — all particles in there are familiar to us and all interactions are familiar. That is why we have phrased: We declare that we are living in the quantum 4-dimensional Minkowski space-time with the force-fields gauge-group structure $SU_c(3) \times SU_L(2) \times U(1) \times SU_f(3)$ built-in from the very beginning. This "overall background" could see the lepton world, of atomic sizes, of the $SU_L(2) \times U(1) \times SU_f(3)$ symmetry (i.e., the other (123) symmetry). It could also see the quark world, of the nuclear sizes, of the $SU_c(3) \times SU_L(2) \times U(1)$ symmetry [i.e., the (123) symmetry].

The other (123) symmetry in the lepton world is required for the mathematical consistency reason (to eliminate Landau ghosts and to give the other asymptotic freedom) as well as for the experimental needs (i.e., neutrino oscillations and the generation problem, at least).

This is how we describe the *smallest* units of matter. Our world is a world that has the *smallest* units of matter. The language is consistent, and appears to be complete.

Maybe it is essential to remind ourselves that a star of five solar-star mass would be an aggregate of 10^{60} smallest units of matter, a really gigantic large number of the smallest units of matter. Thus, a possible rescaling, in view of the unknown quantity of dark matter, of the Newton's gravitational law should not be a surprise from a theorist's point of view.

Basically, the macroscopic Newton's laws yield

$$m_\nu a = force = G'_N \frac{m_\nu M}{r^2}, \tag{96}$$

or,

$$a = G'_N \frac{M}{r^2}, \tag{97}$$

independent of m_ν. So, since the early Universe, each neutrino halo would follow the visual ordinary-matter object, such as a planet or a star, as five times in weight the invisible dark matter.

Neutrino halos cannot split or fracture by themselves, since the neutrinos have *only tiny* mass, compared to all the other particles (except the photon) in the Standard Model. In the early Universe, they would follow the formation of the (visual) ordinary-matter objects, at the end, each such ordinary-matter object would have one neutrino halo.

Further evolution of neutrino halos, including the reactions which we discussed above, appears only as a very small perturbation. In terms of many billions of years, the part of neutrino halos appears to be stable.

As emphasized in the early paper, the visual ordinary-matter object is an aggregate of typically 10^{60} smallest units of matter, really macroscopic or gigantic. This factor of 10^{60} is bewildering, so much complexities at different levels and, yet, so much simplicities and symmetries.

Thus, CBν's are the *only long-lived* particles, provided that the smallest units of matter are described by the Standard Model. It is deduced that, in our World, CBν's must be the 25% dark matter, in the form of neutrino halos.

Neutrino halos are Fermi-Dirac gases, incompressible because of Pauli's exclusion principle, having the sizes far bigger than the Schwarzschild size of the visual ordinary-matter objects (such as stars). In other words, the point-like structure, such as the Schwarzschild black hole, would never be formed in our Universe. The destiny of our World is that they are filled up with neutrino halos, each accompanied by a (previously visual) dead star or a macroscopic object.

So, our World (i.e., our Universe) is basically to begin with the 3 K cosmic microwave background (CMB), with massless photons almost uniformly distributed coupled with the various neutrino halos, each of them having a visual ordinary-matter macroscopic object, such as planets, stars, etc. The incompressible neutrino halos sort of control the final destiny of the star systems.

It might be easier to detect the Fermi-Dirac sphere of a neutrino halo, such as using *IceCube*, via the Fermi-Dirac sphere of the neutrino halo of the Earth. Yes or not; the Standard Model is closed in the beautiful mathematics, consistently and (maybe) completely, hoping that this would be the last chapter of our Universe. Eventually we would realize that our Universe is described by a beautiful mathematics as well as an elegant physics - the abstract of the metaphysics on the one hand and the reality on the other.

Under the Standard Model, we understand everything in our Universe, from the smallest units of matter to atoms, to molecules, and to the Universe itself. That is why we could name it "the Standard Model of All Centuries", anticipating that the Newton's classical era would soon be replaced, and that the Standard Model would stay there for a while and maybe forever.

In addition, we can safely conclude that black holes do not exist in our Universe. Neutrino halos have helped to change the path of the gigantic evolution of the knowledge.

References

Fock, V. A., Zeits. F. Physik **75**, 622 (1932).

Jordan, P. and Klein, O., *ibid.* **45**, 751 (1927), for systems obeying Bose-Einstein statistics.

Jordan, P. and Winger, E., *ibid.* **47**, 631 (1928), for systems obeying Fermi-Dirac statistics.

E. Mishkin, *Lecture Notes on Relativistic Quantum Mehanics* (1964), Poly. Tech. Brooklyn.

Chapter 8. Quantization of Free Fields

We shall quantize the scalar field Φ of the Klein-Gordon equation, the vector electromagnetic field A_μ and the spinor field Ψ of the Dirac equation, all uncoupled to any source of charges or currents.

8.1 Klein-Gordon, Real (Pseudo) Scalar Field $\phi(x_\mu)$

The solutions of the Klein-Gordon equation Eq. (67), Ch. 6, are

$$e^{ik_\nu x_\nu} = e^{i(\mathbf{k}\cdot\mathbf{r}-k_0 x_0)}, \tag{1}$$

which form a complete set of functions.[1] We expand $\phi(x_\nu)$ in the form

$$\phi(x_\nu) = \frac{1}{(2\pi)^3}\int d^4k\,\delta(k_\nu^2+\mu^2)Ca_{\mathbf{k}}e^{ik_\nu x_\nu}, \tag{3}$$

where $\delta(k_\nu^2+\mu^2)$ is given in (A2) of the appendix at the end of this chapter, C is a constant and $a_{\mathbf{k}}$ is an operator yet to be determined. As $\phi(x_\nu)$ is a real field, $\phi(x_\nu)$ is a hermitian operator. Changing k_ν into $-k_\nu$ changes $a_{\mathbf{k}}$ into $a_{-\mathbf{k}}$. It will be seen that $a_{-\mathbf{k}}$ is the adjoint of $a_{\mathbf{k}}$ [see Eq. (14) below]. $\phi(x_\nu)$ is the Fourier transform of $\delta(k_\nu^2+\mu^2)Ca_{\mathbf{k}}$.

On using the step function $\vartheta(x_0)$ in (A11) and the last relation of (A12), one can write

$$\begin{aligned}\phi(x_\lambda) &= \frac{1}{(2\pi)^3}\int d^4k[\vartheta(k_0)+\vartheta(-k_0)]\delta(k_\nu^2+\mu^2)Ca_{\mathbf{k}}e^{ik_\nu x_\nu}\\ &= \phi_+(x_\nu)+\phi_-(x_\nu),\end{aligned} \tag{4}$$

where

$$\phi_+(x_\lambda) = \frac{1}{(2\pi)^3}\int d^4k\vartheta(k_0)\delta(k_\nu^2+\mu^2)Ca_{\mathbf{k}}a^{ik_\nu x_\nu}$$

$$\phi_-(x_\lambda) = \frac{1}{(2\pi)^3}\int d^4k\vartheta(k_0)\delta(k_\nu^2+\mu^2)Ca_{-\mathbf{k}}e^{-ik_\nu x_\nu} \tag{5}$$

---

[1] *For notations, see the appendix at the end of this chapter,*

$$\begin{aligned}&x_4 = ix_0 = it,\\ &k_4 = ik_0, \quad k_0 = E_{\mathbf{k}} = \omega_{\mathbf{k}} = \sqrt{\mathbf{k}^2+\mu^2}, \quad \mu = m_0,\\ &k_\nu^2 = k_\nu k_\nu = \mathbf{k}^2 - \mathbf{k}_0^2 = -\mu^2,\\ &\mathbf{k}^2+\mu^2 = k_0^2.\end{aligned} \tag{2}$$

Or, in ordinary units,

$$\begin{aligned}&x_4 = ix_0 = ict,\\ &k_4 = ik_0, \quad k_0 = \frac{E_{\mathbf{k}}}{\hbar c} = \frac{\omega_{\mathbf{k}}}{c} = \sqrt{\mathbf{k}^2+\mu^2}, \quad \mu = m_0c.\end{aligned} \quad \text{(A1a–A1c)}$$

From (A2), (A11) and $k_0 = \omega_{\mathbf{k}}$ in Eq. (2), it is seen that[2]

$$\begin{aligned}\vartheta(k_0)\delta(k_\nu^2 + \mu^2) &= \vartheta(k_0)\delta(k_0^2 - \omega_{\mathbf{k}}^2)\\ &= \vartheta(k_0)\frac{1}{2\omega_{\mathbf{k}}}[\delta(k_0 - \omega_{\mathbf{k}}) + \delta(k_0 + \omega_{\mathbf{k}})]\\ &= \frac{1}{2\omega_{\mathbf{k}}}\delta(k_0 - \omega_{\mathbf{k}}),\end{aligned} \tag{6}$$

so that[3]

$$\begin{aligned}\phi_+(x_\lambda) &= \frac{1}{2(2\pi)^3}\int d^3k \frac{1}{\omega_{\mathbf{k}}} C a_{\mathbf{k}} e^{i(\mathbf{k}\cdot\mathbf{r} - \omega_{\mathbf{k}} t)},\\ \phi_-(x_\lambda) &= \frac{1}{2(2\pi)^3}\int d^3k \frac{1}{\omega_{\mathbf{k}}} C a_{-\mathbf{k}} e^{-i(\mathbf{k}\cdot\mathbf{r} - \omega_{\mathbf{k}} t)}.\end{aligned} \tag{7}$$

From Eq. (69) of Ch. 6, the conjugate to $\phi(x_\lambda)$ is

$$\Pi(x_\lambda) = \frac{\partial}{\partial t}\phi(x_\lambda).$$

One writes

$$\Pi(x_\lambda) = \Pi_+(x_\lambda) + \Pi_-(x_\lambda), \tag{8}$$

$$\Pi_+(x_\lambda) = -\frac{i}{2(2\pi)^3}\int d^3k C a_{\mathbf{k}} e^{i(\mathbf{k}\cdot\mathbf{r} - \omega_{\mathbf{k}} t)},$$

$$\Pi_-(x_\lambda) = \frac{i}{2(2\pi)^3}\int d^3k C a_{-\mathbf{k}} e^{-i(\mathbf{k}\cdot\mathbf{r} - \omega_{\mathbf{k}} t)}. \tag{9}$$

We shall take a (large cubic) volume of the field $V = L^3$ and assume periodic conditions

$$k_i = \frac{2\pi}{L} n_i, \quad i = x, y, z, \quad n_i = \pm \text{ integers},$$

which quantize k_i. We choose[4]

$$C = \sqrt{2\omega_{\mathbf{k}} L^3}. \tag{10}$$

---

[2]

$$\vartheta(k_0)\delta(k_\nu^2 + \mu^2) = \frac{c}{2\omega_{\mathbf{k}}}\delta(k_0 - \frac{\omega_{\mathbf{k}}}{c}). \tag{6'}$$

[3] *In ordinary units, Eqs. (7), (9) have a factor c on the right.*

[4]

$$C = \frac{1}{c}\sqrt{2\hbar\omega_{\mathbf{k}} L^3}. \tag{10'}$$

For large L, the integrals over k may be replaced by sums over discrete k,

$$\frac{1}{(2\pi)^3}\int d^3k \quad\rightarrow\quad \frac{1}{L^3}\sum_{\mathbf{k}}$$

and[5]

$$\phi(x_\lambda) = \sqrt{\frac{1}{2L^3}}\sum\frac{1}{\sqrt{\omega_{\mathbf{k}}}}\{a_{\mathbf{k}}e^{i(\mathbf{k}\cdot\mathbf{r}-\omega_{\mathbf{k}}t)} + a_{-\mathbf{k}}e^{-i(k\cdot r-\omega_{\mathbf{k}}t)}\},$$
$$\Pi(x_\lambda) = i\sqrt{\frac{1}{2L^3}}\sum\sqrt{\omega_{\mathbf{k}}}\{-a_{\mathbf{k}}e^{i(\mathbf{k}\cdot\mathbf{r}-\omega_{\mathbf{k}}t)} + a_{-\mathbf{k}}e^{-i(\mathbf{k}\cdot\mathbf{r}-\omega_{\mathbf{k}}t)}\}. \tag{11}$$

These are the general solutions of the Klein-Gordon equation.

Quantization of the field may be accomplished by application of Dirac correspondence principle, which states that, to an elementary Poisson bracket $\{a(t), b(t)\} = c$ (which c some c-number) in a classical system, there corresponds an elementary commutator $[a(t), b(t)] = ic$ in the corresponding quantized system. In the present case, we find, for $t' - t = 0$,

$$[\phi(\mathbf{r}),\ \phi(\mathbf{r}')] = 0, \tag{12a}$$

$$[\phi(\mathbf{r}),\ \Pi(\mathbf{r}')] = i\delta(\mathbf{r}-\mathbf{r}'), \tag{12b}$$

$$[\Pi(\mathbf{r}),\ \Pi(\mathbf{r}')] = 0. \tag{12c}$$

It is to be noted that these last three relations are not Lorentz invariant since they hold only for $t = t'$. That $\delta(\mathbf{r}-\mathbf{r}')$ appears in Eq. (12b) has a simple reason. Two points spatially separated cannot communicate by any instantaneous signal so that $\phi(\mathbf{r}), \Pi(\mathbf{r}')$ are independent and can both be measured with arbitrary accuracy.

It is straightforward to demonstrate that Eqs. (12a)-(12c) hold if and only if the following relations are assumed:

$$[a_{\mathbf{k}}, a_{\mathbf{k}'}] = [a_{-\mathbf{k}}, a_{-\mathbf{k}'}] = 0,$$
$$[a_{\mathbf{k}}, a_{-\mathbf{k}'}] = \delta_{\mathbf{k}\mathbf{k}'}. \tag{13}$$

Let us calculate the commutator

$$[\phi(\mathbf{x}_\lambda), \phi_(\mathbf{x}'_\lambda)] = \phi(\mathbf{x}_\lambda)\phi(\mathbf{x}'_\lambda) - \phi(\mathbf{x}'_\lambda)\phi(\mathbf{x}_\lambda).$$

Let us write for brevity

$$R \equiv \mathbf{k}\cdot\mathbf{r} - \omega_{\mathbf{k}}t = k_\mu x_\mu, \qquad R' \equiv \mathbf{k}'\cdot\mathbf{r}' - \omega_{\mathbf{k}'}t = k'_\mu x'_\mu. \tag{14}$$

Then[6]

$$[\phi(\mathbf{x}_\lambda), \phi(\mathbf{x}'_\lambda)] = \frac{1}{2L^3}\sum_{k,k'}\frac{1}{\sqrt{\omega_{\mathbf{k}}\omega_{\mathbf{k}'}}}\{[a_{\mathbf{k}}, a_{\mathbf{k}'}]e^{i(R+R')} + [a_{-\mathbf{k}}, a_{-\mathbf{k}'}]e^{-i(R+R')}$$
$$+ [a_{\mathbf{k}}, a_{-\mathbf{k}'}]e^{i(R-R')} - [a_{\mathbf{k}'}, a_{-\mathbf{k}}]e^{-i(R-R')}\}. \tag{15}$$

---

[5] *Eqs. (11) have a factor $\sqrt{\hbar}$ on the right.*

[6] *In ordinary units, Eq. (15) has $\hbar$ on the right; Eqs. (17) and (18) have $\frac{\hbar}{c}$ as a factor on the right; Eq. (19) $\frac{\hbar}{c^2}$ on the right.*

Using Eq. (13), we find

$$[\phi(\mathbf{x}_\lambda), \phi(\mathbf{x}'_\lambda)] = \frac{1}{2L^3}\sum \frac{1}{\omega_{\mathbf{k}}}\{e^{i(R-R')} - e^{-i(R-R')}\}$$
$$= \frac{1}{2(2\pi)^3}\int \frac{d^3k}{\omega_{\mathbf{k}}}\{e^{i(k_\nu y_\nu)} - e^{-i(k_\nu y_\nu)}\}$$
$$= -\frac{1}{(2\pi)^3}\int d^3k \frac{\sin k_\nu y_\nu}{k_4}, \quad (16)$$

where

$$y_\nu = x_\nu - x'_\nu, \qquad k_4 = i\omega_{\mathbf{k}}.$$

Comparison with the $\triangle(x)$ function of (A-6b) now leads to

$$[\phi(x_\lambda), \phi(x'_\lambda)] = \frac{1}{i}\triangle(x_\lambda - x'_\lambda), \quad (17)$$

which is a Lorentz invariant commutation relation.
Similarly,

$$[\phi(x_\lambda), \Pi(x'_\lambda)] = \frac{i}{(2\pi)^3}\int d^3k \cos k_\nu y_\nu$$
$$= -\frac{1}{i}\frac{\partial}{\partial t}\triangle(x_\lambda - x'_\lambda), \quad (18)$$

$$[\Pi(x_\lambda), \Pi(x'_\lambda)] = -\frac{1}{(2\pi)^3}\int d^3k k_4 \sin k_\nu y_\nu$$
$$= -i\frac{\partial^2}{\partial t^2}\triangle(x_\lambda - x'_\lambda). \quad (19)$$

Let us come back to the relations Eq. (13). They are identical with the commutation relations for the creation and annihilation operators $a^\dagger_{\mathbf{k}}$ and $a_{\mathbf{k}}$ if we identify them as follows:

$$a_{\mathbf{k}} = a_{\mathbf{k}}, \qquad a_{-\mathbf{k}} = a^\dagger_{\mathbf{k}}. \quad (20)$$

To fully bring out the meaning of $a_{\mathbf{k}}$ and $a_{-\mathbf{k}}$ in Eq. (11), let us take the Heisenberg equations of motion[7]

$$\frac{d\phi}{dt} = -i[\phi, H], \qquad \frac{d\Pi}{dt} = -i[\Pi, H] \quad (21)$$

where

$$[A, B] = AB - BA.$$

[7] *In ordinary units, Eqs. (21), (22), (25), and (26) have a factor $\hbar$ on the left; (23), (27) a factor $\hbar$ on the right.*

From Eq. (11), one obtains

$$\sum_{\mathbf{k}} \sqrt{\omega_{\mathbf{k}}}[a_{\mathbf{k}}e^{iR} - a_{-\mathbf{k}}e^{-iR}] = \sum_{\mathbf{k}} \frac{1}{\sqrt{\omega_{\mathbf{k}}}}[(a_{\mathbf{k}}e^{iR} + a_{-\mathbf{k}}e^{-iR}),\, H], \tag{22}$$

where for brevity as in Eq. (12), $R \equiv \mathbf{k}\cdot\mathbf{r} - \omega_{\mathbf{k}}t$.
From Eq. (21), one obtains

$$\begin{aligned}[a_{\mathbf{k}},\, H] &= \omega_{\mathbf{k}}a_{\mathbf{k}},\\ [a_{-\mathbf{k}},\, H] &= -\omega_{\mathbf{k}}a_{-\mathbf{k}}.\end{aligned} \tag{23}$$

The momentum $P_j, j = 1, 2, 3$, is, from Eq. (74) of Ch. 6,

$$P_j = \int_{V_3} \frac{1}{2}\{\Pi(x_\lambda)\partial_j\phi(x_\lambda) + (\partial_j\phi(x_\lambda))\Pi(x_\lambda)\}d^3x. \tag{24}$$

By means of the commutation relations Eq. (13), it can be shown that for any function $F(\phi, \pi)$ of $\phi(x_\lambda)$ and $\pi(x_\lambda)$, the following relation holds

$$\nabla F = -i[\mathbf{P}, F]. \tag{25}$$

In particular, for $F = \phi(x_\lambda)$ of Eq. (11), this relation gives

$$\sum \mathbf{k}(a_{\mathbf{k}}e^{iR} - a_{-\mathbf{k}}e^{-iR}) = -\sum_{\mathbf{k}} [\mathbf{P}, (a_{\mathbf{k}}e^{iR} + a_{-\mathbf{k}}e^{-iR})]. \tag{26}$$

From this, one gets, in a way similar to Eq. (23),

$$\begin{aligned}[a_{\mathbf{k}}, \mathbf{P}] &= \mathbf{k}a_{\mathbf{k}},\\ [a_{-\mathbf{k}}, \mathbf{P}] &= -\mathbf{k}a_{-\mathbf{k}}.\end{aligned} \tag{27}$$

Let $E_0, \mid E_0\rangle, \mathbf{P}_0, \mid \mathbf{P}_0\rangle$ be the eigenvalue and eigenstate of H and $\mathbf{P}$

$$(H - E_0) \mid E_0\rangle = 0, \quad (\mathbf{P} - \mathbf{P}_0) \mid \mathbf{P}_0\rangle = 0. \tag{28}$$

From Eq. (27) and Eq. (28), one obtains[8]

$$\begin{aligned} H(a_{\mathbf{k}} \mid E_0\rangle) &= (E_0 - \omega_{\mathbf{k}})(a_{\mathbf{k}} \mid E_0\rangle),\\ \mathbf{P}(a_{\mathbf{k}} \mid \mathbf{P}_0\rangle) &= (\mathbf{P}_0 - \mathbf{k})(a_{\mathbf{k}} \mid \mathbf{P}_0\rangle),\\ H(a_{-\mathbf{k}} \mid E_0\rangle) &= (E_0 + \omega_{\mathbf{k}})(a_{-\mathbf{k}} \mid E_0\rangle),\\ \mathbf{P}(a_{-\mathbf{k}} \mid \mathbf{P}_0\rangle) &= (\mathbf{P}_0 + \mathbf{k})(a_{-\mathbf{k}} \mid \mathbf{P}_0\rangle).\end{aligned} \tag{29}$$

These show that $a_{\mathbf{k}}$ lowers the eigenvalue of H by $\omega_{\mathbf{k}}$ and the eigenvalue of $\mathbf{P}$ by k, while $a_{-\mathbf{k}}$ raises that of H by $\omega_{\mathbf{k}}$ and that of $\mathbf{P}$ by k. This completes the identification Eq. (20) of $a_{-\mathbf{k}}$ with the creation operator $a^\dagger_{\mathbf{k}}$, and $a_{\mathbf{k}}$ with the annihilation operator $a_{\mathbf{k}}$ of Eqs. (73a, b), Eq. (11) then brings out the particle aspect of the field $\phi(x_\lambda)$.

[8] *In ordinary units, $\omega_{\mathbf{k}}, k$ on the right in (29) stand for $\hbar\omega_{\mathbf{k}}, \hbar k$ respectively.*

If one uses the occupation number representation of Eq. (34) in Ch. 7, then

$$| n_1, n_2, n_3, ...\rangle$$

is a state in which n_i oscillators have the momentum

$$\mathbf{k}_i$$

and energy

$$E_{k_i} = \sqrt{\mathbf{k}_i^2 + m_0^2}, \quad i = 1, 2, 3, ...$$

From Eqs. (70) and (71) of Ch. 7, one has

$$\begin{aligned} a_{k_i} \mid n_1, n_2, ..., n_i, ...\rangle &= \sqrt{n_i} \mid n_1, n_2, ..., n_i - 1, ...\rangle, \\ a_{k_i}^\dagger \mid n_1, n_2, ..., n_i, ...\rangle &= \sqrt{n_i + 1} \mid n_1, n_2, ..., n_i + 1, ...\rangle, \end{aligned} \tag{30}$$

with the occupation-number operator $N_{\mathbf{k}}$ of Eq. (76) in Ch. 7

$$N_{\mathbf{k}} = a_{\mathbf{k}}^\dagger a_{\mathbf{k}}, \tag{31}$$

one has, for the energy and momentum operator,[9]

$$\begin{aligned} H &= \sum_{\mathbf{k}} E_{\mathbf{k}} N_{\mathbf{k}} = \sum_{\mathbf{k}} E_{\mathbf{k}} a_{\mathbf{k}}^\dagger a_{\mathbf{k}}, \\ \mathbf{P} &= \sum_{\mathbf{k}} \mathbf{P}_{\mathbf{k}} N_{\mathbf{k}} = \sum_{\mathbf{k}} \mathbf{k} a_{\mathbf{k}}^\dagger a_{\mathbf{k}}. \end{aligned} \tag{32}$$

and

$$\begin{aligned} H \mid n_1, n_2, ...\rangle &= (\sum_{\mathbf{k}} n_{\mathbf{k}} E_{\mathbf{k}}) \mid n_1, n_2, ...\rangle, \\ \mathbf{P} \mid n_1, n_2, ...\rangle &= (\sum_{\mathbf{k}} n_{\mathbf{k}} \mathbf{k}) \mid n_1, n_2, ...\rangle. \end{aligned} \tag{33}$$

From Eq. (24), one finds that, for $E_2 - E_1 > 0$,

$$(E_2 - E_1 - \omega_{\mathbf{k}})\langle E_2 \mid a_{\mathbf{k}}^\dagger \mid E_1\rangle = 0,$$

so that

$$\langle E_2 \mid a_{\mathbf{k}}^\dagger \mid E_1\rangle = 0 \;\text{ unless }\; \omega_{\mathbf{k}} = E_2 - E_1. \tag{34a}$$

Similarly, for $E_1 - E_2 > 0$,

$$\langle E_2 \mid a_{\mathbf{k}} \mid E_1\rangle = 0 \;\text{ unless }\; \omega_{\mathbf{k}} = E_1 - E_2. \tag{34b}$$

[9] *In ordinary units, k in Eqs. (32) and (33), $\omega_{\mathbf{k}}$ in Eqs. (34a) and (34b) stand for $\hbar k, \hbar\omega_{\mathbf{k}}$ respectively.*

The real, scalar Klein-Gordon field $\phi(x_\lambda)$, when quantized, may be applied to the neutral pion π^0 which is experimentally known to have zero spin, and is a pseudo scalar [i.e., $\phi(x_\lambda)$ changes sign upon space inversion $\mathbf{r} \to -\mathbf{r}$].

To close this section, it is useful to note that the momentum operator given by Eq. (32) differs from that in Eq. (24) by a c-number. We write

$$P_j = \int_{V_3} : \frac{1}{2}\{\Pi(x_\lambda)\partial_j\phi(x_\lambda) + (\partial_j\phi(x_\lambda))\Pi(x_\lambda)\} : d^3x, \tag{35}$$

where the double colon " : : " indicates "normal ordering" of the operators, meaning that creation operators always precede (i.e., are to the left of) all annihilation operators. In general, we have, for an arbitrary operator $\hat{Q}$,

$$\langle 0 \mid : \hat{Q} : \mid 0 \rangle = 0. \tag{36}$$

Analogously, by normal ordering the quantized Noether's current, we obtain the energy operator as given by Eq. (32).

8.2 Klein-Gordon Complex, Scalar Field

The quantization of the real scalar fields $\psi_1(x_\lambda), \psi_2(x_\lambda)$ of Eq. (78) in Ch. 6 follows the same procedure as in Eqs. (10)-(19) of Ch. 7. Thus for the hermitian operators ψ_1, ψ_2, analogously to Eq. (11), we now have, on writing

$$a(\mathbf{k}),\, a^\dagger(\mathbf{k}) \;\text{ for }\; a_\mathbf{k},\, a_\mathbf{k}^\dagger, \tag{37}$$

$$\psi_j(x_\lambda) = \sqrt{\frac{1}{2L^3}} \sum_\mathbf{k} \frac{1}{\sqrt{\omega_\mathbf{k}}}\{a_j(\mathbf{k})e^{iR} + a_j^\dagger(\mathbf{k})e^{-iR}\},$$
$$j = 1, 2, \quad R \equiv k_\nu x_\nu = \mathbf{k}\cdot\mathbf{r} - \omega_\mathbf{k} t, \quad R' \equiv \mathbf{k}'\cdot\mathbf{r}' - \omega_{\mathbf{k}'}t'. \tag{38}$$

$$\begin{aligned}
&[a_j(\mathbf{k}),\, a_j(\mathbf{k}')] = [a_j^\dagger(\mathbf{k}),\, a_j^\dagger(\mathbf{k}')] = 0,\\
&[a_j(\mathbf{k}), a_j^\dagger(\mathbf{k}')] = \delta_{\mathbf{k}\mathbf{k}'}.
\end{aligned} \tag{39}$$

Now we introduce

$$\begin{aligned}
a_\mathbf{k} &= \frac{1}{\sqrt{2}}(a_1(\mathbf{k}) + ia_2(\mathbf{k})),\\
a_\mathbf{k}^\dagger &= \frac{1}{\sqrt{2}}(a_1^\dagger(\mathbf{k}) - ia_2^\dagger(\mathbf{k})),\\
b_\mathbf{k} &= \frac{1}{\sqrt{2}}(a_1(\mathbf{k}) - ia_2(\mathbf{k})),\\
b_\mathbf{k}^\dagger &= \frac{1}{\sqrt{2}}(a_1^\dagger(\mathbf{k}) + ia_2^\dagger(\mathbf{k})),
\end{aligned} \tag{40}$$

and by analogy with Eq. (20),

$$a_{-\mathbf{k}} = b^\dagger_{\mathbf{k}}, \qquad b_{-\mathbf{k}} = a^\dagger_{\mathbf{k}}. \tag{41}$$

Then from Eqs. (39) and (40), one obtains

$$\begin{aligned}
&[a_{\mathbf{k}}, b_{\mathbf{k}'}] = [a^\dagger_{\mathbf{k}}, b^\dagger_{\mathbf{k}'}] = [a_{\mathbf{k}}, b^\dagger_{\mathbf{k}'}] = [b_{\mathbf{k}}, a^\dagger_{\mathbf{k}'}] = 0,\\
&[a_{\mathbf{k}}, a^\dagger_{\mathbf{k}'}] = [b_{\mathbf{k}}, b^\dagger_{\mathbf{k}'}] = \delta_{\mathbf{k}\mathbf{k}'}.
\end{aligned} \tag{42}$$

In terms of the $a_{\mathbf{k}}$, $b_{\mathbf{k}}$, $a^+_{\mathbf{k}}$, $b^+_{\mathbf{k}}$, one obtains for the non-hermitian operators $\Psi(x_\lambda)$, $\Psi^+(x_\lambda)$, $\Pi(x_\lambda)$, $\Pi^+(x_\lambda)$ corresponding to Ψ, Ψ^*, Π, Π^* in Eqs. (76) and (79) of Ch. 6, the expressions[10]

$$\begin{aligned}
\Psi(x_\lambda) &= \sqrt{\frac{1}{2L^3}} \sum_{\mathbf{k}} \frac{1}{\sqrt{\omega_{\mathbf{k}}}} \{a_{\mathbf{k}} e^{iR} + b^\dagger_{\mathbf{k}} e^{-iR}\},\\
\Psi^\dagger(x_\lambda) &= \sqrt{\frac{1}{2L^3}} \sum_{\mathbf{k}} \frac{1}{\sqrt{\omega_{\mathbf{k}}}} \{b_{\mathbf{k}} e^{iR} + a^\dagger_{\mathbf{k}} e^{-iR}\},\\
\Pi(x_\lambda) &= i\sqrt{\frac{1}{2L^3}} \sum_{\mathbf{k}} \sqrt{\omega_{\mathbf{k}}} \{-b_{\mathbf{k}} e^{iR} + a^\dagger_{\mathbf{k}} e^{-iR}\},\\
\Pi^\dagger(x_\lambda) &= i\sqrt{\frac{1}{2L^3}} \sum_{\mathbf{k}} \sqrt{\omega_{\mathbf{k}}} \{-a_{\mathbf{k}} e^{iR} + b^+_{\mathbf{k}} e^{-iR}\}.
\end{aligned} \tag{43}$$

Using the commutation relation (17) for the two real field $\psi_1(x_\lambda), \psi_2(x_\lambda)$, one can show that

$$[\Psi(x_\lambda),\ \Psi^\dagger(x'_\lambda)] = \frac{1}{i}\triangle(x_\lambda - x'_\lambda) \tag{44}$$

which is Lorentz invariant, and

$$[\Psi(x_\lambda),\ \Psi(x'_\lambda)] = [\Psi^\dagger(x_\lambda),\ \Psi^\dagger(x'_\lambda)] = 0. \tag{45}$$

Corresponding to Eqs. (12a), (12b), and (12c) for a space-like separation $\mathbf{r} - \mathbf{r}'$, (i.e., $t - t' = 0$), one obtains

$$[\Psi(\mathbf{r}, t),\ \Psi^\dagger(\mathbf{r}', t)] = [\Pi(\mathbf{r}, t),\ \Pi^+(r', t)] = 0, \tag{46a}$$

$$[\Psi(\mathbf{r}, t),\ \Psi^\dagger(\mathbf{r}', t)] = [\Psi^\dagger(x_\lambda),\ \Pi(x_\lambda)] = 0, \tag{46b}$$

$$[\Psi(\mathbf{r}, t),\ \Pi(\mathbf{r}', t)] = -[\Psi^\dagger(\mathbf{r}, t),\ \Pi^\dagger(\mathbf{r}', t)] = -\frac{1}{i}\delta(\mathbf{r} - \mathbf{r}'). \tag{46c}$$

he same remark following Eq. (12b) applies here to Eq. (46c).

---

[10] *In ordinary units, a factor* $\sqrt{\hbar}$ *on the right in Eq. (43);* $\frac{\hbar}{c}$ *on the right in (44);* $\hbar$ *on the right in (46c);* $\frac{1}{\hbar}$ *on the right in* Q*, in (47).*

That the $a_{\mathbf{k}}, b_{\mathbf{k}}, a_{\mathbf{k}}^{\dagger}, b_{\mathbf{k}}^{\dagger}$ have the meaning of annihilation and creation operators can be seen by following through the same procedure as in Eqs. (23)–(34).

From Eq. (89) of Ch. 6, we have the total electric charge of the field in a volume V_3

$$Q = \int_{V_3} \rho d^3x = -ie \int_{V_3} (\Psi \dot{\Psi}^* - \Psi^* \dot{\Psi}) d^3x,$$

upon symmetrizing,

$$Q = -\frac{ie}{2} \int_{V_3} (\Psi \Pi + \Pi \Psi - \Psi^\dagger \Pi^\dagger - \Pi^\dagger \Psi^\dagger) d^3x, \tag{47}$$

and on using Eq. (41)

$$Q = e \sum_{\mathbf{k}} (a_{\mathbf{k}}^\dagger a_{\mathbf{k}} - b_{\mathbf{k}}^\dagger b_{\mathbf{k}}). \tag{47a}$$

This suggests the interpretation that $a_{\mathbf{k}}^\dagger, a_{\mathbf{k}}$ are the creation and annihilation operator respectively of positively charged particle of the quantized field, $b_{\mathbf{k}}^\dagger, b_{\mathbf{k}}$ are similarly for the negatively charge particle of the quantized field.

Analogously to Eq. (31), we define the occupation number operators for + and - charged particles

$$N_{k(+)} = a_{\mathbf{k}}^\dagger a_{\mathbf{k}}, \quad N_{k(-)} = b_{\mathbf{k}}^\dagger b_{\mathbf{k}},$$

and

$$N_+ = \sum_{\mathbf{k}} N_{k(+)}, \quad N_- = \sum_{\mathbf{k}} N_{k(-)}. \tag{48}$$

In terms of these operators,

$$Q = e(N_+ - N_-). \tag{47b}$$

From Eqs. (40) and (45a), we obtain

$$\begin{aligned} [a_{\mathbf{k}}, Q] &= e a_{\mathbf{k}}, & [a_{\mathbf{k}}^\dagger, Q] &= -e a_{\mathbf{k}}^\dagger, \\ [b_{\mathbf{k}}, Q] &= -e b_{\mathbf{k}}, & [b_{\mathbf{k}}^\dagger, Q] &= e b_{\mathbf{k}}^\dagger. \end{aligned} \tag{49}$$

The Heisenberg equations of motion of Ψ and Ψ^+ are[11]

$$\frac{d}{dt}\Psi = -i[\Psi, H], \qquad \frac{d}{dt}\Psi^+ = -i[\Psi^\dagger, H], \tag{50}$$

and analogously to Eq. (25),

$$\nabla \Psi = -i[\mathbf{P}, \Psi], \qquad \nabla \Psi^\dagger = -i[\mathbf{P}, \Psi^\dagger]. \tag{51}$$

[11] *In ordinary units, a factor* $\frac{1}{\hbar}$ *on the right in (50), (51).*

From these equations, one obtains[12]

$$\begin{aligned}
[a_{\mathbf{k}}, H] &= E_{\mathbf{k}} a_{\mathbf{k}}, & [a^\dagger_{\mathbf{k}}, H] &= -E_{\mathbf{k}} a^\dagger_{\mathbf{k}}, \\
[b_{\mathbf{k}}, H] &= E_{\mathbf{k}} b_{\mathbf{k}}, & [b^\dagger_{\mathbf{k}}, H] &= -E_{\mathbf{k}} b^\dagger_{\mathbf{k}}, \\
[a_{\mathbf{k}}, \mathbf{P}] &= \mathbf{k} a_{\mathbf{k}}, & [b^\dagger_{\mathbf{k}}, \mathbf{P}] &= -\mathbf{k} b^\dagger_{\mathbf{k}}, \\
[b_{\mathbf{k}}, \mathbf{P}] &= \mathbf{k} b_{\mathbf{k}}, & [b^\dagger_{\mathbf{k}}, \mathbf{P}] &= -\mathbf{k} b^\dagger_{\mathbf{k}}.
\end{aligned} \tag{52}$$

Let $\mid 0\rangle$ be the vacuum state

$$a_{\mathbf{k}} \mid 0\rangle = b_{\mathbf{k}} \mid 0\rangle = 0,$$

$$H \mid 0\rangle = 0, \quad \mathbf{P} \mid 0\rangle = 0. \tag{53}$$

Then one gets

$$\begin{aligned}
H &= \sum_{\mathbf{k}} E_{\mathbf{k}} (N_{k(+)} + N_{k(-)}) \\
&= \sum_{\mathbf{k}} E_{\mathbf{k}} (a^\dagger_{\mathbf{k}} a_{\mathbf{k}} + b^\dagger_{\mathbf{k}} b_{\mathbf{k}}), \\
\mathbf{P} &= \sum_{\mathbf{k}} \mathbf{k} (a^\dagger_{\mathbf{k}} a_{\mathbf{k}} + b^\dagger_{\mathbf{k}} b_{\mathbf{k}}).
\end{aligned} \tag{54}$$

The charged Klein-Gordon fields Ψ, Ψ^* , or ψ_1, ψ_2 can be taken to apply to the Π^+ and Π^- pions which have experimintally been found to have spin zero.

Again, it is useful to note that the operators Q, H, and P [Eqs. (47b) and (54)] are related to the quantized conserved Noether's currents [cf. §6.3.] by normal ordering. In other words, Noether's theorem gives rise to conserved dynamical invariants which, upon quantization and normal ordering, describe the corresponding quantities in the quantized theory.

8.3 Electromagnetic Fields

The general procedure of the quantization of fields has been illustrated by the examples of the scalar fields of the Klein-Gordon equation in the two preceding sections. For the electromagnetic field A_μ of the Maxwell equations, the general method is the same, but there are a few special considerations.[13]

(i) The rest mass m_0 of the quantized field particles (photons) is zero.

(ii) In the Lorentz gauge, the condition Eq. (94) of Ch. 6.

$$\partial_\mu A_\mu = 0, \tag{55}$$

imposes one relation among the A_μ so that only three of the four A_μ's are independent.

---

[12] *In ordinary units, k on the right in (52), (54) stands for $\hbar k$.*

[13] *See, e.g., I. J. R. Aitchison, An Informal Introduction to Gauge Field Theories (Cambridge University Press, London, 1982).*

Furthermore, if one chooses, in any Lorentz frame, a gauge in which $A_4 = 0$, one has Eq. (100) of Ch. 6,

$$div\, A = \partial_{\mathbf{k}} A_{\mathbf{k}} = 0 \tag{56}$$

so that only two of the A_μ are independent.
(iii) Since the (physical) fields **E** and **B** are invariant under the gauge transformation (97), Ch. 6, it is necessary to see the relation of this to the definition of the operators and the physical state of an electromagnetic field.
(1) Without the Lorentz condition For reasons that will become clear later in the section, we need to start *without* the Lorentz condition (55) so that the A_μ are independent. We use the following notations

$$A(A_1, A_2, A_3, A_4 = iA_0), \tag{57}$$

A_1, A_2, A_3, A_0 are hermitian operators, so that A_4 is anti-hermitian, i.e.,

$$\begin{aligned} A_j^\dagger(x) &= A_j(x), & j &= 1, 2, 3, \\ A_0^\dagger(x) &= A_0(x), & A_4^\dagger(x) &= -A_4(x). \end{aligned} \tag{58}$$

The energy-momentum relation is

$$k^2 = \mathbf{k}^2 - k_0^2 = 0, \tag{59a}$$

where[14]

$$E_{\mathbf{k}} = \mid \mathbf{k} \mid = \omega_{\mathbf{k}} = k_0. \tag{59b}$$

The solutions of Eq. (101) of Ch. 6,

$$\partial_\nu \partial_\nu A_\mu(x) = 0, \tag{60}$$

namely

$$e^{ikx}, \quad kx = k_\nu x_\nu, \tag{61}$$

form a complete set, so that one may expand $A_\mu(x)$ as in Eq. (3)

$$A_\mu(x) = \frac{1}{(2\pi)^3} \int d^4k\, \delta(k_\nu^2) C a_\mu(\mathbf{k}) e^{ikx}, \tag{62}$$

which, on account of the $\delta(k_\nu^2)$ function, is really a 3-dimensional integral. We choose, as in Eq. (10),[15]

$$C = \sqrt{2\omega_{\mathbf{k}} L^3}. \tag{63}$$

14

$$E_{\mathbf{k}} = c\hbar \mid \mathbf{k} \mid = \hbar\omega_{\mathbf{k}} = c\hbar k_0. \tag{59b$'$}$$

15

$$C = \frac{1}{2}\sqrt{2\omega_{\mathbf{k}} L^3}. \tag{63}$$

On account of the hermitian properties Eq. (58), we have

$$a_j(\mathbf{k}) = a_j^\dagger(-\mathbf{k}), \qquad a_0(\mathbf{k}) = a_0^\dagger(-\mathbf{k}). \tag{64}$$

From Eq. (6), we have[16]

$$\delta(k^2) = \frac{1}{2\omega_\mathbf{k}}(\delta(k_0+\omega_\mathbf{k}) + \delta(k_0-\omega_\mathbf{k})). \tag{65}$$

Hence

$$A_\mu(x) = \frac{1}{2(2\pi)^3}\int d^3k \frac{1}{\omega_\mathbf{k}} C(a_\mu(\mathbf{k})e^{iR} + a_\mu^\dagger(\mathbf{k})e^{-iR}) \tag{66}$$

$$R = \mathbf{k}\cdot\mathbf{r} - \omega_\mathbf{k} t. \tag{67}$$

Again, as in Eq. (11), we quantize k by taking the field in a large volume L^3 and assuming periodic conditions

$$k_j = \frac{2\pi}{L}n_j, \qquad n_j = \pm\,\text{integers}, \tag{68}$$

and replacing

$$\frac{1}{(2\pi)^3}\int d^3k \quad \text{by} \quad \frac{1}{L^3}\sum_\mathbf{k} \tag{69}$$

so that Eq. (6) becomes[17]

$$\mathbf{A}(x) = \sqrt{\frac{1}{2L^3}}\sum_\mathbf{k}\frac{1}{\sqrt{\omega_\mathbf{k}}}\{a(\mathbf{k})e^{iR} + a^\dagger(\mathbf{k})e^{-iR}\} \tag{70a}$$

$$\equiv \mathbf{A}_+(x) + \mathbf{A}_-(x), \tag{70b}$$

$$A_0(x) = \sqrt{\frac{1}{2L^3}}\sum_\mathbf{k}\frac{1}{\sqrt{\omega_\mathbf{k}}}\{a_0(\mathbf{k})e^{iR} + a_0^\dagger(\mathbf{k})e^{-iR}\} \tag{71a}$$

$$= A_{0+}(x) + A_{0-}(x). \tag{71b}$$

From Eqs. (105) and (102) of Ch. 6, we have

$$\begin{aligned}\dot{A}_j(x) &= \Pi_j(x) + i\partial_j A_4(x),\\ \dot{A}_4(x) &= \Pi_4(x) - i\partial_j A_j(x)\end{aligned} \tag{72a}$$

---

[16] *In ordinary units, $\omega_\mathbf{k}$ stands for $\frac{\omega_\mathbf{k}}{c}$ in (65), (66).*

[17] A factor $\sqrt{\hbar}$ on the right in (70a), (71a); a factor $\frac{\hbar}{c}$ on the right in (74), (74a).

$$\Pi_j(x) = -i(-\partial_4 A_j + \partial_j A_4),$$
$$\Pi_4(x) = i(\partial_4 A_4 + \partial_j A_j). \tag{72b}$$

The $a_j(\mathbf{k}), a_j^\dagger(\mathbf{k}), a_0(\mathbf{k}), a_0^\dagger(\mathbf{k})$, $j = 1, 2, 3$, operators satisfy the relations

$$\begin{aligned}
&[a_i(\mathbf{k}), a_j(\mathbf{k}')] = [a_i^\dagger(\mathbf{k}), a_j^\dagger(\mathbf{k}')] = 0, \qquad i, j = 1, 2, 3,\\
&[a_0(\mathbf{k}), a_0(\mathbf{k}')] = [a_0^\dagger(\mathbf{k}), a_0^\dagger(\mathbf{k}')] = 0,\\
&[a_i(\mathbf{k}), a_j^\dagger(\mathbf{k}')] = \delta_{ij}\delta_{\mathbf{kk'}},\\
&[a_0(\mathbf{k}), a_0^\dagger(\mathbf{k}')] = -\delta_{\mathbf{kk'}},
\end{aligned} \tag{73}$$

all other commutators vanishing.[18]

From Eqs. (70a), (71a), and (73), calculations give[19]

$$[A_\mu(x), A_\nu(x')] = -\frac{1}{(2\pi)^3}\int d^3k \frac{\sin k_\lambda (x-x')_\lambda}{k_4}\delta_{\mu\nu} \tag{74}$$

$$= \frac{1}{i}D(x-x')\delta_{\mu\nu}, \tag{74a}$$

where $D(x-x')$ is the invariant function in (A16a). The above commutator relation is Lorentz invariant.

For $t = t'$, the above relation reduces to

$$[A_\mu(\mathbf{r}, t), A_\nu(\mathbf{r}', t)] = 0, \quad \text{by } (A21). \tag{74b}$$

Similarly[20]

$$\begin{aligned}
[A_j(x), \Pi_\ell(x')] &= \frac{i}{(2\pi)^3}\int d^3k \cos(k_\nu(x-x')_\nu)\\
&= i\partial_t D(x-x')\delta_{j\ell}, \quad j, \ell = 1, 2, 3,
\end{aligned} \tag{75}$$

$$\begin{aligned}
[A_j(x), \Pi_4(x')] &= [A_4(x), \Pi_j(x')]\\
&= -\partial_j D(x-x'),
\end{aligned} \tag{76}$$

$$[A_4(x), \Pi_4(x')] = i\partial_t D(x-x'), \tag{77}$$

$$[A_j(\mathbf{r}, t), \Pi_\ell(\mathbf{r}', t)] = i\delta(\mathbf{r}-\mathbf{r}')\delta_{j\ell}, \tag{75a}$$

[18] *The negative sign in the last of the relations in Eq. (73) comes from the anti-hermiticity of A_4, namely $A_4^\dagger = -A_4$, in Eq. (58). On account of this negative sign, one may regard $a_0^\dagger$, a_0 as the annihilation and creation operators respectively.*

[19] *In ordinary units, a factor $\frac{\hbar}{c}$ on the right in (74), (74a).*

[20] *In ordinary units, a factor $\hbar\epsilon_0$ on the right of (76), (75a), (77a); a factor $\frac{\hbar\epsilon_0}{c}$ on the right of (75), (77).*

$$[A_4(\mathbf{r},t), \Pi_4(\mathbf{r}',t)] = i\delta(\mathbf{r}-\mathbf{r}'). \tag{77a}$$

All the above results have been obtained without making use of the Lorentz condition Eq. (55),

$$\partial_\mu A_\mu = 0.$$

(2) Polarization of photons From[21]

$$\begin{aligned} &\mathbf{B} = \nabla \times \mathbf{A}, && \mathbf{E} = -(\partial_0 \mathbf{A} + \nabla A_0), \\ &\partial_0 = \frac{\partial}{\partial x_0} = \frac{\partial}{\partial t}, && \end{aligned} \tag{78}$$

and

$$\mathbf{B}(x) \equiv \mathbf{B}_+(x) + \mathbf{B}_-(x)$$

$$\mathbf{E}(x) \equiv \mathbf{E}_+(x) + \mathbf{E}_-(x), \tag{79}$$

one obtains from Eq. (70a)[22]

$$\begin{aligned} \mathbf{B}_+(x) &= i\sqrt{\frac{1}{2L^3}} \sum_{\mathbf{k}} \frac{1}{\sqrt{\omega_{\mathbf{k}}}} [\mathbf{k} \times \mathbf{a}(\mathbf{k})] e^{iR}, \\ \mathbf{B}_-(x) &= -i\sqrt{\frac{1}{2L^3}} \sum_{\mathbf{k}} \frac{1}{\sqrt{\omega_{\mathbf{k}}}} [\mathbf{k} \times \mathbf{a}^\dagger(\mathbf{k})] e^{-iR}, \end{aligned} \tag{80}$$

$$\begin{aligned} \mathbf{E}_+(x) &= i\sqrt{\frac{1}{2L^3}} \sum_{\mathbf{k}} \frac{1}{\sqrt{\omega_{\mathbf{k}}}} \{|\,\mathbf{k}\,|\,\mathbf{a}(\mathbf{k}) - \mathbf{k}\ \mathbf{a}_0(\mathbf{k})\} e^{iR}, \\ \mathbf{E}_-(x) &= -i\sqrt{\frac{1}{2L^3}} \sum_{\mathbf{k}} \frac{1}{\sqrt{\omega_{\mathbf{k}}}} \{|\,\mathbf{k}\,|\,\mathbf{a}^\dagger(\mathbf{k}) - \mathbf{k}\ \mathbf{a}_0^\dagger(\mathbf{k})\} e^{-iR}. \end{aligned} \tag{81}$$

Owing to the unusual sign for the last equation of (73), we must introduce the physical vacuum state $|\,0\rangle$, in a careful manner.
From Eqs. (80) and (81), it is seen that $|\,0\rangle$ may be defined by

$$\mathbf{B}_+(x)\,|\,0\rangle = 0, \tag{82}$$

[21]

$$\begin{aligned} \mathbf{E} &= -c(\partial_0 \mathbf{A} + \nabla A_0), \\ \partial_0 &= \frac{\partial}{\partial x_0} = \frac{1}{c}\frac{\partial}{\partial t}. \end{aligned} \tag{78$'$}$$

[22] *In ordinary units, a factor $c\sqrt{\hbar}$ on the right in (80); a factor $c^2\sqrt{\hbar}$ on the right in (81).*

$$\mathbf{E}_+(x) \mid 0\rangle = 0, \tag{83}$$

i.e., for any $\mathbf{k}$, one must have

$$[\mathbf{k} \times \mathbf{a}(\mathbf{k})] \mid 0\rangle = 0, \tag{84}$$

$$(\mid \mathbf{k} \mid \mathbf{a}(\mathbf{k}) - \mathbf{k}\ \mathbf{a}_0(\mathbf{k})) \mid 0\rangle = 0. \tag{85}$$

Let us choose a rectangular coordinate system with the x_3-axis along $\mathbf{k}$, and $a_1(\mathbf{k}), a_2(\mathbf{k})$ are transverse to $\mathbf{k}$, whereas $a_3(\mathbf{k})$ is "longitudinal" along $\mathbf{k}$. In this coordinate system, $k_1 = k_2 = 0$, and Eq. (84) gives for the transverse parts:

$$a_1(\mathbf{k}) \mid 0\rangle = 0, \qquad a_2(\mathbf{k}) \mid 0\rangle = 0 \tag{86}$$

and Eq. (85) gives for the longitudinal part

$$(a_3(\mathbf{k}) - a_0(\mathbf{k})) \mid 0\rangle = 0. \tag{87}$$

From Eq. (73), one finds

$$[a_3(\mathbf{k}) - a_0(\mathbf{k}),\ a_3^\dagger(\mathbf{k}) - a_0^\dagger(\mathbf{k})] = 0. \tag{88}$$

From Eqs. (87) and (88), it is seen that[23]

$$(a_3^\dagger(\mathbf{k}) - a_0^\dagger(\mathbf{k})) \mid 0\rangle = 0, \tag{89}$$

$$(a_3^\dagger(\mathbf{k}) - a_0^\dagger(\mathbf{k}))^n \mid 0\rangle = 0, \qquad n = 2, 3, ... \tag{90}$$

We introduce

$$\begin{aligned} a_1^\dagger(\mathbf{k}) \mid 0 > &= \mid 1, 0\rangle, \\ a_2^\dagger(\mathbf{k}) \mid 0 > &= \mid 0, 1\rangle, \end{aligned} \tag{91}$$

which represent states, respectively, with one photon with momentum $\mathbf{k}$ along x_1 and x_2 axis. One says that the photon is polarized along the x_1 or x_2 axis. The notation

$$\mid n, m\rangle = \frac{1}{\sqrt{n!m!}}((a_1^\dagger(\mathbf{k}))^n (a_2^\dagger(\mathbf{k}))^m \mid 0\rangle, \tag{92}$$

represents a state with n photons polarized along the x_1 - axis and m photons polarized along the x_2 - axis.

From

$$[a_1^\dagger,\ a_3 - a_0] = 0, \qquad [a_2^\dagger,\ a_3 - a_0] = 0,$$

---

[23] *Eq. (88) yields*

$$\langle 0 \mid (a_3 - a_0)(a_3^\dagger - a_0^\dagger) \mid 0\rangle = \langle 0 \mid (a_3^\dagger - a_0^\dagger)(a_3 - a_0) \mid 0\rangle = 0,$$

so that Eq. (89) follows as a consequence.

it follows that

$$[(a_1^\dagger)^n,\ a_3 - a_0] = [(a_2^\dagger)^m,\ a_3 - a_0] = 0,$$

and

$$[(a_1^\dagger)^n (a_2^\dagger)^m, a_3 - a_0] = 0,$$

and

$$(a_1^\dagger)^n (a_2^\dagger)^m (a_3 - a_0) = (a_3 - a_0)(a_1^\dagger)^n (a_2^\dagger)^m,$$

and

$$(a_1^\dagger)^n (a_2^\dagger)^m (a_3 - a_0) \mid 0\rangle = (a_3 - a_0)(a_1^\dagger)^n (a_2^\dagger)^m \mid 0\rangle.$$

By Eqs. (87) and (92), it is seen that

$$(a_3(\mathbf{k}) - a_0(\mathbf{k})) \mid n, m\rangle = 0, \tag{93}$$

for arbitrary state $\mid n, m\rangle$. From the definition of $A_{\mu+}(x)$ in Eqs. (70) and (71), it is seen that Eq. (93) is equivalent to

$$\partial_\mu A_{\mu+}(x) \mid n, m\rangle = 0, \tag{94}$$

for any state $\mid n, m\rangle$. This is a condition on the operators $A_{\mu+}$ of the quantized electromagnetic field. It is called the "subsidiary condition", and it takes the place of the Lorentz condition (55)

$$\partial_\mu A_\mu = 0, \tag{55}$$

in the classical theory.[24]

From Eqs. (50) and (51),

$$\begin{aligned} \partial_t A_\mu(x) &= -i[A_\mu(x), H], \\ \nabla A_\mu(x) &= -i[\mathbf{P}, A_\mu(x)], \end{aligned} \tag{95}$$

and Eqs. (70) and (71) for $A_\mu(x)$, one obtains, as Eqs. (23) and (27),

$$\begin{aligned} [\mathbf{a}(\mathbf{k}), H] &= E_{\mathbf{k}}\mathbf{a}(\mathbf{k}), & [\mathbf{a}^\dagger(\mathbf{k}), H] &= -E_{\mathbf{k}}\mathbf{a}^\dagger(\mathbf{k}), \\ [a_j(\mathbf{k}), \mathbf{P}] &= \mathbf{k}a_j(\mathbf{k}), & [a_j^\dagger(\mathbf{k}), \mathbf{P}] &= -\mathbf{k}a_j^\dagger(\mathbf{k}). \end{aligned} \tag{96}$$

The vacuum state $\mid 0\rangle$ satisfies

$$H \mid 0\rangle = 0, \qquad \mathbf{P} \mid 0\rangle = 0. \tag{97}$$

[24] *Following the step leading to Eq. (89), we see that*

$$(a_3^\dagger(\mathbf{k}) - a_0^\dagger(\mathbf{k})) \mid n, m\rangle = 0. \tag{93a}$$

The contributions due to time-like photons are always cancelled by those due to longitudinal photons.

Let the occupation numbers for states with polarization along x_1 and x_2 axis be N_1 and N_2.

$$N_1(\mathbf{k}) = a_1^\dagger(\mathbf{k})a_1(\mathbf{k}), \qquad N_2(\mathbf{k}) = a_2^\dagger(\mathbf{k})a_2(\mathbf{k}). \tag{98}$$

Then, analogously to Eq. (32), one has

$$\begin{aligned} H &= \sum_{\mathbf{k}} E_{\mathbf{k}}(N_1(\mathbf{k}) + N_2(\mathbf{k})), \\ \mathbf{P} &= \sum_{\mathbf{k}} \mathbf{k}(N_1(\mathbf{k}) + N_2(\mathbf{k})). \end{aligned} \tag{99}$$

Only a_1, $a_1^\dagger$, a_2, $a_2^\dagger$, the "transverse" operators, appear in H and P; there are no "longitudinal" photons.

As the time-like and longitudinal photons are included, H and P contain terms proportional to

$$N_3(\mathbf{k}) - N_0(\mathbf{k}) = a_3^\dagger(\mathbf{k})a_3(\mathbf{k}) - a_0^\dagger(\mathbf{k})a_0(\mathbf{k}).$$

The expectation value is, from Eq. (93),

$$\begin{aligned} &\langle n, m \,|[a_3^\dagger a_3 - a_0^\dagger a_0] \mid n, m\rangle \\ &= \langle n, m \mid [a_3^\dagger - a_0^\dagger]a_3 \mid n, m\rangle \\ &= 0, \end{aligned} \tag{100}$$

indicating that the expectation value involves only transverse photons. In other words, the physically allowed states may contain certain number of time-like and longitudinal photons but effects due to these two types of photons cancel each other. A gauge transformation changes the admixture of these "unphysical" photons such that cancellation always holds.

(3) Gauge invariance The condition for the Lorentz condition $\partial_\mu A_\mu = 0$ for the gauge transformation

$$A_\mu \;\rightarrow\; A_\mu + \partial_\mu \chi(x), \tag{101a}$$

is

$$\partial_\nu \partial_\nu \chi(x) = 0, \tag{101b}$$

where $\chi(x)$ is a real, scalar function $\chi(x_\mu)$, i.e., $\chi(x)$ is a hermitian operator. Let the solution $\chi(x)$ of $\partial_\nu\partial_\nu\chi(x) = 0$ be Fourier analyzed as the $A_\mu(x)$ in Eq. (66), and, with the relation (20)

$$\mathbf{a}(-\mathbf{k}) = \mathbf{a}^\dagger(\mathbf{k}),$$

and the hermiticity of χ, similar to Eqs. (7) and (9) ,

$$\chi(-\mathbf{k}) = \chi^\dagger(\mathbf{k}), \tag{102}$$

one obtains the following transformation relations of the $a(\mathbf{k})$, $a^\dagger(\mathbf{k})$, $a_0(\mathbf{k})$, $a_0^\dagger(\mathbf{k})$ operators

$$\begin{aligned}
\mathbf{a}(\mathbf{k}) &\rightarrow \mathbf{a}(\mathbf{k}) + i\mathbf{k}\chi(\mathbf{k}),\\
\mathbf{a}^\dagger(\mathbf{k}) &\rightarrow \mathbf{a}^\dagger(\mathbf{k}) - i\mathbf{k}\chi^\dagger(\mathbf{k}),\\
a_0(\mathbf{k}) &\rightarrow a_0(\mathbf{k}) + i \mid \mathbf{k} \mid \chi(\mathbf{k}),\\
a_0^\dagger(\mathbf{k}) &\rightarrow a_0^\dagger(\mathbf{k}) - i \mid \mathbf{k} \mid \chi^\dagger(\mathbf{k}).
\end{aligned} \tag{103}$$

From these relations and $k_1 = k_2 = 0$ (by the choice of the coordinate system in Eq. (86), it is seen that

$$\begin{aligned}
a_1(\mathbf{k}) &\rightarrow a_1(\mathbf{k}), & a_2(\mathbf{k}) &\rightarrow a_2(\mathbf{k}),\\
a_1^\dagger(\mathbf{k}) &\rightarrow a_1^\dagger(\mathbf{k}), & a_2^\dagger(\mathbf{k}) &\rightarrow a_2^\dagger(\mathbf{k}),
\end{aligned} \tag{104}$$

i.e., the "transverse" operators are invariant under gauge transformation Eq. (101), and, since $k_3 =\mid \mathbf{k} \mid$,

$$a_3(\mathbf{k}) - a_0(\mathbf{k}) \ \text{ is also invariant.} \tag{105}$$

This, on account of Eq. (87)

$$(a_3(\mathbf{k}) - a_0(\mathbf{k})) \mid 0\rangle = 0, \tag{106}$$

means that a gauge transformation Eq. (101) does not affect the vacuum state and energy and momentum of the field Eq. (99).

In a formal way, the above theory of the quantized free electromagnetic field is successful, by the creation and annihilation operators, in bringing out the photon (particle) aspect

$$E_{\mathbf{k}} = \omega_{\mathbf{k}}, \qquad \mathbf{p}_{\mathbf{k}} = \mathbf{k} \tag{107}$$

which are the Einstein relations.

There is a troublesome point in the theory, namely, each degree of freedom has a zero point energy $\frac{1}{2}\omega_{\mathbf{k}}$, and since a field has an infinite number of degrees of freedom, the zero point energy of the field is also infinite.

8.4 Dirac Electron-Positron Field[25]

As mentioned in Chapter 3, Sect. 4, in the Majorana representation the 4-component Dirac wave function can be separated into two pairs of mutually complex-conjugated functions.

$$\begin{pmatrix} \Psi^{(1)} \\ \Psi^{(2)} \end{pmatrix} \qquad \text{and} \qquad \begin{pmatrix} \Psi^{(3)} = \Psi^{*(1)} \\ \Psi^{(4)} = \Psi^{*(2)} \end{pmatrix} \tag{108}$$

[25] *See Exercise 8.1. on quantization of the Dirac particle in the Pauli-Dirac repsentation.*

For a free electron field, the solutions are plans waves

$$\Psi^{(s)}(x) = u_s e^{ikx}, \qquad s = 1, 2,$$
$$\Psi^{(t)}(x) = u_t e^{-ikx}, \qquad t = 3, 4, \tag{109}$$

$$\Psi^{(3)} = \Psi^{*(1)}, \qquad \Psi^{(4)} = \Psi^{*(2)}. \tag{110}$$

The u_s, u_t and their normalization have been given in Eqs. (39, 40a, b, c, d), Ch. 3.

On expanding $\Psi(x)$, $\Psi^*(x)$ in the complete set of plane waves Eqs. (88), (89), Ch. 3, one has

$$\Psi(x) = \frac{1}{\sqrt{L^3}} \sum_{\mathbf{k}} \{\sum_s a_s u_s e^{iR} + \sum_t b_t u_t e^{-iR}\},$$
$$\Psi^*(x) = \frac{1}{\sqrt{L^3}} \sum_{\mathbf{k}} \{\sum_s a_s^* u_s^* e^{-iR} + \sum_t b_t^* u_t^* e^{iR}\}, \tag{111}$$

where a_s, a_s^*, a_t, a_t^* are all complex numbers (i.e., not operators), and the summation of s is from 1 to 2; of t from 3 to 4, and

$$R \equiv k_\mu x_\mu = \mathbf{k} \cdot \mathbf{r} - \omega_{\mathbf{k}} t. \tag{112}$$

The energy-momentum relation is

$$k_\mu^2 = k_\mu k_\mu = \mathbf{k}^2 - k_0^2 = -m_0^2$$

$$\pm k_0 = E_{\mathbf{k}} = \omega_{\mathbf{k}} \tag{113}$$

$$E_{\mathbf{k}} = \sqrt{\mathbf{k}^2 + m_0^2},$$

and

$$u_s^\dagger u_s = 1, \qquad u_t^\dagger u_t = 1. \tag{114}$$

Note that the indices $s = 1, 2$, $t = 3, 4$ do *not* refer to the spinor components, but to the 4-component u. Thus $u_s, s = 1$, is a column 4-component spinor, $u_s^\dagger$ a row 4-component spinor, etc. In Majorana's representation (110), $u_1, u_2, u_3 = u_1^*$, $u_4 = u_2^*$ are four column 4-component spinors.

To quantize the field Ψ, Ψ^*, the first step is to replace the functions Ψ, Ψ^* by non-hermitian operators $\Psi(x), \Psi^\dagger(x)$, and the a_s, a_s^*, b_t, b_t^* by annihilation and creation operators a_s, $a_s^\dagger$, b_t, $b_t^\dagger$

$$\Psi(x) = \frac{1}{\sqrt{L^3}} \sum_{\mathbf{k}} \sum_s \{a_s(\mathbf{k}) u_s(\mathbf{k}) e^{iR} + b_s^\dagger(\mathbf{k}) u_s^*(\mathbf{k}) e^{-iR}\}$$
$$\Psi^\dagger(x) = \frac{1}{\sqrt{L^3}} \sum_{\mathbf{k}} \sum_s \{a_s^\dagger(\mathbf{k}) u_s^\dagger(\mathbf{k}) e^{-iR} + b_s(\mathbf{k}) \tilde{u}_s(\mathbf{k}) e^{iR}\}. \tag{115}$$

where u_s, u_s^* are column matrices, $u_s^\dagger \equiv \tilde{u}_s^*$, $\tilde{u}_s$ are row matrices. The summation over s is from 1 to 2.

The meaning of Eq. (115) is as follows:

$a_s(\mathbf{k})$, $a_s^\dagger(\mathbf{k})$ are the annihilation and creation operators respectively of electron in the state of charge $-e$, momentum $\mathbf{p}$ and energy $E_{\mathbf{k}}$.

$b_s(\mathbf{k})$, $b_s^\dagger(\mathbf{k})$ are the annihilation and creation operators respectively of positron ($+e$, $\mathbf{p}$, $E_{\mathbf{k}}$ state).

These operators obey the anti-commutation relations,

$$\begin{aligned}
\{a_s(\mathbf{k}), a_{s'}(\mathbf{k}')\}_+ &= \{b_s(\mathbf{k}), b_{s'}(\mathbf{k}')\}_+ = 0,\\
\{a_s^\dagger(\mathbf{k}), a_{s'}^\dagger(\mathbf{k}')\}_+ &= \{b_s^\dagger(\mathbf{k}), b_{s'}^\dagger(\mathbf{k}')\}_+ = 0,\\
\{a_s(\mathbf{k}), b_{s'}^\dagger(\mathbf{k}')\}_+ &= \{a_s^\dagger(\mathbf{k}), b_{s'}(\mathbf{k}')\}_+ = 0,\\
\{a_s(\mathbf{k}), b_{s'}(\mathbf{k}')\}_+ &= \{a_s^\dagger(\mathbf{k}), b_{s'}^\dagger(\mathbf{k}')\}_+ = 0,\\
\{a_s(\mathbf{k}), a_{s'}^\dagger(\mathbf{k}')\}_+ &= \{b_s(\mathbf{k}), b_{s'}^\dagger(\mathbf{k}')\}_+ = \delta_{ss'}\delta_{\mathbf{k}\mathbf{k}'},
\end{aligned} \tag{116}$$

where

$$\{A, B\}_+ = AB + BA. \tag{117}$$

We introduce the electron and positron occupation number $N_{s-}(\mathbf{k}), N_{s+}(\mathbf{k})$

$$N_{s-}(\mathbf{k}) = a_s^\dagger(\mathbf{k})a_s(\mathbf{k}), \qquad N_{s+}(\mathbf{k}) = b_s^\dagger(\mathbf{k})b_s(\mathbf{k}) \tag{118}$$

The vacuum $\mid 0\rangle$ is such that

$$a_s(\mathbf{k}) \mid 0\rangle = 0, \qquad b_s(\mathbf{k}) \mid 0\rangle = 0, \tag{119}$$

and therefore

$$N_{s-}(\mathbf{k}) \mid 0\rangle = 0, \qquad N_{s+}(\mathbf{k}) \mid 0\rangle = 0. \tag{120}$$

From Eq. (116), one obtains the commutation relations

$$\begin{aligned}
[N_{s\pm}(\mathbf{k}), N_{s'\mp}(\mathbf{k}')]_- &= 0,\\
[N_{s-}(\mathbf{k}), a_{s'}(\mathbf{k}')]_- &= -\delta_{ss'}\delta_{\mathbf{k}\mathbf{k}'}a_s(\mathbf{k}),\\
[N_{s+}(\mathbf{k}), b_{s'}(\mathbf{k}')]_- &= -\delta_{ss'}\delta_{\mathbf{k}\mathbf{k}'}b_s(\mathbf{k}),\\
[N_{s-}(\mathbf{k}), a_{s'}^\dagger(\mathbf{k}')]_- &= \delta_{ss'}\delta_{\mathbf{k}\mathbf{k}'}a_s^\dagger(\mathbf{k}),\\
[N_{s+}(\mathbf{k}), b_{s'}^\dagger(\mathbf{k}')]_- &= \delta_{ss'}\delta_{\mathbf{k}\mathbf{k}'}b_s^\dagger(\mathbf{k}).
\end{aligned} \tag{121}$$

If in the equation (115) for $\Psi(x)$, $\Psi^+(x)$ one makes the interchange

$$\begin{aligned}
a_s(\mathbf{k}) &\longleftrightarrow b_s(\mathbf{k}), & a_s^\dagger(\mathbf{k}) &\longleftrightarrow b_s^\dagger(\mathbf{k}),\\
u_s(\mathbf{k}) &\longleftrightarrow u_s^\dagger(\mathbf{k}), & (i.e., u_s^*(\mathbf{k}) &\longleftrightarrow \tilde{u}_s(\mathbf{k})),
\end{aligned} \tag{122}$$

then in Eq. (115),

$$\begin{Bmatrix} \Psi(x) \\ \Psi^\dagger(x) \end{Bmatrix} \quad \text{goes into} \quad \begin{Bmatrix} \Psi^\dagger(x) \\ \Psi(x) \end{Bmatrix}, \tag{123}$$

i.e., the electron goes into the positron and vice versa. This is the charge conjugation transformation. In a free electron field (i.e., no electromagnetic interaction), the quantized field theory is symmetrical in the electron-positron interchange.

From Chapter 6, Eqs. (115)-(118) for the Hamiltonian, the 4-current and the total electric charge, one has, on using the $\Psi, \Psi^\dagger$ of Eq. (115), and the anticommutation relations (116),

$$H = \sum_{\mathbf{k}} \sum_{s=1}^{2} E_s[a_s^\dagger(\mathbf{k})a_s(\mathbf{k}) + b_s^\dagger(\mathbf{k})b_s(\mathbf{k})], \tag{124a}$$

$$Q = -e \sum_{\mathbf{k}} \sum_{s=1}^{2} [a_s^\dagger(\mathbf{k})a_s(\mathbf{k}) - b_s^\dagger(\mathbf{k})b_s(\mathbf{k})], \tag{125a}$$

and if $| \, 0 >$ is the vacuum state of Eqs. (119) and (120),

$$H = \sum_{\mathbf{k}} \sum_{s=1}^{2} E_{\mathbf{k}}(N_{s-}(\mathbf{k}) + N_{s+}(\mathbf{k})), \tag{124b}$$

$$Q = -e \sum_{\mathbf{k}} \sum_{s=1}^{2} (N_{s-}(\mathbf{k}) - N_{s+}(\mathbf{k})). \tag{125b}$$

From this and the preceding three sections on the Klein-Gordon, electromagnetic field, one summarizes the following result:

For complex conjugate field [Klein-Gordon complex $\Psi(x)$, $\Psi^*(x)$, Dirac $\Psi(x)$, $\Psi^*(x)$], the non-hermitian operators $\Psi(x)$, $\Psi^\dagger(x)$ correspond, respectively, to particle and anti-particle (π^+ and π^- for the Klein-Gordon field; the electron-positron for the Dirac field).

For real field [Klein-Gordon real, scalar field $\phi(x)$ and the electromagnetic field $A_\mu(x)$], the operators $\Psi(x), \Psi^\dagger(x)$ are hermitian, and the antiparticle of the field is the particle itself (π^0 of the Klein-Gordon field and the photon of the electromagnetic field).

8.5 Dirac's Theory of Emission and Absorption of Radiation

In the preceding sections of the present chapter, we have treated a few examples of the quantization of 'pure' fields, for example, electromagnetic fields in regions of space free of electric charges and currents.

We now proceed to consider the following physical situation: an atom with one electron, in an electromagnetic field.

In the Schrödinger theory, the system is described by the equation[26]

$$i\hbar\frac{\partial\Psi}{\partial t} = (H_0 + H_1)\Psi, \tag{126}$$

where

$$H_0 = -\frac{\hbar}{2m}\nabla^2 + V(r) + H_f$$

$$H_1 = -\frac{e}{mc}(\mathbf{A}\cdot\mathbf{p}). \tag{127}$$

the first two terms in H_0 are the kinetic and potential energy of the electron in the atom; H_f is the Hamiltonian of the free electromagnetic field. H_1 is the interaction between the electron and the field.

From Eq. (32), one has

$$H_f = \sum_{k\lambda}\hbar\omega_{\mathbf{k}}a^\dagger_\lambda(\mathbf{k})a_\lambda(\mathbf{k}), \tag{128}$$

where λ is to denote the state of polarization of the photon. The vector potential A is, from Eq. (70a)

$$\mathbf{A}(x_\mu) = \sum_{k,\lambda}c\sqrt{\frac{2\pi h}{L^3\omega_{\mathbf{k}}}}\mathbf{S}_\lambda(\mathbf{k})[a_\lambda(\mathbf{k})e^{iR} + a^\dagger_\lambda(\mathbf{k})e^{-iR}], \tag{129}$$

$$R = \mathbf{k}\cdot\mathbf{r} - \omega_{\mathbf{k}}t,$$

where $\mathbf{S}_\lambda(\mathbf{k})$ is a unit vector with $\lambda = 1, 2$ for the two directions of polarizations as discussed in Eq. (70a).[27]

For the system H_0, the solution of the Schrödinger is

$$\Phi_j = v_j(r)\psi_j(x_\mu), \tag{130}$$

where

$$(-\frac{\hbar}{2m}\nabla^2 + V(r) - \epsilon_j)v_j(r) = 0, \tag{131}$$

and $\psi_j(x_\mu)$ is the state function of the electromagnetic field which in the occupation number representation is

$$\begin{aligned} a^\dagger_\lambda(\mathbf{k})\psi_{\ldots n_{k\lambda}\ldots} &= \sqrt{n_{k\lambda}+1}\,\psi_{\ldots n_{k\lambda}+1\ldots}, \\ a_\lambda(\mathbf{k})\psi_{\ldots n_{k\lambda}\ldots} &= \sqrt{n_{k\lambda}}\,\psi_{\ldots n_{k\lambda}-1\ldots}. \end{aligned} \tag{132}$$

---

[26] *The interaction* $-\frac{e}{mc}(\mathbf{A}\cdot\mathbf{p})$ *with the electromagnetic field represented by a vector potential* $\mathbf{A}$ *comes from Eq. (16) of Chapter 5. For this section, we shall not impose the* $\hbar = c = 1$ *unit system.*

[27] *This expression for* $A(x)$ *differs from that in Eq. (70a) by a factor* $\sqrt{4\pi}$ *for a change to the e.s.u system.*

$n_{k\lambda}$ being the number of photons of momentum $\hbar\mathbf{k}$ and polarization $\lambda\,(\lambda = 1, 2)$.

The interaction H_1 is

$$H_1 = -\frac{e}{m}\sqrt{\frac{2\pi\hbar}{L^3}}\sum_{k\lambda}\frac{1}{\sqrt{\omega_{\mathbf{k}}}}(\mathbf{S}_\lambda\cdot\mathbf{p})\{a_\lambda(\mathbf{k})e^{iR} + a_\lambda^\dagger(\mathbf{k})e^{-iR}\}. \tag{133}$$

On account of this interaction, the Ψ in Eq. (126) is not a stationary state but is time dependent. Let the Ψ be expanded in the complete set of $v_j\psi_j$,

$$\Psi = \sum_j c_j(t)v_j\psi_j\ \exp(-\frac{i}{\hbar}\epsilon_j t). \tag{134}$$

From Eq. (126), one obtains

$$\frac{dc_\ell}{dt} = -\frac{i}{\hbar}\sum_j(\Phi_\ell, H_1\Phi_j)c_j(t)\ \exp(i\omega_{\ell j}t), \tag{135}$$

$$\hbar\omega_{\ell j} = \epsilon_\ell - \epsilon_j. \tag{136}$$

To integrate the equation for c_ℓ, let the initial state be an atom in a state v_u and the field with $n_{k\lambda}$ photons in state $\mathbf{k}, \lambda$,

$$\Phi = v_u\psi_{\ldots n_{k\lambda}\ldots}, \tag{137}$$

so that[28]

$$c_u(t=0) = 1, \qquad c_j(t=0) = 0, \qquad j \neq u. \tag{138}$$

Then from Eqs. (133) and (132),

$$(\Phi_\ell, H_1\Phi_u) = \frac{e}{m}\sqrt{\frac{2\pi\hbar}{L^3}}\sum_{k\lambda}\sqrt{\frac{n_{\mathbf{k}}+1}{\omega_{\mathbf{k}}}}(v_\ell, (\mathbf{S}_\lambda\cdot\mathbf{p})e^{-i\mathbf{k}\cdot\mathbf{r}}v_u)e^{i\omega_{\mathbf{k}}t}. \tag{139}$$

The probability that the system (atom + radiation field) has gone to state v_ℓ in time t is

$$|\ c_\ell(t)\ |^2$$

provided t is not too long and $|\ c_\ell(t)\ |^2$ is still very small compared with 1.

$$|\ c_\ell(t)\ |^2 = \frac{1}{\hbar L^3}(\frac{2\pi e}{m})^2\sum_{k\lambda}\frac{n_{\mathbf{k}}+1}{\omega_{\mathbf{k}}}\ |\ (v_\ell, (\mathbf{S}_\lambda\cdot\mathbf{p})e^{-i\mathbf{k}\cdot\mathbf{r}}v_n)\ |^2\ \frac{\sin^2\xi}{\xi^2}(\frac{t^2}{2\pi}), \tag{140}$$

where

$$\xi = \frac{\omega_{\ell u} + \omega_{\mathbf{k}}}{2}t.$$

[28] *The perturbation theory method is that of Chapter 7, Sect. 2(1), Vol. I ("Quantum Mechanics"). See (VII-32) - (VII-36), etc., in Vol. I.*

Consider the situation when the initial state u is such that the atom is in an excited state υ_u , and υ_ℓ is a lower state,

$$\epsilon_u - \epsilon_\ell = \hbar\omega_{u\ell} > 0, \tag{141}$$

so that

$$\xi = (\frac{\omega_{\mathbf{k}} - \omega_{u\ell}}{2})t. \tag{142}$$

We are then seeking the probability that an atom in an excited state υ_u makes a transition to a lower state υ_ℓ increasing a photon $\hbar\omega_{\mathbf{k}}$ in the field.

The function $\frac{\sin^2\xi}{\xi^2}$ has a very sharp maximum at $\xi = 0$, and

$$\frac{1}{\pi}\int_{-\infty}^{\infty}\frac{\sin^2\xi}{\xi^2}d\xi = 1.$$

We shall replace

$$\frac{1}{\pi}\frac{\sin^2\xi}{\xi^2} \quad \text{by} \quad \delta(\xi) = \frac{2}{t}\delta(\omega_{\mathbf{k}} - \omega_{u\ell}). \tag{143}$$

To evaluate the matrix element in Eq. (140), we make the dipole approximation (as in (VII-47)-(VII-50), Vol. I, for $kr = \frac{2\pi r}{\lambda} \ll 1$, i.e., for wavelengths long compared with atomic size) and obtain (by using (VII-50), Vol. I.)

$$\begin{aligned}\mathbf{S}_\lambda \cdot (\upsilon_\ell, \mathbf{p}e^{-ik\cdot r}\upsilon_u) &\simeq \mathbf{S}_\lambda \cdot (\upsilon_\ell, \mathbf{p}\upsilon_u)\\ &= im\omega_{\ell u}(\mathbf{S}_\lambda \cdot \mathbf{r}_{\ell u}),\end{aligned} \tag{144}$$

where

$$\mathbf{r}_{\ell u} = \int \upsilon_\ell^* \mathbf{r} \upsilon_u d^3x. \tag{145}$$

The polarization factor $\mathbf{S}_\lambda \cdot \mathbf{r}$ can be evaluated by choosing the coordinate system of Eq. (86), namely, with the x_3 -axis along $\mathbf{k}$.
The $\mathbf{S}_1, \mathbf{S}_2$ are unit vectors of polarization transverse to $\mathbf{k}$. We shall now take $\mathbf{S}_1$ to be in the plane of $\mathbf{k}$ and $\mathbf{r}$, and $\mathbf{S}_2$ normal to this plane. If ϑ is the angle between $\mathbf{k}$ and $\mathbf{r}$, then

$$(\mathbf{S}_1 \cdot \mathbf{r}_{\ell u}) = \sin\vartheta \mathbf{r}_{\ell u}, \qquad (\mathbf{S}_2 \cdot \mathbf{r}_{\ell u}) = 0. \tag{146}$$

The summation over the discrete $\mathbf{k}$ in Eq. (140) is again replaced, for a large volume L^3, by an integral

$$\begin{aligned}\sum_{\mathbf{k}} &\to (\frac{L}{2\pi})^3 \int d^3k\\ &= \frac{1}{c^3}(\frac{L}{2\pi})^3 \int d\Omega \omega_{\mathbf{k}}^2 d\omega_{\mathbf{k}},\end{aligned}$$

where $d\Omega$ is the element of solid angle for d^3k. Thus

$$| c_\ell(t) |^2 = t\frac{e^2}{2\pi\hbar c}\int d\Omega \int d\omega_{\mathbf{k}}(n_{\mathbf{k}}+1)\omega_{\mathbf{k}}^3 \sin^2\theta \mid \langle \ell \mid \mathbf{r} \mid u\rangle \mid^2 \delta(\omega_{\mathbf{k}} - \omega_{u\ell}). \tag{147}$$

(1) Spontaneous emission coefficient $A(u \to \ell)$

There are two terms in Eq. (147). Consider the term not containing $n_{\mathbf{k}}$, i.e., independent of the presence or absence of photons $n_{\mathbf{k}}$ in the field. Upon integrating over $\omega_{\mathbf{k}}$ and $\sin^2\vartheta d\cos\vartheta d\varphi$ (directions of photon $\mathbf{k}$), one obtains for the probability per unit time $\frac{1}{t} \mid c_\ell(t) \mid^2$,

$$P = \frac{1}{t} \mid c_\ell(t) \mid^2 = \frac{4\omega_{u\ell}^3}{3\hbar c^3} \mid \langle \ell \mid e\mathbf{r} \mid u\rangle \mid^2 . \tag{148}$$

This transition probability, being independent of the $n_{\mathbf{k}}(E_{\mathbf{k}} = \hbar\omega_{\mathbf{k}})$ of the electromagnetic field, can be identified with Einstein's spontaneous emission coefficient $A(u \to \ell)$. In fact, the expression in Eq. (148) agrees with the expression (VII-54), Vol. I, for $A(u \to \ell)$,

$$A(u \to \ell) = \frac{64\pi^4\nu^3}{3hc^3} \mid \langle \ell \mid e\mathbf{r} \mid u\rangle \mid^2, \tag{149}$$

which is there obtained indirectly from the Einstein relation

$$A(u \to \ell) = \frac{8\pi h\nu^3}{c^3} B(u \leftarrow \ell) \tag{150}$$

and the absorption coefficient $B(u \leftarrow \ell)$ is obtained from the perturbation theory.

(2) Induced emission and absorption coefficient, $B(u \to \ell)$ and $B(u \leftarrow \ell)$

Consider the term containing $n_{\mathbf{k}}$ in the integral in Eq. (147). This term, being proportional to the number $n_{\mathbf{k}}$ of quanta of energy $\hbar\omega_{\mathbf{k}}$, can be identified with the induced emission process of Einstein. Let $\rho(\nu)d\nu$ be the energy density of the radiation field having frequendy between ν and $\nu + d\nu$. The $n_{\mathbf{k}}$ in Eq. (147) or (140), is

$$n_{\mathbf{k}} = \frac{L^3\rho(\nu_{\mathbf{k}})d\nu_{\mathbf{k}}}{2h\omega_{\mathbf{k}}}, \tag{151}$$

the factor 2 in the denominator coming from the fact that in an unpolarized field of radiation, half the energy is associated with each direction of polarization. Hence from Eqs. (140), (141), and (146), one obtains

$$\begin{aligned} P = \frac{1}{t} \mid c_\ell(t) \mid^2 &= \frac{2\pi}{3\hbar^2} \mid \langle \ell \mid e\mathbf{r} \mid u\rangle \mid^2 \rho(\nu) \\ &\equiv B(u \to \ell)\rho(\nu), \end{aligned} \tag{152}$$

which gives indeed the Einstein $B(u \to \ell)$ coefficient for induced emission (or, negative absorption) as obtained in (VII-53), Vol. I, and in Chapter 1, Sect. 6, Vol. I.

For absorption, we take the second equation in Eq. (132), i.e., $a_\lambda(\mathbf{k})\psi$. This leads to $n_{\mathbf{k}}$ replacing $n_{\mathbf{k}} + 1$ in Eqs. (132), (138), and (147). All the steps of calculation above are the same. The result is, for the absorption of one photon $\hbar\omega_{\mathbf{k}} = \hbar\omega_{u\ell}$ by the atom, the probability per second is

$$\begin{aligned} P(u \leftarrow \ell) &= \frac{2}{3\hbar^2} \mid \langle \ell \mid e\mathbf{r} \mid u \rangle \mid^2 \rho(\nu) \\ &\equiv B(u \leftarrow \ell)\rho(\nu), \end{aligned} \tag{153}$$

so that

$$B(u \leftarrow \ell) = B(u \rightarrow \ell),$$

as obtained by Einstein from the equilibrium radiation energy distribution (1917).[29]

The above theory, of treating the emission and absorption of radiation by the method of quantized field, is due to Dirac (1927). This is the beginning of the theory of quantized field, which is extended immediately by P. Jordan, O. Klein, E. Wigner, W. Pauli in 1928.

[29] *A Einstein, Physik Zeit.* **18**, *121 (1917); Mitt. d. Phys. Ges. Zurich,* **18** *(1916); Verhandlungen der Deutschen Physik Gesells* **18**, *318 (1916).*

Appendix

Green's Functions $\triangle$ and D

We shall obtain here the Green's functions of the equations

$$\partial_\nu \partial_\nu A_\mu(x) = 0, \qquad x = (x_1, x_2, x_3, x_4),$$
$$(\partial_\nu \partial_\nu - \mu^2)\phi(x) = 0, \qquad x_4 = ix_0 = ict,$$

which are the classical equations, respectively, of the 4-potentials of electromagnetic field and the scalar field of the Klein-Gordon equation. These equations are in the relativistic form, and their Green's functions are Lorentz invariant. The Green's functions appear in the commutation relations when the fields are quantized.

We shall summarize our notations:

$$\begin{aligned} x &= (x_1, x_2, x_3, x_4) = (\mathbf{r}, ict = ix_0), \\ p &= (p_1, p_2, p_3, p_4) = (\mathbf{p}, p_4 = \frac{iE}{c} \equiv ip_0), \\ p^2 &= p_\nu^2 = \mathbf{p}^2 + p_4^2 = \mathbf{p}^2 - \frac{E^2}{c^2} = -m_0^2 c^2, \end{aligned} \tag{A1a}$$

$$\mathbf{k} = \frac{1}{\hbar}\mathbf{p}, \qquad k_4 = ik_0, \tag{A1b}$$

$$\mu = \frac{m_0 c}{\hbar}, \quad (\frac{1}{\mu} = \frac{1}{2\pi}\lambda_c, \quad \lambda_c = \text{Compton wave length})$$

Hence

$$\begin{aligned} k^2 &= k_\nu^2 = \mathbf{k}^2 + k_4^2 \\ &= \mathbf{k}^2 - k_0^2 = -\mu^2 \\ E_{\mathbf{k}} &= c\sqrt{\mathbf{p}^2 + m_0^2 c^2} \\ &= \hbar c\sqrt{\mathbf{k}^2 + \mu^2} \end{aligned} \tag{A1c}$$

In the following, we shall use the "natural units" in which

$$\hbar = c = 1.$$

Thus

$$\begin{aligned} k_\nu^2 + \mu^2 &= \mathbf{k}^2 + \mu^2 - k_0^2 \\ &= E_{\mathbf{k}}^2 - k_0^2 \end{aligned} \tag{A1d}$$

$k_\nu^2 + \mu^2 = 0$ is a Lorentz invariant. Thus the Dirac δ function $\delta(k_\nu^2 + \mu^2)$ is a Lorentz invariant operator. It is expressible in the form[30]

$$\delta(k^2 + \mu^2) = \frac{1}{2E_{\mathbf{k}}}(\delta(k_0 + E_{\mathbf{k}}) + \delta(k_0 - E_{\mathbf{k}})) \tag{A2}$$

We define a step function $\epsilon(k_0)$ of the sign of k_0

$$E(k_0) = \begin{cases} 1, & k_0 > 0 \\ -1, & k_0 < 0 \end{cases} \tag{A3}$$

$k_4\,(ik_0)$ is the time-component of the 4-vector k. Under proper Lorentz transformations inside the light cone

$$r^2 - c^2t^2 < 0,$$

the sign of t, and of k_0, does not change. Hence $\epsilon(k_0)$ is a Lorentz invariant function.

From (A2) and (A3), we have, since $E_{\mathbf{k}}$ is always positive,

$$\epsilon(k_0)\delta(k_\nu^2 + \mu^2) = \frac{1}{2E_{\mathbf{k}}}\{\delta(k_0 - E_{\mathbf{k}}) - \delta(k_0 + E_{\mathbf{k}})\}. \tag{A4}$$

(1) $\triangle(x)$ function

Let us define the Lorentz invariant $\triangle(x)$ function[31]

$$\triangle(x) = \frac{i}{(2\pi)^3}\int d^4k e^{ik_\nu x_\nu}\epsilon(k_0)\delta(k_\nu^2 + \mu^2), \tag{A5}$$

where

$$d^4k = d^3k dk_0, \qquad (ik_0 = k_4), \tag{A5a}$$

---

[30] *From*

$$\delta(ax) = \frac{\delta(x)}{\mid a \mid}$$

one sees

$$\int_{-\infty}^{\infty} f(x)\delta[(x-a)(x-b)]dx = \frac{1}{\mid a-b \mid}[f(a) + f(b)]$$

Alse

$$\int_{-\infty}^{\infty} f(x)\frac{\delta(x-a) + \delta(x-b)}{\mid a-b \mid}dx = \frac{1}{\mid a-b \mid}[f(a) + f(b)].$$

Hence (A2).

[31] *The $\triangle(x)$ function can also be defined by the integral*

$$\triangle(x) = \frac{1}{(2\pi)^4}\int_C d^4k \frac{e^{ik_\nu x_\nu}}{k_\nu^2 + \mu^2}. \tag{A7}$$

where the contour C is a large circle about $k_0 = 0$ on the complex k_0 plane and two small circles, all in the counter clockwise sense, about the two poles $k_0 = +E_{\mathbf{k}}$ of $k_\nu^2 + \mu^2 = (E_{\mathbf{k}} - k_0)(E_{\mathbf{k}} + k_0)$ in (A1d). The $\triangle(x)$ so defined is identical with (A6a).

i.e., $\triangle(x)$ is the Fourier transform of $\epsilon(k_0)\delta(k_\nu^2+\mu^2)$. On account of the $\delta(k_\nu^2+\mu^2)$ function, the integral is really over a 3-dimensional surface of the 4-dimensional k space. On using (A-4), we have ($x_0 = ct,\quad x_4 = ix_0$)

$$\begin{aligned}\triangle(x) &= \frac{i}{(2\pi)^3}\int d^3k e^{ik\cdot r}\frac{1}{2E_{\mathbf{k}}}\int_{-\infty}^{\infty} dk_0 e^{-k_0x_0}\{\delta(k_0-E_{\mathbf{k}})-\delta(k_0+E_{\mathbf{k}})\} \quad &(A6)\\ &= \frac{1}{(2\pi)^3}\int d^3k e^{ik\cdot r}\frac{\sin x_0\sqrt{k^2+\mu^2}}{\sqrt{k^2+\mu^2}} \quad &(A6a)\\ &= -\frac{i}{(2\pi)^3}\int d^3k\frac{\sin k_\nu x_\nu}{k_4}. \quad &(A6b)\end{aligned}$$

The $\triangle(x)$ above defined has the following properties:

(i)

$$\triangle(x)_{x_0=0} = 0, \qquad \text{from (A-6a).} \tag{A8}$$

(ii)

$$\begin{aligned}\frac{\partial}{\partial t}\triangle(x)\mid_{t=0} &= \frac{1}{(2\pi)^3}\int d^3k e^{i\mathbf{k}\cdot\mathbf{r}}\\ &= \delta(r), \qquad (\delta(r)=\delta(x_1)\delta(x_2)\delta(x_3)).\end{aligned} \tag{A9}$$

(iii) $\triangle(x)$ satisfies the Klein-Gordon equation

$$(\partial_\nu\partial_\nu - m_0^2)\triangle(x) = 0, \tag{A10}$$

as is seen from (A5).

(iv) Let us introduce the step function

$$\vartheta(x_0) = \begin{cases}1, & x_0 > 0;\\ 0, & x_0 < 0.\end{cases} \tag{A11}$$

For the same reason as for $\epsilon(k_0)$ in (A3), $\vartheta(x_0)$ is an invariant function under proper Lorentz transformations. This $\vartheta(x_0)$ has the following properties:

$$\begin{aligned}\epsilon(x_0) &= 2\vartheta(x_0) - 1,\\ \epsilon^2(x_0) &= 1,\\ \vartheta^2(x_0) &= \vartheta(x_0),\\ \vartheta(x_0)\vartheta(-x_0) &= 0,\\ \vartheta(x_0)\pm\vartheta(-x_0) &= \begin{cases}1\\ \epsilon(x_0).\end{cases}\end{aligned} \tag{A12}$$

In view of the last relation, $\triangle(x)$ in (A5) can be written

$$\triangle(x) = \triangle_+(x) - \triangle_-(x), \tag{A13}$$

where

$$\triangle_+(x) = -\frac{i}{(2\pi)^3}\int d^4k e^{ik_\nu x_\nu}\vartheta(k_0)\delta(k_\nu^2+\mu^2)$$
$$= \frac{1}{(2\pi)^4}\int_{C_+} d^4k \frac{e^{ik_\nu x_\nu}}{k_\nu^2+\mu^2}, \qquad d^4k = dkdk_0, \tag{A13a}$$

$$\triangle_- = \frac{1}{(2\pi)^4}\int_{C_-} d^4k \frac{e^{ik_\nu x_\nu}}{k_\nu^2+\mu^2}, \qquad d^4k = dkdk_0, \tag{A13b}$$

where C_+ is a closed curve, in the counter clockwise sense, on the complex k_0-plane about the pole $k_0 = E_{\mathbf{k}}$, and C_- is a closed curve, in the clockwise sense, about the pole $k_0 = -E_{\mathbf{k}}$.

(2) D(x) function

For fields with $m_0 = 0 \quad (\mu = \frac{m_0 c}{\hbar} = 0)$, (A1d) is

$$k_\nu^2 = \mathbf{k}^2 + k_4^2 = \mathbf{k}^2 - k_0^2 = 0, \tag{A14a}$$

$$E_{\mathbf{k}} = \mid \mathbf{k} \mid . \tag{A14b}$$

and (A4) is now

$$\epsilon(k_0)\delta(k_\nu^2) = \frac{1}{2\mid \mathbf{k} \mid}\{\delta(k_0 - \mid \mathbf{k} \mid) - \delta(k_0 + \mid \mathbf{k} \mid)\} \tag{A15}$$

and D(x) is similarly defined as the (negative) Fourier transform of $\epsilon(k_0)\delta(k^2)$, i.e., [from (A6), (A6a)],

$$D(x) = -\frac{i}{(2\pi)^3}\int d^4k e^{ik_\nu x_\nu}\epsilon(k_0)\delta(k_\nu^2) \tag{A16}$$

$$= -\frac{1}{(2\pi)^3}\int d^3k e^{i\mathbf{k}\cdot\mathbf{r}}\frac{\sin x_0 \mid \mathbf{k} \mid}{\mid \mathbf{k} \mid}. \tag{A16a}$$

Similarly to (A7), D(x) can also be defined by the integral

$$D(x) = \frac{1}{(2\pi)^3}\int_C d^4k \frac{e^{ik_\nu x_\nu}}{k_\nu^2}, \qquad k_4 = i \mid \mathbf{k} \mid \tag{A17}$$

where the contour is similar to that in (A7), the two poles here being similar to that in (A7), the poles being $k_0 = \pm \mid \mathbf{k} \mid$ (or $\pm E_{\mathbf{k}}$). From (A17), one obtains (A16a).

To evaluate D(x) in (A16a), let us use spherical polar coordinates for $\mathbf{k}$, and denote $\mid \mathbf{k} \mid$ by K, so that

$$d^3k = K^2 dK d\cos\theta d\varphi,$$

and

$$D(x) = -\frac{1}{2\pi^2 r}\int_0^\infty \sin x_0 K \sin rK dK$$
$$= -\frac{1}{4\pi r}\int_0^\infty \{\cos K(x_0 - r) - \cos K(x_0 + r)\}dK$$
$$= -\frac{1}{4\pi r}\{\delta(x_0 - r) - \delta(x_0 + r)\} \qquad \text{(A18a)}$$
$$= -\frac{1}{2\pi}\epsilon(x_0)\delta(x^2), \quad \text{using (A15).} \qquad \text{(A18)}$$

AS

$$\delta(x^2) = \delta(\mathbf{r}^2 - x_0^2) = \delta(\mathbf{r}^2 - t^2),$$

it is seen that D(x) = 0 except on the light cone $x^2 - t^2 = 0$.

Thus D(x) has the following properties

(i)

$$D(x) = 0 \qquad \text{except on the light cone.} \qquad \text{(A19)}$$

(ii) From (A16) and (A18),

$$D(x) \text{ is its own Fourier transform.} \qquad \text{(A20)}$$

(iii)

$$D(x)\mid_{x_0=0} = 0, \qquad \text{from (A16a).} \qquad \text{(A21)}$$

(iv)

$$\frac{\partial}{\partial t}D(x)\mid_{x_0=0} = -\delta(\mathbf{r}) \qquad \text{(A22)}$$

(v) D(x) satisfies the d'Alembertian equation

$$\partial_\nu\partial_\nu D(x) = 0, \qquad \text{from (A14a).} \qquad \text{(A23)}$$

(vi) From (A18) and the last relation in (A12),

$$D(x) = -\frac{1}{2\pi}\epsilon(x_0)\delta(x^2)$$
$$= \frac{1}{2\pi}\{\vartheta(-x_0) - \vartheta(x_0)\}\delta(x^2). \qquad \text{(A24)}$$

We define the retarded and advanced D(x)

$$D_{ret}(x) \equiv \frac{1}{2\pi}\vartheta(x_0)\delta(x^2) = \frac{1}{4\pi r}\delta(x_0 + r),$$
$$D_{adv}(x) \equiv \frac{1}{2\pi}\vartheta(-x_0)\delta(x^2) = \frac{1}{4\pi r}\delta(x_0 - r). \qquad \text{(A25)}$$

$D_{adv}(x), D_{ret}(x)$ both satisfy the equation

$$\partial_\nu\partial_\nu D(x) = -\delta^4(x) = -\delta^3(\mathbf{r})\delta(x_0). \qquad \text{(A26)}$$

References

Dirac, P. A. M., Proc. Roy. Soc. London **A114**, 243, 710 (1927). The general theory of quantization of fields as qualitatively outlined in Sect. 1, and the theory of emission and absorption of radiation (the Einstein $A_{(m\to n)}$ coefficient), is first given by Dirac.

Jordan, P. and Klein, O., Zeits. f. Physik **45**, 751 (1927), for systems satisfying Bose-Einstein statistics.

Jordan P. and Wigner, E., ibid. **47**, 631 (1928), for systems satisfying Fermi-Dirac statistics.

Heisenberg, W. and Pauli, W., ibid. **56**, 1 (1929), for canonical formalism of field quantization.

Pauli, W. and Weisskopf, V., Helvetica Phys. Acta **7**, 709 (1934), for the Klein-Gordon field.

Jordan, P. and Pauli, W., Zeits. f. Physik **47**, 151, (1928), for the $D(x_\nu)$ function.

Dirac, P. A. M., Proc. Cambridge Phil. Soc. **30**, 150 (1934), for the $\triangle(x_\nu)$ and $D(x_\nu)$ functions.

Heitler, W., *The Quantum Theory of Radiation*, 3rd ed. (Oxford Univ. Press, London, 1954).

Wentzel, G., *Quantum Theory of Fields* (Interscience Publ., New York, 1949).

Jauch, J. M. and Rohlich, F., *The Theory of Photons and Electrons* (Addison-Wesley, Mass., 1955).

Bjorken. J. D. and Drell, S. D., *Relativistic Quantum Field* (McGraw Hill, N. Y., 1967).

Itzykson, C. and Zuber, J. -B., *Quantum Field Theory* (McGraw-Hill, New York, 1980).

Lee, T. D., *Particle Physics and Introduction to Field Theory* (Harwood, Chur-London-New York, 1981).

Aitchison, I. J. R., *An Informal Introduction to Gauge Field Theories* (Cambridge University Press, London, 1982).

Cheng, T. P., and Li, L. -F., *Gauge Theory of Elementary Particle Physics* (Clarendon Press, Oxford, 1984).

Exercises: *Chapter 8*

1. **(i)** Show that Eq. (13) is equivalent to Eqs. (12a)-(12c).

(ii) In the case of the electromagnetic field, write down the elementary equal-time commutators and show that your results are equivalent to Eqs. (73).

(iii) The "normal-mode" expansion for the Dirac field may also be specified by

$$\psi(\mathbf{x},t) = \int \frac{d^3p}{(2\pi)^{\frac{3}{2}}} \sum_s \{a_s^{(+)}(\mathbf{p})u(\mathbf{p},s)e^{ip\cdot x} + a_s^{(-)*}(\mathbf{p})v(\mathbf{p},s)e^{-ip\cdot x}\}, \tag{154}$$

where u- and v-spinors are already given in Ch. 2 (Exercise 1). The momentum conjugate to $\psi(\mathrm{x})$ is then specified by

$$\pi(\mathbf{x},t) = \frac{\partial\mathcal{L}}{\partial(\partial_0\psi)} = +i\bar{\psi}\gamma_4 = i\psi^\dagger(\mathbf{x},t). \tag{155}$$

Quantization of the Dirac field is accomplished by imposing the following equal-time anti-commutation relations:

$$\begin{aligned}
&\{\psi_i(\mathbf{x},t),\ \ \psi_j(\mathbf{y},t)\} = 0,\\
&\{\psi_i^\dagger(\mathbf{x},t),\ \ \psi_j^\dagger(\mathbf{y},t)\} = 0,\\
&\{\psi_i(\mathbf{x},t),\ \ i\psi_j^\dagger(\mathbf{y},t)\} = i\delta_{ij}\delta^3(\mathbf{x}-\mathbf{y})
\end{aligned} \tag{156}$$

We introduce the following set of anti-commutation relations,

$$\begin{aligned}
&\{a_s(+)(\mathbf{p}),\ \ a_s(+)*(\mathbf{p}')\} = \delta_{ss'}\delta^3(\mathbf{p}-\mathbf{p}'),\\
&\{a_s(-)(\mathbf{p}),\ \ a_s(-)*(\mathbf{p}')\} = \delta_{ss'}\delta^3(\mathbf{p}-\mathbf{p}'),\\
&\text{all other elementary anticommutators} = 0.
\end{aligned} \tag{157}$$

Demonstrate that Eqs. (156) and Eqs. (157) are completely equivalent to each other, showing that quantization of the fields may also be accomplished equally well in the Pauli-Dirac representation.

2. We define

$$\begin{aligned}
T(\varphi_i(x)\varphi_j(y)) \equiv &\varphi_i(x)\varphi_j(y), \quad \text{if} \quad x_0 > y_0;\\
&\pm\varphi_j(y)\varphi_i(x), \quad \text{if} \quad y_0 > x_0.\\
&(+ \text{ for bosons } \& - \text{ for fermions.})
\end{aligned} \tag{158}$$

The chronological pairing, or T-pairing, is defined to be the vacuum expectation value of $T(\varphi_i(x)\varphi_j(y))$.

(i) Use Eqs. (70)–(73) to show that, up to $\delta_{\mu\nu} \to \delta_{\mu\nu} - \lambda\frac{\kappa_\mu\kappa_\nu}{k^2}$,

$$\begin{aligned} A_\mu(x)A_\nu(y) &\equiv -i\delta_{\mu\nu}D_0^c(x-y) \\ &= \frac{\delta_{\mu\nu}}{(2\pi)^4 i}\int d^4k e^{-ik\cdot(x-y)}\frac{1}{k^2 - i\varepsilon} \end{aligned} \tag{159}$$

(ii) Use Eqs. (111)–(116) to show that

$$\begin{aligned} \psi(x)\bar{\psi}(y) &\equiv -iS^c(x-y) \\ &= \frac{1}{(2\pi)^4 i}\int d^4p e^{ip\cdot(x-y)}\frac{m - i\gamma\cdot p}{m^2 + p^2 - i\varepsilon} \end{aligned} \tag{160}$$

Chapter 9. Quantum Electrodynamics I: S-Matrix Elements

As already mentioned earlier (Ch. 0 and Ch. 6), the theory of quantized electromagnetic fields began with the work of Dirac in 1927, followed immediately by the work of Jordan and Wigner, and by Fermi in 1930. The quantum theory of a pure electron field presents no difficulties, nor does the quantized theory of a free electromagnetic field. For a coupled system of an electron interacting with an electromagnetic field, however, a relativistic quantum field theory, known as "quantum electrodynamics" or QED, presents serious deep-rooted difficulties arising from the persistent presence of infinities in the results of calculations of physical amplitudes. A major breakthrough came in the mid 1940's with the work of Tomonaga in Japan and Schwinger, Feynman, and Dyson in the United States. As shall be explained later in this and the next chapters, the theory succeeds in "subtracting" these infinities away in a systematic manner so that finite results are obtained, which have been found to be in remarkable agreement with the observed Lamb shifts and the "g anomaly".

Recent developments in particle physics have led to the general acceptance of the $SU(3)$ gauge theory of strong interactions among quarks and gluons, known as quantum chromodynamics or QCD, and the Glashow-Salam-Weinberg (GSW) $SU(2)_L \times U(1)$ theory of electroweak interactions as the standard model. The GSW theory not only contains QED as one of its important components, but uses QED as its prototype in the sense that both gauge principle and renormalizability are served as the guideline for constructing the theory. In a similar vein, QCD has been proposed, and tested to some extent, as the candidate theory of strong interactions. Despite the amazing successes of the standard model which we shall summarize at the end of this volume, we need to look for the deeper reasons why theories of this kind work so well, in which presence of infinities is a persistent feature. Till now, there seems not yet a satisfactory answer to this very question.

In this chapter, we introduce the calculational scheme for QED, leaving discussion of renormalization to the subsequent chapter. This will then pave the way for our presentation of the standard model in the last part of the book (Part III). A pedestrian, rather than formal, approach is adopted in our introduction to this important subject. Thus, treatments, such as through the path-integral formulation, which are technical in nature, will either be relegated to appendices, or omitted entirely. By giving up the opportunity of phrasing the theory perhaps in a more elegant manner, we wish to present the subject in terms of the concepts which we had become familiar from lessons in classical dynamics and ordinary quantum mechanics. The conventional pedestrian approach has both the benefit of learning to know how to make specific predictions from the theory and that of gaining strong intuition (and perhaps better insights) toward the subject.

9.1 The Evolution Operator and the S-Matrix

Consider a physical system described by the Hamiltonian,

$$H = H_0 + H', \tag{1}$$

where H_0 is the unperturbed Hamiltonian and H' is a small perturbation. It is assumed that the perturbation H' is turned on during the period $-T_0 < t < T_0$ with T_0 some large time. Consider the state $| t_0 >$ which is assumed to be an eigenstate of H at time t_0. The state will evolve with time:

$$| t\rangle = U(t,t_0) | t_0\rangle, \tag{2}$$

where $U(t,t_0)$ is referred to as the "evolution operator". For the sake of simplicity, we suppress the known time-dependent phase factor e^{-iE_0t} related to H_0 so that $U(t,t_0)$ is determined entirely by H'.

It is clear that the evolution operator $U(t',t)$ satisfies the following properties:

i.

$$U(t_2,t_0) = U(t_2,t_1)U(t_1,t_0), \qquad \text{for } t_2 > t_1 > t_0. \tag{3a}$$

It arises because $| t_2\rangle = U(t_2,t_0) | t_0\rangle$ and $| t_2\rangle = U(t_2,t_1) | t_1\rangle = U(t_2,t_1)(U(t_1,t_0) | t_0\rangle)$.

ii.

$$U^{-1}(t_1,t_0) = U(t_0,t_1), \qquad \text{(Existence of the inverse operator)}. \tag{3b}$$

iii.

$$U^{\dagger}(t_1,t_0)U(t_1,t_0) = 1, \qquad \text{(unitarity)}. \tag{3c}$$

It comes from $\langle t_1 | t_1\rangle = \langle t_0 | t_0\rangle$ and $| t_1\rangle = U(t_1,t_0) | t_0\rangle$.

iv.

$$U(t,t) = 1, \qquad \text{(Existence of the identity operator)}. \tag{3d}$$

We find

$$\begin{aligned} i\frac{d}{dt}U(t,t_0) &\equiv i \lim_{\triangle t\to 0} \frac{U(t+\triangle t,t_0) - U(t,t_0)}{\triangle t} \\ &= i \lim_{\triangle t\to 0} \frac{U(t+\triangle t,t)U(t,t_0) - U(t,t_0)}{\triangle t} \\ &= i\{ \lim_{\triangle t\to 0} \frac{U(t+\triangle t,t) - 1}{\triangle t}\}U(t,t_0). \end{aligned} \tag{4}$$

We define the interaction Hamiltonian:

$$H_{int}(t) \equiv i \lim_{\triangle t\to 0} \frac{U(t+\triangle t,t) - 1}{\triangle t}, \tag{5}$$

so that $H_{int}(t)$ charaterizes how the system evolves with time at t. It follows that

$$i\frac{d}{dt}U(t,t_0) = H_{int}(t)U(t,t_0). \tag{6}$$

Note that

$$i\frac{d}{dt}U(t,t_0) | t_0\rangle = H_{int}(t)U(t,t_0) | t_0\rangle,$$

or,

$$i\frac{d}{dt} | t\rangle = H_{int}(t) | t\rangle, \tag{7}$$

which suggests that $H_{int}(t) = H'$ with H' given by Eq. (1).

Eq. (6) can readily be solved:

$$\begin{aligned} U(t,t_0) &= 1 - i\int_{t_0}^{t} dt_1 H_{int}(t_1)U(t_1,t_0) \\ &= 1 - i\int_{t_0}^{t} dt_1 H_{int}(t_1) \\ &\quad +(-i)^2 \int_{t_0}^{t} dt_1 \int_{t_0}^{t_1} dt_2 H_{int}(t_1)H_{int}(t_2) \\ &\quad +... \end{aligned} \tag{8}$$

Define the chronological or T-product:

$$T(A(t_1)B(t_2)) = \begin{cases} A(t_1)B(t_2) & \text{for } t_1 > t_2 \ ; \\ B(t_2)A(t_1) & \text{for } t_1 < t_2 \ . \end{cases} \tag{9}$$

Here $A(t)$ and $B(t)$ are *even* functions in fermion operators; that is, the number of fermion operators which enter $A(t)$ or $B(t)$ is even. An extra minus sign should be introduced in Eq. (9) for $t_1 < t_2$ if $B(t_2)A(t_1)$ differs from $A(t_1)B(t_2)$ by *odd* permutations of fermion operators. Note that

$$\begin{aligned} &\int_{t_0}^{t} dt_1 \int_{t_0}^{t_1} dt_2 H_{int}(t_1)H_{int}(t_2) \\ &\quad = \frac{1}{2}\int_{t_0}^{t} dt_1 \int_{t_0}^{t} dt_2 T(H_{int}(t_1)H_{int}(t_2)), \end{aligned} \tag{10a}$$

and, more generally,

$$\begin{aligned} &\int_{t_0}^{t} dt_1 \int_{t_0}^{t_1} dt_2 ... \int_{t_0}^{t_{n-1}} dt_n H_{int}(t_1)H_{int}(t_2)...H_{int}(t_n) \\ &= \frac{1}{n!}\int_{t_0}^{t} dt_1 \int_{t_0}^{t} dt_2 ... \int_{t_0}^{t} dt_n T(H_{int}(t_1)H_{int}(t_2)...H_{int}(t_n)). \end{aligned} \tag{10b}$$

Therefore, Eq. (8) becomes

$$\begin{aligned} &U(t,t_0) \\ &= T(\exp(-i)\int_{t_0}^{t} dt H_{int}(t)) \\ &\equiv 1 + \sum_{n=1}^{\infty} \frac{(-i)^n}{n!} \int_{t_0}^{t} dt_1 \int_{t_0}^{t} dt_2 ... \int_{t_0}^{t} dt_n T(H_{int}(t_1)H_{int}(t_2)...H_{int}(t_n)). \end{aligned} \tag{11}$$

The scattering operator (matrix), or S-matrix, is defined by

$$S = \lim_{t\to\infty} \lim_{t_0\to -\infty} U(t,t_0), \tag{12}$$

provided that the double limit exists in a certain operator sense. Eqs. (1)–(12) provide an intuitive and simple way to introduce the concept of the scattering matrix which is tied closely to the evolution operator. The major problem which we encounter in practice is that, for a system of interacting fields, the limiting procedures appearing in these equations such as Eqs. (12) and (5) can hardly be categorized in a mathematically rigorous manner. Consider

$$\begin{aligned} H_{int}(t) &= \int d^3x H_{int}(x,t) \\ &\equiv -\int d^4x \mathcal{L}_{int}(x,t), \end{aligned} \tag{13}$$

so that Eqs. (12) and (11) yield

$$S = T(\exp i \int d^4x \mathcal{L}_{int}(x)). \tag{14}$$

It is customary to introduce the transition operator, or T-matrix, as follows:

$$\langle f \mid S \mid i \rangle = \delta_{fi} + i(2\pi)^4\delta^4(\sum_f p_f - \sum_i p_i)T_{fi}. \tag{15}$$

For the scattering problem, H_0 is often taken to be the free Hamiltonian so that the initial state (at $t < -T_0$) and the final state (at $t > T_0$) are plane waves. For instance, the initial and final states in Bhabha scattering, $e^+e^- \to e^+e^-$, are specified by

$$\mid i \rangle = b^+_{s_1}(\mathbf{p}_1)a^+_{s_2}(\mathbf{p}_2) \mid 0 \rangle, \tag{16a}$$

$$\mid f \rangle = b^+_{s'_1}(\mathbf{p}'_1)a^+_{s'_2}(\mathbf{p}'_2) \mid 0 \rangle. \tag{16b}$$

Here $a^+_s(\mathbf{p})$ [$b^+_s(\mathbf{p})$] is the creation operator for an electron [positron] of spin s and three-momentum $\mathbf{p}$. The cross section, which is the transition probability per unit volume per unit time, is given by

$$\sigma = \frac{1}{f}\int \frac{d^3p'_1}{N_0}\int \frac{d^3p'_2}{N_0} \bar{\sum} \mid \langle f \mid S \mid i \rangle \mid^2 \cdot \frac{1}{VT}, \tag{17}$$

where f is the flux factor ($=\mid v_1 - v_2 \mid$ in the case of two-particle scattering), N_0 is some phase-space normalization factor (a specific power of 2π in our normalization), and $\bar{\Sigma}$ denotes suitable summation and averaging over internal indices such as spins. V and T are the total volume and time, respectively, so that

$$\begin{aligned} & \mid (2\pi)^4\delta^4(\sum p_f - \sum p_i) \mid^2 \\ = & \mid \int d^4x \exp ix \cdot (\sum p_f - \sum p_i) \mid^2 \\ = & \int d^4x \exp ix \cdot (\sum p_f - \sum p_i) \cdot (2\pi)^4\delta^4(\sum p_f - \sum p_i) \\ = & VT(2\pi)^4\delta^4(\sum p_f - \sum p_i). \end{aligned} \tag{18}$$

Accordingly, we find, for $f \neq i$,

$$d\sigma = \frac{1}{f}(\frac{d^3p'_1}{N_0})(\frac{d^3p'_2}{N_0})(2\pi)^4\delta^4(p'_1 + p'_2 - p_1 - p_2)\bar{\sum} \mid T_{fi} \mid^2. \tag{19}$$

The phase-space factor N_0 depends on the choice for the various normalization factors associated with wave functions and others. Later in this chapter, we shall set up Feynman rules for calculating T_{fi} in such a way that all powers of 2π are suitably taken into account by an appropriate choice of N_0 . In this way, we obtain

$$N_0 = (2\pi)^3. \tag{20}$$

On the flux factor, $f = 1$ for a high energy fixed target scattering experiment while $f = 2$ for a collider experiment.

Finally, it should be stressed that the formulae obtained in this section may be modified slightly in order to describe bound-state problems such as Lamb shifts or the "g anomaly". We shall turn to these problems later in Ch. 10.

9.2 S-Matrix Elements and Feynman Rules

Consider the interaction of the electron (positron) field $\psi(x)$ with the electromagnetic field $A_\mu(x)$:

$$\mathcal{L}_{int}(x) = -ie : \bar{\psi}(x)\gamma_\mu\psi(x)A_\mu(x) : \tag{21}$$

where $: \alpha\beta...\gamma :$ denotes the "normal ordering" of the operator product $\alpha\beta...\gamma$. Normal ordering is specified as follows:

i. If $\alpha_1, \alpha_2, ...\alpha_n$ are creation or annihilation operators, then

$$: \alpha_1\alpha_2...\alpha_n := \eta_N \alpha_{i_1}\alpha_{i_2}...\alpha_{i_n} \tag{22}$$

where $\{i_1, i_2, ..., i_n\}$ is a permutation of $1, 2, ..., n$ such that all creation operators must precede annihilation operators. The phase factor η_N is -1 if the number of permutations for fermions is odd and $+1$ otherwise.

ii. Let $\alpha_1^i, \alpha_2^i, ..., \alpha_n^i$ be creation and annihilation operators and c_i be c-numbers. Then

$$: \sum_i c_i\alpha_1^i\alpha_2^i...\alpha_n^i := \sum_i c_i : \alpha_1^i\alpha_2^i...\alpha_n^i : \tag{23}$$

Accordingly, we have, for the vacuum state $\mid 0\rangle$,

$$\langle 0 \mid : \alpha\beta...\gamma : \mid 0\rangle = 0. \tag{24}$$

It is clear that, for any given operator product $\alpha\beta...\gamma$, the normal ordering $: \alpha\beta...\gamma :$ is uniquely specified since any two creation (annihilation) operators either commute or anticommute.

There is a problem because of elevation of $\mathcal{L}_{int}(x)$ as a c-number in classical field theory (Lagrangian formalism) to a q-number as given by Eq. (21). Derivation of the Euler-Lagrange equation (the field equation) from a Lagrangian density $\mathcal{L}(x)$, or application of Noether's theorem from a given $\mathcal{L}(x)$ in generating physical observables (such as four-momentum and the angular-momentum tensor) involves variation of a functional with respect to a field function. Mathematical rigor is lost as field functions are treated as operators. Normal ordering just introduced makes the situation even worse, since a specific basis for defining creation and annihilation operators must be chosen as a reference point. Thus, the canonical formalism as stressed by T. D. Lee and many others (where field functions are treated as operators from the outset) or the path-integral formulation as initiated by R. P. Feynman provides a more coherent treatment of the problem than the naïve Lagrangian formalism. Nevertheless, the situation seems less problematic when the $\mathcal{L}_{int}(x)$ of Eq. (21) is used only in connection with the S-matrix of Eq. (14).

We write

$$\psi(\mathbf{x},t)=\int\frac{d^3p}{(2\pi)^{\frac{3}{2}}}\sum_s\{a_s^-(\mathbf{p})u(\mathbf{p},s)e^{ip\cdot x}+b_s^+(\mathbf{p})v(\mathbf{p},s)e^{-ip\cdot x}\},\tag{25}$$

which is just Eq. (109), Ch. 8. Using Eq. (114), Ch. 8, we have

$$\{a_s^-(\mathbf{p}),a_{s'}^+(\mathbf{p}')\}=\delta_{ss'}\delta^3(\mathbf{p}-\mathbf{p}'),\tag{26a}$$

$$\{b_s^-(\mathbf{p}),b_{s'}^+(\mathbf{p}')\}=\delta_{ss'}\delta^3(\mathbf{p}-\mathbf{p}'),\tag{26b}$$

$$\text{all other anticommutators}=0.\tag{26c}$$

It is then straightforward to show that

$$\psi(x;pair)\bar{\psi}(y;pair)=\frac{1}{(2\pi)^4i}\int d^4p e^{ip\cdot(x-y)}\frac{m-i\gamma\cdot p}{m^2+p^2-i\varepsilon},\tag{27}$$

where we have introduced the chronological pairing or T-pairing as follows:

$$\psi(x;pair)\bar{\psi}(y;pair)\equiv\langle 0\mid T(\psi(x)\bar{\psi}(y))\mid 0\rangle\equiv -iS^c(x-y).\tag{28}$$

Analogously, we write, for the electromagnetic field,

$$\begin{aligned}A_\mu(x)=\int\frac{d^3k}{(2\pi)^{\frac{3}{2}}}\frac{1}{\sqrt{2\omega}}\sum_{\lambda=1}^{2}\{&c_\lambda(\mathbf{k})\varepsilon_\mu^\lambda(\mathbf{k})e^{ik\cdot x}\\&+c_\lambda^+(\mathbf{k})\varepsilon_\mu^{\lambda*}(\mathbf{k})e^{-ik\cdot x}\},\end{aligned}\tag{29}$$

where $\varepsilon_\mu^\lambda(\mathbf{k})$ with $\lambda=1,2$ describe two possible transverse polarizations:

$$\varepsilon^\lambda\cdot\varepsilon^{\lambda\dagger}=\varepsilon^\lambda\cdot\varepsilon^{\lambda*}-\varepsilon_0^\lambda\varepsilon_0^{\lambda*}=1,\tag{30a}$$

$$k\cdot\varepsilon^\lambda=\mathbf{k}\cdot\varepsilon^\lambda-k_0\varepsilon_0^\lambda=0.\tag{30b}$$

[Cf. Eq. (127), Ch. 8.]

We have

$$[c_\lambda(\mathbf{k}),c_{\lambda'}^+(\mathbf{k}')]=\delta_{\lambda\lambda'}\delta^3(\mathbf{k}-\mathbf{k}'),\tag{31a}$$

$$\text{all other commutators}=0\ .\tag{31b}$$

Accordingly, we find

$$\begin{aligned}A_\mu(x;pair)A_\nu(y;pair)&\equiv -i\delta_{\mu\nu}D_0^c(x-y)\\&=\frac{\delta_{\mu\nu}}{(2\pi)^4i}\int d^4k e^{-ik\cdot(x-y)}\frac{1}{k^2-i\varepsilon}.\end{aligned}\tag{32}$$

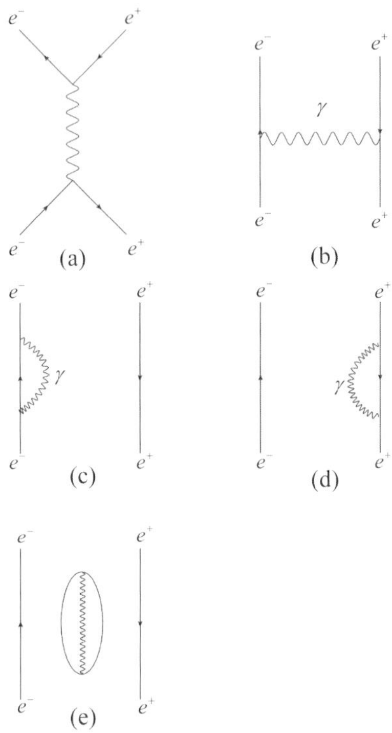

Figure 1: Pictorial representation of the second-order S-matrix element for the Bhabha scattering.

Now, consider Bhabha scattering,

$$e^{+}(p_1) + e^{-}(p_2) \to e^{+}(p_1') + e^{-}(p_2'). \tag{33}$$

The initial and final states are specified by Eqs. (16a) and (16b), respectively. We wish to evaluate the S-matrix element S_{fi}:

$$S_{fi} \equiv \langle f \mid S \mid i \rangle = \sum_{n=0}^{\infty} S_{fi}^{(n)}, \tag{34}$$

with

$$S_{fi}^{(n)} = \langle f \mid \frac{i^n}{n!} \int d^4x_1 ... d^4x_n T(\mathcal{L}_{int}(x_1) ... \mathcal{L}_{int}(x_n)) \mid i \rangle. \tag{35}$$

It is straightforward to obtain

$$S_{fi}^{(0)} = \delta_{fi} \equiv \delta^3(\mathbf{p}_1 - \mathbf{p}_1')\delta_{s_1 s_1'}\delta^3(\mathbf{p}_2 - \mathbf{p}_2')\delta_{s_2 s_2'}, \tag{36}$$

$$S_{fi}^{(1)} = 0. \tag{37}$$

Eq. (37) follows from $\langle 0 \mid A_\mu(x) \mid 0\rangle = 0$. The leading nontrivial contribution $S_{fi}^{(2)}$ is given by

$$
\begin{aligned}
S_{fi}^{(2)} &= \langle f \mid \frac{i^2}{2!}\int d^4x_1 d^4x_2 T(\mathcal{L}_{int}(x_1)\mathcal{L}_{int}(x_2)) \mid i\rangle \\
&= \langle f \mid \frac{i^2\cdot(-ie)^2}{2}\int d^4x_1 d^4x_2 \\
&\quad T(:\bar{\psi}(x_1)\gamma_\mu\psi(x_1)A_\mu(x_1)::\bar{\psi}(x_2)\gamma_\nu\psi(x_2)A_\nu(x_2):)\mid i\rangle \\
&= \frac{i^2\cdot(-ie)^2}{2}\int d^4x_1 d^4x_2 A_\mu(x_1;pair)A_\nu(x_2;pair)\langle 0\mid a_{s_2'}(\mathbf{p}_2')b_{s_1'}(\mathbf{p}_1') \\
&\quad \cdot T(:\bar{\psi}(x_1)\gamma_\mu\psi(x_1)::\bar{\psi}(x_2)\gamma_\nu\psi(x_2):)b_{s_1}^+(\mathbf{p}_1)a_{s_2}^+(\mathbf{p}_2)\mid 0\rangle. \qquad (38)
\end{aligned}
$$

The standard way to simplify the expression (38) is to pair off each creation operator by an annihilation operators until all creation and annihilation operators are paired off. We obtain

$$
\begin{aligned}
S_{fi}^{(2)} = &\frac{i^2\cdot(-ie)^2}{2}\int d^4x_1 d^4x_2 A_\mu(x_1;pair)A_\nu(x_2;pair) \\
&\cdot\{a(p1)b(p2):\bar{\psi}(x_1;p1)\gamma_\mu\psi(x_1;p2)::\bar{\psi}(x_2;p3)\gamma_\nu\psi(x_2;p4):b^+(p3)a^+(p4) \\
&+(\text{1st term with } x_1\leftrightarrow x_2 \text{ and } \mu\leftrightarrow\nu) \\
&+a(p1)b(p2):\bar{\psi}(x_1;p1)\gamma_\mu\psi(x_1;p4)::\bar{\psi}(x_2;p3)\gamma_\nu\psi(x_2;p2):b^+(p3)a^+(p4) \\
&+(\text{3rd term with } x_1\leftrightarrow x_2 \text{ and } \mu\leftrightarrow\nu) \\
&+a(p1)b(p2):\bar{\psi}(x_1;p1)\gamma_\mu\psi(x_1;p3)::\bar{\psi}(x_2;p3)\gamma_\nu\psi(x_2;p4):b^+(p2)a^+(p4) \\
&+(\text{5th term with } x_1\leftrightarrow x_2 \text{ and } \mu\leftrightarrow\nu) \\
&+a(p1)b(p2):\bar{\psi}(x_1;p4)\gamma_\mu\psi(x_1;p2)::\bar{\psi}(x_2;p3)\gamma_\nu\psi(x_2;p4):b^+(p3)a^+(p1) \\
&+(\text{7th term with } x_1\leftrightarrow x_2 \text{ and } \mu\leftrightarrow\nu) \\
&+a(p1)b(p2):\bar{\psi}(x_1;p3)\gamma_\mu\psi(x_1;p4)::\bar{\psi}(x_2;p4)\gamma_\nu\psi(x_2;p3):b^+(p2)a^+(p1)\}.
\end{aligned}
\qquad (39)
$$

Here the *l*st, 3rd, 5th, 7th, and 9th terms may be represented pictorially by Figs. 1(a) - 1(e), respectively.

The last three diagrams, Figs. 1(c)–1(e), represent "renormalization" of the lowest-order graph, i.e., $S_{fi}^{(0)}$, and they do not contain explicitly an interaction between e^+ and e^-.

Equivalently, we start out to define the T-pairing Eq. (9) to avoid the point $t_1 = t_2$ because of the ambiguity at $t_1 = t_2$. If this point is removed completely, then these diagrams, or the fifth to 9th terms, are simply not there. But we have to keep the appropriate limits when approaching to the equal-time limit $t_1 = t_2$, rather than throwing them away completely. Thus, we keep them in some "causal" way and learn how to handle ultraviolet divergences.

Accordingly, we shall focus our attention on the first two diagrams in Fig. 1. Using Eqs. (25) and (26), we obtain

$$
\begin{aligned}
\bar{\psi}(\mathbf{x},t;pair)b_s^+(\mathbf{p},pair) &\equiv \langle 0\mid T(\bar{\psi}(\mathbf{x},t)b_s^+(\mathbf{p};t=-\infty))\mid 0\rangle \\
&= \frac{1}{(2\pi)^{\frac{3}{2}}}\,\bar{v}(\mathbf{p},s)e^{ip\cdot x}, \qquad (40a)
\end{aligned}
$$

$$\psi(x; pair)a_s^+(\mathbf{p}, pair) = \frac{1}{(2\pi)^{\frac{3}{2}}}u(\mathbf{p}, s)e^{ip\cdot x}, \tag{40b}$$

$$a_s(\mathbf{p}, pair)\bar{\psi}(x; pair) = \frac{1}{(2\pi)^{\frac{3}{2}}}\bar{u}(\mathbf{p}, s)e^{-ip\cdot x}, \tag{40c}$$

$$b_s(\mathbf{p}, pair)\psi(x; pair) = \frac{1}{(2\pi)^{\frac{3}{2}}}\upsilon(\mathbf{p}, s)e^{-ip\cdot x}. \tag{40d}$$

Thus, the first four terms of Eq. (39) may be written in the following form:

$$\begin{aligned} S_{fi}^{\prime(2)} &= (\frac{1}{(2\pi)^{\frac{3}{2}}})^4(2\pi)^4\delta^4(p_1' + p_2' - p_1 - p_2)i^2(-ie)^2 \\ &\quad \cdot\{\bar{u}(p_1')\gamma_\mu\upsilon(p_2')\frac{1}{i}\frac{1}{(p_1+p_2)^2 - i\varepsilon}\bar{\upsilon}(p_2)\gamma_\nu u(p_1) \\ &\quad -\bar{u}(p_1')\gamma_\mu u(p_1)\frac{1}{i}\frac{1}{(p_1' - p_2')^2 - i\varepsilon}\bar{\upsilon}(p_2)\gamma_\nu\upsilon(p_2')\}, \end{aligned} \tag{41}$$

where the minus sign related to the second term comes from an odd number of permutations among the fermion operators. It is clear that Eq. (41) may also be obtained via application of a set of Feynman rules in momentum space:

(a) Fermion propagator:

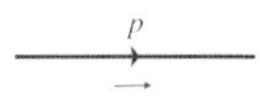

$$\Leftrightarrow \quad \frac{1}{i}\frac{m - i\gamma\cdot p}{m^2 + p^2 - i\varepsilon} \tag{42a}$$

(b) Photon propagator:

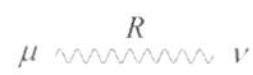

$$\Leftrightarrow \quad \frac{1}{i}\frac{\delta_{\mu\nu}}{k^2 - i\varepsilon} \tag{42b}$$

(c) Vertex:

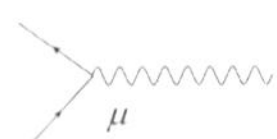

$$\Leftrightarrow \quad i\cdot(-ie)\gamma_\mu \tag{42c}$$

(d) Summation over dummy discrete indices or integration over the internal momentum $[(2\pi)^{-4}\int d^4p]$ is always implied.

(42d)

(e) External lines:

$$\begin{array}{ll} u(\mathbf{p},s) & \text{for incoming spinor,} \\ \bar{u}(\mathbf{p},s) & \text{for outgoing spinor;} \\ \bar{v}(\mathbf{p},s) & \text{for incoming antispinor,} \\ v(\mathbf{p},s) & \text{for outgoing antispinor;} \\ \dfrac{\varepsilon_\mu^\lambda(\mathbf{k})}{\sqrt{2k_0}} & \text{for a photon in the initial state,} \\ \dfrac{\varepsilon_\mu^{\lambda*}(\mathbf{k})}{\sqrt{2k_0}} & \text{for a photon in the final state.} \end{array} \tag{42e}$$

(f) There is an additional sign for each fermion loop, some counting factor for a given set of identical particle (just to ensure the proper normalization).
There is a non-essential factor $(-i)$ in going from S_{fi} to T_{fi} [cf. Eq. (15)]. Finally, factors of 2π are counted altogether at the end such that the cross section (or decay rate) is given by

$$d\sigma = \frac{1}{f}(\prod_i \frac{d^3p_i'}{(2\pi)^3})(2\pi)^4\delta^4(\sum p_i' - \sum p_i)\bar{\sum} \mid T_{fi} \mid^2, \tag{42f}$$

with f the flux factor, $\{p_i'\}$ the final momenta, and $\bar{\Sigma}$ denoting the appropriate summation and averaging over discrete indices.

(g) There is an additional minus sign if an odd number of permutations among the fermion operators of the same kind is required to separate out completely the designated T-pairings.

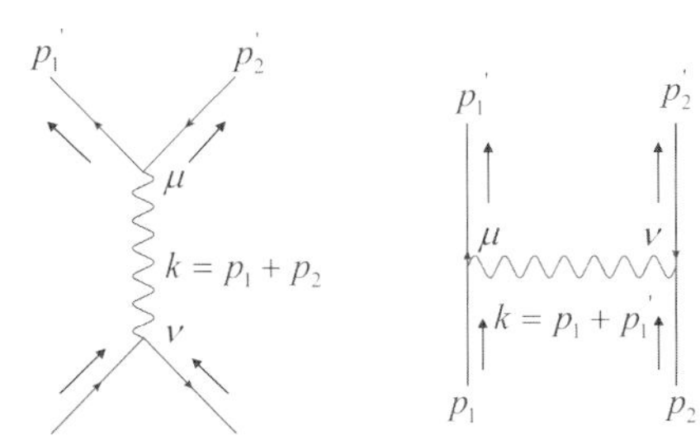

Figure 2: Feynman diagrams for the Bhabha scattering.

Application of Feynman rules to the Bhabha scattering diagrams as illustrated in

Figs. 2(a) and 2(b) yields

$$\begin{aligned}T_{fi} = (-i)\{&\bar{u}(p_1')(e\gamma_\mu)v(p_2')\frac{1}{i}\frac{1}{(p_1+p_2)^2 - i\varepsilon}\,\bar{v}(p_2)(e\gamma_\mu)u(p_1)\\ &- \bar{u}(p_1')(e\gamma_\mu)u(p_1)\frac{1}{i}\frac{1}{(p_1'-p_1)^2 - i\varepsilon}\,\bar{v}(p_2)(e\gamma_\mu)v(p_2')\},\end{aligned} \tag{43}$$

which may be obtained from Eqs. (41) and (15) *except* the known overall factor of (2π) [associated with Eqs. (40a)-(40d)]. The expression (43) is to be used in connection with Eq. (42f).

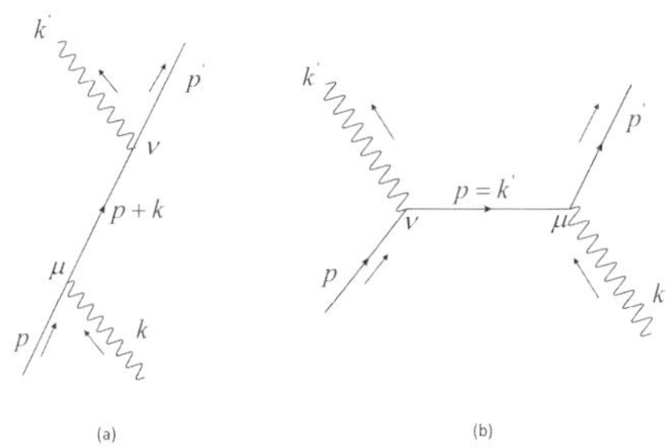

Figure 3: Feynman diagrams for Compton scattering.

Analogously, application of Feynman rules to Compton scattering, as illustrated by Figs. 3(a) and 3(b), yields

$$\begin{aligned}T_{fi} = (-i)\{&\bar{u}(p')(e\gamma_\nu)\frac{\varepsilon_\nu'^{\lambda'*}(k')}{\sqrt{2k_0'}}\frac{1}{i}\frac{m - i\gamma\cdot(p+k)}{m^2+(p+k)^2 - i\varepsilon}\\ &\cdot(e\gamma_\mu)\frac{\varepsilon_\mu^\lambda(k)}{\sqrt{2k_0}}u(p)\\ &+\bar{u}(p')(e\gamma_\mu)\frac{\varepsilon_\mu^\lambda(k)}{\sqrt{2k_0}}\frac{1}{i}\frac{m - i\gamma\cdot(p-k')}{m^2+(p-k')^2 - i\varepsilon}\\ &\cdot(e\gamma_\nu)\frac{\varepsilon_\nu'^{\lambda'*}(k')}{\sqrt{2k_0'}}u(p)\}.\end{aligned} \tag{44}$$

This amplitude will be used later in §.9.3.

In closing this section, we wish to make remarks concerning the last three diagrams in Fig. 1. *First*, we note that the disconnected bubble of Fig. 1(e) appears as a term in S_0:

$$S_0 \equiv < 0 \mid S \mid 0 >= 1 +$$  $$+ \ldots. \tag{45}$$

It is clear that, for any given diagram D, DS_0^i will also be a legitimate Feynman diagram where S_0^i is the i-th diagram in S_0 . These contributions are dropped since S_0 represents vacuum fluctuations that cannot be observed.

Second, an electron line which appears in S_{fi} always looks like

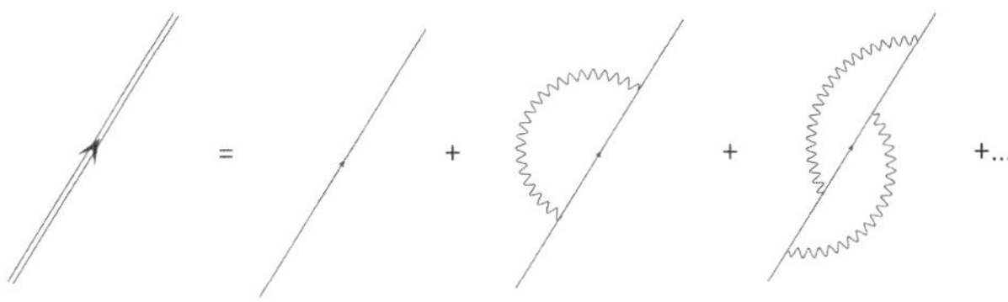

(46)

so that we may take into account Figs. 1(c) and 1(d) by interpreting the full electron line as the "physical" one (i.e., by using the observed electron mass and electric charge in the evaluation of a given diagram).

We shall return to the problem of renormalization in Ch. 10, where questions related to these diagrams will be addressed in some detail.

9.3 Calculation of Cross Sections

(1) $e^+e^- \to e^+e^-$ at High Energies

In the framework of QED, the transition amplitude for $e^+e^- \to e^+e^-$ is given by

$$\begin{aligned} T_{fi} &= e^2\{\frac{1}{s}\bar{u}(p_1')\gamma_\mu v(p_2')\bar{v}(p_2)\gamma_\mu u(p_1) \\ &\quad -\frac{1}{t}\bar{u}(p_1')\gamma_\mu u(p_1)\bar{v}(p_2)\gamma_\mu v(p_2')\}, \end{aligned} \tag{47}$$

where $s \equiv -(p_1+p_2)^2$ and $t \equiv -(p_1'-p_1)^2$ [cf. Eq. (43)]. In an e^+e^- collider experiment, the cross section is determined via the formula,

$$d\sigma = \frac{1}{2}\frac{d^3p_1'}{(2\pi)^3}\frac{d^3p_2'}{(2\pi)^3}(2\pi)^4\delta^4(p_1'+p_2'-p_1-p_2)\overline{\sum} \mid T_{fi}\mid^2 . \tag{48}$$

We find, for $\sqrt{s} \gg m_e$,

$$\frac{d\sigma}{d\Omega} = \frac{1}{4}(\frac{\sqrt{s}}{2})^2\frac{1}{(2\pi)^2}\overline{\sum} \mid T_{fi}\mid^2, \tag{49}$$

where Ω is the solid angle defined by $\mathbf{p}_1'$.

In an experiment in which no polarizations (spins) are detected, we have

$$\begin{aligned} \overline{\sum} \mid T_{fi}\mid^2 = \frac{1}{4}\sum_{\text{all spins}} & e^4\{\frac{1}{s}\bar{u}(p_1')\gamma_\mu v(p_2')\bar{v}(p_2)\gamma_\mu u(p_1) \\ & -\frac{1}{t}\bar{u}(p_1')\gamma_\mu u(p_1)\bar{v}(p_2)\gamma_\mu v(p_2')\} \\ & \cdot\{\frac{1}{s}\bar{u}(p_1')\gamma_\nu v(p_2')\bar{v}(p_2)\gamma_\nu u(p_1) \\ & -\frac{1}{t}\bar{u}(p_1')\gamma_\nu u(p_1)\bar{v}(p_2)\gamma_\nu v(p_2')\}^*. \end{aligned} \tag{50}$$

Note that

$$\sum_s u(p,s)\bar{u}(p,s) = \frac{-i\gamma\cdot p+m}{2E}, \tag{51a}$$

$$\sum_s v(p,s)\bar{v}(p,s) = \frac{-i\gamma\cdot p-m}{2E}. \tag{51b}$$

Thus, we find

$$\begin{aligned}
&\bar{\sum} \mid T_{fi}\mid^2 \\
= \;& e^4\frac{1}{4}\{\frac{1}{s^2}Tr\,\frac{-i\gamma\cdot p_2'-m}{2E_2'}\gamma_\nu\frac{-i\gamma\cdot p_1'+m}{2E_1'}\gamma_\mu \\
&\cdot Tr\,\frac{-i\gamma\cdot p_2-m}{2E_2}\gamma_\mu\frac{-i\gamma\cdot p_1+m}{2E_1}\gamma_\nu \\
&-\frac{1}{st}[Tr\,(\frac{-i\gamma\cdot p_1'+m}{2E_1'}\gamma_\mu\frac{-i\gamma\cdot p_2'-m}{2E_2'}\gamma_\nu\frac{-i\gamma\cdot p_2-m}{2E_2}\gamma_\mu\frac{-i\gamma\cdot p_1+m}{2E_1}\gamma_\nu) \\
&+h.c.] \\
&+\frac{1}{t^2}Tr\,\frac{-i\gamma\cdot p_1'+m}{2E_1'}\gamma_\mu\frac{-i\gamma\cdot p_1+m}{2E_1}\gamma_\nu \\
&\cdot Tr\,\frac{-i\gamma\cdot p_2'-m}{2E_2'}\gamma_\nu\frac{-i\gamma\cdot p_2-m}{2E_2}\gamma_\mu\}.
\end{aligned} \tag{52}$$

Note that $\gamma_\mu\gamma_\nu+\gamma_\nu\gamma_\mu = 2\delta_{\mu\nu}$ and $\gamma_5=\gamma_1\gamma_2\gamma_3\gamma_4$ yield

$$Tr\,\gamma_\mu\gamma_\nu = 4\delta_{\mu\nu}, \tag{53a}$$

$$Tr\,\gamma_\mu\gamma_\nu\gamma_\sigma\gamma_\rho = 4(\delta_{\mu\nu}\delta_{\sigma\rho}-\delta_{\mu\sigma}\delta_{\nu\rho}+\delta_{\mu\rho}\delta_{\nu\sigma}), \tag{53b}$$

$$Tr\,\gamma_5\gamma_\mu\gamma_\nu\gamma_\sigma\gamma_\rho = 4\varepsilon_{\mu\nu\sigma\rho}. \tag{53c}$$

For $\sqrt{s}\gg m_e$, Eq. (52) becomes

$$\begin{aligned}
\bar{\sum}\mid T_{fi}\mid^2 = e^4\cdot\frac{1}{4}\cdot(16E_1E_2E_1'E_2')^{-1} \\
\cdot\{\frac{1}{s^2}Tr\,\gamma\cdot p_2'\gamma_\nu\gamma\cdot p_1'\gamma_\mu\cdot Tr\,\gamma\cdot p_2\gamma_\mu\gamma\cdot p_1\gamma_\nu \\
+\frac{1}{st}[Tr\,\gamma\cdot p_1'\gamma_\mu\gamma\cdot p_2'\gamma_\nu\gamma\cdot p_2\gamma_\mu\gamma\cdot p_1\gamma_\nu + h.c.] \\
+\frac{1}{t^2}Tr\,\gamma\cdot p_1'\gamma_\mu\gamma\cdot p_1\gamma_\nu\cdot Tr\,\gamma\cdot p_2'\gamma_\nu\gamma\cdot p_2\gamma_\mu\}.
\end{aligned} \tag{54}$$

Note that

$$\begin{aligned}
&Tr\,\gamma\cdot p_2'\gamma_\nu\gamma\cdot p_1'\gamma_\mu\cdot Tr\,\gamma\cdot p_2\gamma_\mu\gamma\cdot p_1\gamma_\nu \\
=&4(p_{2\nu}'p_{1\mu}'+p_{2\mu}'p_{1\nu}'-\delta_{\mu\nu}p_2'\cdot p_1') \\
&\cdot 4(p_{2\mu}p_{1\nu}+p_{2\nu}p_{1\mu}-\delta_{\mu\nu}p_2\cdot p_1) \\
=&16(2p_1'\cdot p_2p_1\cdot p_2'+2p_1'\cdot p_1p_2'\cdot p_2),
\end{aligned} \tag{55a}$$

$$\begin{aligned}&Tr\,\gamma\cdot p_1'\gamma_\mu\gamma\cdot p_1\gamma_\nu\cdot Tr\,\gamma\cdot p_2'\gamma_\nu\gamma\cdot p_2\gamma_\mu\\&=16\,(2p_1'\cdot p_2p_1\cdot p_2'+2p_1'\cdot p_2'p_1\cdot p_2).\end{aligned}\tag{55b}$$

Using $(\gamma\cdot a)(\gamma\cdot b)=2a\cdot b-(\gamma\cdot b)(\gamma\cdot a)$ and $\gamma_\mu\gamma_\mu=4$, we find

$$\begin{aligned}&Tr\,\gamma\cdot p_1'\gamma_\mu\gamma\cdot p_2'\gamma_\nu\gamma\cdot p_2\gamma_\mu\gamma\cdot p_1\gamma_\nu\\=\;&2p_{2\mu}'Tr\,\gamma\cdot p_1'\gamma_\nu\gamma\cdot p_2\gamma_\mu\gamma\cdot p_1\gamma_\nu-2\delta_{\mu\nu}Tr\,\gamma\cdot p_1'\gamma\cdot p_2'\gamma\cdot p_2\gamma_\mu\gamma\cdot p_1\gamma_\nu\\&+2p_{2\mu}Tr\,\gamma\cdot p_1'\gamma\cdot p_2'\gamma_\nu\gamma_\mu\gamma\cdot p_1\gamma_\nu-4Tr\,\gamma\cdot p_1'\gamma\cdot p_2'\gamma_\nu\gamma\cdot p_2\gamma\cdot p_1\gamma_\nu\\=\;&2p_{2\mu}'\{2p_{1\nu}'Tr\,\gamma\cdot p_2\gamma_\mu\gamma\cdot p_1\gamma_\nu-4Tr\,\gamma\cdot p_1'\gamma\cdot p_2\gamma_\mu\gamma\cdot p_1\}\\&-2\{2p_{1\mu}Tr\,\gamma\cdot p_1'\gamma\cdot p_2'\gamma\cdot p_2\gamma_\mu-4Tr\,\gamma\cdot p_1'\gamma\cdot p_2'\gamma\cdot p_2\gamma\cdot p_1\}\\&+2p_{2\mu}\{2\delta_{\mu\nu}Tr\,\gamma\cdot p_1'\gamma\cdot p_2'\gamma\cdot p_1\gamma_\nu-2p_{1\nu}Tr\,\gamma\cdot p_1'\gamma\cdot p_2'\gamma_\mu\gamma_\nu\\&+4Tr\,\gamma\cdot p_1'\gamma\cdot p_2'\gamma_\mu\gamma\cdot p_1\}\\&-4\{2p_{2\nu}Tr\,\gamma\cdot p_1'\gamma\cdot p_2'\gamma\cdot p_1\gamma_\nu-2p_{1\nu}Tr\,\gamma\cdot p_1'\gamma\cdot p_2'\gamma\cdot p_2\gamma_\nu\\&+4Tr\,\gamma\cdot p_1'\gamma\cdot p_2'\gamma\cdot p_2\gamma\cdot p_1\}.\end{aligned}\tag{56}$$

Using Eq. (53b), we then obtain

$$Tr\,\gamma\cdot p_1'\gamma_\mu\gamma\cdot p_2'\gamma_\nu\gamma\cdot p_2\gamma_\mu\gamma\cdot p_1\gamma_\nu=-32p_1\cdot p_2'\,p_2\cdot p_1'.\tag{57}$$

Therefore, we find

$$\begin{aligned}\bar{\sum}\mid T_{fi}\mid^2=e^4\cdot\frac{1}{4}\cdot(E_1E_2E_1'E_2')^{-1}\cdot\{&\frac{1}{s^2}(2p_1'\cdot p_2p_1\cdot p_2'+2p_1'\cdot p_1p_2'\cdot p_2)\\&+\frac{1}{st}(-4)p_1\cdot p_2'p_2\cdot p_1'\\&+\frac{1}{t^2}(2p_1'\cdot p_2p_1\cdot p_2'+2p_1'\cdot p_2'p_1\cdot p_2)\}.\end{aligned}\tag{58}$$

Choose the kinematics as in the center-of-mass (CM) frame:

$$\begin{aligned}&\mathbf{p}_1=-\mathbf{p}_2=\mathbf{p},\qquad\qquad \mathbf{p}_1'=-\mathbf{p}_2'=\mathbf{p}',\\&\mathbf{p}_1\cdot\mathbf{p}_1'=\mid\mathbf{p}_1\mid\mid\mathbf{p}_1'\mid\cos\theta;\\&E\cong\mid\mathbf{p}\mid,\qquad\qquad E'\cong\mid\mathbf{p}'\mid.\end{aligned}\tag{59}$$

We obtain, from Eqs. (49) and (58),[1]

$$\frac{d\sigma}{d\Omega}=\frac{\alpha^2}{4s}\{(1+\cos^2\theta)-4\csc^2\frac{\theta}{2}\cos^4\frac{\theta}{2}+2\csc^4\frac{\theta}{2}(1+\cos^4\frac{\theta}{2})\}.\tag{60}$$

Note that the first term in Eq. (60) describes the reation $e^+e^-\to\mu^+\mu^-$ at high energies[2]:

$$\frac{d\sigma}{d\Omega}(e^+e^-\to\mu^+\mu^-)=\frac{\alpha^2}{4s}(1+\cos^2\theta),\tag{61}$$

[1] *See, e.g., Bjorken, J.D. and Drell, S.D., Relativistic Quantum Mechanics (McGraw-Hill, New York, 1964).*

[2] *See, e.g., Halzen, F. and Martin, A.D., Quarks and Leptons: An Introductory Course in Modern Particle Physics (John Wiley & Sons, New York, 1984), Ch. 11.*

so that

$$\sigma(e^+e^- \to \mu^+\mu^-) = \frac{4\pi\alpha^2}{3s}. \tag{62}$$

Analogously, production cross section for quark-antiquark pairs in e^+e^- collisions in given by

$$\sigma(e^+e^- \to \text{hadrons}) = 3\sum_q Q_q^2 \frac{4\pi\alpha^2}{3s}, \tag{63}$$

where 3 is the color factor. Accordingly, the ratio R as measured frequently in e^+e^- machines is given by

$$\begin{aligned} R &\equiv \frac{\sigma(e^+e^- \to \text{hadrons})}{\sigma(e^+e^- \to \mu^+\mu^-)} \\ &= 3\sum_q Q_q^2 \\ &= 2 \quad \text{for} \quad \sqrt{s} < m(\psi' s); \\ &\quad \frac{10}{3} \quad \text{for} \quad m(\psi' s) < \sqrt{s} < m(\Upsilon' s); \\ &\quad \frac{11}{3} \quad \text{for} \quad \sqrt{s} > m(\Upsilon' s), \end{aligned} \tag{64}$$

which agrees with the observed values of R as a function of $\sqrt{s}$, supporting the three-color hypothesis.

(2) Compton Scattering $\gamma e^- \to \gamma e^-$

The transition amplitude for Compton scattering,

$$\gamma(k,\lambda) + e^-(p,s) \to \gamma(k',\lambda') + e^-(p',s'), \tag{65}$$

is given by Eq. (44); see Figs. 3(a) and 3(b). The differential cross section is determined from Eq. (42f) with the flux factor $f = 1$:

$$d\sigma = \frac{d^3p'}{(2\pi)^3}\frac{d^3k'}{(2\pi)^3}(2\pi)^4\delta^4(p'+k'-p-k)\overline{\sum} \mid T_{fi} \mid^2. \tag{66}$$

On integrating over d^3k',

$$d\sigma = \frac{1}{(2\pi)^2} d^3p'\delta(p_0'+k_0'-p_0-k_0)\overline{\sum} \mid T_{fi} \mid^2. \tag{67}$$

Choose the kinematics:

$$\mathbf{p} = 0, \qquad \mathbf{p}' + \mathbf{k}' = \mathbf{k}, \qquad \mathbf{p}'\cdot\mathbf{k} = \mid \mathbf{p}' \mid\mid \mathbf{k} \mid \cos\theta, \tag{68}$$

so that

$$\begin{aligned} d^3p' &= p'^2 dp' d\Omega_e' \\ &= p' p_0' dp_0' d\Omega_e', \end{aligned} \tag{69a}$$

$$\begin{aligned}&\delta(p_0'+k_0'-p_0-k_0)\\&=(1+\frac{p_0'-k_0\frac{|\mathbf{p}'|}{p_0'}\cos\theta}{p_0+k_0-p_0'})^{-1}\,\delta(p_0'-p_{0c}'),\end{aligned} \tag{69b}$$

with p_{0c}' the observed final electron energy.
Thus, we obtain

$$\frac{d\sigma}{d\Omega}=\frac{p'p_0'}{(2\pi)^2}\,\frac{p_0+k_0-p_0'}{p_0+k_0-k_0\frac{|\mathbf{p}'|}{p_0'}\cos\theta}\,\bar{\sum}\mid T_{fi}\mid^2. \tag{70}$$

For the Compton scattering of unpolarized electrons by unpolarized photons, we have

$$\bar{\sum}\mid T_{fi}\mid^2=\frac{1}{4}\sum_s\sum_\lambda\sum_{s'}\sum_{\lambda'}\mid T_{fi}\mid^2. \tag{71}$$

Summation over the photon polarizations λ and λ' may be carried out with the aid of the formula:

$$\sum_\lambda\varepsilon^\lambda(k)\cdot\mathbf{a}\,\varepsilon^{\lambda*}(k)\cdot\mathbf{b}=\mathbf{a}\cdot\mathbf{b}-\mathbf{a}\cdot\hat{k}\,\mathbf{b}\cdot\hat{k}. \tag{72}$$

Here we need to consider only the two physical transverse polizations for the initial (or final) photon.

The rest of the algebra has to do with evaluation of traces of the products of γ-matrices, which is left as an exercise. The final result is the well-known Klein-Nishina-Tamm formula[3]:

$$\frac{d\sigma}{d\Omega_e'}=\frac{\alpha^2}{4m_0^2}\,(\frac{k_0'}{k_0})^2\,\{4cos^2\theta+\frac{k_0}{k_0'}+\frac{k_0'}{k_0}-2\}. \tag{73}$$

(3) Electron-Positron Annihilation $e^+e^-\to\gamma\gamma$

We shall also investigate briefly the process of mutual annihilation of an electron and a positron. The simplest diagram which corresponds to this process is shown by Fig. 4, which is the only first-order diagram.

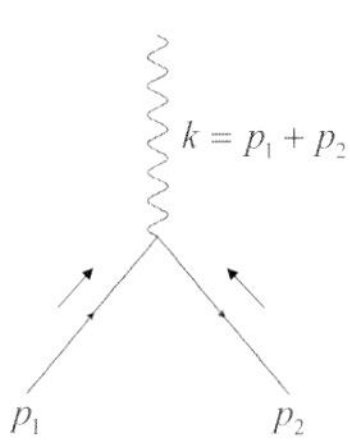

Figure 4: Annihilation of electron and positron into a single photon.

[3] *See, e.g., Bogoliubov, N.N. and Shirkov, D.V., Introduction to the Theory of Quantized Fields, 3rd Edition (John Wiley & Sons, New York, 1980), Ch. 4.; or, Heitler, W., Quantum Theory of Radiation, 3rd edition (Oxford University, 1956), p. 217.*

However, it may readily be seen that the one-photon annihilation described by this diagram is forbidden by the energy-momentum conservation laws. Indeed, the conservation laws yield

$$\mathbf{k} = \mathbf{p}_1 + \mathbf{p}_2,$$
$$|\,\mathbf{k}\,| = \sqrt{|\,\mathbf{p}_1\,|^2 + m_0^2} + \sqrt{|\,\mathbf{p}_2\,|^2 + m_0^2}.$$

By transforming, for example, to the system in which the center of mass of the electron and the positron is at rest ($\mathbf{p}_1 + \mathbf{p}_2 = 0$), we obtain a clear contradiction.

The two-photon annihilation is described by the two second-order diagrams obtained from Figs. (3a) and (3b) by crossing.

We may carry out the calculation in the system in which the center of mass of the electron and the positron is at rest. Then by setting

$$\mathbf{p}_1 = \mathbf{p}, \qquad \mathbf{p}_2 = -\mathbf{p}, \qquad \mathbf{k}_1 = \mathbf{k}, \qquad \mathbf{k}_2 = -\mathbf{k}, \tag{74}$$

We obtain

$$\begin{aligned} electron &: (E = \sqrt{\mathbf{p}^2 + m_0^2}, \mathbf{p}), \\ positron &: (E, -\mathbf{p}), \\ first\,photon &: (k_0 =|\,\mathbf{k}\,|, \mathbf{k}), \\ second\,photon &: (k_0, -\mathbf{k}). \end{aligned} \tag{75}$$

The transition amplitude is (see Eq. (44))

$$\begin{aligned} T_{fi} = (-i)\{&\bar{v}(p_2)(e\gamma_\nu)\frac{\varepsilon_\nu'^{\lambda'*}(k_2)}{\sqrt{2k_{20}}}\frac{1}{i}\frac{m - i\gamma\cdot(p_1 - k_1)}{m^2 + (p_1 - k_1)^2 - i\varepsilon} \\ &\cdot (e\gamma_\mu)\frac{\varepsilon_\mu^{\lambda*}(k_1)}{\sqrt{2k_{10}}}u(p_1) \\ &+ \bar{v}(p_2)(e\gamma_\mu)\frac{\varepsilon_\mu^{\lambda*}(k_1)}{\sqrt{2k_{10}}}\frac{1}{i}\frac{m - i\gamma\cdot(p_1 - k_2)}{m^2 + (p_1 - k_2)^2 - i\varepsilon} \\ &\cdot (e\gamma_\nu)\frac{\varepsilon_\nu'^{\lambda'*}(k_2)}{\sqrt{2k_{20}}}u(p_1)\}. \end{aligned} \tag{76}$$

Using

$$\delta(k_{10} + k_{20} - p_{10} - p_{20}) = \frac{1}{2}\delta(k_{10} - k_0), \tag{77}$$

we find, with the flux factor $f =|\,v_1 - v_2\,|= 2p/E$ in the CM frame,

$$\frac{d\sigma}{d\Omega} = \frac{E}{2p}\frac{1}{(2\pi)^2}\frac{k_0^2}{2}\overline{\sum}\,|\,T_{fi}\,|^2, \tag{78}$$

with $p \equiv|\,\mathbf{p}\,|$.

Again, it is straightforward, albeit tedious, to evaluate the traces associated with $\bar{\Sigma} \mid T_{fi} \mid^2$. For an unpolarized experiment, we obtain the well-known formula for the differential cross section[4]:

$$\frac{d\sigma}{d\Omega} = \frac{\alpha^2}{4k_0 p}\{\frac{k_0^2 + p^2 + p^2 \sin^2\theta}{k_0^2 - p^2\cos^2\theta} - \frac{2p^4 \sin^4\theta}{(k_0^2 - p^2\cos^2\theta)^2}\}, \tag{79}$$

with $cos\theta \equiv \mathbf{p} \cdot \mathbf{k}/(\mid \mathbf{p} \mid \mid \mathbf{k} \mid)$.

[4] *See, e.g., Heitler, W., Quantum Theory of Radiation, 3rd edition (Oxford University, 1956), p. 269.*

References

Wu, T.-Y., *Quantum Mechanics* (World Scientific, Singapore, 1986), p. 293; on the evolution operator.

Bjorken, J.D. and Drell, S.D., *Relativistic Quantum Mechanics* (McGraw-Hill, New York, 1964).

Halzen, F. and Martin, A.D., *Quarks and Leptons: An Introductory Course in Modern Particle Physics* (John Wiley & Sons, New York, 1984).

Bogoliubov, N.N. and Shirkov, D.V., *Introduction to the Theory of Quantized Fields*, 3rd Edition (John Wiley & Sons, New York, 1980).

Heitler, W., *Quantum Theory of Radiation*, 3rd edition (Oxford University, 1956).

Exercises: *Chapter 9*

1. Prove Eq. (11) from Eq. (8). Discuss possible ambiguities in your proof in light of possible singularites at equal times.

2. Use the anticommutation or commutation relations Eqs. (26) and (31) to obtain Eqs. (27), (32), and (40a)-(40d).

3. On Compton scattering, derive the final formula on the differential cross section, Eq. (73), from Eqs. (44), (70), and (71).

4. On electron-positron annihilation $e^+e^- \to \gamma\gamma$, derive the differential cross secion, Eq. (79), from Eqs. (76) and (78).

Chapter 10. Quantum Electrodynamics II: Renormalization

This chapter deals with the unsolved problem regarding ultraviolet divergences of the various sorts, mainly in summarizing the developments in the 20th Century. There is some hope in the 21st Century because most of these divergences can be expressed in terms of some masses and the Standard Model[1] is a dimensionless and massless theory, apart from the "ignition" term in the spontaneous symmetry breaking (SSB). (Notes are given in 2017.)

Let us look into the structure of quantum electrodynamics in greater detail. To this end, we consider the three diagrams illustrated in Fig. 1.

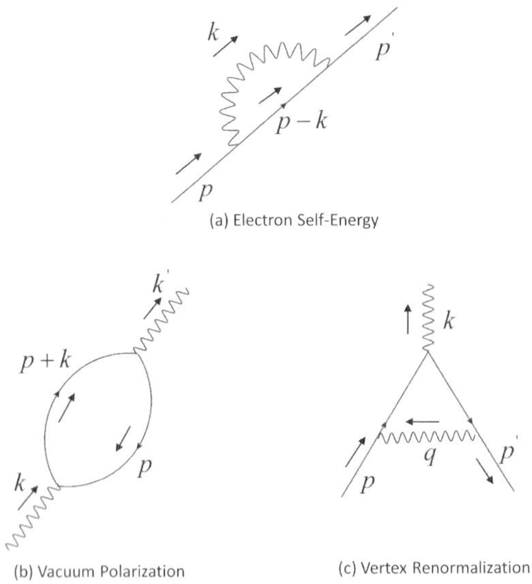

Figure 1: Leading renormalization diagrams in QED: electron self-energy (a), vacuum polarization (b), and vertex renormalization (c).

Specifically, Fig. 1(a) represents the scenario that an electron, as propagating in free space, may emit and reabsorb a virtual photon. Using QED Feynman rules, Eqs. (42a)–(42f) of Ch. 9, we obtain

$$S_a = -i(2\pi)^4\delta^4(p' - p)\delta_{s's}\bar{u}(p')\Sigma(p)u(p), \tag{1}$$

with

$$\Sigma(p) = -ie^2 \int \frac{d^4k}{(2\pi)^4}\gamma_\mu \frac{m - i\gamma\cdot(p-k)}{m^2 + (p-k)^2 - i\varepsilon}\gamma_\mu \frac{1}{k^2 - i\varepsilon}. \tag{2}$$

This contribution is often called "electron self-energy". As it stands, however, the $\Sigma(p)$ as given by Eq. (2) is ill-defined since, at large k, the integrand behaves like k^{-3} and $\int d^4k k^{-3}$

[1] *W-Y. Pauchy Hwang, arXiv:1304.4705v2 [hep-ph] 25 August 2015; "The Standard Model".*

diverges. Taking into account the fact that integration over an expression that is odd in k_μ vanishes identically, we still anticipate that $\Sigma(p)$ diverges logarithmically.

Before considering any resolution to this problem, we write down the S-matrix elements in the case of Figs. 1(b) and 1(c).

$$S_b = -i(2\pi)^4\delta^4(k'-k)\delta_{\lambda\eta}\frac{1}{2k_0}\varepsilon_\nu^{\eta*}(k')\varepsilon_\mu^\lambda(k)\Pi_{\mu\nu}(k), \tag{3}$$

$$S_c = -i(2\pi)^4\delta^4(p'+k-p)\frac{\varepsilon_\mu^\lambda(k)}{\sqrt{2k_0}}(-ie)\bar{u}(p')\Gamma_\mu u(p), \tag{4}$$

with

$$\Pi_{\mu\nu}(k) = ie^2\int\frac{d^4p}{(2\pi)^4}T_r\gamma_\mu\frac{m+i\gamma\cdot p}{m^2+p^2-i\varepsilon}\gamma_\nu\frac{m+i\gamma\cdot(p+k)}{m^2+(p+k)^2-i\varepsilon}, \tag{5}$$

$$\Gamma_\mu = -ie^2\int\frac{d^4q}{(2\pi)^4}\gamma_\alpha\frac{m-i\gamma\cdot(p'+q)}{m^2+(p'+q)^2-i\varepsilon}\gamma_\mu\frac{m-i\gamma\cdot(p+q)}{m^2+(p+q)^2-i\varepsilon}\gamma_\alpha\frac{1}{q^2-i\varepsilon}. \tag{6}$$

Simple power counting indicates that $\Pi_{\mu\nu}(k)$ and Γ_μ are also divergent. It is customary to refer to $\Pi_{\mu\nu}$ and Γ_μ ad "vacuum polarization" and "vertex renormalization", respectively.

What goes wrong? Physics is an empirical science. In principle, we can use only a well-defined mathematical model (without infinities) to model or describe an observed phenomenon. The fact that the diagrams illustrated by Fig. 1, if taken at the face value, are divergent must have critical implications for local field theories as a whole. As stated in the introductory chapter (Ch. 0), serious attempts were made, during the decade from the mid 1950's to the mid 1960's, in searching for an alternative means of describing interactions among elementary particles in terms of the S-matrix approach, in which one tries to determine the S-matrix elements using general principles (such as unitarity and microscopic causality) and a minimal set of dynamical assumptions. However, efforts to find an alternative approach were not very fruitful while gauge field theories have scored an amazing comeback since early 1970's. Successes of the standard model, which is phrased in terms of local gauge field theories (with QED as a prototype), suggest strongly the usefulness of the concept of a local field theory, despite the presence of infinities associated with the diagrams such as those in Fig. 1. The question concerning what goes wrong with these infinities is thus transcended to the questions as to why infinities are there and how we can make sense out of them. These questions are what we wish to investigate in the rest of this chapter, although we shall see that affirmative answers remain to be very elusive and in some sense rather mysterious.

An important clue comes from the observation that the diagrams such as those in Fig. 1. may become well-defined if the assumption of a 4-dimensional Minkowski space-time with usual topology can somehow be altered. As the first illustrative example, consider possible existence of a maximum momentum scale (or a minimum length scale) for which the theory is applicable. The electron self-energy, $\Sigma(p)$ as given by Eq. (2), is finite if the integration over d^4k is carried out only up to some cut-off momentum k_{max}. This is Feynman's cut-off regularization method. To do it in a Lorentz-covariant manner, we may use the Pauli-Villars regularization method and replace the photon propagator by

$$reg\, D_0^c(k) = \frac{1}{k^2-i\varepsilon} + \sum_M C_M\frac{1}{M^2+k^2-i\varepsilon}, \tag{7}$$

and the fermion propagator by

$$reg\, S^c(p) = (m - i\gamma \cdot p)\{\frac{1}{m^2 + p^2 - i\varepsilon} + \sum_M C_M \frac{1}{M^2 + p^2 - i\varepsilon}\}. \tag{8}$$

Here continuity of the regularized function, together with all its derivatives up to order $n-1$ inclusive, reguires the following conditions:

$$\sum_i C_i = 0, \qquad \sum_i C_i M_i^2 = 0, \; ..., \qquad \sum_i C_i M_i^{2n} = 0, \tag{9}$$

where the summation over i includes $i = 0$ with $C_0 = 1$ and $M_0 = m$. Eqs. (7)-(9) constitute the basis for the Pauli-Villars regularization scheme, which we wish to elaborate on later in §.10.1. Taking $n = 1$, we find, from Eqs. (2), (7), (8), and (9),

$$\begin{aligned} reg\, \Sigma(p) &= -ie^2 \int \frac{d^4k}{(2\pi)^4} \gamma_\mu \{m - i\gamma \cdot (p-k)\} \cdot \\ &\cdot \{\frac{1}{m^2 + (p-k)^2 - i\varepsilon} - \frac{1}{M^2 + (p-k)^2 - i\varepsilon}\}\gamma_\mu \\ &\cdot \{\frac{1}{k^2 - i\varepsilon} - \frac{1}{M^2 + k^2 - i\varepsilon}\}, \end{aligned} \tag{10}$$

which is finite for $0 < m \ll M < \infty$.

An alternative method for regularization of gauge fields was introduced by 't Hooft and Veltman in 1972. They considered an analytic continuation of the S-matrix elements in the complex n-plane, where n is a variable that for positive integer values equals the dimension of the space involved with respect to loop quantities. The physical situation corresponds to $n = 4$. The generalized S-matrix elements so defined are analytic in n and the infinities of perturbation theory manifest as poles at $n = 4$. The "dimensional regularization" of 't Hooft and Veltman will be introduced in some detail in §.10.2.

Of course, there are additional ways to redefine the self-energy $\Sigma(p)$ and other divergences so that the "regularized" expressions, which reproduce original resulsts in a specific limit, are finite. In all cases, there is an additional parameter, say μ^2, which may be chosen to have the dimension of $(mass)^2$. It is clear that the theory is defined as we know μ^2 in addition to the parameters m, e, and others, which appear in the lagrangion. Changing the scale μ^2 from μ_0^2 to μ_1^2, the parameters (m, e) change from (m_0, e_0) to (m_1, e_1). The structure of the theory remains to be the same, as reflected by the renormalization-group analysis. The renormalization-group concept will be introduced in §.10.3. As an application, the concept of a running coupling constant is also described there.

10.1 Pauli-Villars Regularization

Consider the electron self-energy in the Pauli-Villars regularization scheme i.e., Eq. (10). Using $\gamma_\mu\gamma_\mu = 4$ and $\gamma_\mu i\gamma\cdot p\gamma_\mu = -2i\gamma\cdot p$, we obtain

$$\begin{aligned} reg\,\Sigma(p) &= -ie^2\int\frac{d^4k}{(2\pi)^4}\{4m+2i\gamma\cdot(p-k)\} \\ &\quad\cdot\{\frac{1}{m^2+(p-k)^2-i\varepsilon}-\frac{1}{M^2+(p-k)^2-i\varepsilon}\} \\ &\quad\cdot\{\frac{1}{k^2-i\varepsilon}-\frac{1}{M^2+k^2-i\varepsilon}\}. \end{aligned} \tag{11}$$

The following mathematical identities are useful:

$$\frac{1}{m^2+k^2-i\varepsilon} = i\int_0^\infty d\alpha e^{-i\alpha(m^2+k^2-i\varepsilon)}, \tag{12a}$$

$$\int_{-\infty}^{\infty} dt e^{i(at^2+2bt)} = \frac{1+i}{\sqrt{2}}\sqrt{\frac{\pi}{a}}e^{-\frac{ib^2}{a}}, \quad (a>0); \tag{12b}$$

$$\int_{-\infty}^{\infty} dt e^{i(at^2+2bt)} = \frac{1-i}{\sqrt{2}}\sqrt{\frac{\pi}{\mid a\mid}}e^{-\frac{ib^2}{a}}, \quad (a<0); \tag{12c}$$

$$\int d^4k e^{-i(ak^2+2bk)} = \frac{\pi^2}{ia^2}e^{+\frac{ib^2}{a}}, \quad (a>0); \tag{12d}$$

$$\int d^4k e^{-i(ak^2+2bk)}k_\mu = \frac{ib_\mu}{a}(\frac{\pi}{a})^2 e^{+\frac{ib^2}{a}}, \quad (a>0); \tag{12e}$$

$$\int d^4k e^{-i(ak^2+2bk)}k_\mu k_\nu = \frac{a\delta_{\mu\nu}-2ib_\mu b_\nu}{2a^2}(\frac{\pi}{a})^2 e^{+\frac{ib^2}{a}}, \quad (a>0); \tag{12f}$$

$$\int d^4k e^{-i(ak^2+2bk)}k^2 = \frac{2a-ib^2}{a^2}(\frac{\pi}{a})^2 e^{+\frac{ib^2}{a}}, \quad (a>0). \tag{12g}$$

Eq. (11) becomes

$$\begin{aligned} reg\,\Sigma(p) &= -ie^2\int\frac{d^4k}{(2\pi)^4}\{4m+2i\gamma\cdot(p-k)\} \\ &\quad\cdot i\int_0^\infty d\beta e^{-i\beta((p-k)^2-i\varepsilon)}\{e^{-i\beta m^2}-e^{-i\beta M^2}\} \\ &\quad\cdot i\int_0^\infty d\alpha e^{-i\alpha(k^2-i\varepsilon)}\{1-e^{-i\alpha M^2}\} \\ &= \frac{e^2}{8\pi^2}\int_0^\infty d\alpha\int_0^\infty d\beta\frac{e^{-\varepsilon(\alpha+\beta)}}{(\alpha+\beta)^2}e^{-i(\frac{\alpha\beta}{\alpha+\beta}p^2)}(2m+i\gamma\cdot p\frac{\alpha}{\alpha+\beta}) \\ &\quad\cdot(1-e^{-i\alpha M^2})(e^{-i\beta m^2}-e^{-i\beta M^2}). \end{aligned} \tag{13}$$

We introduce

$$\alpha = \xi\lambda, \qquad \beta = (1-\xi)\lambda, \tag{14}$$

so that

$$\frac{\partial(\alpha,\beta)}{\partial(\xi,\lambda)} = \lambda. \tag{15}$$

We obtain

$$reg\ \Sigma(p) = \frac{e^2}{8\pi^2}\int_0^1 d\xi(2m + i\gamma\cdot p\xi)J_\varepsilon(\xi, M), \tag{16}$$

with

$$\begin{aligned} J_\varepsilon(\xi, M) &= \int_0^\infty \frac{d\lambda}{\lambda} e^{-\lambda\varepsilon - i\lambda\xi(1-\xi)p^2}(1 - e^{-i\xi\lambda M^2}) \\ &\quad \cdot(e^{-i(1-\xi)\lambda m^2} - e^{-i(1-\xi)\lambda M^2}). \end{aligned} \tag{17}$$

Using

$$\int_0^\infty \frac{d\lambda}{\lambda}(e^{iA\lambda} - e^{iB\lambda})e^{-\varepsilon\lambda} = \ln\frac{B + i\varepsilon}{A + i\varepsilon}, \tag{18}$$

we obtain

$$J_0(\xi, M) = \ln\frac{M^2 + \xi p^2}{m^2 + \xi p^2} + \ln\frac{\xi M^2 + (1-\xi)m^2 + \xi(1-\xi)p^2}{M^2 + \xi(1-\xi)p^2}. \tag{19}$$

Substituting Eq. (19) back into Eq. (16), we write

$$reg\,\Sigma(p) = \Sigma_{div}(p) + \Sigma'(p), \tag{20}$$

where

$$\Sigma'(p) = \frac{e^2}{8\pi^2}\int_0^1 d\xi(2m + i\gamma\cdot p\xi)\ln\frac{m^2}{m^2 + \xi p^2}, \tag{21}$$

$$\begin{aligned} \Sigma_{div}(p) &= \frac{e^2}{8\pi^2}\int_0^1 d\xi(2m + i\gamma\cdot p\xi) \\ &\quad \cdot\ln\{\xi\frac{M^2 + \xi p^2}{m^2}\cdot\frac{\xi M^2 + (1-\xi)m^2 + \xi(1-\xi)p^2}{\xi M^2 + \xi^2(1-\xi)p^2}\}. \end{aligned} \tag{22}$$

For sufficiently large M, we find

$$\Sigma_{div}(p) = \frac{e^2}{(4\pi)^2}\{\ln\frac{M^2}{m^2}(4m + i\gamma\cdot p) + (-\frac{i}{2}\gamma\cdot p - 4m)\}. \tag{23}$$

By going over to configuration space,

$$\Sigma(x-y) = \frac{1}{(2\pi)^4}\int d^4p e^{ip\cdot(x-y)}\Sigma(p), \tag{24}$$

we find, for sufficiently large M,

$$\begin{aligned} reg\,\Sigma(x) &= \frac{e^2}{(4\pi)^2}\{\ln\frac{M^2}{m^2}(4m - \gamma_\mu\partial_\mu) \\ &\quad +(\frac{1}{2}\gamma_\mu\partial_\mu - 4m)\}\delta^4(x) + \Sigma'(x), \end{aligned} \tag{25}$$

with

$$\Sigma'(x) \equiv \frac{1}{(2\pi)^4}\int d^4p e^{ip\cdot x}\Sigma'(p). \tag{26}$$

Note that

$$\lim_{M\to\infty} \; reg\ \Sigma(x) = \Sigma'(x) \quad everywhere \quad except \quad x = 0. \tag{27}$$

To understand the meaning of these results, we consider the Feynman regularization method. Instead of Eq. (11), we introduce the Feynman cutoff factor,

$$\begin{aligned} reg\,\Sigma_F(p) &= -ie^2\int\frac{d^4k}{(2\pi)^4}\{4m + 2i\gamma\cdot(p-k)\} \\ &\quad \cdot\frac{1}{m^2+(p-k)^2-i\varepsilon}\cdot\frac{1}{k^2-i\varepsilon}\cdot\frac{M^2}{M^2+k^2} \\ &= -ie^2\int\frac{d^4k}{(2\pi)^4}\{4m + 2i\gamma\cdot(p-k)\} \\ &\quad \cdot\frac{1}{m^2+(p-k)^2-i\varepsilon}\{\frac{1}{k^2-i\varepsilon}-\frac{1}{M^2+k^2-i\varepsilon}\}. \end{aligned} \tag{28}$$

Following the same procedure, we find, for sufficiently large M,

$$\begin{aligned} reg\,\Sigma_F(x) &= \frac{e^2}{(4\pi)^2}\{\ln\frac{M^2}{m^2}(4m-\gamma_\mu\partial_\mu) + (\frac{1}{2}\gamma_\mu\partial_\mu - 4m)\}\delta^4(x) \\ &\quad +\Sigma'_F(x), \end{aligned} \tag{29}$$

where Σ'_F differs from Σ' by a quantity,

$$\frac{e^2}{(4\pi)^2}(2m-\frac{3}{4}\gamma_\mu\partial_\mu)\delta^4(x). \tag{30}$$

Accordingly, we may write the general expression for $\Sigma'(p)$ in the form,

$$\begin{aligned} \Sigma'(p) &= \frac{e^2}{8\pi^2}\{\int_0^1 d\xi(2m+i\gamma\cdot p\xi)\ln\frac{m^2}{m^2+\xi p^2} \\ &\quad +C_1(i\gamma\cdot p+m)+C_2 m\}, \end{aligned} \tag{31}$$

where C_1 and C_2 are finite constants which depend on the method of regularization. Nevertheless, the expression obtained by subtracting from reg $\Sigma(p)$ the first two terms of Maclaurin's series,

$$reg\ \Sigma(p) - reg\ \Sigma(0) - \frac{\partial \text{reg}\ \Sigma(p)}{\partial p_\mu}\Big|_{p=0}\cdot p_\mu,$$

does not depend on the method of regularization.

We proceed to consider the regularization of the vacuum polarization $\Pi_{\mu\nu}(k)$ as given by Eq. (5). We obtain

$$\begin{aligned} reg\,\Pi_{\mu\nu}(k) &= ie^2\int\frac{d^4p}{(2\pi)^4}T_r\gamma_\mu(m+i\gamma\cdot p)\gamma_\nu(m+i\gamma\cdot(p+k))\cdot \\ &\quad \cdot(\frac{1}{m^2+p^2-i\varepsilon}-\frac{1}{M^2+p^2-i\varepsilon}) \\ &\quad \cdot(\frac{1}{m^2+(p+k)^2-i\varepsilon}-\frac{1}{M^2+(p+k)^2-i\varepsilon}). \end{aligned} \tag{32}$$

Evaluating the trace and using Eqs. (12a)-(12g), we find

$$\begin{aligned} reg\,\Pi_{\mu\nu}(k) &= -\frac{ie^2}{4\pi^2}\int_0^\infty d\alpha \int_0^\infty d\beta e^{-\varepsilon(\alpha+\beta)-i\frac{\alpha\beta}{\alpha+\beta}k^2}\cdot \\ &\quad \cdot(e^{-i\alpha m^2}-e^{-i\alpha M^2})(e^{-i\beta m^2}-e^{-i\beta M^2}) \\ &\quad \cdot\frac{1}{(\alpha+\beta)^2}\{i\frac{\alpha\beta}{(\alpha+\beta)^2}(k^2\delta_{\mu\nu}-2k_\mu k_\nu) \\ &\quad -\delta_{\mu\nu}(im^2-\frac{1}{\alpha+\beta})\}. \end{aligned} \tag{33}$$

Using Eq. (14), we find

$$\begin{aligned} reg\,\Pi_{\mu\nu}(k) &= -\frac{ie^2}{4\pi^2}\int_0^1 d\xi \int_0^\infty d\lambda \frac{e^{-\varepsilon\lambda}}{\lambda}\cdot \\ &\quad \cdot\{f(\lambda)[i\xi(1-\xi)(k^2\delta_{\mu\nu}-2k_\mu k_\nu)-im^2\delta_{\mu\nu}] \\ &\quad +\frac{1}{\lambda}f(\lambda)\delta_{\mu\nu}\}, \end{aligned} \tag{34}$$

with

$$\begin{aligned} f(\lambda) = e^{-i\lambda\xi(1-\xi)k^2}&(e^{-i\xi\lambda m^2}-e^{-i\xi\lambda M^2})\cdot \\ &\cdot(e^{-i(1-\xi)\lambda m^2}-e^{-i(1-\xi)\lambda M^2}). \end{aligned} \tag{35}$$

Noting that, with $f(\lambda)<\infty$ as $\lambda\to\infty$ and $f(\lambda)\to c\lambda^2$ as $\lambda\to 0$,

$$\begin{aligned} \int_0^\infty \frac{d\lambda}{\lambda^2}f(\lambda)e^{-\varepsilon\lambda} &= \int_0^\infty \frac{d\lambda}{\lambda}\frac{\partial}{\partial\lambda}(e^{-\varepsilon\lambda}f(\lambda)) \\ &\to \int_0^\infty \frac{d\lambda}{\lambda}e^{-\varepsilon\lambda}\frac{\partial f(\lambda)}{\partial\lambda} \quad \text{as} \quad \varepsilon\to 0, \end{aligned} \tag{36}$$

we obtain

$$\begin{aligned} reg\,\Pi_{\mu\nu}(k) &= -\frac{ie^2}{4\pi^2}\int_0^1 d\xi \int_0^\infty \frac{d\lambda}{\lambda}e^{-\varepsilon\lambda}\cdot \\ &\cdot\{f(\lambda)[i\xi(1-\xi)(k^2\delta_{\mu\nu}-2k_\mu k_\nu)-im^2\delta_{\mu\nu}]+\frac{\partial f(\lambda)}{\partial\lambda}\delta_{\mu\nu}\}. \end{aligned} \tag{37}$$

A straightforward computation yields, in the limit of sufficiently large M,

$$reg\,\Pi_{\mu\nu}(k) = -\frac{e^2}{8\pi^2}\delta_{\mu\nu}(M^2-m^2)+(k^2\delta_{\mu\nu}-k_\mu k_\nu)\frac{e^2}{12\pi^2}\ln\frac{M^2}{m^2}+\Pi'_{\mu\nu}(k), \tag{38}$$

with

$$\Pi'_{\mu\nu}(k) = -\frac{e^2}{2\pi^2}(k^2\delta_{\mu\nu}-k_\mu k_\nu)\int_0^1 d\xi\xi(1-\xi)\ln\frac{m^2+\xi(1-\xi)k^2}{\xi(1-\xi)m^2}. \tag{39}$$

In configuration space, we have

$$\begin{aligned} reg\,\Pi_{\mu\nu}(x) &= -\frac{e^2}{8\pi^2}\delta_{\mu\nu}(M^2-m^2)\delta^4(x) \\ &\quad -\frac{e^2}{12\pi^2}\ln\frac{M^2}{m^2}(\delta_{\mu\nu}\partial_\alpha\partial_\alpha-\partial_\mu\partial_\nu)\delta^4(x)+\Pi'_{\mu\nu}(x), \end{aligned} \tag{40}$$

so that $reg\,\Pi_{\mu\nu}(x)$ converges to $\Pi'_{\mu\nu}(x)$ everywhere except at $x = 0$.

At this juncture, it is useful to consider the question related to gauge invariance. Eq. (3) yields

$$S_b \propto (2\pi)^4 \delta^4(k' - k) A_\mu(k) \Pi_{\mu\nu}(k) A_\nu(k). \tag{41}$$

All physical observables must be invariant under the gauge transformation,

$$A_\mu(x) \ \rightarrow \ A'_\mu(k) = A_\mu(x) + \partial_\mu \chi(x), \tag{42a}$$

or,

$$A_\mu(k) \ \rightarrow \ A'_\mu(k) = A_\mu(k) + ik_\mu \chi(k). \tag{42b}$$

Thus, we have

$$k_\mu \Pi_{\mu\nu}(k) = k_\nu \Pi_{\mu\nu}(k) = 0. \tag{43}$$

It is clear that Eq. (38) is not in a gauge-invariant form, indicating that the Pauli-Villars regularization method is not gauge invariant. Dropping the first term in Eq. (38) and carrying out the integration associated with $\Pi'_{\mu\nu}(k)$, we obtain

$$reg\,\Pi_{\mu\nu}(k) = (k^2 \delta_{\mu\nu} - k_\mu k_\nu)\left(\frac{e^2}{12\pi^2} \ln \frac{M^2}{m^2} - \frac{e^2}{60\pi^2} \frac{k^2}{m^2}\right). \tag{44}$$

To investigate the meaning of Eq. (44), we consider Rutherford scattering as illustrated by Fig. 2.

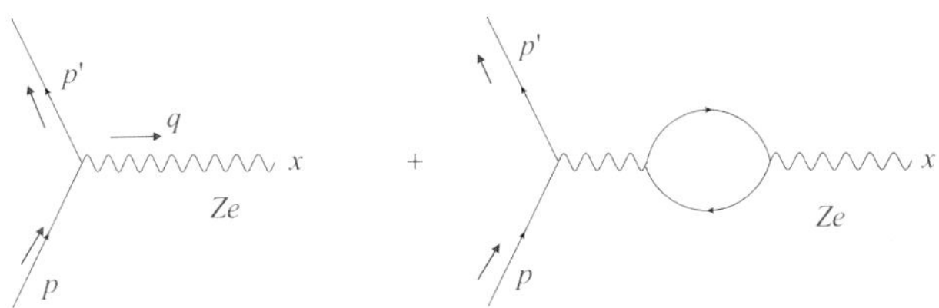

Figure 2: Rutherford scattering including effects due to vacuum polarization.

The transition amplitude looks like

$$\begin{aligned} &\bar{u}(p')(e\gamma_\mu)u(p) \cdot \frac{1}{i}\frac{1}{q^2} \cdot ZeJ_\mu(q) \\ &+ \bar{u}(p')(e\gamma_\mu)u(p) \cdot \frac{1}{i}\frac{1}{q^2} \cdot (-i)\Pi_{\mu\nu}(q) \cdot \frac{1}{i}\frac{1}{q^2} \cdot ZeJ_\nu(q). \end{aligned} \tag{45}$$

Using Eq. (44) and $q_\mu J_\mu(q) = 0$, we find

$$\begin{aligned} T &\propto -ei\bar{u}(p')\gamma_\mu u(p) \cdot ZeJ_\mu(q) \cdot \frac{1}{q^2}\{1 - \frac{e^2}{12\pi^2} \ln \frac{M^2}{m^2} + \frac{e^2}{60\pi^2} \frac{q^2}{m^2}\} \\ &= -Ze_R^2 i\bar{u}(p')\gamma_\mu u(p) \cdot J_\mu(q) \cdot \frac{1}{q^2}\{1 + \frac{e_R^2}{60\pi^2} \frac{q^2}{m^2}\}. \end{aligned} \tag{46}$$

Here we have introduced

$$e_R \equiv e\left(1 - \frac{e^2}{12\pi^2} \ln \frac{M^2}{m^2}\right)^{\frac{1}{2}}, \tag{47}$$

which is to be identified with the "physical" or "renormalized" charge. The Coulomb potential is obtained by Fourier-transforming the first term of Eq. (46).

$$V_0(\mathbf{r}) = \int \frac{d^3q}{(2\pi)^3} e^{i\mathbf{q}\cdot\mathbf{r}} \left(-\frac{Ze_R^2}{\mathbf{q}^2}\right) = -\frac{Ze_R^2}{4\pi r}. \tag{48}$$

Taking into account both terms in Eq. (46), we find

$$V(\mathbf{r}) = -\frac{Ze_R^2}{4\pi r} - \frac{Ze_R^4}{60\pi^2 m^2}\delta^3(\mathbf{r}). \tag{49}$$

The second term contributes $-27\ MHz$ to the total Lamb shift of $+1057\ MHz$ between the $2S_{\frac{1}{2}}$ and $2P_{\frac{1}{2}}$ levels of the hydrogen atom, which has been measured to an accuracy of about 0.01%.

We proceed to consider the vertex renormalization diagram of Fig. 1(c). It is a routine exercise to extract the "finite" contribution out of Eqs. (4) and (6). We may write

$$-ei\bar{u}(p')\{\gamma_\mu[1 - \frac{\alpha}{3\pi}\frac{q^2}{m^2}(\ln\frac{m^2}{\mu^2} - \frac{3}{8})] - \frac{\alpha}{2\pi}\frac{1}{2m}\sigma_{\mu\nu}q_\nu\}u(p), \tag{50}$$

where the leading term refers to the vertex diagram without any loop and the parameter μ is the small cutoff parameter associated with the removal of infrared divergences. Now, the $O(\alpha)$ correction associated with γ_μ can be combined with Eq. (49), yielding a potential:

$$V(\mathbf{r}) = -\frac{Ze_R^2}{4\pi r} + \frac{Ze_R^4}{12\pi^2}\frac{1}{m^2}(\ln\frac{m^2}{\mu^2} - \frac{3}{8} - \frac{1}{5})\delta^3(\mathbf{r}). \tag{51}$$

The second term accounts for the observed value of the Lamb shift ($+1057\ MHz$). The first two terms are from the vertex renormalization while the third one if from the previous vacuum polarization ($\frac{1}{60}$).

The $\sigma_{\mu\nu}$ term in Eq. (50) also has its own story. The Dirac magnetic moment of an electron is given by

$$\mu = -\frac{e}{2m}\sigma = -g\frac{e}{2m}\mathbf{S}, \tag{52}$$

with

$$g = 2. \tag{53}$$

The last term in Eq. (50) yield a correction:

$$\mu = -\frac{e}{2m}(1 + \frac{\alpha}{2\pi})\sigma. \tag{54}$$

Or, we have, quoting the most recent result for the reason of accuracy (Kinoshita and Lindquist 1981),

$$\begin{aligned} \frac{g-2}{2} &= \frac{\alpha}{2\pi} - 0.328478966(\frac{\alpha}{\pi})^2 + (1.1765 \pm 0.0013)(\frac{\alpha}{\pi})^3 \\ &\quad + (-0.8 \pm 2.5)(\frac{\alpha}{\pi})^4 + ... \\ &= (1159652.460 \pm 0.127 \pm 0.075) \times 10^{-9}, \end{aligned} \tag{55}$$

where the first error (± 0.127) is due to the uncertainty associated with the fine-structure constant and the second error (± 0.075) is purely theoretical. The experimental value of the electron's anomalous magnetic moment, as from the 1988 publication of Particle Data Group (Phys. Lett. **B204**, p.1), is

$$(\frac{g-2}{2})_{exp} = (1159652.193 \pm 0.010) \times 10^{-9}, \tag{56}$$

which is in excellent agreement with Eq. (55). The experimental number of 2016 is even more precise, adding a few significant digits. Theory-wise, it is more important to add the contributions from the other gauge bosons, such as the family gauge bosons in the Standard Model of the 21st Centuries, anticipated to become that of all centuries.

To sum up, the *finite* contributions associated with Figs. 1(a)–1(b), i.e., those contributions which do not behave like $\delta^4(x)$ or derivatives of $\delta^4(x)$ in configuration space [cf. Eqs. (25) and (40)], are linked to the observable effects such as the Lamb shift and the electron anomalous magnetic moment. These contributions do not depend on the method of regularization. This fact is of specific importance for any future attempt to understand ultraviolet divergences in QED.

To sum up on the Pauli-Villars regularization, the method tries to take care of the causality $-i\epsilon$ requirement; although the introduction of the M^2 terms seems to be "extra" due to infinities, the resultant finite terms might make some sense.

10.2 Dimensional Regularization

The dimensional regularization method was first adopted in 1972 by 't Hooft and Veltman in their attempt to prove the renormalizability of nonabelian gauge field theories. To describe the central ideas underlying the method, we wish to follow closely the presentation given in their original paper.

The dimensional regularization procedure was based originally on the observation that Ward identities, i.e., identites arising from gauge invariance, hold irrespective of the dimension of the space involved. By introducing a fictitious fifth dimension and attributing a very large momentum inside the loop to the fifth dimension, we may then formulate suitable regulator diagrams which are gauge invariant. It is clear that the procedure breaks down for diagrams containing two or more closed loops since then the "fifth" component of loop momentum may be distributed over the various internal lines. It was suggested that more dimensions would have to be introduced, and thus the idea of continuation in the number of dimensions emerged. In practice, suitable definitions may be introduced such that a slight continuation from dimension 4 into fractal dimensions allows for realization of the dimensional regularization procedure.

Infrared difficulties associated with massless particles are not the subject of our discussion here.

As an example we take a charged scalar field $\phi(x)$ interacting with the electromagnetic field $A_\mu(x)$. The gauge invariant lagrangian is specified by

$$\mathcal{L} = -[D_\mu\phi]^\dagger[D_\mu\phi] - m^2\phi^\dagger\phi, \tag{57a}$$

with $D_\mu \equiv \partial_\mu - ieA_\mu(x)$ the gauge invariant derivative. Thus the interaction terms are given by

$$\mathcal{L}_{int} = +ie[\partial_\mu\phi]^\dagger A_\mu\phi - ieA_\mu\phi^\dagger[\partial_\mu\phi] - e^2 A_\mu A_\mu \phi^\dagger\phi. \tag{57b}$$

We wish to illustrate the central ideas by considering the one-loop graphs with the lowest order photon self-energy diagrams as the specific example. Extensions to two loops or more will be mentioned at the end. The diagrams are illustrated in Fig. 3.

Figure 3: The lowest order photon self-energy diagrams for a charged scalar particle.

Using the Feynman rules which can easily be obtained from Eq. (57b), we find that the S-matrix element for Fig. 3 is proportional to the integral specified by

$$J_{\mu\nu} = e^2 \int d_n p[\frac{(2p+k)_\mu(2p+k)_\nu}{(p^2+m^2)((p+k)^2+m^2)} - \frac{2\delta_{\mu\nu}}{p^2+m^2}]. \tag{58}$$

where p and k are the n component loop momentum and the external momentum, respectively. It is straightforward to see that we may recast Eq. (58) in a slightly different form:

$$J_{\mu\nu} = e^2 \int_0^1 dx \int d_n p \frac{4p_\mu p_\nu + 2p_\mu k_\nu + 2k_\mu p_\nu + k_\mu k_\nu - 2((p+k)^2+m^2)\delta_{\mu\nu}}{(p^2+2pkx+k^2x+m^2)^2}.$$

Evaluating the integral with the aid of the formulae listed in the Appendix at the end of the chapter and noting that in the end terms odd in $(1-2x)$ may be dropped, we obtain, with $d_n p = d(ip_0)d_{n-1}p$ used here and in the Appendix,

$$J_{\mu\nu} = e^2 i\pi^{\frac{1}{2}n}\Gamma(2-\frac{1}{2}n)\int_0^1 dx \frac{(1-2x)^2(k_\mu k_\nu - k^2\delta_{\mu\nu})}{(m^2+k^2x(1-x))^{2-\frac{1}{2}n}}. \tag{59}$$

This expression is manifestly gauge invariant, since $k_\mu J_{\mu\nu} = k_\nu J_{\mu\nu} = 0$. In the complex n plane there are simple poles for $n = 4, 6, 8$, etc. Note that gauge invariance holds for any n. This is the property mentioned earlier that Ward identities do not involve the dimensionality of space.

To carry out the regularization procedure, we subtract from (59) the pole and its residue at the dimension $n = 4$:

$$e^2 i\pi^2 \frac{2}{4-n}(k_\mu k_\nu - k^2\delta_{\mu\nu})\int_0^1 dx(1-2x)^2, \tag{60}$$

which is a polynomial in the external momentum, and is again gauge invariant. Subtracting (60) from (59) and taking the limit $n = 4$, we obtain the customary result:

$$\begin{aligned} J_{\mu\nu}^{reg} &= -ie^2\pi^2(k_\mu k_\nu - k^2\delta_{\mu\nu})\int_0^1 dx(1-2x)^2 \ln(m^2+k^2x(1-x)) \\ &\quad +C(k_\mu k_\nu - k^2\delta_{\mu\nu}), \end{aligned} \tag{61}$$

where C is a constant related to the n dependence other than in the exponent of the denominator. The constant C is in fact undetermined, as may be seen as follows: Suppose that in (59) we replace e^2 by e^2M^{4-n}, where M is an arbitrary mass. Thus, the dimension of Eq. (59) is $(mass)^2$ which is independent of the dimension n. However, C in (61) is changed by a term proportional to $\ln M$ (which is arbitrary). It is useful to note that such ambiguity also occurs if we apply the Pauli-Villars regularization scheme to Fig. 3.

The above simple heuristic derivation already displays many of the features of the dimensional regularization method. It is clear that, in practical calculations for one loop diagrams, the method provides a very simple scheme for computing gauge invariant results. It could for instance be used to show cancellation of divergencies in the manifestly unitary set of Feynman diagrams.

However, there are serious objections to the above manipulations. First of all, our starting point Eq. (58) is meaningless for $n \geq 2$. In order to obtain a sensible result, one must (i) extend the Feynman rules such that for non-integer n all diagrams give rise to well-defined expressions, and (ii) define a suitable limiting procedure for $n \to 4$, which restores originally convergent diagrams to their original values while originally divergent diagrams are given a meaning consistent with unitarity, etc.

Thus, we shall first come up with a possible redefinition of the S-matrix. Let us consider again Eq. (58). First we split the n-dimensional space in a 4 dimensional (physical) space and an $n-4$ dimensional subspace:

$$\int d_n p \to \int d_4 \underline{p} \int d_{n-4} P. \tag{62}$$

Multiplying (58) with two arbitrary physical four vectors $e_{1\mu}$ and $e_{2\nu}$ we see that (58) depends on the direction of $\underline{p}$ but not on the direction of P. Here we have used $(pk) = (\underline{pk})$, $(e_1p) = (e_1\underline{p})$, $(e_2p) = (e_2\underline{p})$ and $p^2 = \underline{p}^2 + \omega^2$, with ω the magnitude of P in the $n-4$ dimensional subspace. Introducing polar coordinates in P space and integrating over angles one finds:

$$J = \int d_4\underline{p} \int d\omega\omega^{n-5} \frac{2(\pi)^{\frac{1}{2}(n-4)}}{\Gamma(\frac{1}{2}(n-4))} f(\underline{p}, \omega^2). \tag{63}$$

where the dependence on the external vectors e_1, e_2 and k is not shown explicitly. Note that (63) may still quite meaningless, since the second integral in (63) contains an infrared divergence for $n < 4$. Thus, we continue our formal manipulations until we arrive at an expression that can be given a meaning. This divergence is superficial and may be removed by partial integration:

$$\int_0^\infty d\omega\omega^{n-5} f(\underline{p}, \omega^2) = -\frac{2}{n-4} \int_0^\infty d\omega\omega^{n-3} \frac{\partial}{\partial\omega^2} f(\underline{p}, \omega^2),$$

where surface terms have been neglected. Doing this λ times on Eq. (63), we obtain

$$\frac{\pi^{\frac{1}{2}(n-4)}2}{\Gamma(\frac{1}{2}(n-4)+\lambda)} \int d_4\underline{p} \int_0^\infty d\omega\omega^{n-5+2\lambda}(-\frac{\partial}{\partial\omega^2})^\lambda f(\underline{p}, \omega^2) \tag{64}$$

For the quadratically divergent diagrams of Fig. 3 (in 4 dimensional space), this is a well defined formula for $4-2\lambda < n < 2$. *Eq. (64) with sufficiently large λ defines the contribution* *of one loop diagrams to the generalized S-matrix elements in a finite region of the complex n-plane. This region is the domain of convergence of the integrals in Eq. (64).*

By taking a sufficiently large λ the domain of convergence extends to arbitrarily small n. Furthermore, the degree of convergence is $2-n$ as far as ultraviolet behavior is concerned and $n-4+2\lambda$ for the infrared behavior. It is clear that, by choosing a suitable λ and n, one has a representation of the generalized diagrams in some region of the n-plane in terms of arbitrarily convergent integrals.

If a diagram is convergent in 4-dimensional space then the redefinition Eq. (64) exists for $n < n_0$ with $n_0 > 4$. Moreover, for $n = 4$, Eq. (64) equals the result evaluated in the conventional way, as may be seen by taking $\lambda = 1$ and setting $n = 4$. Thus, for finite diagrams our prescription gives the conventional results in the limit $n = 4$. For divergent diagrams, Eq. (64) will be meaningless for $n = 4$. However, as will be shown, Eq. (64) may be continued in the complex n-plane to large n values. The result will in general have a pole at $n = 4$. In order to make sense in the limit $n = 4$, one must introduce counterterms in the perturbation expansion, and those counterterms must be chosen to cancel the poles appearing at $n = 4$. Whether this can be done in a consistent manner is a separate and difficult issue, which is related to the concept of "renormalization" to be discussed in §.10.3.

For values of n outside the domain of convergence of the integrals in Eq. (64), the contribution to the generalized S-matrix elements is to be defined as the analytic continuation of Eq. (64).

It turns out to be possible to construct explicitly this analytic continuation toward larger n values. The method is as follows: By means of partial integration which is valid inside the domain of convergence of Eq. (64) we may derive a new formula, which is identical to Eq. (64) inside the domain of convergence of Eq. (64) but is analytic in n in an enlarged domain. By the principles of analytic continuation, this new formula is then used to define the analytic continuation of Eq. (64) in this enlarged domain.

In view of the importance of this construction, it is of importance to formulate the concept as clearly as possible. We consider the integral specified by

$$I = \int d_k p \frac{p_a^{\lambda 1} p_b^{\lambda 2} ... p_c^{\lambda j}}{((p+k_1)^2 + m_1^2)^{\alpha_1}((p+k_2)^2 + m_2^2)^{\alpha_2} ... ((p+k_\ell)^2 + m_1^2)^{\alpha_\ell}}. \tag{65}$$

Here p_a, p_b (etc.) are components a, b (etc.) of p. The α_j are positive integers and can be larger than 1. The exponents $\lambda_i ... \lambda_j$ are not necessarily integers. Eq. (64) is of the form Eq. (65) with $k = 5$, where the integration over p_5 in Eq. (65) is nothing but the ω-integration in Eq. (64). Thus p_5 occurs with an n-dependent exponent in the numerator. Also $\underline{p}_1$, $\underline{p}_2$, etc. may occur in the numerator, they are contained in Eq. (64) in the function f. Note that the differentiations with respect to ω^2 in Eq. (64) have as net effect an increase of the exponents of the factors in the denominator.
The integral in Eq. (65) is convergent if

$$\begin{aligned} &\lambda_1 > -1,\ \lambda_2 > -1,\ ...,\ \lambda_j > -1; \\ &k + \lambda_1 + \lambda_2 + ...\lambda_j - 2(\alpha_1 + \alpha_2 + ...\alpha_\ell) < 0. \end{aligned} \tag{66}$$

Next we insert in Eq. (65) the expression, which is identical to unity:

$$\frac{1}{k}\sum_{i=1}^{k}\frac{\partial p_i}{\partial p_i}. \tag{67}$$

Within the region (66) we may perform partial integrations with respect to p_i. After some trivial algebra we obtain:

$$I = -\frac{\lambda_1+\lambda_2+...\lambda_j}{k}I + \frac{2(\alpha_1+\alpha_2+...\alpha_\ell)}{k}I - \frac{1}{k}I',$$

with

$$\begin{aligned} I' &= \int d_k p p_a^{\lambda 1}...p_c^{\lambda j}\{\frac{2\alpha_1(m_1^2+k_1^2+(pk_1))}{((p+k_1)^2+m_1^2)^{\alpha_1+1}(\,)^{\alpha_2}...(\,)^{\alpha_\ell}} \\ &\quad +\frac{2\alpha_2(m_2^2+k_2^2+(pk_2))}{(\,)^{\alpha_1}(\,)^{\alpha_2+1}...(\,)^{\alpha_\ell}}+...+\frac{2\alpha_\ell(m_\ell^2+k_\ell^2+(pk_\ell))}{(\,)^{\alpha_1}(\,)^{\alpha_2}...(\,)^{\alpha_\ell+1}}\}, \end{aligned} \tag{68}$$

or

$$I = -\frac{1}{(k+\lambda_1+\lambda_2+...\lambda_j-2\alpha_1-2\alpha_2-...-2\alpha_\ell)}I'. \tag{69}$$

The integral I' converges if

$$\begin{aligned} &\lambda_1>-1,\ \lambda_2>-1,\ ...,\ \lambda_j>-1 \\ &k+\lambda_1+\lambda_2+...\lambda_j-2(\alpha_1+\alpha_2+...\alpha_\ell)<1, \end{aligned} \tag{70}$$

which is a domain larger than Eq. (66). The right hand side of (69) is the explicit representation of the analytic continuation of I into this domain.

For one loop diagrams, the variable n appears linearly in some exponent λ, so that one obtains an explicit representation valid in an arbitrarily large domain in the complex n-plane. With the prescription (69) one may now evaluate the integrals in the example Eq. (58). The result is of course precisely Eq. (60).

For diagrams with two closed loops one may proceed in a similar way. There will be two n-fold integrals, and one writes:

$$\int d_n p \int d_n p' \to \int d_4\underline{p}\int d_4\underline{p}'\int d_{n-4}P\int d_{n-4}P'. \tag{71}$$

In the P' integral the fifth axis is taken in the direction of the $(n-4)$ vector P:

$$(71) \to \int d_4\underline{p}\int d_4\underline{p}'\int d_{n-4}P\int dp_5'\int d_{n-5}P'.$$

The integrands will be independent of the P and P' directions. The integration over angles may be performed:

$$\begin{aligned} (71) \quad \to \quad & \frac{4\pi^{\frac{1}{2}(2n-9)}}{\Gamma(\frac{1}{2}(n-4))\Gamma(\frac{1}{2}(n-5))}\cdot \\ & \cdot\int d_4\underline{p}\int d_4\underline{p}'\int_0^\infty d\omega\,\omega^{n-5}\int_{-\infty}^{\infty}dp_5'\int_0^\infty d\omega'\,\omega'^{n-6}. \end{aligned} \tag{72}$$

The argument of such integrals will be a function of the components $\underline{p}$ and $\underline{p}'$, of ω^2, $p_5'^2+\omega'^2$ and $(p_5'+\omega)^2+\omega'^2$ (arising from p^2, p'^2 and $(p+p')^2$).
Eq. (72) may be written in an elegant from by introducing a two dimensional space, and the vectors

$$q=\begin{pmatrix}\omega\\0\end{pmatrix}, \qquad q'=\begin{pmatrix}p_5'\\\omega'\end{pmatrix} \tag{73}$$

We have, with ϵ_{ij} the completely antisymmetric tensor in two dimensions ($\epsilon_{12}=1$),

$$\begin{aligned}&\int d_np\int d_np'\, f(p^2,p'^2,(p+p')^2)\\&=\frac{2\pi^{\frac{1}{2}(2n-9)-1}}{\Gamma(\frac{1}{2}(n-4))\Gamma(\frac{1}{2}(n-5))}\int d_4\underline{p}\int d_4\underline{p}'\int d_2q\int d_2q'\,(\epsilon_{ij}q_iq_j')^{n-6}\cdot\\&\cdot\theta(\epsilon_{ij}q_iq_j')f(q^2,q'^2,(q+q')^2).\end{aligned} \tag{74}$$

The step-function θ is needed because of the lower limit $\omega'>0$ in the ω' integration in Eq. (72). We have dropped explicit indication of the dependence on the components of $\underline{p}$ and $\underline{p}'$.

The equivalent of Eq. (64) may now be obtained by partial integrations. To this purpose, one observes that

$$(\epsilon_{ij}q_iq_j')^\alpha=\frac{1}{(\alpha+2)(\alpha+1)}\epsilon_{ab}\frac{\partial}{\partial q_a}\frac{\partial}{\partial q_b'}(\epsilon_{ij}q_iq_j')^{\alpha+1}. \tag{75}$$

Applying Eq. (75) λ times together with subsequent partial integrations, one obtains an expression similar to Eq. (64):

$$\begin{aligned}&\frac{2\pi^{\frac{1}{2}(2n-11)}}{\Gamma(\frac{1}{2}(n-4)+\lambda)\Gamma(\frac{1}{2}(n-5)+\lambda)}\\&\times\int d_4\underline{p}\int d_4\underline{p}'\int d_2q\int d_2q'\,(\epsilon_{ij}q_iq_j')^{n-6+2\lambda}\\&\times\ \theta(\epsilon_{ij}q_iq_j')\,(\frac{\partial^2}{\partial q^2\partial q'^2}+\frac{\partial^2}{\partial q^2\partial(q+q')^2}+\frac{\partial^2}{\partial q'^2\partial(q+q')^2})^\lambda\\&\times\ f(q^2,q'^2,(q-q')^2).\end{aligned} \tag{76}$$

Again, Eq. (76) and its analytic continuation to larger n define the contribution of the two-loop diagrams to the generalized S-matrix elements. Explicit representations for large n may be obtained by operations similar to Eqs. (67)–(70) described earlier. Note that we need four such operations in the two loop case.
It is clear that the procedure to redefine the generalized S-matrix elements for the three or more closed loop cases may be worked out in an analogous manner.

The above prescription applies if all loop particles are scalars. To complete our prescription to cover vector fields we note that indices that are part of the propagators contained in the loops now take the values 1 to n for integer n. This is because polarization vectors corresponding to internal lines become n-component vectors. The only practical consequence of this fact is that in doing the vector algebra of all occurring loop indices one must use the rule, in continuing into larger n (including fractal n),

$$\delta_{\mu\mu} = n. \tag{77}$$

After that, one has expressions that can be used to define the diagrams for non-integer n. In establishing Ward-dentities, one sees that there is an interplay between these factors n and the factors n occurring in association with averaging over all directions in p space of factors like $p_\mu p_\nu$. [See Eqs. (A7) and (A8) in the Appendix.]

Extensions to fermions

The extension to fermions is based on the following observation: Everything may be formulated such that only traces of strings of γ-matrices occur. If there are external fermion lines this may be done through the use of suitable projection operators. These traces must then be evaluated according to the rules generalized to n dimensions:

$$\{\gamma_\mu, \gamma_\nu\} = 2\delta_{\mu\nu}, \tag{78}$$

$$Tr(S) = 0 \quad if \quad S \text{ is an odd string of } \gamma's, \tag{79}$$

$$Tr(I) = 4. \tag{80}$$

Remember $\delta_{\mu\mu} = n$. As far as γ-matrices is concerned, any Ward identity relying only on Eq. (78) (as in quantum electrodynamics) will be satisfied for every n.

Note that there is no place for the pseudo-scalar γ_5 (in conventional notation) in Eq. (78), as there is no place for the pseudo tensor $\epsilon_{\mu\nu\alpha\beta}$. This places certain limitations on the regularization method. (See the orginal paper of 't Hooft and Veltman.)

The rule Eq. (80) can be satisfied by finite matrices only for $n = 4$, but this is of no importance because we are only interested in a consistent algebra for $n \neq 4$. Or, in n-dimensional space one will have

$$Tr(I) = f(n)$$

where f is a function of n only. We need only $f(n) = 4$, and the deviations of $f(n)$ from $f(4)$ are never important for Ward identities because one always compares diagrams with an equal number of traces.

As an expample we consider the lowest order photon self-energy diagram in quantum electrodynamics as illustrated by Fig. 1(b). According to Eq. (5), the contribution is proportional to, in the n dimensional space,

$$I_{\mu\nu} = \int d_n p \frac{Tr[\gamma_\mu(i\gamma(p+k)+m)\gamma_\nu(i\gamma p + m)]}{((p+k)^2+m^2)(p^2+m^2)}$$

The trace may be evaluated using Eqs. (78), (79) and (80). Taking denominators together, one obtains:

$$I_{\mu\nu} = 4\int_0^1 dx \int d_n p \frac{\delta_{\mu\nu}(m^2+p^2+pk) - 2p_\mu p_\nu - p_\mu k_\nu - k_\mu p_\nu}{(p^2+2pkx+k^2x+m^2)^2}$$

Using the formulae listed in the Appendix, one obtains

$$I_{\mu\nu} = \frac{4i\pi^{\frac{1}{2}n}}{\Gamma(2)}\Gamma(2-\frac{1}{2}n)\int_0^1 dx \frac{2x(1-x)(k_\mu k_\nu - \delta_{\mu\nu}k^2)}{(m^2+k^2x(1-x))^{2-\frac{1}{2}n}},$$

which is manifestly gauge invariant. It is of some interest to compare this result with what we have obtained earlier in the Pauli-Villars regularization method. [See Eqs. (38) and (39).]

10.3 Introduction to Renormalization

Consider a physical process as illustrated by Fig. 4.

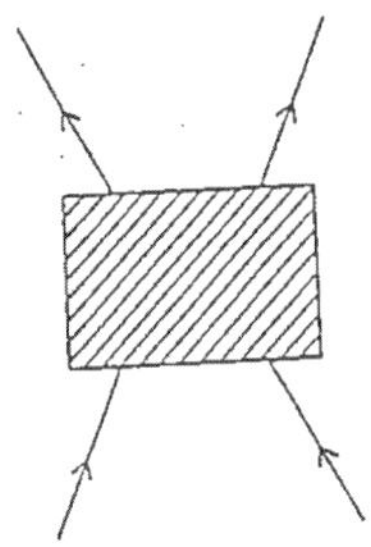

Figure 4: A schematic view of a typical physical process.

Let $\{p_j\}$, m, and g be, respectively, the external momenta, the mass parameter, and the dimensionless coupling constant. (Our discussion refers to an arbitrary physical process.) The amplitude, i.e. the S-matrix element, can be represented by a vertex function:

$$\Gamma(-\frac{p_j^2}{\mu^2}, \frac{m^2}{\mu^2}, g),$$

where μ^2 is the renormalization parameter required in the treatment of ultraviolet divergent diagrams. For instance, μ^2 can be taken to be the cut-off mass squared M^2 in the Pauli-Villars regularization scheme or μ^2 in the dimensional regularization scheme. Now, suppose that we change the renormalization point from μ^2 to $\bar{\mu}^2$. Note that

$$g \to \bar{g}(\frac{\bar{\mu}^2}{\mu^2}, \frac{m^2}{\mu^2}, g), \tag{81a}$$

$$m \to \bar{m}(\frac{\bar{\mu}^2}{\mu^2}, \frac{m^2}{\mu^2}, g). \tag{81b}$$

Thus, the consistency of the theory demands that the amplitude calculated at the new renormalization point $\bar{\mu}^2$ must be related to that obtained at μ^2 in a simple manner:

$$\Gamma(-\frac{p_j^2}{\bar{\mu}^2}, \frac{\bar{m}^2}{\bar{\mu}^2}, \bar{g}) = Z_\Gamma(\frac{\bar{\mu}^2}{\mu^2}, \frac{m^2}{\mu^2}, g)\Gamma(-\frac{p_j^2}{\mu^2}, \frac{m^2}{\mu^2}, g), \tag{81c}$$

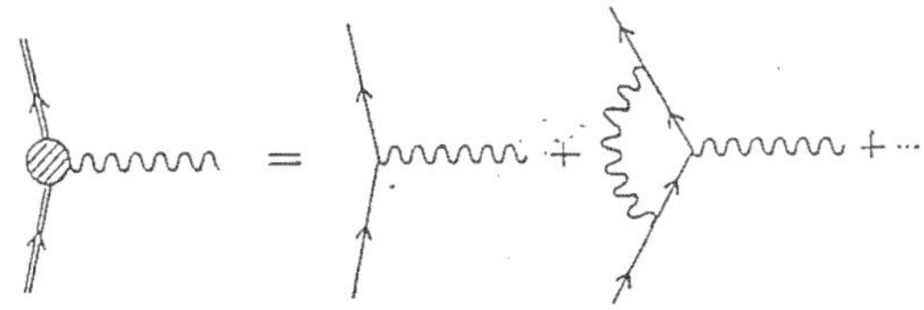

Figure 5: (a). A pictorial representation of Eq. (81a).

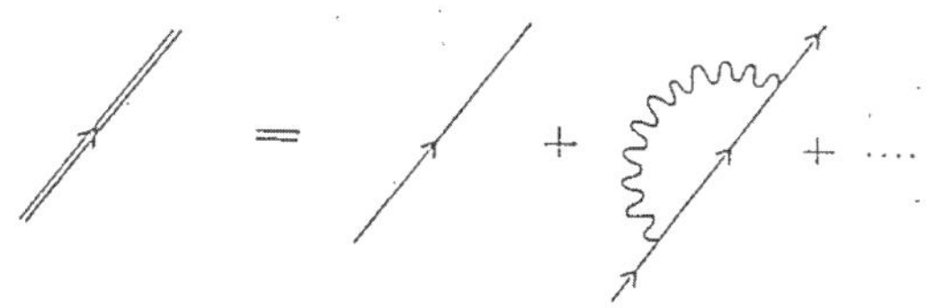

Figure 5: (b). A pictorial representation of Eq. (81b).

where Z_Γ is a characteristic function for the given vertex Γ.

Differentiating Eq. (81c) with respect to $\bar{\mu}^2$ and then setting $\bar{\mu}^2 = \mu^2$, we obtain

$$\{\mu^2 \frac{\partial}{\partial \mu^2} + \beta(\frac{m^2}{\mu^2}, g)\frac{\partial}{\partial g} + \gamma_m(\frac{m^2}{\mu^2}, g)m^2 \frac{\partial}{\partial m^2} - \gamma_\Gamma(\frac{m^2}{\mu^2}, g)\}\Gamma(-\frac{p_j^2}{\mu^2}, \frac{m^2}{\mu^2}, g) = 0, \tag{82a}$$

with

$$\beta(\frac{m^2}{\mu^2}, g) \equiv \bar{\mu}^2 \frac{\partial}{\partial \bar{\mu}^2} \bar{g}(\frac{\bar{\mu}^2}{\mu^2}, \frac{m^2}{\mu^2}, g) \mid_{\bar{\mu}=\mu}, \tag{82b}$$

$$\begin{aligned} \gamma_m(\frac{m^2}{\mu^2}, g) &\equiv \frac{\bar{\mu}^2}{\bar{m}^2} \frac{\partial}{\partial \bar{\mu}^2} \bar{m}^2(\frac{\bar{\mu}^2}{\mu^2}, \frac{m^2}{\mu^2}, g) \mid_{\bar{\mu}=\mu}, \\ &= \bar{\mu}^2 \frac{\partial}{\partial \bar{\mu}^2} ln\, \bar{m}^2(\frac{\bar{\mu}^2}{\mu^2}, \frac{m^2}{\mu^2}, g) \mid_{\bar{\mu}=\mu}, \end{aligned} \tag{82c}$$

$$\gamma_\Gamma(\frac{m^2}{\mu^2}, g) \equiv \bar{\mu}^2 \frac{\partial}{\partial \bar{\mu}^2} ln\, Z_\Gamma(\frac{\bar{\mu}^2}{\mu^2}, \frac{m^2}{\mu^2}, g) \mid_{\bar{\mu}=\mu}. \tag{82d}$$

Eq. (82a) is known by several names: Ovsyannikov's equation, Gell-Mann-Low equation, or Callan-Symanzik equation. Although it is clearly beyond the scope of this book to elucidate how these renormalization-group (RG) equations [Eqs. (82)] should be used in connection with gauge field theories, we wish to stress the point that presence of infinities would render a field theory meaningless if it were not possible to treat these infinities *consistently* in the framework of RG equations. A field theory for which infinities can be handled *consistently* in the framework of RG equations will be referred to as a "renormalizable field theory". Otherwise, it is a non-renormalizable field theory. [*Caution:* This strict definition of "renormalizability", albeit the most appropriate one, is not adopted in field theory books because it may be extremely difficult to prove that a field theory is renormalizable.]

More precisely, presence of infinities, if taken at its face value, already renders the theory meaningless mathematically. However, a renormalizable field theory makes sense because it makes sense for an arbitrary renormalization point $\bar{\mu}^2\,(<\infty)$ and the theories at two different renormalization points, say μ^2 and $\bar{\mu}^2$ (both $<\infty$), are correlated via RG equations. In other words, there is a specific structure, which exists at all finite μ^2 and so at $\mu^2 \to \infty$, that makes a renormalizable field theory a meaningful theory.

It is a rather tedious task to show how, in a given regularization scheme, those ultraviolet infinities which we encounter in QED can be absorbed consistently into redefined couplings including the charge e and the mass m. This task is of course very important since, otherwise, the renormalization step illustrated by Eqs. (45)-(54) to interpret the finite parts of divergent graphs as being physical would be completely unfounded. Since it is not the objective of this book to expose ourselves to the technical complexities associated with the actual renormalization program for QED, we shall close our discussions by turning our attention to another very interesting consequence of the renormalization group equations, namely, that the coupling constant is running, i.e., depends on Q^2 that we are probing.

We may determine the beta function using Eqs. (81a) and (82b). We quote the result for QED (Baker and Johnson 1969):

$$\beta(\alpha) = \frac{\alpha^2}{3\pi} + \frac{\alpha^3}{4\pi^2} + 0(\alpha^4). \tag{83}$$

Note that the first term is just α times the coefficient of the $ln\,\frac{M^2}{m^2}$ term in Eq. (44). To make the physical meaning of Eq. (83) a little more transparent, we need to consider the RG equation for $\bar{g}(\frac{\bar{\mu}^2}{\mu^2}, g)$ for, say, a massless theory. Let us define

$$t \equiv \frac{\bar{\mu}^2}{\mu^2}. \tag{84}$$

Choosing $t'' = t't$, we have

$$g'' \equiv \bar{g}(t', g') \quad with \quad g' \equiv \bar{g}(t, g),$$

or,

$$\bar{g}(t't, g') = \bar{g}(t', \bar{g}(t, g)).$$

Differentiating with respect to t' and then setting $t' = 1$, we find

$$t\frac{\partial}{\partial t}\bar{g}(t, g) = \beta(\bar{g}). \tag{85}$$

Note that Eq. (85) is more general than Eq. (82b), the latter defining $\beta(g)$.

Eq. (85) can readily be solved for $\bar{g}$. We obtain

$$\int_{t=1}^{t} \frac{d\bar{g}}{\beta(\bar{g})} = ln\ t. \tag{86}$$

Consider the case of QED to lowest order:

$$\beta(\alpha) = \frac{\alpha^2}{3\pi} \tag{87a}$$

We find

$$3\pi(\frac{1}{\alpha(\mu^2)} - \frac{1}{\alpha(\bar{\mu}^2)}) = ln\frac{\bar{\mu}^2}{\mu^2},$$

or,

$$\alpha(\bar{\mu}^2) = \frac{\alpha(\mu^2)}{1 - \frac{1}{3\pi}\alpha(\mu^2)ln\frac{\bar{\mu}^2}{\mu^2}}. \tag{87b}$$

In an approximate sense, the renormalization parameter μ^2 can be identified also as the momentum squared Q^2 which defines the scale of the physics that we are probing. In other words, we may rewrite Eq. (87b) in a more familiar form:

$$\alpha(Q^2) = \frac{\alpha(Q_0^2)}{1 - \frac{1}{3\pi}\alpha(Q_0^2)ln\frac{Q^2}{Q_0^2}}. \tag{88}$$

At $Q_0^2 \sim (1GeV)^2$, we have $\alpha(Q_0^2) = \frac{1}{137}$. We note that

$$\alpha(Q^2) \rightarrow \infty \quad as \quad 1 - \frac{1}{3\pi}\alpha(Q_0^2)ln\frac{Q^2}{Q_0^2} \rightarrow 0, \tag{89}$$

which is the well-known Landau ghost problem.

Appendix

Some Useful Formulae For Dimensional Regularization

The n-dimensional integration is defined by

$$\int d_n x f(x) = \int f(x) r^{n-1} dr\, sin^{n-2}\theta_{n-1} d\theta_{n-1}\, sin^{n-3}\theta_{n-2} d\theta_{n-2} ... d\theta_1, \tag{A1}$$

with $0 \leq \theta_i \leq \pi$, except $0 \leq \theta_1 \leq 2\pi$.

If $f(x)$ depends only on $r = \sqrt{x_1^2 + ... x_n^2}$ one may perform the integration over angles using

$$\int_0^\pi sin^m \theta d\theta = \sqrt{\pi} \frac{\Gamma(\frac{1}{2}(m+1))}{\Gamma(\frac{1}{2}(m+2))}, \tag{A2}$$

leading to

$$\int d_n x f(r) = \frac{2\pi^{\frac{1}{2}n}}{\Gamma(\frac{1}{2}n)} \int f(r) r^{n-1} dr. \tag{A3}$$

Eqs. (A1)–(A3) define the basis of the dimensional regularization. Note that the causality $-i\epsilon$ requirement is *not* there, i.e., not in the definition.

Another useful formula is given by

$$\int_0^\infty dx \frac{x^\beta}{(x^2 + M^2)^\alpha} = \frac{1}{2} \frac{\Gamma(\frac{1}{2}(\beta+1))\Gamma(\alpha - \frac{1}{2}(\beta+1))}{\Gamma(\alpha)(M^2)^{\alpha - \frac{1}{2}(\beta+1)}} \tag{A4}$$

Keeping the prescriptions and definitions of §.10.2. in mind, the following equations hold for arbitrary n:

$$\int d_n p \frac{1}{(p^2 + 2kp + m^2)^\alpha} = \frac{i\pi^{\frac{1}{2}n}}{(m^2 - k^2)^{\alpha - \frac{1}{2}n}} \frac{\Gamma(\alpha - \frac{1}{2}n)}{\Gamma(\alpha)}. \tag{A5}$$

$$\int d_n p \frac{p_\mu}{(p^2 + 2kp + m^2)^\alpha} = \frac{i\pi^{\frac{1}{2}n}}{(m^2 - k^2)^{\alpha - \frac{1}{2}n}} \frac{\Gamma(\alpha - \frac{1}{2}n)}{\Gamma(\alpha)} (-k_\mu). \tag{A6}$$

$$\int d_n p \frac{p_\mu p_\mu}{(p^2 + 2kp + m^2)^\alpha} = \frac{i\pi^{\frac{1}{2}n}}{(m^2 - k^2)^{\alpha - \frac{1}{2}n}} \cdot \frac{1}{\Gamma(\alpha)} \{\Gamma(\alpha - \frac{1}{2}n) k^2 + \Gamma(\alpha - 1 - \frac{1}{2}n) \frac{1}{2} n (m^2 - k^2)\}. \tag{A7}$$

$$\int d_n p \frac{p_\mu p_\nu}{(p^2 + 2kp + m^2)^\alpha} = \frac{i\pi^{\frac{1}{2}n}}{(m^2 - k^2)^{\alpha - \frac{1}{2}n}} \cdot \frac{1}{\Gamma(\alpha)} \{\Gamma(\alpha - \frac{1}{2}n) k_\mu k_\nu + \Gamma(\alpha - 1 - \frac{1}{2}n) \frac{1}{2} \delta_{\mu\nu} (m^2 - k^2)\}. \tag{A8}$$

$$\int d_n p \frac{p_\mu p_\nu p_\lambda}{(p^2+2kp+m^2)^\alpha} = \frac{i\pi^{\frac{1}{2}n}}{(m^2-k^2)^{\alpha-\frac{1}{2}n}} \cdot \frac{1}{\Gamma(\alpha)}\{-\Gamma(\alpha-\frac{1}{2}n)k_\mu k_\nu k_\lambda$$
$$-\Gamma(\alpha-1-\frac{1}{2}n)\frac{1}{2}(\delta_{\mu\nu}k_\lambda+\delta_{\mu\lambda}k_\nu+\delta_{\nu\lambda}k_\mu)(m^2-k^2)\}. \tag{A9}$$

$$\int d_n p \frac{p^2 p_\mu}{(p^2+2kp+m^2)^\alpha} = \frac{i\pi^{\frac{1}{2}n}}{(m^2-k^2)^{\alpha-\frac{1}{2}n}} \cdot \frac{1}{\Gamma(\alpha)}(-k_\mu)\{\Gamma(\alpha-\frac{1}{2}n)k^2$$
$$+\Gamma(\alpha-\frac{1}{2}n-1)\frac{1}{2}(n+2)(m^2-k^2)\}. \tag{A10}$$

The above equations contain indices μ, ν, λ. These indices are understood to be contracted with arbitrary n-vectors q_1, q_2 etc. In computing the integrals one first integrates over the part of n-space orthogonal to the vectors k, q_1, q_2 etc., using (A1)–(A4). After that, the expressions are meaningful also for non-integer n. Note that formally (A6)-(A10) may be obtained from (A5) by differentiation with respect to k, or by using $p^2 = (p^2+2pk+m^2) - 2pk - m^2$.

As is well-known, the Feynman parameter method for non-integer exponents is often very useful:

$$\frac{1}{a^\alpha b^\beta} = \frac{\Gamma(\alpha+\beta)}{\Gamma(\alpha)\Gamma(\beta)} \int_0^1 dx \frac{x^{\alpha-1}(1-x)^{\beta-1}}{(ax+b(1-x))^{\alpha+\beta}}, \tag{A11}$$

valid for $\alpha > 0, \beta > 0$. If one needs this formula for α in the neighbourhood of 0 one may write

$$\frac{1}{a^\alpha b^\beta} = \frac{a}{a^{\alpha+1}b^\beta}$$

and then use (A11). The generalization of eq. (A11) for many factors is also useful:

$$\frac{1}{a_1^{\alpha_1}a_2^{\alpha_2}...a_m^{\alpha_m}} = \frac{\Gamma(\alpha_1+\alpha_2+...\alpha_m)}{\Gamma(\alpha_1)\Gamma(\alpha_2)...\Gamma(\alpha_m)} \cdot \int_0^1 dx_1 \int_0^{x_1} dx_2 ... \int_0^{x_{m-2}} dx_{m-1}$$
$$\times \frac{x_{m-1}^{\alpha_i-1}(x_{m-2}-x_{m-1})^{\alpha_2-1}...(1-x_1)^{\alpha_m-1}}{[a_1 x_{m-1}+a_2(x_{m-2}-x_{m-1})+....+a_m(1-x_1)]^{\alpha_1+\alpha_2+...\alpha_m}}. \tag{A12}$$

References

't Hooft, G. and Veltman, M., Nucl. Phys. **B44**, 189 (1972); on dimensional regularization.

Kinoshita, T. and Lindquist, W.B., Phys. Rev. Lett. **47**, 1573, (1981); on higher-order calculation of the $g-2$.

Baker, M. and Johnson, K., Phys. Rev. **183**, 1292 (1969); on the β function for QED.

Bjorken, J.D. and Drell, S.D., *Relativistic Quantum Fields* (McGraw-Hill, New York, 1965).

Halzen, F. and Martin, A.D., *Quarks and Leptons: An Introductory Course in Modern Particle Physics* (John Wiley & Sons, New York, 1984).

Bogoliubov, N.N. and Shirkov, D.V., *Introduction to the Theory of Quantized Fields*, 3rd Edition (John Wiley & Sons, New York, 1980).

Itzykson, C., and Zuber, J., *Introduction to Quantum Field Theory* (McGraw-Hill, New York, 1980).

Cheng, T.-P., and Li, L.-F., *Gauge Theory of Elementary Particle Physics* (Clarendon Press, Oxford, 1984).

Exercises: *Chapter 10*

1. On the Pauli-Villars regularization scheme, work out the following problems.

(i) Prove Eqs. (12a)-(12g).

(ii) Examine the step leading to Eq. (37) starting from Eq. (32).

(iii) Derive Eq. (38) from Eq. (37).

2. On the dimensional regularization scheme, work out the following problems.

(i) Prove Eqs. (A3)-(A6), (A8), and (A11).

(ii) Fill the missing steps leading to Eq. (60).

(iii) Obtain the last equation in §.10.2.

3. Consider possible renormalization effects due to the diagrams illustrated in Figs. 5(a) and 5(b). Follow the step leading to Eq. (82a) and try to construct the set of renormalization group equations for the charge e, the mass m, etc. Could you justify Eqs. (45)-(54)?

Part C. The Standard Model

Chapter 11. Symmetries, Transformations, and Invariants

In the last part of this book, we wish to describe in detail the Standard Model[1], starting with quantum chromodynamics (QCD), an $SU(3)$ gauge field theory of strong interactions, and the Glashow-Salam-Weinberg (GSW) $SU(2) \times U(1)$ gauge field theory of electroweak interactions. In the Standard Model, we declare that we are living in the quantum 4-dimensional Minkowski space-time with the force-fields gauge-group structure $SU_c(3) \times SU_L(2) \times U(1) \times SU_f(3)$ built-in from the very beginning. From this overall background, we can see the quark world of nuclear sizes and also the lepton world of atomic sizes.

The force-fields gauge-group structure has to built-in from the very beginning because the involved forces sort of determine the displacements from one point to another in the (Minkowski) space-time via the gauge principle.

Hermann Weyl in 1919 introduced the concept of "scale invariance", or "gauge invariance" as Weyl himself called it, in his attempt to unify gravity and eletromagnetism. In the quantum mechanical description of a charged particle, one makes the substitution:

$$p_\mu - \frac{e}{c}A_\mu \rightarrow -i\hbar\{\frac{\partial}{\partial x_\mu} - i\frac{e}{\hbar c}A_\mu\}.$$

In 1927, Fock observed that one could obtain quantum electrodynamics using this operator. London (1927) soon pointed out that Fock's observation amounts to adding an additional phase i to Weyl's notion of "gauge invariance". Thus, the concept of "gauge invariance", as it is called nowadays, is in fact "phase invariance" of some sort. Another breakthrough along this line had to wait until 1954 as Yang and Mills[2] attempted to describe nucleon-nucleon interactions in terms of *local* $SU(2)$ isospin phase symmetry. The idea was revolutionary since, in the proposed non-abelian gauge theory as often referred to as "Yang-Mills theory" nowadays, the way of identifying, or distinguishing between, protons and neutrons, varies from one space-time point to the other. Analogously, the color of a quark, as described by an $SU(3)$ Yang-Mills theory (i.e., QCD), varies with the space-time point under local $SU(3)$ gauge transformations.

In the preceding chapters, we have seen, to a certain extent, the important role played by the concepts of the invariance of physical laws under certain transformations, as may be summarized by Noether's theorem formulated in §.6.3. According to Noether's theorem, these concepts are related to the concepts of symmetry and laws, known as "conservation laws".

It is often useful to differentiate between *exact symmetries* and *approximate symmetries*. Invariance under Lorentz transformations (which form a Lie group called "Lorentz group") is believed to be an "exact symmetry". Lorentz group contains the rotational $SU(2)$ symmetry group as a subgroup. On the other hand, the isospin symmetry, an $SU(2)$ symmetry which arises from the almost equal mass for the proton and the neutron, is regarded as an "approximate symmetry". In the framework of present-day gauge field theories, gauge principle is taken as an exact symmetry whereas conservation of "lepton number" or "baryon number" in fundamental reactions is assumed to be only "approximate", although such distinction is yet to be confirmed further by experiments.

---

[1] *W-Y. Pauchy Hwang, arXiv:1304.4705v2 [hep-ph] 25 August 2015.*

[2] *Yang, C.N., and Mills, R. L., Phys. Rev.* **96**, *191 (1954).*

The revolutionary suggestion to extend the notion of "gauge invariance" from the $U(1)$ group to a non-abelian group such as $SU(2)$ or $SU(3)$ is to be elucidated in §.12.1. To this end, we wish in this chapter to introduce Lie groups $SU(2)$ and $SU(3)$ as groups of symmetries of some kind. In particular, we review in §.11.1., respectively, rotation symmetry as an *exact* $SU(2)$ symmetry and isospin symmetry as an *approximate* $SU(2)$ symmetry. In §.11.2., we consider flavor $SU(3)$ symmetry, a generalization of isospin symmetry to include strangeness, as an example of *approximate* $SU(3)$ symmetry. As a by-product, we introduce the naïve quark model for low-lying mesons and baryons. In §.14.3., we mention briefly additional symmetries often encountered in particle and nuclear physics. In the Appendix, we discuss briefly potentials and phase in quantum mechanics.

11.1 $SU(2)$ Symmetries in Particle Physics

(1) Invariance under Spatial Rotations as an Exact $SU(2)$ Symmetry

Consider rotations in configuration space of a physical system that is characterized by the Hamiltonian H. The states of the system are labeled by $\mid \psi\rangle$, $\mid \phi\rangle$, etc.

The set of rotations of a system form a group, each rotation being an element of the group. Two successive rotations R_1 followed by R_2 (written as the "product" R_2R_1) are equivalent to a single rotation (that is, to another group element). The set of rotations is closed under "group multiplication." There is an identity element (no rotation), and every rotation has an inverse (rotate back again). The "product" is not necessarily commutative, $R_1R_2 \neq R_2R_1$, but the associative law $R_3(R_2R_1) = (R_3R_2)R_1$ always holds. The rotation group is a continuous group in that each rotation can be labeled by a set of continuously varying parameters $(\theta_1, \theta_2, \theta_3)$. These can be regarded as the components of a vector θ directed along the axis of rotation with magnitude given by the angles of rotation. See §.4.5. (2) for the specification of a rotation.

Thus, the theory of groups is the mathematics that arise naturally in treatment of symmetries in physics.

The rotation group is a Lie group. Every rotation can be expressed as the product of a succession of infinitiesimal rotations (rotations arbitarily close to the identity). The group is thus completely defined by the "neighborhood of the identity."

To ensure that an experimental result does not depend on the specific laboratory orientation of the system which we are measuring, rotations must form a symmetry group of a system. The rotations must leave the transition probabilities of the system invariant. Suppose that under a rotation R the states of a system transform as

$$\mid \psi\rangle \to\mid \psi'\rangle = U \mid \psi\rangle. \tag{1}$$

The probability that a system described by $\mid \psi\rangle$ will be found in state $\mid \phi\rangle$ must be unchanged by R,

$$\mid \langle\phi \mid \psi\rangle \mid^2 =\mid \langle\phi' \mid \psi'\rangle \mid^2 =\mid \langle\phi \mid U^\dagger U \mid \psi\rangle \mid^2, \tag{2}$$

so that U must be a unitary operator. The operators $U(R_1)$, $U(R_2)$ ...., form a group with exactly the same structure as the original group R_1, R_2, ... That is, they form a unitary representation of the rotation group.

It is assumed that the Hamiltonian H is unchanged by a symmetry operator R of the system and the matrix elements are preserved:

$$\langle \phi' \mid H \mid \psi' \rangle = \langle \phi \mid U^\dagger H U \mid \psi \rangle = \langle \phi \mid H \mid \psi \rangle,$$

so that

$$H = U^\dagger H U, \qquad \text{or,} \qquad [U, H] \equiv UH - HU = 0. \tag{3}$$

The equation of motion,

$$i\frac{d}{dt} \mid \psi(t)\rangle = H \mid \psi(t)\rangle, \tag{4}$$

is not affected by the designated rotation. Accordingly, the expectation value of U is a constant of motion:

$$i\frac{d}{dt}\langle \psi(t) \mid U \mid \psi(t)\rangle = \langle \psi(t) \mid UH - HU \mid \psi(t)\rangle = 0, \tag{5}$$

where it is assumed that U does not have an explicit time dependence.

As a basic property of a Lie group, all the rotation group properties are in fact defined through infinitesimal rotations in the neighborhood of the identity. To see this, we consider a rotation through an infinitesimal angle ε about the 3- (or z) axis. We may write, to first order in ε,

$$U = 1 - i\varepsilon J_3. \tag{6}$$

The operator J_3 is called the generator of rotations about the 3-axis. We have

$$\begin{aligned} 1 = U^\dagger U &= (1 + i\varepsilon J_3^\dagger)(1 - i\varepsilon J_3) \\ &= 1 + i\varepsilon(J_3^\dagger - J_3) + O(\varepsilon^2). \end{aligned}$$

so that J_3 is hermitian (an important property for a physical observable). Note that the factor i has been introduced in Eq. (6) to ensure the hermiticity of the operator J_3.

To understand the physical significance of the operator J_3, we consider the effect of a rotation on the wave function $\psi(\mathbf{r})$ describing the system. As already familiar in quantum mechanics, one should distinguish between two points of view: Either we can keep the axes fixed and rotate the system (the active viewpoint) or we may rotate the axes and keep the physical system fixed (the passive viewpoint). The viewpoints are equivalent; a rotation of the axes through an angle θ is the same as a rotation of the physical system by $-\theta$. We adopt the active viewpoint and rotate the physical system. The wave function ψ' describing the rotated state at $\mathbf{r}$ is then equal to the original function ψ at the point $R^{-1}\mathbf{r}$, which is transformed into $\mathbf{r}$ under the rotation R:

$$\psi'(\mathbf{r}) = \psi(R^{-1}\mathbf{r}). \tag{7}$$

This determines the correspondence between ψ' and ψ:

$$\psi' = U\psi. \tag{8}$$

Thus, we obtain, for an infinitesimal rotation ε about the z axis,

$$\begin{aligned} U\psi(x,y,z) &= \psi(R^{-1}\mathbf{r}) = \psi(x+\varepsilon y, y-\varepsilon x, z) \\ &\simeq \psi(x,y,z) + \varepsilon(y\frac{\partial\psi}{\partial x} - x\frac{\partial\psi}{\partial y}) \\ &= (1 - i\varepsilon(xp_y - yp_x))\psi. \end{aligned} \tag{9}$$

Comparing Eq. (9) with Eq. (6), we identify the *generator*, J_3, of rotations about the 3- (or z) axis with the third-component of the angular momentum operator.

Eq. (5) indicates that the eigenvalues of the observable J_3 are constants of motion. A symmetry of the system has led to a conservation law. The fact that experiments performed with an apparatus of different orientations yield identical physics results (rotational symmetry) has led to the conservation of angular momentum. This is in fact a special case of Noether's theorem introduced earlier in §.6.3. Of course, Noether's theorem may be realized for an arbitrary symmetry transformation that leaves the system invariant.

A rotation through a finite angle θ may be built up from a succession of n infinitesimal rotations

$$U(\theta) = (U(\varepsilon))^n = (1 - i\frac{\theta}{n}J_3)^n \to e^{-i\theta J_3}, \quad \text{as } n \to \infty. \tag{10}$$

Upon introducing similar hermitian generators of rotations, J_1 and J_2, respectively about the 1- and 2-axes, we obtain the commutator algebra of the generators as follows:

$$[J_i, J_j] = i\varepsilon_{ijk}J_k, \tag{11}$$

where $\varepsilon_{ijk} = +1(-1)$ if ijk is a cyclic (anticyclic) permutation of 1 2 3 and $\varepsilon_{ijk} = 0$ otherwise. The set of relations (13), which is said to form a Lie algebra, completely define the group properties; the ε_{ijk} coefficients are called the structure constants of the group.

For the rotation group, the combination

$$J^2 = J_1^2 + J_2^2 + J_3^2, \tag{12}$$

commutes with all three generators of the group,

$$[J^2, J_i] = 0 \qquad \text{with } i = 1,2,3. \tag{13}$$

Such nonlinear function of the generators which commute with all the generators is called an *invariant* or *Casimir operator*. It follows that we can construct simultaneous eigenstates $\mid jm\rangle$ of J^2 and one of the generators, say J_3. Using only Eq. (11), we may show that

$$\begin{aligned} J^2 \mid jm\rangle &= j(j+1) \mid jm\rangle, \\ J_3 \mid jm\rangle &= m \mid jm\rangle, \end{aligned} \tag{14}$$

with $m = -j, -j+1, ..., j$, and where j can take one of the values 0, $\frac{1}{2}$, 1, $\frac{3}{2}$, ... Introducing the "raising" and "lowering" operators:

$$J_\pm = J_1 \pm iJ_2, \tag{15}$$

we find

$$J_{\pm} \mid jm\rangle = (j(j+1) - m(m \pm 1))^{1/2} \mid j, m \pm 1\rangle. \tag{16}$$

A state $\mid jm\rangle$ is transformed under a rotation through an angle θ about the 2-axis into a linear combination of the $2j+1$ states $\mid jm'\rangle$, with $m' = -j, -j+1, ..., j$:

$$e^{-i\theta J_2} \mid jm\rangle = \sum_{m'} d^j_{m'm}(\theta) \mid jm'\rangle, \tag{17}$$

where the coefficients $d^j_{m'm}$ are frequently called *rotation matrices*. Eq. (17) indicates that, although all the $2j+1$ states are mixed by rotations, the states with the same j and all possible m transform among themselves under rotations. That is, they form the basis of a $(2j+1)$-dimensional *irreducible representation* of the rotation group. The set of such states is called a *multiplet*.

The generators in the lowest-dimensional nontrivial representation of the rotation group ($J = \frac{1}{2}$) may be written as follows:

$$J_i = \frac{1}{2}\sigma_i, \qquad \text{with } i = 1, 2, 3, \tag{18}$$

where σ_i are the Pauli matrices

$$\sigma_1 = \begin{pmatrix} 0 & 1 \\ 1 & 0 \end{pmatrix}, \qquad \sigma_2 = \begin{pmatrix} 0 & -i \\ i & 0 \end{pmatrix}, \qquad \sigma_3 = \begin{pmatrix} 1 & 0 \\ 0 & -1 \end{pmatrix}. \tag{19}$$

The basis for this representation is conventionally chosen to be the eigenvectors of σ_3,

$$\begin{pmatrix} 1 \\ 0 \end{pmatrix} \quad \text{and} \quad \begin{pmatrix} 0 \\ 1 \end{pmatrix},$$

which describe a spin-$\frac{1}{2}$ particle of spin projection up ($m = +\frac{1}{2}$ or $\uparrow$) and spin projection down ($m = -\frac{1}{2}$ or $\downarrow$) along the 3-axis, respectively.

The Pauli matrices σ_i are hermitian, and the transformation matrices

$$U(\theta_i) = e^{-i\theta_i \sigma_i/2}, \tag{20}$$

are unitary. The set of all unitary 2×2 matrices is known as the group $U(2)$. However, $U(2)$ is larger than the group of matrices $U(\theta_i)$, since the generators σ_i all are traceless. For any hermitian traceless matrix σ, we can show that

$$det\,(e^{i\sigma}) = e^{iTr\,\sigma} = 1. \tag{21}$$

Since the unit determinant is preserved in matrix multiplication, the set of traceless unitary 2×2 matrices form a special subgroup, $SU(2)$, of $U(2)$. $SU(2)$ is used to denote the special unitary group in two dimensions. The set of transformation matrices $U(\theta_i)$ therefore form an $SU(2)$ group. The $SU(2)$ algebra is just the algebra of the generators J_i, as given by Eq. (11). There are $1, 2, 3, 4, ...$ dimensional representations of $SU(2)$ corresponding to $j = 0, \frac{1}{2}, 1, \frac{3}{2}, ...$, respectively. The two-dimensional representation specified by the σ-matrices is called the *fundamental representation* of $SU(2)$, since from which all other representations can be constructed.

(2) Isospin Symmetry

Owing to the fact that the proton and neutron masses are nearly equal ($m_p = 938.27\,MeV$ versus $m_n = 939.57\,MeV$), we may choose to view the two particles as two substates of one and the same particle called the "nucleon," in analogy with the two spin substates of an electron. We note that electrons of spin *up* and *down* are thought of as one, not two, particles. Accordingly, the mathematical structure used to discuss the similarity of the neutron and the proton is just a xerox copy of spin, and is called "*isospin*".

Consider the description of the two-nucleon system. Each nucleon has spin $\frac{1}{2}$ (with spin states $\uparrow$ and $\downarrow$) and, following the rules for the addition of angular momenta, the composite system may have total spin $S = 1$ or $S = 0$. The composition of these spin triplet and spin singlet states is given by

$$\begin{cases} |\, S = 1,\, M_s = 1\rangle = \uparrow\uparrow \\ |\, S = 1,\, M_s = 0\rangle = \sqrt{\frac{1}{2}}\,(\uparrow\downarrow\, +\, \downarrow\uparrow) \\ |\, S = 1,\, M_s = -1\rangle = \downarrow\downarrow \end{cases} \tag{22a}$$

$$|\, S = 0,\, M_s = 0\rangle = \sqrt{\frac{1}{2}}(\uparrow\downarrow\, -\, \downarrow\uparrow). \tag{22n}$$

Analogously, each nucleon is postulated to have isospin $I = \frac{1}{2}$, with $I_3 = \pm\frac{1}{2}$ for protons and neutrons, respectively. $I = 1$ and $I = 0$ states of the nucleon-nucleon system are then given by

$$\begin{cases} |\, I = 1, \quad I_3 = 1\rangle = pp \\ |\, I = 1, \quad I_3 = 0\rangle = \sqrt{\frac{1}{2}}(pn + np) \\ |\, I = 1, \quad I_3 = -1\rangle = nn \end{cases} \tag{23a}$$

$$|\, I = 0,\, I_3 = 0\rangle = \sqrt{\frac{1}{2}}(pn - np). \tag{23b}$$

The three-dimensional isospin space is so defined that its relation to isospin is identical to that of configuration space (x, y, z) to spin. A rotation in isospin space induces transformations on isospin states, called "isospin transformations". There is much evidence to show that the nuclear force is invariant under isospin transformations. In other words, the interaction is independent of the value of I_3 in the $I = 1$ multiplet of Eq. (23a).

Nowadays, we may attribute the success of $SU(2)$ isospin symmetry to the fact that the u and d quarks have almost equal masses while strong interactions (or QCD) are flavor independent.

11.2 Flavor $SU(3)$ Symmetry: Isospin and Strangeness

(1) $SU(3)$ Group

The group $SU(3)$ consists of all unitary 3×3 matrices with $det\, U = 1$. The generators for the group may be taken to be any $3^2 - 1 = 8$ linearly independent traceless hermitian 3×3 matrices. Following Gell-Mann,[3] we may take

$$\lambda_i = \begin{pmatrix} \tau_i & 0 \\ 0 & 0 \end{pmatrix}, \quad i = 1, 2, 3; \tag{24a}$$

$$\lambda_4 = \begin{pmatrix} 0 & 0 & 1 \\ 0 & 0 & 0 \\ 1 & 0 & 0 \end{pmatrix}, \quad \lambda_5 = \begin{pmatrix} 0 & 0 & -i \\ 0 & 0 & 0 \\ i & 0 & 0 \end{pmatrix}; \tag{24b}$$

$$\lambda_6 = \begin{pmatrix} 0 & 0 & 0 \\ 0 & 0 & 1 \\ 0 & 1 & 0 \end{pmatrix}, \quad \lambda_7 = \begin{pmatrix} 0 & 0 & 0 \\ 0 & 0 & -i \\ 0 & i & 0 \end{pmatrix},$$

$$\lambda_8 = \frac{1}{\sqrt{3}} \begin{pmatrix} 1 & 0 & 0 \\ 0 & 1 & 0 \\ 0 & 0 & -2 \end{pmatrix}. \tag{24c}$$

With τ_i $(i = 1, 2, 3)$ three 2×2 Pauli matrices. It is possible to have only two of these traceless matrices diagonal which is the maximum number of mutually commuting generators. This number is called the *rank* of the group so that $SU(3)$ is of rank 2. For a Lie group, the number of Casimir operators is equal to the rank of the group. In the case of $SU(3)$, the Casimir operators are taken to be F^2 and G^3. [It is J^2 in the case of $SU(2)$.]

The fundamental representation of $SU(3)$ is a triplet. The three color substates of a quark, red (R or x), yellow (Y or y), and blue (B or z) [as introduced in §.0.1.], constitute the basis for the triplet fundamental representation of $SU(3)$.

$$x = \begin{pmatrix} 1 \\ 0 \\ 0 \end{pmatrix}, \quad y = \begin{pmatrix} 0 \\ 1 \\ 0 \end{pmatrix}, \quad z = \begin{pmatrix} 0 \\ 0 \\ 1 \end{pmatrix}. \tag{25}$$

An arbitrary state φ,

$$\varphi = ax + by + cz = \begin{pmatrix} a \\ b \\ c \end{pmatrix}, \tag{26a}$$

transforms under $SU(3)$ as follows:

$$\varphi \to \varphi' = U(\theta)\varphi, \tag{26b}$$

[3] *Gell-Mann, M., Phys. Rev.* **125**, *1067 (1962).*

with

$$U(\theta) = exp(-i\sum_{j=1}^{8}\theta_j\lambda_j/2). \tag{26c}$$

Here $\{\theta_j\}$ are eight real numbers which characterize some "rotation" in the 8-dimensional $SU(3)$ adjoint space. [Using $SU(2)$ as an analogue, the adjoint space may be viewed as the ordinary space and $\{\theta_1, \theta_2, \theta_3\}$ characterize an ordinary rotation so that the transformation $\varphi \to \varphi'$ describes the effect of an ordinary rotation on the electron spinor (which is a linear combination of Pauli spinors).]

The generators $\{\lambda_i\}$ of the $SU(3)$ group satisfy the Lie algebra,

$$[\lambda_i, \lambda_j] = 2if_{ijk}\lambda_k, \tag{27}$$

with f_{ijk} the completely antisymmetric structure coefficients. Using the Gell-Mann matrices listed in Eqs. (24), we have, for nonvanishing f_{ijk}'s,

$$\begin{aligned} f_{ijk} =& 1 \ \text{ for } (ijk) = (123); \\ & \frac{1}{2} \ \text{ for } (ijk) = (147), (246), (257), \ \text{ or } (345); \\ & -\frac{1}{2} \ \text{ for } (ijk) = (156) \ \text{ or } (367); \\ & \frac{\sqrt{3}}{2} \ \text{ for } (ijk) = (458) \ \text{ or } (678). \end{aligned} \tag{28}$$

Finally, we have

$$\begin{aligned} \{\lambda_i, \lambda_j\} &\equiv \lambda_i\lambda_j + \lambda_j\lambda_i \\ &= 2d_{ijk}\lambda_k + \frac{4}{3}\delta_{ij}, \end{aligned} \tag{29}$$

with

$$\begin{aligned} d_{ijk} =& \frac{1}{2}, \ \text{ for } (ijk) = (146), (157), (256), (344), \ \text{ or } (355); \\ & -\frac{1}{2}, \ \text{ for } (ijk) = (247), (366), \ \text{ or } (377); \\ & \frac{1}{\sqrt{3}}, \ \text{ for } (ijk) = (118), (228), \ \text{ or } (338); \\ & -\frac{1}{\sqrt{3}}, \ \text{ for } (ijk) = (888); \\ & -\frac{1}{(2\sqrt{3})}, \ \text{ for } (ijk) = (448), (558), (668), \ \text{ or } (778). \end{aligned} \tag{30}$$

Note that the structure constants d_{ijk} are completely symmetric. Note also that those d_{ijk} which are not listed in Eq. (30) vanish identically.

(2) Flavor SU(3) Symmetry

Since the pion was discovered in 1947, the nucleon has lost its unique role in particle physics. Subsequently, many more strongly interacting particles (hadrons) have been identified. Some of the new particles were surprisingly long-lived on the time scale of strong interactions, despite being massive enough to decay into lighter objects without violating the conservation of charge or baryon number. For instance, a Σ^- is readily produced by the strong interaction $\pi^- p \to K^+\Sigma^-$ and yet decays only weakly via $\Sigma^- \to n\pi^-$. Gell-Mann and, independently, Nishijima, took this as the hint for the existence of a new additive quantum number called "strangeness" S. They assigned to each hadron an integer value of strangeness,

$$\begin{aligned} S &= 0: \quad && \pi, N, \Delta, ..., \\ S &= 1: \quad && K^+, ..., \\ S &= -1: \quad && \Lambda, \Sigma, ..., \end{aligned} \tag{31}$$

with $-S$ for their antiparticles. Assuming that strong and electromangnetic interactions are forbidden unless S is conserved by the reaction, Gell-Mann and Nishijima were able to account for the strong production and the weak decay of the Σ. Indeed, the total strangeness is conserved in the reaction $\pi^- p \to K^+\Sigma^-$, accounting for the production of the Σ-particle by the strong interaction. Subsequently, the Σ^- can only decay by the strangeness-violating weak interaction $\Sigma^- \to n\pi^-$, yielding its long lifetime. The Gell-Mann and Nishijima scheme was confirmed by observations of the properties of the large number of strange particles that were subsequently discovered.

Now, owing to the existence of a second additive quantum number S in addition to I_3, it was natural to attempt to enlarge isospin symmetry to a larger group, say, a group of rank 2. This new symmetry group must allow for grouping of the hadrons with similar properties naturally into its multiplet representations. Since no strange particles exist that are close in mass to the nucleon, the appropriate grouping was relatively difficult to identify and the choice of group was not obvious at all. As $SU(3)$ was originally proposed in 1961, it groups the n, p, Σ^+, Σ^0, Σ^-, Λ, Ξ^0, and Ξ^-, with a mass spread of nearly $400\,MeV$, into an $SU(3)$ octet representation. In addition, it groups the lightest mesons into an octet, with the K-meson belonging to the same representation as the much lighter π. The extra symmetry linking strange and non-strange particles is thus much more approximate than is isospin.

Noting that the $SU(3)$ multiplet structure of elementary particles was quite similar to the grouping of chemical elements in Mendeleev's periodic table, successes of the $SU(3)$ classification strongly hinted at the existence of a substructure. The $SU(3)$ incorporating the three lightest flavors of quark can only be an approximate symmetry, which is called "flavor $SU(3)$". We may use such symmetry to enumerate low-lying hadronic states. The concept of flavor $SU(3)$ symmetry was not firmly established until 1964. Note that "flavor $SU(3)$" has nothing to do with the concept of "color $SU(3)$", which is believed to be an exact symmetry of fundamental origin. (See Chapter 12.)

Nowadays, the role of the $SU(3)$ group of isospin and strangeness is mostly historical: It sets the stage for the entry of quarks into particle physics.

(3) Quark Model

According to naïve quark models, hadrons are made up of a small variety of more basic entities, called quarks and gluons, bound together in different ways. The fundamental representation of flavor $SU(3)$, the multiplet (u, d, s) from which all other multiplets can be built, is a triplet. Each quark is assigned spin $\frac{1}{2}$ and baryon number $B = \frac{1}{3}$. In naïve quark models, baryons are treated as systems of three quarks (qqq) with quarks confined to the region characterized by the baryon size while mesons are quark-antiquark pairs $(q\bar{q})$ with the quark (antiquark) content restricted to a small region. The additive quantum number "hypercharge" Y $(\equiv B + S)$ rather than the strangeness S is introduced so that the charge Qe is given by

$$Q = I_3 + \frac{Y}{2} \tag{32}$$

Mesons as Quark-Antiquark Pairs

Mesons are quark-antiquark pairs confined to within the region defined by the meson size. Considering the subspace defined by only two flavors, $q = u$ or d, we obtain the $q\bar{q}$ bound-state wave functions by making the substitutions $p \to u$ and $n \to d$ in Eqs. (2). This results in an isotriplet and an isosinglet of mesons[4]:

$$\begin{cases} \mid I = 1, \quad I_3 = 1 >= -u\bar{d} \\ \mid I = 1, \quad I_3 = 0 >= \sqrt{\frac{1}{2}}(u\bar{u} - d\bar{d}) \\ \mid I = 1, \quad I_3 = -1 >= d\bar{u} \end{cases}$$

$$\mid I = 0, I_3 = 0 >= \sqrt{\frac{1}{2}}(u\bar{u} + d\bar{d}).$$

Adding the s quark to the picture, we have five more states:

$$u\bar{s}, d\bar{s};$$
$$s\bar{d}, -s\bar{u};$$
$$s\bar{s}.$$

Thus, there are nine possible $q\bar{q}$ combinations. We may divide the nine states into an $SU(3)$ octet and an $SU(3)$ singlet. Under operations of the $SU(3)$ group, the eight states transform among themselves and do not mix with the singlet state.

Among the nine $q\bar{q}$ states, we note that there are three states which have $I_3 = Y = 0$. These are linear combinations of $u\bar{u}$, $d\bar{d}$, and $s\bar{s}$ states. The singlet combination, η_1, must contain each quark flavor on an equal footing[5]:

$$\eta_1 = \sqrt{\frac{1}{3}}(u\bar{u} + d\bar{d} + s\bar{s}).$$

[4] *Note that, if (u, d) is an isospin doublet, then $(-\bar{d}, \bar{u})$ is also an isospin doublet. The extra sign in $-\bar{d}$ ia essential here.*

[5] *Here we use particle names to denote the various $q\bar{q}$ combinations.*

The second state may taken to be a member of the isospin triplet,

$$\pi^0 = \sqrt{\frac{1}{2}}(u\bar{u} - d\bar{d}).$$

By requiring orthogonality to both η_1 and π^0, the isospin singlet state is found to be

$$\eta_8 = \sqrt{\frac{1}{6}}(u\bar{u} + d\bar{d} - 2s\bar{s}).$$

Like any quantum-mechanical bound system, the $q\bar{q}$ pair should have a discrete energy level spectrum corresponding to the different modes of $q\bar{q}$ excitations. Since the quark has spin $\frac{1}{2}$, the total intrinsic spin of the $q\bar{q}$ pair can be either $S = 0$ or 1. The spin J of the composite meson is then the vector sum of this spin S and the relative orbital angular momentum L between q and $\bar{q}$. The parity of the meson, which specifies the meson property under space inversion, is thus given by

$$P = -(-1)^L. \tag{33}$$

Here the minus sign arises beacuse the q and $\bar{q}$ have opposite intrinsic parity and $(-1)^L$ is due to the space inversion property of the $q\bar{q}$ wavefunction $Y_{LM}(\theta, \phi)$. A neutral $q\bar{q}$ system is an eigenstate of the particle-antiparticle conjugation operator C.

$$C = -(-1)^{S+1}(-1)^L = (-1)^{L+S}, \tag{34}$$

where S is the total intrinsic spin of the $q\bar{q}$ pair. Note that, in Eq. (34), the minus sign arises from interchanging fermions, the factor $(-1)^{S+1}$ is determined from the symmetry of the $q\bar{q}$ spin states [cf. Eqs. (22)], and the factor $(-1)^L$ from the symmetry of $Y_{LM}(\theta, \phi)$.

For the $J^P = 0^-$ $q\bar{q}$, we have two isospin doublets

$$K^0(d\bar{s}), \qquad K^+(u\bar{s}) \quad \text{with } Y = 1$$

$$K^-(s\bar{u}), \qquad \bar{K}^0(s\bar{d}) \quad \text{with } Y = -1.$$

These pseudoscalar mesons form an octet along with Y=0 isotriplet (the π^+, π^0, π^- states) and the $(Y = 0, I = 0)$ state (the η meson). The $SU(3)$ singlet state is identified with η' meson.[6]

To sum up, we have, in group notations,

$$3 \otimes \bar{3} = 8 \oplus 1, \tag{35a}$$

where η_0 is the only member forming the singlet representation ("nonet") while the eight members of the octet representation are listed below:

$$\eta_8, \pi^+, \pi^0, \pi^-, K^+, K^0, \bar{K}^0, K^-. \tag{35b}$$

[6] *In practice, the $\eta_1 - \eta_8$ mixing is of importance since flavor $SU(3)$ symmetry breaking is not negligible.*

Looking over the particle data table, there are also nine lowlying vector mesons ($J^P = 1^-$) which may be grouped into an octet and nonet representations of flavor $SU(3)$ symmetry:

$$\omega, \rho^+, \rho^0, \rho^-, K^{*+}, K^{*0}, \bar{K}^{*0}, K^{*-}, \phi. \tag{36}$$

Here ω and ϕ are states obtained by mixing "ideally" the ω_1 and ω_8 (analogues of η_1 and η_8 described above) such that ϕ is an almost pure $s\bar{s}$ state.

Finally, we remark on the color wave function for mesons. In view of color confinement, it was conjectured earlier that all observed hadrons are colorless, or in the singlet representation of the color $SU(3)$ group. Using the result which we have just learned for flavor $SU(3)$, we find that the color wave function for a meson is unique:

$$\Phi_M^c = \sqrt{\frac{1}{3}}(x\bar{x} + y\bar{y} + z\bar{z}), \tag{37}$$

where x, y, and z are three color substates introduced earlier in Eq. (3), §.0.1.

Baryons as Three-Quark Systems

The flavor $SU(3)$ decomposition of the 27 possible qqq combinations is a little more involved than that for mesons. Nevertheless, the quark content of baryons can be readily obtained using the same techniques. We first combine the first two quarks and arrange the nine qq combinations into two SU(3) multiplets,

$$3 \otimes 3 = 6 \oplus \bar{3}, \tag{38}$$

where the 6 is symmetric and the $\bar{3}$ is antisymmetric under interchange of the two quarks. The members of the **6** representation ("sextet") are listed below:

$$\begin{aligned} &uu, \quad dd, \quad ss, \\ &(ds)_S \equiv \frac{1}{\sqrt{2}}(ds + sd), \qquad (su)_S, \qquad (ud)_S. \end{aligned} \tag{39a}$$

The members of the antitriplet representation are given by

$$(ds)_A \equiv \frac{1}{\sqrt{2}}(ds - sd), \qquad (su)_A, \qquad (ud)_A. \tag{39b}$$

Adding the third quark to the triplet, we obtain the decomposition:

$$\begin{aligned} 3 \otimes 3 \otimes 3 &= (6 \otimes 3) \oplus (\bar{3} \otimes 3) \\ &= 10 \oplus 8 \oplus 8 \oplus 1. \end{aligned} \tag{40}$$

The decuplet states are totally symmetric under interchange of quarks, as evidenced by the uuu, ddd, and sss members. The symmetric combination of "uud" is given by

$$\Delta_S \equiv \frac{1}{3}[uud + (ud + du)u]. \tag{41}$$

The two other *uud* states orthogonal to Δ are

$$p_s = \sqrt{\frac{1}{6}}[(ud + du)u - 2uud], \tag{42}$$

$$p_a = \sqrt{\frac{1}{2}}(ud - du)u. \tag{43}$$

The states p_s and p_a have mixed symmetry, where the subscripts are used to indicate the symmetry property under interchange of the first two quarks. The quark structrue of the other states can be obtained in a similar manner. The completely antisymmetric three-quark state is

$$(qqq)_{singlet} = \frac{1}{6}(uds - usd + sud - sdu + dsu - dus). \tag{44}$$

Replacing $u \rightarrow\uparrow$ and $d \rightarrow\downarrow$ to Eqs. (41)–(43), we immediately obtain three possible spin states for a quark triplet:

$$\begin{aligned} \chi(S) &= \sqrt{\frac{1}{3}}(\uparrow\uparrow\downarrow + \uparrow\downarrow\uparrow + \downarrow\uparrow\uparrow) \\ \chi(M_s) &= \frac{1}{6}(\uparrow\downarrow\uparrow + \downarrow\uparrow\uparrow - 2\uparrow\uparrow\downarrow) \\ \chi(M_A) &= \sqrt{\frac{1}{2}}(\uparrow\downarrow\uparrow - \downarrow\uparrow\uparrow). \end{aligned} \tag{45}$$

To enumerate the baryons expected in the quark model, we must combine the $SU(3)$ flavor wave function such as Eqs. (41)–(43) with the spin wave function, Eq. (45).

There is a problem with this symmetry of the ground state. For example, a Δ^{++} of $J_3 = \frac{3}{2}$ is described by the symmetric wave function,

$$u\uparrow u\uparrow u\uparrow, \tag{46}$$

in violation with Pauli exclusion principle for a system of identical fermions. As already noted earlier, the explanation is that the quarks possess an additional internal degree of freedom, called "color", which can take three possible values, x, y, or z. The quarks form a fundamental triplet of an $SU(3)$ color symmetry which is believed to be exact. All observed hadrons are postulated to be colorless. That is, they belong to singlet representations of the $SU(3)$ color group. The color wavefunction for a baryon is given by[7]

$$(qqq)_{color\ singlet} = \sqrt{\frac{1}{6}}(xyz - xzy + yzx - yxz + zxy - zyx). \tag{47}$$

The required antisymmetric property of the total wavefunction under particle exchange is then ensured: It is overall symmetric in space, spin, and flavor structure and completely antisymmetric in color space. Note that this precludes the choice of Eq. (44) as the flavor wave function.

[7] *See Eq. (44).*

As an example of an explicit quark model wavefunction, we have, for a spin-up proton,

$$\begin{aligned} \mid p \uparrow\rangle =& \sqrt{\frac{1}{18}}[uud(\uparrow\downarrow\uparrow + \downarrow\uparrow\uparrow - 2\uparrow\uparrow\downarrow) + udu(\uparrow\uparrow\downarrow + \downarrow\uparrow\uparrow - 2\uparrow\downarrow\uparrow) \\ & + duu(\uparrow\downarrow\uparrow + \uparrow\uparrow\downarrow - 2\downarrow\uparrow\uparrow)] \\ =& \sqrt{\frac{1}{18}}[u\uparrow u\downarrow d\uparrow + u\downarrow u\uparrow d\uparrow - 2u\uparrow u\uparrow d\downarrow + permutations]. \end{aligned} \tag{48}$$

The three quarks have zero orbital angular momentum in ground-state baryons, so that the parity of the state, $(-1)^{\ell+\ell'}$, is positive for lowlying baryons.

We may now look over the particle data table on lowlying baryons. There are eight $J^P = \frac{1}{2}^+$ baryons which may be grouped together to form an octet representation:

$$\begin{gathered} p, \quad n, \\ \Sigma^+, \quad \Sigma^0, \quad \Sigma^-, \quad \Lambda, \\ \Xi^+, \quad \Xi^-. \end{gathered} \tag{49}$$

In addition, there are ten lowlying $J^P = \frac{3}{2}^+$ baryons which can be grouped into a decuplet representation:

$$\begin{gathered} \Delta^{++}, \quad \Delta^+, \quad \Delta^0, \quad \Delta^-, \\ \Sigma^{*+}, \quad \Sigma^{*0}, \quad \Sigma^{*-}, \\ \Xi^{*+}, \quad \Xi^{*-}, \\ \Omega^-. \end{gathered} \tag{50}$$

As a matter fact, the existence of the last member in the decuplet, namely Ω^-, was a prediction made by Gell-Mann as he tried to identify members for a possible decuplet representation under flavor $SU(3)$. The prediction was soon confirmed by experiments, strengthening considerably the idea of flavor $SU(3)$ symmetry.

There are quark models which provide detailed descriptions of the hadron wave functions, which include wave functions in configuration space, spin space, flavor space, and color space and sometime even with specific background fields related to confinement. A general survey of the various quark models may be found in the literature.

11.3 Additional Symmetries in Particle Physics

In addition to $SU(2)$ and $SU(3)$ symmetries which were elucidated in §.11.1. and §.11.2., similar discussions may easily be extended to many other symmetries.

(1) For a system possessing translational symmetry in time, i.e., invariant under translation in time, there is the law of conservation of energy.[8]

[8] *We wish to insert a remark here to emphasize the real significance of the use of tensors and vectors. The Lorentz transformation (or the general coordinate transformation in the general relativity theory) defines the tensors (and vectors) by their transformation properties, so that a tensor equation is automatically invariant in form under the transformation, and any physical law expressible in the form of a tensor equation is automatically guaranteed to obey the relativity theory.*

(2) For a dynamical system possessing translational symmetry in a certain direction in space, dynamical laws are invariant under linear translations of coordinate systems in that direction, leading to the law of conservation of momentum in that direction.

(3) For systems in uniform relative motion, the physical laws are invariant under the Lorentz transformation. This gives (or, is the expression of) the relativity principles.

Note that the ten dynamical operators, including (1) the generator for time translation ("*Hamiltonian*"), i.e., for item (1) above, (2) the three generators for spatial translations ("*momenta*"), i.e., for item (2) above, (3) the three *Lorentz boost* operators, as in item (3) above, and (4) the *angular momenta* (for generating rotaions in configuration space, as discussed earlier in §.11.1.), form a Lie group, called "proper Lorentz group". The Lie algebra may be called "Poincaré algebra". (Dirac 1949)

(4) Parity symmetry, charge conjugation symmetry, and time-reversal symmetry. See §.4.4. for an introduction to these important concepts.

It has been intuitively taken for granted that physical laws have a symmetry with respect to a mirroring operation (i.e., a reflection of the coordinate system on the $X - Y$ plane, say.). This intuitive feeling has been justified by the discovery of the empirical law (Laporte, 1924) for dipole radiation from the analysis of the spectrum lines of iron atom. That this left-right symmetry is not universal was proposed by T. D. Lee and C. N. Yang in 1956 and soon experimentlally verified by C. S. Wu and others for the interactions involved in nuclear and neutron β decays, $\pi \to \mu$ decay, and $\mu \to e$ decay.[9]

It was found that the combined symmetry of charge conjugation and space inversion (CP) was broken in the case of the 2π decay of the K_L^0 meson, by Christenson et al. in 1964.[10] If the CTP theorem is assumed (as is the case for a local field theory), then the symmetry of time (with respect to reversal) was also broken.

(5) The invariance under gauge transformation as introduced in §.1.1. for a Klein-Gordon particle and in §.4.2. for a Dirac particle.

It can be shown that from the requirement of the invariance of the electromagnetic laws under the gauge transformation follows the conservation of electric charge, which is of course not a new result but a familiar empirical fact.

As already mentioned earlier in this chapter, the concept of "gauge invariance" was first introduced by Weyl (1919), revised slightly by Fock (1927) and London (1927),[11] and finally generalized to the case of a non-abelian gauge group by Yang and Mills (1954). Generalization of "$U(1)$ gauge symmetry" to a non-abelian group will be

[9] *See, e.g., Commins, E.D. and Bucksbaum, P.H., Weak Interactions of Leptons and Quarks (Cambridge University Press, 1983).*

[10] *Christenson, J.H., Cronin, J.W., Fitch, V.L., and Turlay, R., Phys. Rev. Lett.* **13**, *138 (1964).*

[11] *In ordinary quantum mechanics, the equation of an electron in an electromagnetic field is invariant under the gauge transformation if it is accompanied by a suitable transformation in the phase of the wave function. For the meaning of the phase in quantum mechanics and its possible relation to gauge symmetry, consult the Appendix for further discussions.*

described, in the case of QCD or an $SU(3)$ Yang-Mills theory, at the beginning of the next chapter (§.12.1.). Here we note that the notion of "gauge symmetry" under a non-abelian gauge group has become the backbone of the modern theory of particle physics, namely, the "Standard Model" consisting of an $SU(3)$ Yang-Mills theory of strong interactions and the Glashow-Salam-Weinberg $SU(2) \times U(1)$ Yang-Mills theory of eletroweak interactions.

(6) There is a permutation symmetry for a system of many particles (fermions or bosons). As elucidated in detail in Ch. 7, permutation symmetry determines the statistics of the many-body system.

To sum up, it is fair to say that, in the present form of elementary particle physics, the concepts of symmetry has assumed a very fundamental role.

Appendix

Potentials and Phase in Quantum Mechanics

To shed light on the meaning of the gauge symmetry, we append here a few notes[12] which we believe may be relevant for the problem. The notes are organized as follows: In §.*A*.1. and §.*A*.2., we contrast the role of field and potential, respectively, in classical physics and in quantum mechanics. In §.*A*.3., we introduce Dirac's theory of the magnetic monopole. In §.*A*.4., aspects related to the Aharonov-Bohm experiment are briefly discussed. For geometrization of gauge field theories, the lecture note given by C. N. Yang (1975) at the Sixth Hawaii Topical Conference in Particle Physics provides an in-depth treatment of the problem while, for an introduction to the interesting question of Berry's phase (Berry 1984), the articles by Berry himself (1988) and by R. Jackiw (1988) may be helpful.

A.1. Field and Potential in Classical Physics[13]

In classical dynamics, the force $\mathbf{F}\,(F_x, F_y, F_z)$ appears explicitly in the equations of motion of a particle,

$$m\frac{d^2\mathbf{r}}{dt^2} = \mathbf{F}(\mathbf{r}).$$

It is true that a potential can be introduced

$$\mathbf{F} = -\nabla V,$$

but all potentials differing from V by a constant give the same $\mathbf{F}$ so that V is not uniquely determined by $\mathbf{F}$.

In classical electromagnetic theory the basic laws (Maxwell's equations, together with Lorentz's force expression) are all explicitly expressed in terms of the fields $\mathbf{E}$, $\mathbf{D}$, $\mathbf{B}$, $\mathbf{H}$. One may introduce the potentials $(\mathbf{A}, \phi)$ such that

$$\mathbf{E} = -\nabla\phi - \frac{1}{c}\frac{\partial \mathbf{A}}{\partial t}, \qquad \mathbf{H} = \nabla \times A, \tag{A1}$$

but under the gauge transformation to new potentials $(\mathbf{A'}, \phi')$

$$\mathbf{A}' = \mathbf{A} + \nabla\chi, \qquad \phi' = \phi - \frac{1}{c}\frac{\partial\chi}{\partial t}, \qquad \partial_\mu\partial_\mu\chi = 0, \tag{A2}$$

the fields $\mathbf{E}$, $\mathbf{H}$ and the Maxwell equations (together with the Lorentz gauge condition) remain invariant.

---

[12] *This appendix is prepared using the colloquium lectures given by one of us (T.-Y. Wu) some ten years ago both at the State University of New York at Buffalo and at National Tsing-hua University on Dirac's theory of the magnetic monopole and in that connection also on the Aharonov-Bohm theory.*

[13] *In view of the nature of the materials to be discussed, the natural unit system ($\hbar = c = 1$) is not adopted in this specific appendix.*

Thus, in classical dynamics and electromagnetic theory, the potentials do not have definite meaning; they are "mathematical tools"; only the fields **F**, **E**, **H**, etc. are measurable and have physical meaning.

A.2. Field and Potential in Quantum Mechanics

(i) In quantum mechamics, the situation regarding the roles of the fields and the potentials are different!

In the non-relativistic Schrödinger equation,

$$(\frac{1}{2m}\mathbf{p}^2 + V - E)\psi = 0,$$

$$-\frac{\hbar}{i}\frac{\partial}{\partial t}\psi = (\frac{1}{2m}\mathbf{p}^2 + V)\psi,$$

or the relativistic Dirac equation,

$$\gamma_\mu(\frac{\hbar}{i}\frac{\partial}{\partial x^\mu} - \frac{e}{c}A_\mu)\psi = imc\psi, \tag{A3}$$

$$x^\mu = (x, y, z, ict), \qquad A_\mu = (A_x\, A_y\, A_z, i\phi),$$

the potentials V, A, ϕ appear and not the fields.

The next question is "how do the potentials affect the wave function of a system?" Or, do the potentials have observable effects?

(ii) Introduction of potential A_μ; non-integrable phase

Let us start with a free electron whose equation is

$$(\gamma_\mu \frac{\hbar}{i}\frac{\partial}{\partial x_\mu} - imc)\psi = 0. \tag{A4}$$

Let

$$\psi = \psi'\, exp(-\frac{ie}{\hbar c}\int A_\mu dx_\mu). \tag{A5$'$}$$

Then ψ' satisfies the equation

$$exp(-\frac{ie}{\hbar c}\int A_\mu dx_\mu)[\gamma_\mu(\frac{\hbar}{i}\frac{\partial}{\partial x_\mu} - \frac{e}{c}A_\mu) - imc]\psi' = 0,$$

i.e., ψ' satisfies the equation for an electron in the potential $(A, i\phi)$, and ψ' is related to ψ by

$$\psi' = \psi\, exp(\frac{ie}{\hbar c}\int A_\mu dx_\mu). \tag{A5}$$

Now the phase

$$\beta = \frac{e}{\hbar c}\int A_\mu dx_\mu, \tag{A6}$$

is not an integrable function of the point x_μ, but depends on the path of integration C from some initial point to x_μ.[14]
Note that the β as specified by Eq. (A6) is a property of the potential alone, and is independent of ψ.
The difference in phase between two points $x_\mu^{(1)}$ and $x_\mu^{(2)}$ depends on the path from $x_\mu^{(1)}$ to $x_\mu^{(2)}$!

The phase $\beta = \frac{e}{\hbar c}\int A_\mu dx_\mu$ is then said to be non-integrable.

Thus we obtain

Theorem I: The effect of a potential is to introduce a non-integrable phase in the wave function ψ of the potential-free particle.

The relationship between gauge transformation of the electromagnetic potentials and the phase of wave functions is first shown by H. Weyl in 1929. The possibility of using non-integrable phases to characterize geometrically a gauge theory was explored by C. N. Yang in 1974.

(iii) Gauge invariance and phase

The Dirac equation for an electron in the potential $(\mathbf{A}, i\phi)$ is (A3),

$$[\gamma_\mu(\frac{\hbar}{i}\frac{\partial}{\partial x_\mu} - \frac{e}{c}A_\mu) - imc]\psi = 0. \tag{A3}$$

On making the gauge transformation (A2)

$$\mathbf{A}' = \mathbf{A} + \nabla\chi, \qquad \phi' = \phi - \frac{1}{c}\frac{\partial\chi}{\partial t}, \qquad \partial_\mu\partial_\mu\chi = 0, \tag{A2}$$

and at the same time the transformation

$$\psi' = \psi\, exp(-\frac{ie}{\hbar c}\chi), \tag{A9}$$

Eq. (A3) becomes

$$[\gamma_\mu(\frac{\hbar}{i}\frac{\partial}{\partial x_\mu} - \frac{e}{c}A'_\mu) - imc]\psi' = 0. \tag{A10}$$

---

[14] *By Stokes' theorem,*

$$\oint A_\mu dx_\mu = \int\int_\sigma curl\,\mathbf{A}\cdot d\mathbf{S}. \tag{A7}$$

Unless $Curl\,\mathbf{A} = 0$, i.e., unlesss

$$A_\mu = \frac{\partial\chi}{\partial x_\mu}, \qquad \chi = a\ scalar\ function, \tag{A8}$$

the integral $\oint A_\mu dx_\mu$ does not vanish.

Thus the effect of a gauge transformation (A2) is to introduce a phase factor $exp(-\frac{ie}{\hbar c}\chi)$ to ψ. The phase

$$\beta = \frac{e}{\hbar c}\chi(x_\mu), \tag{A11}$$

is a function of the 4-point x_μ, and the change in phase in going from $x_\mu^{(1)}$ to $x_\mu^{(2)}$ is

$$\beta^{(1)} - \beta^{(2)} = \frac{e}{\hbar c}(\chi(x_\mu^{(2)}) - \chi(x_\mu^{(1)})), \tag{A12}$$

which is independent of the path joining $x_\mu^{(1)}$ and $x_\mu^{(2)}$. The change in phase in going around a closed curve is therefore zero.

Such a phase as (A11) is said to be integrable.

Theorem II: In quantum mechanics, a gauge transformation of the potential $A_\mu(\mathbf{A}, i\phi)$ causes a phase change which is integrable.

A.3. Dirac's Theory of the Magnetic Monopole

(i) Phase of wave function in a magnetic field

As seen from Eqs. (A5′) and (A6), it is possible to generate the desired equation for a Dirac particle moving in an electromagnetic field from the free Dirac equation by substituting the wave function ψ by ψ':

$$\psi' = \psi e^{i\beta}, \qquad \beta = \frac{e}{\hbar c}\int A_\mu dx_\mu. \tag{A13}$$

The phase β is determined by the potential A_μ and is the same for all wave functions. Thus, even though β is not integrable, it still has the following property, namely, the change $\Delta\beta$ when a wave function is carried around any given closed curve (in space-time) is the same for all ψ. This property follows from the following considerations. Let φ_m, ψ_n be any two arbitrary wave functions. The integral

$$\int \varphi_m^* \psi_n d\tau = \text{a complex number}, \tag{A14}$$

is a measure of the "identity" (sameness) of the two states φ_m and ψ_n. If both φ_m and ψ_n are carried around an arbitrary closed curve Γ, the change $\Delta(-\beta_m + \beta_n)$ in $\varphi_m^*\psi_n$ must be zero (or $2\pi n$) in order that the integral in Eq. (A14) have a definite value. From this it follows that

$$\Delta\beta_m = \Delta\beta_n, \qquad \text{or} \quad (\pm 2\pi n). \tag{A15}$$

The indeterminacy $2\pi n$ can be eliminated on the following consideration: Let the closed curve Γ shrink continuously. On account of the continuous character of the ψ's, the phases $\Delta\beta_m$ and $\Delta\beta_n$ must decrease continuously to zero, respectively. Hence $\Delta\beta_m = \Delta\beta_n$; there cannot be any difference $2\pi n$.

But the argument based on continuity would fail in the following case: The wave function ψ has a *nodal line* on which $\psi = 0$. When $\psi = 0$, the phase itself has no meaning (undefined).

In this case, taking ψ around a nodal line (in 3-dimensional space), we have no continuity argument for requiring $\Delta\beta$ to be "small"; we may only say that $\Delta\beta$ approximates to $2\pi n$, with n some integer.[15] The numerical value of n is a property of the nodal line; the sign of n depends on the sense of revolution around the nodal line.

At this point, Dirac introduced an important new idea, namely:

Let ψ_0 be the wave function that has a nodal line. Along a small closed curve Γ around its nodal line, the change in β_0 of ψ_0 is $\Delta\beta_0$. The difference between $\Delta\beta_0$ and the nearest value $2\pi n$ (n being a positive or negative integer) is the same as the change $\triangle\beta$ in phase β, around the same closed curve Γ, of any wave function ψ not having that nodal line,

$$\triangle\beta_0 - 2\pi n = \triangle\beta. \tag{A16}$$

This is a new idea, an assumption, and cannot be proved, nor derived. It is a conjecture of the overall consistency among all the solutions to the same equation. As we shall see immediately, it leads to the quantization rule for the strength of the Dirac magnetic monopole.

From (A13) or (A6), we then obtain, for Γ in 3-space,

$$\triangle\beta_0 - 2\pi n = \frac{e}{\hbar c}\int_\sigma \mathbf{H}\cdot d\sigma, \tag{A17}$$

where σ is any surface whose boundary is Γ.

Let Γ shrink so that σ becomes a "bottle"; in the limit of Γ shrinking to a point, σ becomes a closed surface, a bubble, and $\triangle\beta_0$ approaches zero, so that

$$-2\pi n = \frac{e}{\hbar c}\oint_\sigma \mathbf{H}\cdot d\sigma. \tag{A18}$$

The righthand side $\frac{e}{\hbar c}\oint \mathbf{H}\cdot d\sigma$ has nothing to do with the nodal line, by the original assumption (A16). Hence the value of n is independent of the wave function, and can only be a property of the field.

For example, take $n = \pm$ integer and $n \neq 0$. The righthand side ordinarily vanishes,

$$\begin{aligned}\oint \mathbf{H}\cdot d\sigma &= \iiint \nabla\cdot\mathbf{H}\,d\tau,\\ &= 0, \qquad \text{since} \quad \nabla\cdot\mathbf{H} = 0.\end{aligned}$$

There is then a contradiction!

This contradiation can be avoided by assuming the presence of a magnetic pole of magnetic charge g (in the same unit as electric charge e), so that

$$\oint \mathbf{H}\cdot d\sigma = 4\pi g, \tag{A19}$$

and Eq. (A18) becomes

$$2\pi n = \frac{e}{\hbar c}4\pi g,$$

[15] *n must be an integer because of the single-valued nature of the quantum mechanical wave function.*

or

$$g = \frac{\hbar c}{2e} n, \qquad n = 0, \pm 1, \pm 2, ... \tag{A20}$$

If $n = 1$, then the smallest magnetic monopole g is

$$\frac{g}{e} = \frac{\hbar c}{2e^2} = \frac{1}{2\alpha} = \frac{137}{2}. \tag{A21}$$

The importance of the relation (A20) is this: If the quantization of electric charge (the universal unit e) is accepted, then (A20) is the law of quantization of the magnetic pole strength.

Then, for $n \neq 0$, the nodal line have an endpoint inside the closed surface σ; for, otherwise, a nodal line cutting the surface σ an even number of times will have their total contributions to $2\pi n$ equal to zero, n being positive or negative depending on the direction of the nodal line and the sense of the curve Γ around it. The endpoint of the nodal line is the seat of the magnetic monopole as we shall see in subsection (ii) below.

Following Dirac's conjecture, we see that, unless there exists a magnetic monopole (or something else that is similar), it is not possible to have a wave function that has a nodal line.

If ψ_0 is taken along a closed curve Γ around a number of nodal lines, we shall make small closed curves around each of them. The assumption (A16) is now

$$\triangle\beta_0 - 2\pi \sum n = \triangle\beta, \tag{A16a}$$

where $\triangle\beta$ is $\frac{e}{\hbar c}\iint H \cdot d\sigma$ calculated for any surface whose boundary is Γ, and $\sum n$ in the sum of n over the contributions of the various nodal lines. (A18) is now replaced by

$$-2\pi \sum n = \frac{e}{\hbar c} \oint_\sigma \mathbf{H} \cdot d\sigma. \tag{A18a}$$

Each nodal line having an endpoint inside the closed surface σ contributes a $2\pi n$.

These endpoints being the same for all wave functions are, according to the conclusion following (A18), not the properties of the wave function; they are the singular properties of the potential $\mathbf{A}$.

(ii) Singularities of potential and monopole

We have seen, following (A21), that the endpoints in σ are properties of the potential and not of the wave functions.

It remains to show that the vector potential $\mathbf{A}$ due to a monopole precisely has a line of singularities. [And along the line of singularities in $\mathbf{A}$, there is then a nodal line in the wave function of an electron!]. Assume that a monopole of strength g in placed at the origin. The magnetic field (in Gauss system) is

$$\mathbf{H} = \frac{g}{r^3}\mathbf{r} \tag{A22}$$

and the vector potential A such that $\mathbf{H} = curl\, \mathbf{A}$ is

$$A_x = -\frac{gy}{r(r+z)}, \qquad A_y = \frac{gx}{r(r+z)}, \qquad A_z = 0, \tag{A23}$$

or, in polar coordinates

$$A_r = 0, \qquad A_\vartheta = 0, \qquad A_\varphi = \frac{g}{r}\tan\frac{\vartheta}{2} \tag{A24}$$

Let C be a latitude circle with latitude θ, and S be the surface of the cap above **C**. Then

$$\begin{aligned} \oint_C \mathbf{A}\cdot ds &= \oint_S \mathbf{H}\cdot d\sigma = g\oint_S \frac{1}{r^3}\mathbf{r}\cdot d\sigma, \\ &= 2\pi g(1-cos\theta), \\ &= \begin{cases} 0, & \text{for } \theta = 0, \\ 4\pi g, & \text{for } \theta = \pi. \end{cases} \end{aligned} \tag{A25}$$

When C shrinks to an infinitesimally small circle as $\theta \to \pi$, the last equality of (A25) becomes $4\pi g$ and the l.f.s. would approach zero if **A** is finite. In that case, we have a contradiction $0 = 4\pi g$.

But from (A24), it is seen that at $\vartheta = \pi$, A_φ is infinite!

Since the radius r in Eqs. (A23)-(A25) is entirely arbitrary, it is seen that that the half-line from $(x, y, z) = (0, 0, 0)$ to $(0, 0, -\infty)$ is a line of singular **A**. The magnetic monopole is at the *end* $(0, 0, 0)$ of the line of singularities.

A.4. The Aharonov-Bohm Experiment

(i) The principle of the Aharonov-Bohm experiment

The principle of the experiment is as follows:

Since $A_\mu\, dx_\mu$ is Lorentz invariant, its value can be calculated in a Lorentz frame in which $dx_4 = 0$, and

$$A_\mu\, dx_\mu \to \mathbf{A}\cdot d\mathbf{x},$$

$$\int A_\mu\, dx_\mu \to \sum \mathbf{A}\cdot d\mathbf{x} \to \int \mathbf{A}\cdot d\mathbf{x},$$

$$\oint A_\mu\, dx_\mu \to \oint \mathbf{A}\cdot d\mathbf{x} = \oint \mathbf{H}\cdot d\sigma.$$

Imagine a long thin solenoid (perpendicular to the plane of this paper). A coherent beam of electrons is split at A into two beams, along path C_1 and C_2 (both on the plane of this paper, encircling the long thin solenoid), which are reunited at B. It is assumed that there is a nonzero magnetic field lines **H** cutting through the surface (which is the surface of this paper) expanded by the two paths C_1 and C_2 but $\mathbf{H} = 0$ on the paths and elsewhere outside the solenoid.

The difference in phase of the two beams at B is[16]

$$\int_{C_2} \mathbf{A} \cdot ds - \int_{C_1} \mathbf{A} \cdot ds = \oint \mathbf{A} \cdot ds = \oint \mathbf{H} \cdot d\sigma \neq 0.$$

The difference in phase can be detected by the interference effect at B. This experiment was carried out by Chambers, and a positive result was found!

(ii) The Essence of the Aharonov-Bohm experiment

The important point, as emphasized by Aharonov and Bohm, is the following: The electrons, along the paths C_1 and C_2, move in a region of space where the field $\mathbf{H}$ is zero, so that the electrons nowhere have experienced any "force" (Lorentz force). On the other hand, the potential A cannot be zero everywhere; otherwise $\oint \mathbf{A} \cdot ds$ would have been zero. Hence the experiment shows that, in quantum mechanics, the potential plays a role (i.e., in determining some observable effect such as interference) not expected on classical theories! As stressed by C.N. Yang (1974), the knowledge on the field strengths $F_{\mu\nu}$ under-describes electromagnetism while the knowledge on the potentials A_μ over-describes the situation, leading to the attempt (C. N. Yang 1974) of using the non-integrable phases to characterize a gauge theory.

(iii) Remarks

(1) It is to be noted that, while the potential A affects the phase $\frac{e}{\hbar c} \int A_\mu \, dx_\mu$, this effect is gauge invariant so that all potentials connected by the gauge transformation (A2) are indistinguishable in the experiment of Chambers. This must be remembered when one is thinking of regarding the potentials as quantum mechanical observables.

(2) In the experimental arrangement of Chambers, the phase difference $\oint A_\mu dx_\mu = \oint_C \mathbf{A} \cdot dx$ depends on the curve C. If C does not enclose the solenoid, then $\oint_C \mathbf{A} \cdot ds = \oint \mathbf{H} \cdot d\sigma = 0$. Thus the space around a solenoid (or a magnet) is not simply-connected; it can be made simply-connected by a cut. That this property of space in connection with a magnetic field towards the phase of a wave function is not the first instance of its kind.

In 1931, Dirac, on consideration of the phase of the wave function, arrived at the new idea of a magnetic monopole. We have given a report on the theory in Subsection **A.3.** above.

In 1984, M. V. Berry made an interesting new observation by noting that there may be nontrivial phases for a quantum mechanical system that is influenced adiabatically by the background system with (periodic) slow motion. The surprising discovery of Berry's phases

[16] *The changes in phase due to the bending of the electron beams can be calculated and allowed for in the change in the interference fringe due to the enclosed* $\mathbf{H}$,

$$\frac{e}{\hbar c} \oint \mathbf{H} \cdot d\sigma,$$

which is a "quantum effect".

in the well-known framework of the adiabatic approximation adds a new dimension to the question of phases and angles in quantum mechanics. The articles by Berry himself (1988) and by R. Jackiw (1988) provide an excellent introduction to the subject.

We add the subsection *A*.5. in view of its potential fundamental importance. The excerpts are taken from *The Universe*, **4-3**, 4 (2016) while the author B.H.J. McKellar is one of the author of the HMW phase.

A.5. The He-McKellar-Wilkens (HMW) Phase

Phase is one of the many scientific words with more than one meaning — even in physics. I (McKellar) am using *phase* as the phase of a complex number, and not as a phase of matter. In this sense phases are an inescapable part of doing quantum mechanics, the essential ingredients of which are the complex amplitudes used to calculate the probability of a particular process. The superposition of quantum mechanical amplitudes leads us to be able to observe the relative phases of those amplitudes, although the probabilities calculated in quantum mechanics are invariant under a change of phase of all of the amplitudes. Long ago it was realised that invariance under local (*i.e.*, position dependent) phase changes was possible for charged particles. The contribution to the wave equation from the differentiation of the phase is compensated by a gauge transformation of the vector potential. The wave function after transformation satisfies just the same wave equation as the untransformed wave function. These gauge interactions now form the keystone on which our understanding of much of the fundamental physics of the universe is built.

In these circumstances one can imagine the surprise, and to some extent disbelief, in the physics community when Aharonov and Bohm [1] pointed out that these were circumstances when the position dependent phase, added up as a charged particle moved in a circuit around a magnetic field, could be observed. The Aharonov-Bohm phase was soon observed, and it was also shown to be a particular example of the *topological phases* introduced by Berry [2].

The phases I want to emphasise can be regarded as duals of the Aharonov-Bohm phases, in a sense I will explain below. The Aharonov-Bohm phase is a result of the motion of an electric charge in the vicinity of, but not in, a magnetic field. In those circumstances there is no Lorentz force on the particle, and the existence of an observable effect in the absence of a classical force was one of the surprising and controversial results of Aharonov and Bohm. The first dual phase proposed and measured is due to the motion of a magnetic dipole in an electric field, and was introduced by Aharonov and Casher [3].

When a magnetic dipole $\boldsymbol{\mu}_m$ is moving in an electric field, we can transform to the rest frame of the dipole, in which there is a *effective magnetic* field $-\boldsymbol{v} \times \boldsymbol{E}$. There is then an interaction energy of $\boldsymbol{\mu}_m \cdot \boldsymbol{v} \times \boldsymbol{E}$ which ultimately gives rise to the Aharonov-Casher phase. In a similar way an electric dipole $\boldsymbol{d}$ moving in a magnetic field feels a *effective electric* field $\boldsymbol{v} \times \boldsymbol{B}$ and has an interaction energy $\boldsymbol{d} \cdot \boldsymbol{v} \times \boldsymbol{B}$. The resulting phase was independently discovered by He and McKellar [4] in 1993 and by Wilkens [5] in 1994. Both the Aharonov-Casher and He-McKellar-Wilkens phases are intrinsically relativistic effects, involving the Lorentz transformation of the electromagnetic field from the laboratory frame to the rest frame of the moving dipole, in spite of the fact that the dipole speeds are low, and working to second order in $|\boldsymbol{v}|$ suffices to describe the experimental situation.

These dipole phases are often described as *topological* or *geometric*, in other words the phase is independent of the detailed path of the dipole and dependent only on the topology of the path. The phases proposed by Aharonov and Casher and by He and McKellar and Wilkens are certainly topological when particular conditions on the trajectories and fields are satisfied. When experimental realisations are considered it is important to determine whether or not these conditions are satisfied before labelling the measured phase as "topological".

While the AC experiment can be done with neutrons or with atoms with permanent intrinsic magnetic dipole moments, the HMW experiment is done with atoms which have induced electric dipole moments. As implied by the 1995 paper of Wei, Han and Wei [6], the electric field required to induce the electric dipole moment significantly changes the conditions on the fields, the trajectory and the test particle for the phase to be topological. It becomes necessary to understand this carefully to decide on the topological nature of the phase in any particular experiment.

This same paper also discussed the possibility of using the HMW effect to drive a superfluid current in liquid helium. Recently my collaborators and I have pointed out the the HMW effect may be able to drive persistent current in a Bose-Einstein condensate of molecular dipoles using experimentally realisable electric and magnetic fields.

In this review paper I concentrate on describing the physics of the various phases and effects described above. For those who wish to explore the details, derivations and more detailed discussions are available in the references I provide. I also refer the reader to my recent, more technical, reviews of this subject [7, 8] Although the Aharonov-Bohm phase is not the subject of this review, it underlies the AC and HMW phases which are, so I begin my review by discussing its main features.

A.5.a. Phases with Dipole Moments

The analysis of the dipole phases is different depending on whether the dipoles are intrinsic or induced. Induced dipoles require an additional field to induce them, and this additional field changes the analysis. For intrinsic dipoles in composite particles, we can still model the motion of the particle using relativistic wave equations. For example we can describe the neutrons used in the first demonstration of the AC effect [9] using the Dirac equation, in spite of the fact that the neutron is a composite particle — the energies we are using are not probing that structure. This device allows the calculations to be relativistic from the beginning, which is useful given the central part that relativistic transformations of the electromagnetic fields play in understanding the physics of the effect. To avoid complications here I will continue to use the Dirac equation. The demonstration of the AC and HMW phases for spin 1 has been given using the Proca equation [10] and any value of the spin has been given using the Bargmann-Wigner equations, and can be looked up by those interested.

A.5.b. The AC Effect

To be explicit look at the AC case — a spin half magnetic dipole of magnitude μ in an electric field described by the field tensor $F^{\mu\nu}$. The Dirac equation is

$$\left(i\gamma^\mu\partial_\mu - \frac{1}{2}\mu\sigma^{\mu\nu}F^{\mu\nu} - m\right)\psi = 0. \tag{A26}$$

Here $\sigma_{\mu\nu} = \frac{1}{2i}\left[\gamma_\mu, \gamma_\nu\right]$ is the spin tensor.

It is immediately clear that we must impose special conditions if we are to transform this equation into the desired form

$$\left(i\gamma^\mu\left[\partial_\mu + i\mu S_\mu\right] - m\right)\psi = 0, \tag{A27}$$

with some *effective vector potential* S_μ, to have any chance of having a topological phase. The model used by Aharonov and Casher was a magnetic dipole going around a line of charges. This suggests simplifying the problem by making the line of charges infinite, imposing translational symmetry parallel to the line of charges, which is taken as the z direction. We now need to work with the Dirac equation in 2 spatial dimensions, or $2+1$ total dimensions. In this space use as the metric tensor and anti-symmetric tensor.

$$g_{\mu\nu} = \text{diag}(1, -1, -1) \qquad \text{and} \qquad \epsilon_{012} = +1. \tag{A28}$$

In $2+1$ dimensions we need only 3 independent Dirac matrices, and so a 2×2 representation of the Dirac matrices in terms of Pauli matrices may be used. The product of two different Pauli matrices is proportional to the remaining Pauli matrix, so the Clifford Algebra generated by the Dirac matrices of the $2+1$ dimensional Dirac equation has just 4 basis operators, the unit operator and the 3 two dimensional Dirac matrices, with the defining equation

$$\gamma^\mu\gamma^\nu = g^{\mu\nu} + is\epsilon^{\mu\nu\lambda}\gamma_\lambda \quad \text{where} \quad s = \pm 1. \tag{A29}$$

The two inequivalent[17] representations of the Clifford Algebra are distinguished by the value of a parameter s. The s values ± 1 correspond to spin up and spin down in the "hidden" third spatial dimension. Possible representations of the two inequivalent sets of basis operators are

$$\gamma^0 = \sigma_3, \quad \gamma^1 = si\sigma_2, \quad \text{and} \quad \gamma^2 = i\sigma_1. \tag{A30}$$

Note that the final results of any calculation of observables are independent of the representation.[18]

Now one can see how the proof of the topological nature of the AC phase in $2+1$ dimensions will work without doing any more calculations. The spin tensor is proportional to the product of two different Dirac matrices, and in $2+1$ dimensions this is not a new matrix, it is just the other Dirac matrix. The interaction term in equation (A26) is then indeed proportional to the product of a Dirac matrix and an *effective vector potential.* This permits us to transform equation (A26) into the free particle equation by a phase transformation, giving rise to the topological AC phase.

Doing the calculations the effective vector potential S_μ is found to be the $2+1$ dimensional dual of the electromagnetic field strength tensor

$$S_\mu = (1/2)\epsilon_{\mu\alpha\beta}F^{\alpha\beta} = (-\boldsymbol{k}\cdot\boldsymbol{B}, \boldsymbol{E}\times\boldsymbol{k}). \tag{A31}$$

[17] *inequivalent* in the sense that they cannot be unitarily transformed into each other.

[18] If one feels uncomfortable with handling the two inequivalent representations one can use 4-spinors and 4×4 Dirac matrices as was done by He and McKellar. The 4×4 representation equivalent to eq (A30) has the $s=+1$ representation in the $(1,1)$ place and the $s=-1$ representation in the $(2,2)$ place in the 2×2 block form of the Dirac matrices.

where $\boldsymbol{k}$ is a unit vector in the z direction i.e., the direction of the magnetic moment. When the magnetic field vanishes, then

$$S_\mu = (0, \boldsymbol{S}) \quad \text{with} \quad \boldsymbol{S} = \boldsymbol{E} \times \boldsymbol{k}. \tag{A32}$$

The AC phase is topological in the planar spatial geometry. It is then important to ask whether the necessary deviations of the experimental geometry from the planar geometry in which the electric fields and the motion of the dipole are both normal to the magnetic moment are such as to render the phase non-topological. Of course no experimental apparatus is infinite and translation invariant in the z direction, and one needs to consider whether that intrusion of the real world renders the phase non-topological.

But the two dimensional geometry is not enough for the phase to be topological, we must also have

(1) $\operatorname{curl} \boldsymbol{S} = 0$ in the interference region

(2) $\operatorname{curl} \boldsymbol{S} \neq 0$ in the excluded region

As $\operatorname{curl} \boldsymbol{S} = \boldsymbol{k}(\operatorname{div} \boldsymbol{E}) - (\boldsymbol{k} \cdot \boldsymbol{\nabla})\boldsymbol{E}$, the simplest way for the excluded region to generate a non-vanishing contribution to the phase is for the path to encircle some charges, giving rise to $\operatorname{div} \boldsymbol{E} \neq 0$. To preserve the $2+1$ dimensional geometry, the charges should be extended uniformly and infinitely in the z direction. The simplest such charge configuration, that chosen by Aharonov and Casher, is a line of charges on the z axis, with a linear electric charge density λ_e.

Implicit in the planar geometry are the conditions

- the magnetic moment $\boldsymbol{\mu}_m = \mu_m \boldsymbol{k}$ is normal to the plane defined by the electric field and the direction of motion of the dipole,
- and that the magnetic moment is constant in magnitude and direction.

The phase, relative to the configuration without any field, developed in the wave function when the particle travels along a closed path which encircles the line of charge is

$$\begin{aligned} \chi_{AC} &= s\mu_m \oint \boldsymbol{S} \cdot d\boldsymbol{r} \\ &= -s\mu_m \int_S (\boldsymbol{\nabla} \cdot \boldsymbol{E})\boldsymbol{k} \cdot d\boldsymbol{S} \\ &= -s\mu_m \lambda_e, \end{aligned} \tag{A33}$$

which is just the result of Aharonov and Casher.

The Dirac Equation was used for the calculations because because that is the simplest and most elegant structure. The proof can also be constructed using the Schrödinger equaion (as done by Hagen [11]) and relativistically for higher spins the Bargmann-Wigner equation (as done by He and McKellar). All of these approaches to the calculation give the same conditions for a topological phase.

A.5.c. Comments on AC Experiments

The original observation of the AC phase, due to Cimmino *et al*, used cold neutrons and followed the original AC geometry. While the precision was not high, with about 24% accuracy, it was certainly measuring a topological phase.

Sangster, Hinds, *et al* [12] used atomic beams and a completely different geometry. The original AC geometry is Fig. 3(a), they used Fig. 3(c). As they point out themselves, the magnetic moment in this case is not constant in magnitude or direction, and so the phase they measure is not topological. The same geometry was used in almost all of the subsequent atomic beam measurements before 2014 — they did not measure a topological phase either, but most of the authors did not remember the caveat provided by their model. A recent example is the 2001 work of Yanagimachi *et al* [13].

The first atomic beam experiment to measure a **topological** AC phase was the 2014 experiment of the Toulouse group [14].

A.5.d. The HMW Effect

The electromagnetic dual of an equation is obtained by interchanging the electric and magnetic fields, electric charges and magnetic monopoles, and electric and magnetic dipoles. Notwithstanding the non-observation, even the non-existence, of magnetic monopoles, this is a mathematically consistent operation. Just as a magnetic dipole moving through in an electric field which has the appropriate geometry may acquire a topological quantum phase, the Aharonov-Casher (AC) phase, a electric dipole in a magnetic field which has the appropriate geometry may also acquire a topological quantum phase. This electromagnetic dual phenomenon was pointed out by He and McKellar [4] in 1993, and independently by Wilkens [5] in 1994. Dowling, Williams and Franson in 1999 named it the He-McKellar-Wilkens (HMW) phase.

The derivation of the quantum topological phase acquired by a magnetic dipole carries over with the obvious changes to the derivation of the quantum topological phase acquired by an electric dipole. In this way we obtain the 2+1 dimensional equation for electric dipoles in an magnetic field.

The interaction of a spin half electric dipole of magnitude d with the electromagnetic field is obtained by substituting the magnitude of the electric dipole for that of the magnetic dipole and the dual electromagnetic field tensor $\tilde{F}^{\mu\nu}$ for the electromagnetic field tensor $F^{\mu\nu}$ in equation (A26):

$$\mu_m \Longrightarrow d \qquad F^{\mu\nu} \Longrightarrow \tilde{F}^{\mu\nu} = (1/2)\epsilon^{\mu\nu\lambda\kappa} F_{\lambda\kappa}. \tag{A34}$$

The "effective vector potential" T_μ in this case is then $2+1$ dimensional dual of the $3+1$ dimensional dual $\tilde{F}^{\alpha\beta}$ of the electromagnetic field strength tensor $F^{\mu\nu}$

$$T_\mu = (1/2)\epsilon_{\mu\alpha\beta}\tilde{F}^{\alpha\beta}. \tag{A35}$$

In the HMW configuration, the electric field vanishes and B_1, B_2 are constant in time. Then

$$\begin{aligned} T_\mu &= (0, \boldsymbol{T}) = (0, T_1, T_2) \\ &= (0, B_2, -B_1) = (0, \boldsymbol{B} \times \boldsymbol{k}), \end{aligned} \tag{A36}$$

where $\boldsymbol{k}$ is a unit vector in the z direction, i.e. the direction of the electric dipole moment. As expected the electric dipole moment is interacting with the *magnetic* field.

A phase transformation of the wave function

$$\psi' = \exp\left[-isd \int^{r} \boldsymbol{T}(\boldsymbol{r}') \cdot \boldsymbol{ds}'\right] \psi \tag{A37}$$

gives a transformed wave function ψ' which satisfies the free Dirac equation

$$(i\gamma^{\mu}\partial_{\mu} - m)\,\psi' = 0. \tag{A38}$$

The phase $sd\int_{\mathcal{C}} \boldsymbol{T}(\boldsymbol{r}') \cdot \boldsymbol{ds}'$ is a topological phase in the $2+1$ dimensional world as long as

(1) curl $\boldsymbol{T} = 0$ in the interference region

(2) curl $\boldsymbol{T} \neq 0$ in the excluded region

$$\operatorname{curl}\boldsymbol{T} = \operatorname{curl}(\boldsymbol{B}\times\boldsymbol{k}) = \boldsymbol{k}(\operatorname{div}\boldsymbol{B}) - (\boldsymbol{k}\cdot\boldsymbol{\nabla})\boldsymbol{B}.$$

The simplest way for the excluded region to generate a non-vanishing contribution to the phase is for it to contain some magnetic charges, giving rise to div $\boldsymbol{B} \neq 0$. The simplest charge configuration preserving the $2+1$ dimensional geometry is a line of magnetic monopoles on the z axis, with a linear magnetic monopole charge density λ_m. In this configuration, curl $\boldsymbol{T} \neq 0$ in the excluded region.

It is also possible to obtain curl $\boldsymbol{T} \neq 0$ in the excluded region without magnetic monopoles with a suitably shaped magnetic field, such that $(k\cdot\nabla)B \neq 0$ there. An experimental realisation of this possibility was suggested by McKellar, He and Klein [15].

Returning to the electric dipoles encircling a line of magnetic monopoles, linear monopole density λ_m, the phase, relative to the configuration without any field, developed in the wave function when the particle travels along a closed path $\mathcal{C} = \partial\mathcal{S}$ which encircles the line of magnetic charge just once is

$$\begin{aligned}\chi_{HMW} &= sd\oint_{\mathcal{C}} \boldsymbol{T}\cdot\boldsymbol{dr} \\ &= -sd\int_{\mathcal{S}} (\boldsymbol{\nabla}\cdot\boldsymbol{B})\boldsymbol{k}\cdot d\mathcal{S} \\ &= -s\mu_e\lambda_m,\end{aligned} \tag{A39}$$

as found by He and McKellar.

Since no magnetic monopoles have yet been been found this manifestation of the HMW phase is not capable of experimental observation. Indeed in the original paper He and McKellar [4] did not even suggest a possible experiment. Wilkens [5], and Dowling *et. al.* did make some suggestions for observations, but the real breakthrough came in 1995 when Wei, Han and Wei [6] emphasised that the experimental realisation would require an **induced** electric dipole, and thus an electric field. The induced dipole is proportional to the electric field and so in the HMW phase,

$$\chi_{HMW} = sd\oint_{\mathcal{C}} \boldsymbol{T}\cdot\boldsymbol{dr} = s\int_{\mathcal{S}} \boldsymbol{\nabla}\times(\boldsymbol{B}\times[d\boldsymbol{k}]), \tag{A40}$$

we would expect that we simply replace

$$d\boldsymbol{k} \Longrightarrow \alpha \boldsymbol{E}, \tag{A41}$$

where α is the electric polarisability of the particle. Then

$$\begin{aligned} & \boldsymbol{\nabla} \times (\boldsymbol{B} \times [\alpha \boldsymbol{E}]) \\ = \; & \alpha \, (\boldsymbol{B} \, \mathbf{div} \, \boldsymbol{E} - \boldsymbol{E} \, \mathbf{div} \, \boldsymbol{B} \\ & + (\boldsymbol{B} \cdot \boldsymbol{\nabla}) \boldsymbol{E} - (\boldsymbol{E} \cdot \boldsymbol{\nabla}) \boldsymbol{B}) \, . \end{aligned} \tag{A42}$$

Now electric charges can generate the topological phase, since we can have $\mathbf{div}\, \boldsymbol{E} \neq 0$ in the excluded region.

This intuitive argument has the flaw that it is ultimately based on the Dirac equation, which holds for intrinsic dipoles and not necessarily for induced dipoles. The analysis should be based on induced dipoles from the start, which is what Wei, Han and Wei actually did. Their analysis, as did Wilkens', followed from the effective electric field seen by a particle moving in an magnetic field — called the Röntgen field $\boldsymbol{E}_{\mathrm{R}} = \boldsymbol{v} \times \boldsymbol{B}$, which is added to the laboratory electric field $\boldsymbol{E}$ to obtain the electric field $\boldsymbol{E}_0 = (\boldsymbol{E} + \boldsymbol{v} \times \boldsymbol{B})$ experienced by the dipole in its rest frame. If the polarisability of the atom is α its electric dipole moment is

$$\boldsymbol{d} = \alpha(\boldsymbol{E} + \boldsymbol{v} \times \boldsymbol{B}), \tag{A43}$$

the Lagrangian is then

$$\mathcal{L} = \frac{1}{2} m \boldsymbol{v}^2 + \frac{1}{2} \alpha (\boldsymbol{E} + \boldsymbol{v} \times \boldsymbol{B})^2. \tag{A44}$$

The Wei, Han, and Wei form of the HMW phase is obtained by neglecting terms $\alpha \boldsymbol{E}^2$ and $\alpha \boldsymbol{B}^2$, and setting $\boldsymbol{v} \cdot \boldsymbol{B} = 0$. The Schrödinger equation becomes

$$\frac{1}{2m} \left(-i\nabla - \alpha (\boldsymbol{B} \times \boldsymbol{E}) \right)^2 \psi = 0, \tag{A45}$$

which can be transformed to the field free Schrödinger by a phase transformation with the phase

$$\chi_{WHW} = \alpha \int_C \boldsymbol{B} \times \boldsymbol{E} \cdot d\boldsymbol{s}, \tag{A46}$$

just as was conjectured above on intuitive grounds. This is a topological phase if

$$\begin{aligned} & \mathrm{curl}\, (\boldsymbol{B} \times \boldsymbol{E}) \\ = \; & \boldsymbol{B} \, \mathbf{div} \, \boldsymbol{E} - \boldsymbol{E} \, \mathbf{div} \, \boldsymbol{B} \\ & + (\boldsymbol{B} \cdot \boldsymbol{\nabla}) \boldsymbol{E} - (\boldsymbol{E} \cdot \boldsymbol{\nabla}) \boldsymbol{B} \end{aligned} \tag{A47}$$

vanishes in the interference region and is non-zero in the excluded region. Now electric charges can generate the topological phase. Inducing the electric dipole removes the link to magnetic monopoles, and it has the additional condition that $\boldsymbol{v} \cdot \boldsymbol{B} = 0$. So far there is no condition on the electric field. Our previous relativistic analysis required that the magnetic moment be normal to the velocity, which suggests that the electric field should also be normal to the velocity. Why has this not come out in the analysis of Wei, Han, and Wei?

In an attempt to answer this question, it would seem to be more reliable to obtain the relativistic corrections by taking the low velocity limit of a fully relativistic theory, rather than trying to introduce the corrections into the non-relativistic result. That is the approach I now adopt [16].

References

Halzen, F., and Martin, A.D., *Quarks and Leptons: An Introductory Course in Modern Particle Physics* (John Wiley & Sons, New York, 1984), Ch. 2.; on a very nice introduction to symmetries and quarks.

Wu, T.Y., *Physics: Its Development and Philosophy* (The Physical Society of the Republic of China, Taipei, 1989), Ch. V.; on symmetry, transformation, and invariance.

Dirac, P.A.M., *Rev. Mod. Phys.* **21**, 392 (1949); on the Lie algebra for the proper Lorentz group.

Weyl, H., *Ann. Physik* **59**, 101 (1919).

Weyl, H., *Zeits. f. Physik* **56**, 330 (1929). This is the classic paper on gauge invariance (Eichinvariance).

Fock, V., *Zeits. f. Physik* **39**, 226 (1927).

London, F., *Zeits. f. Physik* **42**, 375 (1927).

Dirac, P.A.M., *Proc. Royal Soc. London* **A133**, 60 (1931). This is the first paper on the magnetic monopole. It is followed by a second paper in **Phys. Rev. 74**, 817 (1948).

Ehrenberg, W., and Siday, R.E., *Proc. Phys. Soc. London* **B62**, 8 (1949). This is the first paper pointing out the essential idea of the experiment of Chamber (of Aharonov & Bohm's). This paper has received no attention.

Aharonov, Y., & Bohm, D., *Phys. Rev.* **115**, 485 (1959).

Chambers, R.G., *Phys. Rev. Letters* **5**, 3 (1960).

Yang, C.N., *Gauge Fields*, in the Proceedings of the Sixth Hawaii Topical Conference in Particle Physics (1975), Eds. P.N. Dobson, Jr., S. Pakvasa, V.Z. Peterson, and S.F. Tuan, p. 488; *Phys. Rev. Lett.* **33**, 445 (1974).

Wu, T.-Y., *Theoretical Physics* (in Chinese), Vol. 3, p. 164; Vol. 4, p. 75 and p. 68; on singularites of the potential **A** of a magnetic monopole.

M. V. Berry, Proc. R. Soc. **A392**, 45 (1984); Sci. Am. **259**, No. 6, 46 (1988); R. Jackiw, Comments At. Mol. Phys. **21**, 71 (1988); Intl. Jnl. Mod. Phys. **A3**, 285 (1988).

The references given below are mainly for the HMW phase; as given by B.H.J. McKellar.

References (from Section A.5. onwards)

[1] Y. Aharonov and D. Bohm, *Phys. Rev* **115**, 485-491 (1959).

[2] M. V. Berry, *Proc. Roy. Soc.* A **392**, 4557 (1984).

[3] Y. Aharonov and A. Casher, *Phys. Rev. Letters* **53**, 319-321 (1984).

[4] X.-G. He and B. H. J. McKellar, *Phys, Rev. A* **47**, 3424-3525 (1993).

[5] M. Wilkens, *Phys. Rev. Letters* **72**, 5-8 (1994)

[6] H. Wei, R. Han and X. Wei, *Phys. Rev. Letters* **75**, 2071 (1995).

[7] B. H. J. Mckellar, Int. J. Mod. Phys. A **31**, no. 07, 1630004 (2016).

[8] B. H. J. McKellar, The Universe **2**, no. 4, 6 (2014).

[9] A. Cimmino, G. I. Opat, A. G Klein, H. Kaiser, S. A. Werner, M. Arif and R. Clothier, *Phys. Rev. Letters* **63**, 380-383 (1989).

[10] J.A. Swansson, and B. H. J. McKellar, B.H.J., *Journal of Physics A* **34**, 1051-1061 (2001).

[11] C. R. Hagen, *Phys. Rev. Letters* **64**, 2347-2349 (1990).

[12] K. Sangster, E. A. Hinds, S. M. Barnett, E. Rijs and A. G. Sinclair, *Phys. Rev. A* **51**, 1776-1786 (1995).

[13] S. Yanagimachi, M. Kajiro, M. Machiya, and A. Morinaga, *Phys. Rev. A* **65**, 042104 (2002).

[14] J. Gillot, S. Lepoutre, A. Gauguet, J. Vigu, and M. Bchner *The European Physical Journal D*, **68** (2014).

[15] B. H. J. McKellar, X.-G. He and A. G. Klein, AIP Conference Proceedings 1588, 59 (2014); doi: 10.1063/1.4866924.

[16] X.-G. He and B. H. J. McKellar, arXiv:1610.08569, to be published.

Exercises: *Chapter 11*

1. In analogy with our discussion on the $SU(2)$ algebra, define

$$F^a = \frac{\lambda^a}{2}, \qquad a = 1, ..., 8.$$

(i) Write down the Lie algebra for F^a and prove that $F^2 \equiv F^a F^a$ is a Casimir operator.

(ii) Generalize the definition for the raising and lowering operators (such as Eqs. (17)-(19) for $SU(2)$) to the case of $SU(3)$.

(iii) Discuss how to label the members of a given irreducible representation of $SU(3)$.

2. *Use the results obtained from Exercise 1.*

(i) Write down the wave function for $\Delta^{++}(J_3 = +3/2)$ and generate the wave functions of the highest weight ($J_3 = +3/2$) for all other decuplet baryons by applying successively suitable raising and lowering operators. Fix the relative phases of the wave functions as well as the normalization factors of your raising and lowering operators.

(ii) Repeat the exercise for the case of low-lying octet baryons, now starting from the proton wave function given by Eq. (48) in the text.

3. Assume that the color wave function for a three-quark system is given uniquely by Eq. (47) in the text. Also assume that quarks are spin-$\frac{1}{2}$ particles satisfying Pauli exclusion principle.

(i) Working in the space of the first three flavors of quarks, i.e., u, d, and s with all quarks sitting at the lowest S-orbitals, show that there is only one set of octet wave functions *and* that there is no nonet baryon.

(ii) By substituting the s quark by the c quark in the flavor $SU(3)$ octet and decuplet baryons, describe the spectrum for charmed baryons.

4. Classify the states of mesons which contain at least one charm quark or antiquark. Your scheme should contain (1) those well-known $c\bar{c}$ bound states which have been observed experimentally and (2) those low-lying charmed mesons which contain one charm quark or antiquark. Try to give some description on the wave functions of those mesons which you have considered.

Chapter 12. Quantum Chromodynamics

Quantum chromodynamics, or QCD, is an $SU(3)$ gauge field theory of strong interactions among quarks and gluons. The word "chromodynamics" means "color-dynamics", meaning the dynamics among colors carried by quarks and gluons. As explained earlier in Ch. 11, the concept of "gauge invariance" was first introduced by Weyl (1919), revised slightly by Fock (1927) and London (1927), and finally generalized to the case of a non-abelian gauge group by Yang and Mills (1954). Thus, a non-abelian gauge theory is also referred to as "Yang-Mills theory". QCD is just a Yang-Mills theory with $SU(3)$ as the gauge group. In contrast, quantum electrodynamics (QED), which was treated in detail previously in Ch. 9 and Ch. 10, is an abelian gauge theory with $U(1)$ as the gauge group.

12.1 *QCD* is an $SU(3)$ Gauge Theory.

Consider the free lagrangian for a quark of a given flavor,

$$\mathcal{L}_0(x) = -\bar{\psi}(x)\gamma_\mu \partial_\mu \psi(x) - m\bar{\psi}(x)\psi(x), \tag{1}$$

which is almost identical to the lagrangian for a free electron, except that, in the present case, the quark field $\psi(x)$ has three components in color space. Following the important suggestion made by Yang and Mills (1954), we attempt to introduce a local phase transformation in color space:

$$\psi(x) \to \psi'(x) = U(\theta)\psi(x), \tag{2a}$$

with

$$U(\theta) = exp(-i\sum_{a=1}^{8} \theta_a(x)\lambda_a/2), \tag{2b}$$

where λ_a are eight hermitian, traceless Gell-Mann 3×3 matrices introduced earlier in §.11.2. Note that $\{\theta_a(x)\}$ are eight real functions to ensure that $U(\theta)$ is unitary.

Accordingly, the "gauge principle" may be generalized as follows:

$$\partial_\mu \to D_\mu \equiv \partial_\mu - i\frac{g}{2}\lambda_a G^a_\mu(x), \tag{3}$$

so that the lagrangian (1) becomes

$$\mathcal{L}(x) = -\bar{\psi}(x)\gamma_\mu D_\mu \psi(x) - m\bar{\psi}(x)\psi(x). \tag{4}$$

Here $G^a_\mu(x)$ are eight gauge fields which are called "gluons", meaning the bosons that glue quarks together. To ensure that the lagrangian (4) is invariant under the gauge transformation of Eqs. (2a) and (2b), we require

$$\frac{\lambda_a}{2}G^a_\mu \to \frac{\lambda_a}{2}G'^a_\mu = U(\theta)\{\frac{\lambda_a}{2}G^a_\mu - \frac{i}{g}U^{-1}(\theta)\partial_\mu U(\theta)\}U^{-1}(\theta). \tag{5}$$

The task to write down the lagrangian for the gauge field, which must be invariant under the gauge transformation Eq. (5), was first accomplished by Yang and Mills in 1954.

First, we note that the QED analogue, $\partial_\mu G^a_\nu - \partial_\nu G^a_\mu$, is *not* gauge invariant. This is where the non-abelian nature of $SU(3)$ starts to manifest itself in a certain way. Nevertheless, a suitable gauge-invariant expression may be inferred from the gauge principle as follows:

$$\begin{aligned}&\partial_\mu(\frac{1}{2}\lambda_a G^a_\nu) - \partial_\nu(\frac{1}{2}\lambda_a G^a_\mu)\\ &\to D_\mu(\frac{1}{2}\lambda_a G^a_\nu) - D_\nu(\frac{1}{2}\lambda_a G^a_\mu) = \frac{\lambda_a}{2}G^a_{\mu\nu},\end{aligned} \tag{6a}$$

with

$$G^a_{\mu\nu} \equiv \partial_\mu G^a_\nu - \partial_\nu G^a_\mu + g f_{abc} G^b_\mu G^c_\nu. \tag{6b}$$

It is then straightforward to show that the expression specified by

$$T_r \frac{\lambda_a}{2} G^a_{\mu\nu} \frac{\lambda_b}{2} G^b_{\mu\nu} = \frac{1}{2} G^a_{\mu\nu} G^a_{\mu\nu}, \tag{7}$$

is invariant under the gauge transformation Eq. (5). Fixing up the normalization factor as in QED, we obtain the *QCD* lagrangian,

$$\begin{aligned}\mathcal{L}_{QCD}(x) \;=\; &-\frac{1}{4}G^a_{\mu\nu}(x)G^a_{\mu\nu}(x)\\ &-\bar{\psi}(x)\gamma_\mu D_\mu \psi(x) - m\bar{\psi}(x)\psi(x).\end{aligned} \tag{8}$$

Strictly speaking, this lagrangian defines the so-called "classical chromodynamics". The quantized version of it is called "quantum chromodynamics" or *QCD*. Note that the first term in Eq. (8) contains a piece that describes interactions among three gluons and another piece that requires four gluons to interact. This brings in the nonlinearity which makes *QCD* very difficult to comprehend (or to solve numerically) as a theory of strong interactions.

Quantization of a non-abelian gauge field theory such as Eq. (8) is by no means trivial since application of Dirac's correspondence principle, which maps the Poisson algebra of a classical theory into the commutator algebra of the corresponding quantized theory, is obscured by the difficulty in choosing an appropriate set of independent generalized coordinates. In the case of QED [with $U(1)$ as the gauge group], it is possible to identify the transversal components of the gauge field, i.e., $\mathbf{A}_\perp(x)$, as the generalized coordinates and then to work out a quantized theory. This was explained previously in Chapter 8. It is in fact the procedure followed by many authors.[1] The price which we have to pay in this program is that Lorentz covariance [and gauge invariance] is no longer manifest. In addition, it is not clear whether such program can be of any practical use in the non-abelian case.

In Chapter 8, we outlined another quantization procedure in which one sets out to treat the four components of $A_\mu(x)$ as though they are linearly independent. To this end, one must introduce a term proportional to $(\partial_\mu A_\mu(x))^2$ to ensure that the generalized momentum field conjugate to $A_0(x)$ does not vanish identically. To quantize the electromagnetic field, one imposes an appropriate set of elementary commutation relations. The Lorentz condition is then imposed as an operator condition that ensures exact cancellation between "unphysical" timelike and longitudinal photons. Gauge invariance ensures that such cancellation occurs for all cases, i.e., for all gauges and for an arbitrary number of unphysical photons. Manifest

[1] *See, e.g., Lee, T.D., "Particle physics and introduction to field theory" (Harwood, New York, 1981).*

Lorentz covariance is maintained by this quantization procedure, but application of the procedure to the non-abelian case such as *QCD* is nontrivial.

Thus, it is important to consider the covariant quantization through the path-integral formulation, which can easily be generalized from the $U(1)$ case (such as QED) to the non-abelian case (such as *QCD*). In the path-integral formulation, one sets out to deal with the partition function as a path integral and attempts to deduce all physical quantities from it. In view of the extensive mathematical nature of the formalism, we wish to relegate introduction and discussion of this important subject to Appendix A. Here we wish to say a few words about the underlying ideas.

Conceptually, in the path-integral formulation we treat $\partial_\mu G^a_\mu(x) = C^a(x)$ as additional independent degrees of freedom but attempt to maintain the constraint that physical quantities are invariant under an arbitrary gauge transformation, Eqs. (2) and (5). Gauge invariance requires that two sets of $\{G^a_\mu(x)\}$ which are linked by a gauge transformation are completely equivalent, so that only one set in each equivalent class of $\{G^a_\mu(x)\}$ is needed for the determination of the partition function. Qualitatively speaking, fixing the gauge is equivalent to that the gauge is not fixed but the extra degrees of freedom are removed by suitable cancelling fields (which is "Faddeev-Popov ghost fields"). This procedure has been referred to as "gauge fixing". It is clear that gauge fixing and treating $C^a(x)$ as independent degrees of freedom are two separate issues in quantization of a gauge field theory. For technical details, consult Appendix A at the end of this chapter.

As we shall see in Ch. 14, experimental tests of *QCD* rely primarily on the asymptotical free nature of *QCD*, which says that, for sufficiently large Q^2, the strong coupling $\alpha_s(Q^2)$ ($\equiv g^2/(4\pi)$) becomes sufficiently small to warrent perturbative treatments. The idea of a "running" coupling constant was already introduced in §.10.3. for QED and we shall consider it again in §.12.2. to deduce the asymptotically free property of *QCD*. For the purpose of carrying out perturbative *QCD* calculations, it suffices to summarize final results on Feynman rules ('t Hooft, 1971). Qualitatively speaking, these Feynman rules may be obtained from an effective lagrangian,

$$\mathcal{L}_{eff}(x) = \mathcal{L}(x) - \partial_\mu \varphi^a(x) D_\mu \varphi^a(x), \tag{9}$$

where $\mathcal{L}(x)$ is specified by Eq. (8) and $D_\mu\varphi^a(x)$ is given by

$$D_\mu \varphi^a(x) \equiv \partial_\mu \varphi^a(x) + g f_{abc} G^b_\mu(x) \varphi^c(x). \tag{10}$$

$\varphi^a(x)$ is the scalar field of color a that serves as a "cancelling field" and so satisfies Fermi statistics. The fields $\varphi^a(x)$ are the "Fadde'ev-Popov ghost fields" mentioned above. The gluon propagator takes the form,

$$\frac{1}{i}\frac{\delta_{ab}}{k^2 - i\varepsilon}(\delta_{\mu\nu} - \frac{k_\mu k_\nu}{k^2}), \tag{11}$$

which vanishes identically upon contraction by k_μ or k_ν . This corresponds to $\partial_\mu G^a_\mu(x) = 0$, which defines the Landau gauge. It is always possible to modify the effective action by adding a suitable functional $S'(\partial_\mu G^a_\mu)$ such that the new effective lagrangian no longer contains terms involving $\partial_\mu G^a_\mu(x)$ explicitly. The gluon propagator is then given by

$$\frac{1}{i}\frac{\delta_{ab}}{k^2 - i\varepsilon}\delta_{\mu\nu}, \tag{12}$$

which defines the 't Hooft-Feynman gauge.

Feynman rules for carrying out perturbative *QCD* calculations are summarized immediately below. The reader who is interested in the derivation of these rules may consult Appendix A.

(a) *Fermion propagator*

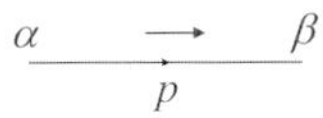

$$\Leftrightarrow \frac{1}{i}\frac{m - i\gamma\cdot p}{m^2 + p^2 - i\varepsilon} \tag{13a}$$

(b) *Gluon propagator*

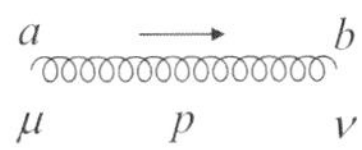

$$\Leftrightarrow \frac{1}{i}\frac{\delta_{ab}}{p^2 - i\varepsilon}[\delta_{\mu\nu} - (1-\xi)\frac{p_\mu p_\nu}{p^2}] \tag{13b}$$

$$\xi = \begin{cases} 0, & \text{Landau gauge;} \\ 1, & \text{'t Hooft-Feynman gauge.} \end{cases}$$

(c) *Ghost propagator*

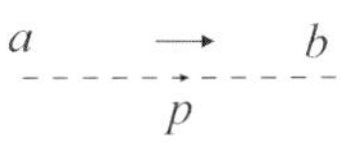

$$\Leftrightarrow \frac{1}{i}\frac{\delta_{ab}}{p^2 - i\varepsilon} \tag{13c}$$

(d) *Fermion vertex*

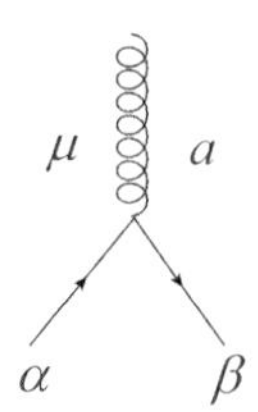

$$\Leftrightarrow -g\gamma_\mu \frac{\lambda^a_{\alpha\beta}}{2} \tag{13d}$$

(e) *Triple gluon vertex*

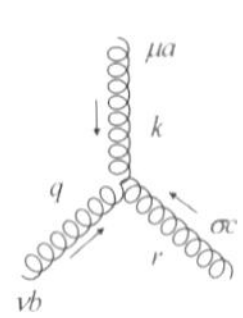

$$\Leftrightarrow \quad -gf_{abc}\{\delta_{\mu\nu}(k-q)_\sigma + \delta_{\nu\sigma}(q-r)_\mu + \delta_{\sigma\mu}(r-k)_\nu\} \tag{13e}$$

(f) *Quartic gluon vertex*

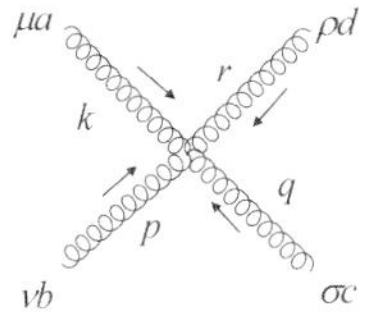

$$\Leftrightarrow -ig^2\{f_{abd}f_{cde}(\delta_{\mu\sigma}\delta_{\nu\rho} - \delta_{\mu\rho}\delta_{\nu\sigma}) + f_{ace}f_{bde}(\delta_{\mu\nu}\delta_{\sigma\rho} - \delta_{\mu\rho}\delta_{\nu\sigma}) + f_{ade}f_{cde}(\delta_{\mu\sigma}\delta_{\nu\rho} - \delta_{\mu\nu}\delta_{\sigma\rho})\} \quad (13f)$$

(g) *Ghost vertex*

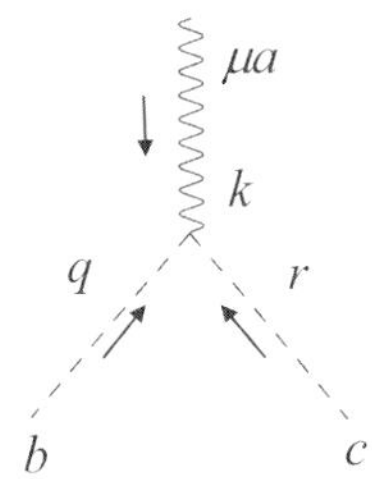

$$\Leftrightarrow +gf_{abc}r_\mu \quad (13g)$$

(h) *External lines*

$$\begin{aligned} &u(\mathbf{p}, s) \text{ for } \textit{incoming spinor}, \\ &\bar{u}(\mathbf{p}, s) \text{ for } \textit{outgoing spinor}; \\ &\bar{v}(\mathbf{p}, s) \text{ for } \textit{incoming antispinor}, \\ &v(\mathbf{p}, s) \text{ for } \textit{outgoing antispinor}; \\ &\frac{\varepsilon_\mu^{a,\lambda}(k)}{\sqrt{2k_0}} \text{ for } \textit{a gluon in the initial state}, \\ &\frac{\varepsilon^{a,\lambda *}(k)}{\sqrt{2k_0}} \text{ for } \textit{a gluon in the final state}. \end{aligned} \quad (13h)$$

(i) Every closed fermion loop or closed ghost loop induces a minus sign. Counting factors are reguired for a set of identical particles to ensure that the state is suitably normalized. There is a factor of $-i$ in going from the S-matrix element S_{fi} to the T-matrix element T_{fi}. Cross sections are computed from T_{fi} as follows:

$$d\sigma = \frac{1}{\mid v_1 - v_2 \mid}\{\prod_f \frac{d^3k_f}{(2\pi)^3}\}(2\pi)^4\delta^4(\sum_f k_f - \sum_i k_i) \cdot \bar{\sum} \mid T_{fi} \mid^2, \quad (13i)$$

where $\bar{\Sigma}$ indicates suitable averaging (over the initial discrete indices) and summation (over the final discrete degress of freedom).

12.2 *QCD* is Asymptotically Free.

Consider the coupling of a gluon to the quark. Such coupling may be described pictorically as in Figure 1.

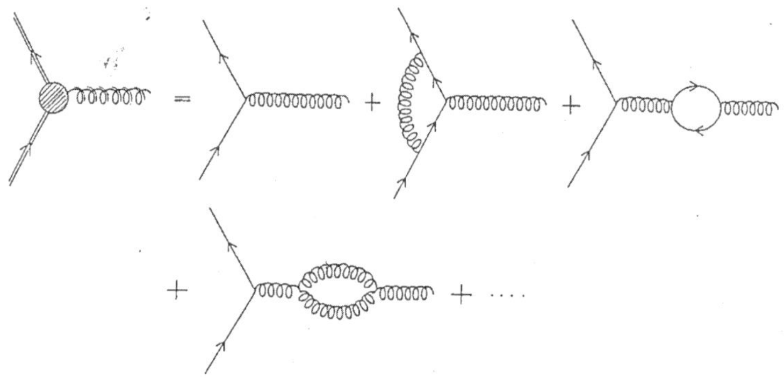

Figure 1:

We may interpret this figure in two different manners:
(A) We may consider the coupling on the left-hand side (LHS) of Fig. 1 as the coupling when the probing scale is characterized by $\bar{\mu}^2$ with $\bar{\mu}$ of the dimension of a mass. In other words, the coupling is $\bar{g}$ when the distance which we are probing is $1/\bar{\mu}$. Now, suppose that we manage to probe the same coupling at a distance $1/\mu$ smaller than $1/\bar{\mu}$. The coupling at μ^2 is g. It is then clear that the coupling $\bar{g}$ at $\bar{\mu}^2$, as looked upon at a smaller distance, is a sum of diagrams as illustrated by the RHS of Fig. 1. We have

$$\bar{g} \equiv \bar{g}(\frac{\bar{\mu}^2}{\mu^2}, \frac{m^2}{\mu^2}, g), \tag{14}$$

provided that the basic parameters of the theory at the scale μ^2 are m^2 and g. In this view, a hierarchy of theories at different values of μ^2 is envisioned and the notion of the "bare coupling" (or "bare mass") is rejected (as μ^2 is arbitrary and the smallest probing distance does not exist).
(B) Suppose that we write down a theory with g and m^2 as the basic parameters. We set out to compute higher order diagrams as described by the RHS of Fig. 1. However, most of them are divergent unless we introduce a regularization of some sort. In other words, we need an additional parameter μ^2 related to the method of regularization. Therefore, the theory makes sense only when the set of the basic parameters include μ^2 in addition to $\{g, m^2\}$. Eq. (14) comes out when we relate theories at two different values of μ^2.

The variation of the coupling constant with respect to the renormalization point μ^2 is charaterized by the beta function:

$$\beta(\frac{m^2}{\mu^2}, g) \equiv \bar{\mu}^2 \frac{\partial}{\partial \bar{\mu}^2} \bar{g}(\frac{\bar{\mu}^2}{\mu^2}, \frac{m^2}{\mu^2}, g) \mid_{\bar{\mu}^2 = \mu^2}, \tag{15}$$

which was considered earlier in §.10.3 for the case of QED.

The β-function in the case of QED is given by (Baker and Johnson, 1969)

$$\beta(\alpha) = \frac{\alpha^2}{3\pi} + \frac{\alpha^3}{4\pi^2} + O(\alpha^4), \tag{16}$$

which was used previously in §.10.3.

The β-function for an $SU(n)_{color}$ gauge theory with N_f flavors of fermions may be obtained from Fig. 1 and is given by (Belavin and Migdal 1974; Caswell 1974; Jones 1974)

$$\begin{aligned} \beta(\alpha_s) &= A\frac{\alpha_s^2}{4\pi} + B\frac{\alpha_s^3}{16\pi^2} + O(\alpha_s^4), \\ A &= -\frac{11}{3}n + \frac{2}{3}N_f, \\ B &= -\frac{34}{3}n^2 + N_f(\frac{13n}{3} - \frac{1}{n}). \end{aligned} \tag{17}$$

To make sense out of Eqs. (15) and (17), we consider again the renormalization group (RG) equation for $\bar{g}(\bar{\mu}^2/\mu^2, g)$, neglecting the mass parameter m^2 for the sake of simplicity. (See §.10.3. for a similar derivation.) We define

$$t \equiv \bar{\mu}^2/\mu^2. \tag{18}$$

Choosing $t'' = t't$, we have

$$g'' \equiv \bar{g}(t', g') \quad with \quad g' \equiv \bar{g}(t, g),$$

or,

$$\bar{g}(t't, g) = \bar{g}(t', \bar{g}(t, g)). \tag{19}$$

Differentiating Eq. (19) with respect to t' and then setting $t' = 1$, we find

$$t\frac{\partial}{\partial t}\bar{g}(t, g) = \beta(\bar{g}), \tag{20}$$

which was just Eq. (85), Ch. 10. Note that Eq. (20) can readily be solved for $\bar{g}$. We obtain

$$\int_{t=1}^{t} \frac{d\bar{g}}{\beta(\bar{g})} = \ln t. \tag{21}$$

Considering the QED case to lowest order, we find

$$3\pi(\frac{1}{\alpha(\mu^2)} - \frac{1}{\alpha(\bar{\mu}^2)}) = \ln\frac{\bar{\mu}^2}{\mu^2},$$

or,

$$\alpha(\bar{\mu}^2) = \frac{\alpha(\mu^2)}{1 - \frac{1}{3\pi}\alpha(\mu^2)\ln\frac{\bar{\mu}^2}{\mu^2}}. \tag{22}$$

Analogously, we obtain, for *QCD* with N_f flavors,

$$\alpha_s(\bar{\mu}^2) = \frac{\alpha_s(\mu^2)}{1 + \frac{1}{4\pi}(11 - \frac{2}{3}N_f)\alpha_s(\mu^2)\ln\frac{\bar{\mu}^2}{\mu^2}}. \tag{23}$$

It is customary to identify the renormalization point μ^2 with the momentum squared Q^2 which defines the scale of the physics that we are probing. In other words, we may rewrite Eqs. (22) and (23) in a more familiar form:

$$\alpha(Q^2) = \frac{\alpha(Q_0^2)}{1 - \frac{1}{3\pi}\alpha(Q_0^2)\ln\frac{Q^2}{Q_0^2}}, \quad \textit{for } QED; \tag{24a}$$

$$\alpha_s(Q^2) = \frac{\alpha_s(Q_0^2)}{1 + \frac{1}{4\pi}(11 - \frac{2}{3}N_f)\alpha_s(Q_0^2)\ln\frac{Q^2}{Q_0^2}}, \quad \textit{for } QCD. \tag{24b}$$

It is known that, at $Q_0^2 \sim (1GeV)^2, \alpha(Q_0^2) = \frac{1}{137}$ and $\alpha_s(Q_0^2) \sim O(1)$. Accordingly, we find

$$\alpha(Q^2) \to \infty \quad as \quad \frac{1}{3\pi}\alpha(Q_0^2)\ln\frac{Q^2}{Q_0^2} \to 1, \tag{25}$$

which occurs at an enormous value of Q^2/Q_0^2. The divergent behavior of $\alpha(Q^2)$ as indicated by Eq. (25) is related to the so-called "Landau ghost problem" which becomes irrelevant as QED gets unified with other interactions well before the Landau ghost appears. On the other hand, we also have

$$\alpha_s(Q^2) \to 0 \quad as \quad Q^2 \to \infty, \tag{26}$$

which means that the interaction becomes weak as Q^2 is large ($\gg Q_0^2$). This phenomenon has been referred to as "asymptotic freedom". It can be shown that inclusion of the next-to-leading order terms, the B term in Eq. (17), does not modify significantly the asymptotically free nature of *QCD*.

It is important to know that *QCD* is asymptotically free. If we use very high energy probes to probe a hadron [which is a system consisting of quarks, antiquarks, and gluons], the large Q^2 implies the smallness of $\alpha_s(Q^2)$ so that interactions among constituents can be treated perturbatively, thereby explaining why parton models are so successful at high energies. Perturbative treatment (Mueller 1981; Reya 1981) of *QCD* effects at high energies allows for specific tests of *QCD* in experiments such as production of jets, decay of quarkonium states, scaling violation in deep inelastic scattering, and Drell-Yan processes.

12.3 Color Confinement

As an important empirical fact, hadrons observed so far are *colorless*, i.e., in a color singlet configuration. This suggests that color fields exist only inside the region specified by the extent of a given hadron. A quark or gluon carries color so that confinement of color explains why quarks and gluons are not observed in isolation. Color confinement also explains why observed hadrons, such as baryons and mesons, appear in the way they are. For example, four-quark clusters (Q^4) must have a net color and so cannot exist in isolation. But, it does not preclude the possibility of having another color-singlet configuration for a multiquark cluster such as ($Q^2\bar{Q}^2$) or Q^6, or having systems consisting only of gluons ("glueballs"), or having systems consisting of both quarks and gluons such as (Q^3g) ("hermaphrodites"), etc., so long as these hadrons are in an overall color-singlet configurations. Search for an unambiguous manifestation of "color" is of critical importance, but it is often obscured by uncertainties related to strong interactions.

A natural way to realize color confinement is to conjecture that the color fields, $\{\mathbf{E}^a, \mathbf{B}^a\}$, exist only in the quark-gluon phase [as characterized by the QCD vacuum or the trivial QCD ground state] that is different from the hadron phase [the physical vacuum or the true QCD ground state]. This has been referred to as the "two-phase picture". In the strict two-phase picture, formation of flux tubes is required if a color constituent is pulled away from a hadron.

By discretizing $SU(2)$ or $SU(3)$ color gauge theory on a lattice [lattice gauge theory], it is possible to investigate whether the two-phase picture emerges. Indeed, extensive early lattice gauge calculations[2], in the so-called "quenched approximation" (where the role of fermions is suppressed) suggested the presence of a deconfinement phase transition at a certain transition temperature T_c, indicating the occurrence of a phase transition from the hadron phase (at $T < T_c$) to the quark-gluon phase (at $T > T_c$). Meanwhile, the restoration phase transition for chiral symmetry is found to take place simultaneously. Although such a simple and nice picture may be only a gross simplification to what is really going on in the real world, it is truly amazing that a non-abelian gauge field theory has so many nontrivial properties.

Of course, the definitive results obtained from lattice gauge calculations do not carry over directly to the case of continuum QCD but this is by far the most convincing evidence that QCD is compatible with the two-phase picture, thereby suggesting confinement of color. The compatibility between QCD and color confinement must be considered as another important reason why QCD is being regarded as the candidate theory of strong interactions among quarks.

The question of how to discretize a gauge field theory on a lattice is a technical issue which is not critical for understanding the subject of our discussion. For gaining some basic insights on the problem, however, an interested reader may consult Appendix B at the end of the present chapter.

To sum up, QCD describes strong interactions among quarks and gluons in a satisfactory manner: At high energies ($Q^2 \gg m_p^2$), it explains why parton models are so successful while, at low energies ($Q^2 \leq m_p^2$), it is compatible with the empirical fact that color is confined. Just like the Glashow-Salam-Weinberg $SU(2) \times U(1)$ electroweak theory, QCD is phrased on the basis of both gauge principle and renormalizability. If these theories sustain future experimental tests, then one must seek for the answer to the basic question as to why the present specific conceptual framework, in which fundamental laws of the nature can be phrased in terms of a gauge field theory, is so powerful and successful.

---

[2] *For a recent review, see, e.g., Kogut, J.B., in "Nuclear and Particle Physics on the Light Cone", Eds. Johnson, M.B., and Kisslinger, L.S. (World Scientific, Singapore, 1989), p. 239 and references therein.*

Appendix A

Method of Path Integrals

Feynman's method of path integrals is a method of formulating a quantum theory without explicit reference to operators and states in Hilbert space. It is a method that can be used for both quantum mechanics and quantum field theory. The method becomes a powerful tool for treating quantized gauge field theories.

A.1. Quantum Mechanics and Path Integrals

As an illustrative example, we consider the scattering of a spinless particle from a potential $V(x)$.[3]

$$i\hbar\frac{\partial\psi}{\partial t} = -\frac{1}{2m}\hbar^2\nabla^2\psi + V(x)\psi \equiv H\psi. \tag{A1}$$

The solution to Eq. (A1) is specified by the wavefunction $\psi(x,t)$ which is a function of x and t. Alternatively, we may relate the solutions at two space-time points, $\psi(x',t')$ and $\psi(x,t)$ $(t' > t)$, through an integral equation:

$$\psi(x',t') = \int K(x',t';x,t)\psi(x,t)dx. \tag{A2}$$

Here the function $K(x',t';x,t)$ is called the "Green's function" or "kernel". It is possible to express K as a path integral. To do so, we first divide the time interval $t'-t$ into $n+1$ small segments of magnitude ϵ. To each value of t, we associate a value of the coordinate x:

$$\begin{aligned} t = t_0 &: x_0 \\ t_1 &: x_1 \\ t_2 &: x_2 \\ &: \\ t_n &: x_n \\ t' = t_{n+1} &: x_{n+1} = x' \end{aligned} \tag{A3}$$

Note that the values x_0 and x' fix the endpoints of a "path" while the intermediate values $x_1, ..., x_n$ determine the rest of the path.

[3] *In §. A.1., we do not set* $\hbar = 1$.

For infinitesimal ϵ, the time evolution is determined from Eq. (A1). For example, we have

$$\begin{aligned}
\psi(x_1,t_1) &\equiv \langle x_1 \mid t_1\rangle \\
&= \langle x_1 \mid U(t_1,t_0) \mid t_0\rangle \\
&= \langle x_1 \mid exp\{-\frac{i}{\hbar}H(x,p)\epsilon\} \mid t_0\rangle \\
&= \int \frac{dx_0 dp}{2\pi\hbar} exp\{-\frac{i}{\hbar}V(\frac{x_1+x_0}{2})\epsilon\}\cdot \\
&\quad \cdot \langle x_1 \mid exp\{-\frac{i}{\hbar}\frac{p^2}{2m}\epsilon\} \mid p\rangle\langle p \mid x_0\rangle\langle x_0 \mid t_0\rangle \\
&= \int \frac{dx_0 dp}{2\pi\hbar} exp\{-\frac{i}{\hbar}V(\frac{x_1+x_0}{2})\epsilon\}\cdot \\
&\quad \cdot exp\{\frac{i}{\hbar}px_1\}exp\{-\frac{i}{\hbar}\frac{p^2}{2m}\epsilon\}exp\{-\frac{i}{\hbar}px_0\}\psi(x_0,t_0).
\end{aligned} \tag{A4}$$

Using the integration formula,

$$\int_{-\infty}^{\infty} dz \exp(iaz^2) = (i\pi/a)^{1/2}, \tag{A5}$$

we obtain, by comparing the result with Eq. (A2),

$$\begin{aligned}
&K(x_1,t_1;x,t_0) \\
&=(\frac{m}{2\pi\hbar i\epsilon})^{\frac{1}{2}} exp\{\frac{i}{\hbar}L(\frac{x_1+x_0}{2},\frac{x_1-x_0}{\epsilon})\epsilon\}.
\end{aligned} \tag{A6}$$

where $L = \frac{m}{2}\dot{x}^2 - V(x)$ is the classical Lagrangian. The operation leading to Eq. (A6) can easily be carried out step by step for the rest of the path. The net result is the desired Green's function:

$$K(x',t';x,t_0) = \lim_{n\to\infty,\epsilon\to 0}(\frac{m}{2\pi\hbar i\epsilon})^{(n+1)/2}\int\prod_{i=1}^{n} dx_i \exp(\frac{i}{\hbar}\int Ldt). \tag{A7}$$

Eq. (A7) is often abbreviated as follows:

$$K = \int [dx] \exp(iS/\hbar), \tag{A8}$$

with $S = \int_{t_0}^{t'} Ldt$ the action for the path. Note that Eq. (A7) is not adequate if the potential is velocity-dependent. In fact, in the case that there are a number of quantized systems corresponding to the same classical mechanical system, it is not clear which quantized system the method of path integrals actually selects out.

As a simple example for obtaining the kernel, we consider the case of a free particle, where $L = L_0 = \frac{m}{2}\dot{x}^2$. The kernel is

$$\begin{aligned}
K_0^{(n)}(x',t';x,t) =&(\frac{m}{2\pi\hbar i\epsilon})^{(n+1)/2}\int_{-\infty}^{\infty} dx_1 ... \int_{-\infty}^{\infty} dx_n \\
&\times \exp[\frac{i}{\hbar}\frac{m}{2\epsilon}(x_1-x_0)^2]\exp[\frac{i}{\hbar}\frac{m}{2\epsilon}(x_2-x_1)^2]... \\
&\times \exp[\frac{i}{\hbar}\frac{m}{2\epsilon}(x_{n+1}-x_n)^2].
\end{aligned} \tag{A9}$$

Introducing the variables,

$$
\begin{aligned}
&x_1 - x_0 = x'_1, \quad x_2 - x_1 = x'_2, \quad ..., \quad x_n - x_{n-1} = x'_n, \\
&x_{n+1} - x_n = (x_{n+1} - x_0) - z = (x' - x) - z, \\
&z = x'_1 + x''_2 + ... + x'_n,
\end{aligned}
\tag{A10}
$$

we rewrite Eq. (9) as follows:

$$
\begin{aligned}
K_0^{(n)}(x', t'; x, t) =& (\frac{m}{2\pi\hbar i\epsilon})^{(n+1)/2} \int dz \int \frac{dp}{2\pi\hbar} \int_{-\infty}^{\infty} dx'_1 ... \int_{-\infty}^{\infty} dx'_n \\
& \times \exp[\frac{i}{\hbar} p \cdot (z - x'_1 - x'_2 - ... - x'_n)] \\
& \times \exp[\frac{i}{\hbar}\frac{m}{2\epsilon}\{x_1'^2 + x_2'^2 + ... + x_n'^2 + (z - (x' - x))^2\}].
\end{aligned}
\tag{A9$'$}
$$

Use Eq. (A5) to carry out the integrations in Eq. (A9$'$) and take into account the fact that in the Schrödinger theory waves propagate forward in time [giving rise to a step function $\theta(t' - t)$ as defined by Eq. (A11), Ch. 8]:

$$
K_0(x', t'; x, t) = [\frac{m}{2\pi\hbar i(t' - t)}]^{1/2} \exp[\frac{im}{2\hbar(t' - t)}(x' - x)^2]\, \theta(t' - t), \tag{A11}
$$

which is a well-known formula in elementary quantum mechanics.

A.2. The Path Integral in Field Theory

It is clear that Eq. (A7) may be generalized readily to field theory. Consider a field ϕ which may have many components. We now divide space-time into four-dimensional cells of volume ϵ^4 and define a "path" by specifying the value of ϕ in each cell. The path integral now consists of summing over all possible values of the field in each cell. Setting $\hbar = 1$, we find a kernel:

$$
K = \int [d\phi] \exp(i \int \mathcal{L}(x) d^4x) \tag{A12}
$$

where $\mathcal{L}$ is the Lagrangian density. In practice, it is convenient to define a generating functional $Z[J]$ by the path integral:

$$
Z[J] = \int [d\phi] \exp\{i \int [\mathcal{L}(x) + J(x)\phi(x)] d^4x\}, \tag{A13}
$$

which is the same as Eq. (A12) except that it contains an additional "source" term in the action. $Z[J]$ plays a central role in the development of quantum field theory according to the method of path integrals since, with the aid of this generating functional, all the standard results of field theory can be obtained. In particular, the n-point Green's function is given by

$$
\begin{aligned}
\frac{\delta^n Z[J]}{\delta J(x_1)\delta J(x_2)...\delta J(x_n)}|_{J=0} &= i^n < 0 \mid T[\phi(x_1)\phi(x_2)...\phi(x_n)] \mid 0 > \\
&\equiv i^n G(x_1, ..., x_n)
\end{aligned}
\tag{A14}
$$

where $\langle 0 \mid T[\phi(x_1)...\phi(x_n)] \mid 0\rangle$ is the vacuum expectation value of the time-ordered product of n fields $\phi(x_i)$.

We may illustrate the procedure by considering the case of a real scalar field ϕ, with Lagrangian $\mathcal{L} = \mathcal{L}_0 + \mathcal{L}_I$, where $\mathcal{L}_0 = -\frac{1}{2}(\partial_\mu\phi\partial_\mu\phi + \mu^2\phi^2)$ and $\mathcal{L}_I = \mathcal{L}_I(\phi)$ is an interaction term which we ignore here but may take into account if necessary. Using integration by parts, we find

$$Z_0[J] = \int [d\phi] \exp\{i \int d^4x[-\frac{1}{2}\phi(-\partial_\mu\partial_\mu + \mu^2 - i\epsilon)\phi + J\phi]\}, \qquad \text{(A15)}$$

where the factor $-i\epsilon$ has been introduced on the ground that the resultant Green function is equivalent to what is expected from the Euclidean-space fields through the imaginary time prescription. Note that the integral in the exponent is the limit of a sum over four-dimensional cells. By labeling the field in cell α as ϕ_α, that in neighboring cell β as ϕ_β, and so on, $\frac{1}{2}(-\partial_\mu\partial_\mu + \mu^2 - i\epsilon)$ may be considered as the limit of a symmetric matrix $A_{\alpha\beta}$ that connects neighboring cells. Note that we have, for symmetric A,

$$\int_{-\infty}^{\infty} \prod_{i=1}^{N} dx_i \exp(-x_i A_{ij} x_j + 2S_k x_k) = (\pi/det\, A)^{1/2} \exp(S_i A_{ij}^{-1} S_j) \qquad \text{(A16)}$$

Employing this result with $S_i = J(x)/2$ and dropping an (infinite but non-essential) multiplicative constant, we obtain

$$Z_0[J] = \exp\{-\frac{1}{2}i \int d^4x d^4y J(x)[-\partial_\mu\partial_\mu + \mu^2 - i\epsilon]^{-1} J(y)\} \qquad \text{(A17)}$$

Using Eq. (14), we then obtain the two-point Green's function:

$$\triangle_F(x) = \int \frac{d^4k}{(2\pi)^4} e^{ik\cdot x} (\frac{1}{k^2 + \mu^2 - i\epsilon}), \qquad \text{(A18)}$$

which yields $(-i)(k^2 + \mu^2 - i\epsilon)^{-1}$ as the Feynman propagator for a scalar meson of mass μ.

A.3. Path Integrals and Quantum Electrodynamics

In the case of quantum electrodynamics (QED), the Green's function for the photon must satisfy

$$(\partial_\nu\partial_\mu - \delta_{\mu\nu}\partial_\eta\partial_\eta)\triangle_{\mu\sigma}(x - y) = \delta_{\nu\sigma}\delta^4(x - y). \qquad \text{(A19)}$$

If we differentiate both side by applying ∂_ν, the left-hand side becomes zero but the right-hand side becomes $\partial_\sigma\delta^4(x - y)$. The problem arises because the inverse of the operator $K_{\mu\nu} = \partial_\nu\partial_\mu - \delta^{\mu\nu}\partial_\eta\partial_\eta$ does not exist. This fact is closely linked to gauge invariance, which states that $F_{\mu\nu}$ is invariant under an arbitrary gauge transformation: $A_\nu \to A_\nu + \partial_\nu\Lambda$. Thus, $K_{\mu\nu}\partial_\nu\Lambda = 0$; or, $K_{\mu\nu}$ has zero eigenvalues and so its inverse is singular.

To obtain the path integral $\int [dA]e^{iS(A)}$ for QED, we integrate over all possible vector potential values in each cell ϵ^4, which include, for each distinct A, all values of A equivalent up to a gauge transformation. That is, we must integrate over all values of Λ. However, for these distinct values of Λ, the lagrangian $\mathcal{L}$ and thus the action S are invariant. Therefore, we are carrying out an integral with a constant integrand over an infinite number of paths. To avoid the problem, we need to select just one path for each gauge-inequivalent A. That this can be done consistently for non-abelian gauge theories such as QCD was first demonstrated by Faddeev and Popov in 1967. Here we wish to illustrate how the solution to the problem results in the introduction of "ghost" fields.

Specifically, we divide the field A_μ into subsets such that, in each subset, none of the members is related to the other members by a gauge transformation. For each member $\tilde{A}_\mu$, we consider all possible gauge transformation Λ. In this way, we have

$$Z = \int [dA] \exp(i \int d^4 x \mathcal{L}) = \int [d\tilde{A}] \exp(i \int d^x \mathcal{L}) \int [d\Lambda], \tag{A20}$$

where we may factor out the integrand for $d\Lambda$ because $\mathcal{L}$ is invariant under gauge transformations.

To remove the unphysical infinite constant $\int [d\Lambda]$, we introduce a factor $exp(-i \int d^4 x C^2/2$ with C some function of A_μ (which helps to fix the gauge), and make the resulting integral independent of the choice of C by multiplying it with the Jacobian $det(\partial C/\partial \Lambda)$. That is, we have

$$\begin{aligned} \int [d\Lambda] \quad &\rightarrow \quad \int [d\Lambda]\, det(\partial C/\partial \Lambda)\, exp(-i \int d^4 x \frac{1}{2} C^2) \\ &= \int dC\, exp(-i \int d^4 x \frac{1}{2} C^2), \end{aligned} \tag{A21}$$

which amounts to multiplying Z by an overall constant. This is permissible since it affects only normalization factors. Thus, the new partition function is given by

$$Z = \int [d\tilde{A}] \int [d\Lambda]\, det(\frac{\partial C}{\partial \Lambda})\, exp[i \int d^4 x (\mathcal{L} - \frac{1}{2} C^2)]. \tag{A22}$$

Under an infinitesimal gauge transformation

$$A_\mu \quad \rightarrow A_\mu \quad + \partial_\mu \Lambda, \tag{A23}$$

the function C is altered as follows:

$$C \quad \rightarrow \quad C + M\Lambda \tag{A24}$$

where M is some operator that may include derivatives. Thus,

$$det(\partial C/\partial \Lambda) = det\, M. \tag{A25}$$

We note that, for any Hermitian matrix A, we have

$$(det\, A)^{-1} = \pi^{-n} i^{-n} \int dz_1 dz_2 ... \int dz_n\, exp(i \langle z \mid A \mid z \rangle), \tag{A26}$$

where $\langle z \mid A \mid z \rangle = z_1^* A z_1 + z_2^* A z_2 + ... + z_n^* A z_n$.

Letting $z = x + iy$ and $dz \equiv dxdy$, we obtain

$$\int dx \exp(-az^*z) = \pi/a,$$

$$\prod_{i=1}^{n} \int dz_i \exp(-a_i z_i^* z_i) = \pi^n/(a_1...a_n). \tag{A27}$$

where the a_i are real.

For a diagonal $n \times n$ matrix A,

$$A = \begin{pmatrix} a_1 & & & & \\ & \cdot & & & \\ & & \cdot & & \\ & & & \cdot & \\ & & & & a_n \end{pmatrix}$$

we have $\sum a_n z_n^* z_n = \langle z \mid A \mid z \rangle$. Substituting each a_j by $-ia_j$, we find

$$\begin{aligned}\int dz_1 \int dz_2 ... \int dz_n \exp(i\langle z \mid A \mid z \rangle) &= i^n \pi^n / a_i ... a_n \\ &= i^n \pi^n / det(A),\end{aligned}$$

which leads to Eq. (26) quoted above. On the other hand, for a nondiagonal hermitian matrix A, we can find a unitary transformation such that

$$z' = Uz, \qquad A' = UAU^{-1}$$

so that $dz_1', ..., dz_n' = dz_1, ..., dz_n$ and Eq. (A26) follows.

Apart from an irrelevant constant factor, Eq. (A26) is the cell equivalent of the equation

$$(det\, M)^{-1} = \int [d\psi] \exp(i \int d^4x \psi^* M \psi). \tag{A28}$$

However, what we really want to know is $det\, M$, whereas Eq. (A28) is a formula for $(det\, M)^{-1}$. The problem may be fixed by assuming that it is possible to represent det M by an equation similar to Eq. (A28):

$$det\, M = \int [d'\phi] \exp(i \int d^4x \phi^* M \phi), \tag{A29}$$

where ϕ is also a complex scalar field and the symbol $[d'\phi]$ is yet to be clarified. We then have

$$\begin{aligned}det\, M (det\, M)^{-1} = 1 = &\int [d\psi] \exp(i \int d^4x \psi^* M \psi) \\ &\times \int [d'\phi] \exp(i \int d^4x \phi^* M \phi).\end{aligned} \tag{A30}$$

Since ψ is a complex scalar field not connected to any sources and $\psi^* M \psi$ appears in every term in the Green's function expansion, the ψ lines must appear in closed loops in all Feynman diagrams. The ϕ lines appear in closed loops for exactly the same reason. However, since the left-hand side of (A30) is unity, the contributions of the ψ and ϕ loops must cancel order by order in the perturbation expansion. This can be achieved if we associate a minus sign with each ϕ loop to cancel the plus sign associated with each ψ loop. This is the significance of the notation $[d'\phi]$. Accordingly, Eq. (A29) may be interpreted as a path integral over the complex scalar ϕ field with a factor of (-1) assigned to every closed loop. This yields Fermi-Dirac statistics, or, the "wrong" statistics for a complex scalar field (leading to the name "ghost").

The net result of the above standard discussion is that the unphysical infinite constant factor associated with gauge invariance has been removed from $Z[J]$. A specific choice of gauge fixes the factor C, which implies the ghost factor $exp(i\int d^4x\phi^* M\phi)$, where M depends on the choice of C. For example, we may choose $C = \partial_\mu A_\mu$. Thus, a gauge transformation $A_\mu \to A_\mu + \partial_\mu \Lambda$ yields $M = \partial_\mu \partial_\mu$ by definition. Therefore, the ghost field in QED does not couple to matter fields and so its role may be ignored.

Consider the $SU(2)$ Yang-Mills field $\mathbf{A}_\mu$. As we recall, the Lagrangian for the gauge field is $\mathcal{L}_{YM} = -\frac{1}{4}\mathbf{F}_{\mu\nu}\cdot\mathbf{F}_{\mu\nu}$, where $\mathbf{F}_{\mu\nu} = \partial_\mu \mathbf{A}_\nu - \partial_\nu \mathbf{A}_\mu - g(\mathbf{A}_\mu \times \mathbf{A}_\nu)$. Note that $\mathcal{L}_{YM}$ is invariant under local gauge transformations of the form

$$\delta\psi = i\epsilon\cdot\frac{\tau}{2}\psi, \qquad \delta\bar{\psi} = -\bar{\psi}i\epsilon\cdot\frac{\tau}{2};$$
$$\delta\mathbf{A}_\mu = -(1/g)\partial_\mu\epsilon - (\epsilon\times\mathbf{A}_\mu).$$

Or, the gauge-invariant derivative is

$$D_\mu = \partial_\mu + igA_\mu\cdot\frac{\tau}{2}.$$

Now, we consider a function corresponding to the Landau gauge condition $\partial_\mu A_\mu = 0$:

$$\mathbf{C} = \partial_\mu \mathbf{A}_\mu. \tag{A31}$$

Thus, we have

$$\begin{aligned}\partial_\mu \mathbf{A}'_\mu &= \partial_\mu \mathbf{A}_\mu - (1/g)\partial_\mu(\partial_\mu\epsilon) - \partial_\mu(\epsilon\times\mathbf{A}_\mu)\\ &= \partial_\mu \mathbf{A}_\mu - (1/g)\partial_\mu D_\mu\epsilon,\end{aligned}$$

which yields

$$C' = C - (1/g)\partial_\mu(D_\mu\epsilon), \tag{A32}$$

or,

$$M = -(1/g)\partial_\mu(D_\mu).$$

Therefore, we have a complex scalar ghost field ϕ with an additional factor:

$$\int [d'\phi]\exp(i\int d^4x\phi^*\partial_\mu D_\mu\phi), \tag{A33}$$

where $[d'\phi]$ signifies that a factor of -1 is associated with each closed loop. The new feature in the massless non-abelian Yang-Mills case is that there is a coupling of the ghost field to the field A_μ, the existence of which would not have been suspected from the original Lagrangian. This coupling cannot be ignored, since otherwise the theory is inconsistent and not renormalizable.

Appendix B

Method of Lattice Gauge Fields

In this appendix, we use Hamiltonian lattice formulation to illustrate how a gauge field theory can be treated on a lattice.

B.1. Formulation

We discretize space into a simple cubic lattice with lattice spacing a. The lattice sites are vectors of the form

$$\mathbf{x} = a(i\hat{\mathbf{e}}_x + j\hat{\mathbf{e}}_y + k\hat{\mathbf{e}}_z), \tag{B1}$$

where i, j, k are integers and $\hat{\mathbf{e}}_i$ are unit vectors along the lattice directions. The directed link from $\mathbf{x}$ to $\mathbf{x} + a\hat{\mathbf{e}}$ will be denoted by $(\mathbf{x}, \hat{\mathbf{e}})$, which may be distinguished from the oppositely directed link $(\mathbf{x} + a\hat{\mathbf{e}}, -\hat{\mathbf{e}})$.

By introducing a lattice, we are forced to give up various space symmetries. Nevertheless, it is possible to maintain gauge invariance. To do so, we wish to require that the theory be invariant under the discrete generalization of the quark field gauge transformation.

It is natural to associate the vector gauge fields with the links. Indeed, we note that, on a lattice, geometry scalars are associated with lattice sites, vectors with directed links, antisymmetric tensors with oriented areas, etc. Our basic unit will be the string operator:

$$U_{\mathbf{x};\hat{\mathbf{e}}} = U(\mathbf{x} + a\hat{\mathbf{e}}, \mathbf{x}; c) = P \exp\ i[\int_{\mathbf{x}}^{\mathbf{x}+a\hat{\mathbf{e}}} \tilde{A} \cdot dz], \tag{B2}$$

with

$$\tilde{A}_\mu \equiv \frac{\lambda^\alpha}{2} g A_\mu^\alpha. \tag{B2a}$$

The integral runs along the link from $\mathbf{x}$ to $\mathbf{x} + a\hat{\mathbf{e}}$. An obvious property, as due to unitarity of U, is

$$U_{\mathbf{x};\hat{\mathbf{e}}}^{-1} = U_{\mathbf{x}+a\hat{\mathbf{e}};-\hat{\mathbf{e}}}.$$

The continuum limit may be achieved by letting a approach zero so that we have, at the classical level,

$$U_{x;\hat{\mathbf{e}}} \approx \exp\{ia\tilde{A}(\mathbf{x}) \cdot \hat{\mathbf{e}}\} \approx 1 + ia\tilde{A}(x) \cdot \hat{\mathbf{e}}, \tag{B3}$$

which may be used to justify the lattice versions of various operators.

U is a unitary 3×3 matrix and thus can be parametrized as an element of $SU(3)$, namely

$$U_{x,\hat{\mathbf{e}}} = exp\{i\frac{\lambda^\alpha}{2} \cdot b_{x,\hat{\mathbf{e}}}^\alpha\}. \tag{B4}$$

Comparing Eq. (B4) with Eq. (B3), we see that the group parameters b are related to the vector potentials A. It is important to note that the b's vary over a compact manifold in distinction to the continuum A's. Integration over the group denoted by $[dU]$ will, in reality, be an integration over the b's. We are now in a position to define a gauge invariant operator whose continuum limit will be related to the color magnetic field B^a. Consider

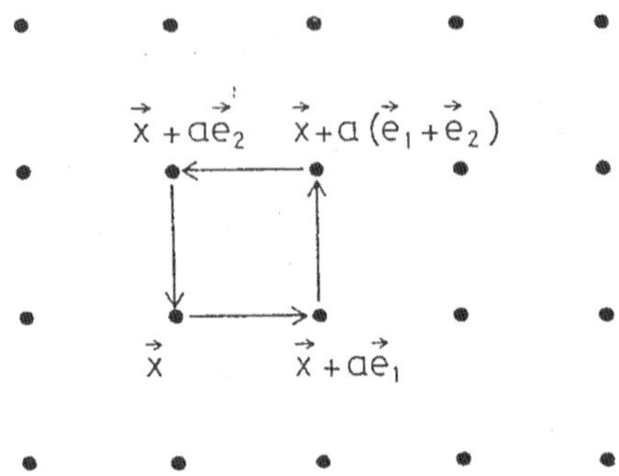

Figure 2: A fundamental plaquette.

the trace of a product of four link operators, $U_{\mathbf{x},\hat{e}}$, along a fundamental lattice square, or plaquette, illustrated in Fig. 2.

$$U_{\mathbf{x};\hat{\mathbf{e}}_1,\hat{\mathbf{e}}_2} = \{U_{\mathbf{x}+a\hat{\mathbf{e}}_2;-\hat{\mathbf{e}}_2} U_{\mathbf{x}+a(\hat{\mathbf{e}}_1+\hat{\mathbf{e}}_2);-\hat{\mathbf{e}}_1} U_{\mathbf{x}+a\hat{\mathbf{e}}_1;\hat{\mathbf{e}}_2} U_{\mathbf{x};\hat{\mathbf{e}}_1}\}. \tag{B5}$$

Using Eq. (B3), we obtain, to order a^2,

$$Tr\, U_{x;\hat{\mathbf{e}}_i;\hat{\mathbf{e}}_j} = \epsilon_{ijk} Tr\, \exp[ia^2 \tilde{B}\cdot\hat{\mathbf{e}}_k] \approx \epsilon_{ijk} Tr\, [1 - \frac{1}{2}a^4(\tilde{B}\cdot\hat{\mathbf{e}}_k)^2], \tag{B6}$$

with

$$F_{\mu\nu} \equiv \frac{1}{g}(\partial_\mu \tilde{A}_\nu - \partial_\nu \tilde{A}_\mu - i[\tilde{A}_\mu, \tilde{A}_\nu]), \tag{B7a}$$

$$\tilde{B}_k \equiv \frac{1}{2} g \epsilon_{ijk} F_{ij}. \tag{B7b}$$

Unless ambiguity may arise, we replace the combination $\mathbf{x}, \hat{\mathbf{e}}_i, \hat{\mathbf{e}}_j$ by p and let

$$U_p = U_{\mathbf{x};\hat{\mathbf{e}}_i,\hat{\mathbf{e}}_j}. \tag{B8}$$

In view of Eq. (B6) a lattice version of

$$H_M = Tr \int d^3x \frac{1}{g^2} B^2(x) \tag{B9}$$

is

$$H_M = \sum_p \frac{1}{g^2 a} Tr\, [2 - U_p - U_p^\dagger]. \tag{B10}$$

We now need a lattice generalization of the electric field. To this end, we note that

$$[\mathbf{E}^\alpha(\mathbf{y})\cdot\hat{\mathbf{e}}_i, U_{\mathbf{x};\hat{\mathbf{e}}_j}] = \delta_{ij}\delta^2(y_\perp) U(\mathbf{x}+a\hat{\mathbf{e}}, \mathbf{y}; c)\frac{\lambda^\alpha}{a} U(\mathbf{y}, \mathbf{x}; c), \tag{B11}$$

if $\mathbf{y}$ coincides with any of the points in the interval $(\mathbf{x}, \mathbf{x}+a\hat{\mathbf{e}})$ and zero otherwise. In the lattice formulation, $U_{\mathbf{x},\hat{\mathbf{e}}}$ must be considered in total and cannot be broken up along the link. We can define an electric field at the start or end of a link. The two fields associated with the link $(\mathbf{x}, \hat{\mathbf{e}})$ are $E^{\alpha}_{\mathbf{x},\hat{\mathbf{e}}}$ and $E^{\alpha}_{\mathbf{x}+a\hat{e};-\hat{\mathbf{e}}}$, and are determined from their commutation relations with link variables $U_{\mathbf{x},\hat{\mathbf{e}}}$. For $\hat{\mathbf{e}}$ positive, we may postulate

$$
\begin{aligned}
[E^{\alpha}_{\mathbf{x};\hat{\mathbf{e}}_i}, U_{\mathbf{y},\hat{\mathbf{e}}_j}] &= \delta_{ij}\delta_{\mathbf{x},\mathbf{y}}U_{\mathbf{x};\hat{\mathbf{e}}_i}\frac{\lambda^{\alpha}}{2},\\
[E^{\alpha}_{\mathbf{x}+a\hat{\mathbf{e}}_i;-\hat{\mathbf{e}}_j}, U_{\mathbf{y};\hat{\mathbf{e}}_j}] &= -\delta_{ij}\delta_{\mathbf{x},\mathbf{y}}\frac{\lambda^{\alpha}}{2}U_{\mathbf{x},\hat{\mathbf{e}}_i}.
\end{aligned} \tag{B12}
$$

The fact that $U^{\dagger}_{\mathbf{x};\hat{\mathbf{e}}} = U_{\mathbf{x}+a\hat{e};-\hat{\mathbf{e}}}$ yields the commutation relations for negative $\hat{\mathbf{e}}$'s. Use of the Jacobi identities yields the electric field commutation relations:

$$
\begin{aligned}
&[E^{\alpha}_{\mathbf{x};\hat{\mathbf{e}}_i}, E^{\beta}_{\mathbf{y};\hat{\mathbf{e}}_j}] = i\delta_{ij}\delta_{\mathbf{x},\mathbf{y}}f^{\alpha\beta\gamma}E^{\gamma}_{\mathbf{x};\hat{\mathbf{e}}_i},\\
&[E^{\alpha}_{\mathbf{x};-\hat{\mathbf{e}}_i}, E^{\beta}_{\mathbf{y};-\hat{\mathbf{e}}_j}] = -i\delta_{ij}\delta_{\mathbf{x},\mathbf{y}}f^{\alpha\beta\gamma}E^{\gamma}_{\mathbf{x};-\hat{\mathbf{e}}_i},\\
&[E^{\alpha}_{\mathbf{x};\hat{\mathbf{e}}_i}, E^{\beta}_{\mathbf{y},-\hat{\mathbf{e}}_j}] = 0.
\end{aligned} \tag{B13}
$$

The two electric fields are not independent of each other; the defining commutation relations provide the connection:

$$\lambda^{\alpha}E^{\alpha}_{\mathbf{x}+a\hat{\mathbf{e}};-\hat{\mathbf{e}}} = -U_{\mathbf{x};\hat{\mathbf{e}}}\lambda^{\beta}E^{\beta}_{\mathbf{x};\hat{\mathbf{e}}}U^{\dagger}_{\mathbf{x};\hat{\mathbf{e}}},$$

which implies that

$$E^2_{\mathbf{x};\hat{\mathbf{e}}} = E^2_{\mathbf{x}+a\hat{\mathbf{e}};-\hat{\mathbf{e}}}. \tag{B14}$$

The electric energy

$$H_E = g^2 Tr\int d^3x\tilde{E}^2 \tag{B15}$$

goes over to

$$H_E = \frac{g^2}{2a}\sum_{\text{links}} E^{\alpha}_{\mathbf{x};\hat{\mathbf{e}}}E^{\alpha}_{\mathbf{x};\hat{\mathbf{e}}}. \tag{B16}$$

Thus, the total lattice QCD Hamiltonian is given by

$$H = \frac{g^2}{2a}\sum_{\text{links}} E^2_{\mathbf{x};\hat{\mathbf{e}}} + \frac{1}{ag^2}Tr\sum_{p}(2 - U_p - U^{\dagger}_p) + \sum_{\mathbf{x}} m_0\bar{q}_{\mathbf{x}}q_{\mathbf{x}}, \tag{B17}$$

where the last term describes the quark mass effect.

The generator of infinitesimal time-independent gauge transformation is specified by

$$\sum_{\mathbf{x}}\{\sum_{\hat{\mathbf{e}}}(E^{\alpha}_{\mathbf{x};\hat{\mathbf{e}}} - E^{\alpha}_{\mathbf{x};-\hat{\mathbf{e}}}) + q^{\dagger}_{\mathbf{x}}\frac{\lambda^{\alpha}}{2}q_{\mathbf{x}}\}, \tag{B18}$$

which annihilates all physical states.

We may rewrite the Hamiltonian of Eq. (B17) as follows

$$
\begin{aligned}
H &= \frac{g^2}{2a}\{\sum_{\text{links}} E^2_{\mathbf{x};\hat{\mathbf{e}}} + xTr\sum_{p}(2 - U_p - U^{\dagger}_p)\} + \sum_{\mathbf{x}} m_0\bar{q}(\mathbf{x})q(\mathbf{x}),\\
x &= 2/g^4.
\end{aligned} \tag{B19}
$$

B.2. An Illustrative Example

In what follows, we wish to consider the quark-antiquark potential, to zeroth, first and second order in x. As we will be interested in the difference in the energies of this configuration and in the vacuum energy, we first calculate the spectrum *without* quarks.

Due to Gauss' law, electric flux lines must close. To zeroth order in x the two lowest lying states are the vacuum, with no electric flux anywhere, and the states where the electric flux is in a 3 or $\bar{3}$ representation along the links of some fundamental plaquette. The first state we identify by $\mid 0\rangle$ with energy $E_0^{(0)} = 0$. The second class of states consists of

$$\begin{aligned} \mid P\rangle &= U_p\mid 0\rangle, \\ \mid \tilde{P}\rangle &= U_p^\dagger\mid 0\rangle. \end{aligned} \tag{B20}$$

Using the definition of U, Eq. (B5), and the commutation relations, Eq. (B12), we obtain

$$\begin{aligned} H_E\mid P\rangle &= E_p^{(0)}\mid P\rangle, \\ H_E\mid \tilde{P}\rangle &= E_p^{(0)}\mid \tilde{P}\rangle, \end{aligned}$$

with

$$E_p^{(0)} = \frac{g^2}{2a}\{4C^{(3)}\} = \frac{16}{3}\frac{g^2}{2a}. \tag{B21}$$

Here $C^{(3)}$ is the quadratic Casimir operator for the 3 or $\bar{3}$ representation.

To first order in x all energies shift by the amount

$$E_{0,p}^{(1)} = 3x\frac{g^2}{2a}N(P), \tag{B22}$$

where $N(P)$ is the number of plaquettes in our system. (Here we assume for the moment a finite world.) The order x^2 correction may be obtained from second-order perturbation theory,

$$E_0^{(2)} = -\frac{g^2}{2a}x^2\sum_p\frac{\mid\langle 0\mid U_p^\dagger\mid P\rangle\mid^2 + \mid\langle 0\mid U_p\mid \tilde{P}\rangle\mid^2}{(16/3)} = -\frac{g^2}{2a}(\frac{3}{8}N(P))x^2. \tag{B23}$$

We may now turn to states with a heavy quark at the origin and a heavy antiquark at $R = na\hat{\mathbf{e}}_1$. Again due to Gauss' law, we must have an electric flux joining the two particles. We expect that the lowest-energy configuration consists of the shortest flux path possible, of which the energy is

$$E^{0a}(x) = \frac{g^2}{2a}\frac{4}{3}n + 2m_0 = \frac{g^2}{2a^2}\frac{4}{3}R + 2m_0. \tag{B24}$$

Accordingly, *strong coupling lattice gauge theories confine quarks.*

To first order in x the perturbing Hamiltonian connects the state with the shortest length to those with an additional plaquette.

Calculating the result and subtracting from it the second order correction to the vacuum, Eq. (B24), we obtain

$$E(R) = \frac{g^2}{2a^2}R[\frac{4}{3} - x^2\frac{11}{153} + ...] + 2m_0. \tag{B25}$$

B.3. Euclidean Lattice Gauge Theories

Lagrangian theories in a path integral formalism may also be transcribed to a lattice, which is four-dimensional:

$$< \prod O_i(\mathbf{x}_i;\hat{\mathbf{e}}_i) > = \frac{1}{Z}\int \prod_{\mathbf{x};\hat{\mathbf{e}}} dU_{\mathbf{x};\hat{\mathbf{e}}} \prod O_i(\mathbf{x}_i;\hat{\mathbf{e}}_i) exp\{\frac{1}{g^2} Tr \sum_p (U_p + U_p^\dagger)\},$$

$$Z = \int \prod_{\mathbf{x};\hat{\mathbf{e}}} dU_{\mathbf{x};\hat{\mathbf{e}}} exp\{\frac{1}{g^2} Tr \sum_p (U_p + U_p^\dagger)\}. \qquad \text{(B26)}$$

This formulation is equivalent to the Hamiltonian formulation *for small g only.* Although the quantum mechanics based on the different formulations is expected to be the same in the continuum limit, various approximate results do depends on the formalism chosen. In many cases, it is useful to compare them.

The operator characterizing the confinement properties of the theory is the Wilson loop integral, of which lattice analogue is

$$W[c] = Tr \prod_{(\mathbf{x},\hat{\mathbf{e}})\in c} U_{\mathbf{x},\hat{\mathbf{e}}}. \qquad \text{(B27)}$$

The product is ordered along the closed curve c.

The strong coupling expansion of the Euclidean theory may be obtained by expanding the exponent in Eq. (B26) as a power series in $1/g^2$. Using the orthogonality property for the representation matrices, we find that the lowest order nonvanishing contribution to $< W[c] >$ is of the order $(1/g^2)^{N(c)}$, where $N(c)$ is the number of plaquettes in the planar area surrounded by the closed curve c. (Bander 1981) Relating the number of plaquettes to the area, i.e., $A = N(c)a^2$, we obtain

$$W[c] = exp(-ln\, g^2\, A/a^2), \qquad \text{(B28)}$$

which results in confinement with the energy of separation

$$E(R) = \frac{1}{a^2} ln\, g^2\, R. \qquad \text{(B29)}$$

Lattice theories, both in the Hamiltonian and in the Euclidean-Lagrangian formulation lead to quark confinement at strong couplings. A crucial question, which we do not address here, is how these results can be extrapolated smoothly to the weak coupling continuum region.

References

Weyl, H., *Ann. Physik* **59**, 101 (1919); Weyl, H., *Zeits. f. Physik* **56**, 330 (1929). The latter is the classic paper on gauge invariance (Eichinvariance).

Fock, V., *Zeits. f. Physik* **39**, 226 (1927).

London, F., *Zeits. f. Physik* **42**, 375 (1927).

Yang, C.N., and Mills, R.L., *Phys. Rev.* **96**, 191 (1954).

Faddeev, L.D. and Popov, V.N., Phys. Lett. **25B**, 29 (1967); on the Faddeev-Popov ghosts.

't Hooft, G., Nucl. Phys. B**33**, 173 (1971); on the Feynman rules for massless Yang-Mill fields.

Baker, M. and Johnson, K., Phys. Rev. **183**, 1292 (1969); on the β-function for QED.

Belavin, A.A. and Migdal, A.A., Pis'ma Zh. Eksp. Teor. Fiz **19**, 317 (1974) [JETP Lett. **19**, 181 (1974)]; Caswell, W.E., Phys. Rev. Lett. **33**, 244 (1974); Jones, D.R.T., Nucl. Phys. B**75**, 531 (1974); on the β-function for $SU(n)$ gauge field theories.

Mueller, A.H., "*Perturbative QCD at high energies*", Phys. Rep. **73C**, 237 (1981); Reya, E., "*Perturbative quantum chromodynamics*", Phys. Rep. **69C**, 195 (1981); on perturbative *QCD.*

Feynman, R.P., and Hibbs, A.R., *Quantum Mechanics and Path Integrals* (McGraw Hill, New York, 1965); Abers, E., and Lee, B.W., *Gauge Theories*, Phys. Rep. **9C**, 1 (1973); on the method of path integrals.

Wilson, K.G., Phys. Rev. D**10**, 2445 (1974); Kogut, J., and Susskind, L., Phys. Rev. D**11**, 395 (1975); Bander, M., Phys. Rep. **75**, 205 (1981); Kogut, J.B., Rev. Mod. Phys. **51**, 659 (1979); *ibid.* **55**, 775 (1983); in *Nuclear and Particle Physics on the Light Cone*, eds. Johnson, M.B., and Kisslinger, L.S. (World Scientific, Singapore, 1989), p. 239; Negele, J.W., in *Quarks, Mesons, and Nuclei I: Strong Interactions*, eds. Hwang, W-Y. P., and Henley, E.M. (World Scientific, Singapore, 1989), p. 1; on the treatment of lattice gauge theories.

Exercises: *Chapter 12*

1. Prove that the lagrangian, Eq. (8), is invariant under gauge transformations, Eqs. (2) and (5).

2. Derive the Feynman rules (13e) and (13f) by calculating $\langle 0 \mid S^{(1)} \mid i \rangle$ with the suitable three-gluon and four-gluon initial states.

3. Use Eq. (17) in connection with Eq. (21). Discuss the asymptotically free behavior of the final result.

Chapter 13. The Glashow-Salam-Weinberg Electroweak Theory

So far we have introduced two very important examples of gauge field theories, QED and QCD. The gauge fields in both cases are massless vector bosons. If both strong and electromagnetic interactions can be described successfully in terms of gauge field theories, it is natural to speculate that weak interactions, or any other fundamental interactions that may exist in nature, can also be phrased in a similar vein. It turns out that this might indeed be the case. Historically, Weinberg and Salam proposed independently in 1967 an $SU(2) \times U(1)$ gauge field theory that unifies the electromagnetic and weak interactions among leptons, before QCD was introduced. The structrue of the electroweak interactions in the Weinberg-Salam $SU(2) \times U(1)$ model agrees to lowest order with what Glashow obtained in 1961 by attempting to unify the electromagnetic and weak interactions, although Weinberg and Salam invoked the concept of the Higgs mechanism (Higgs 1964 and others) in their construction, a key ingredient that 't Hooft used later in 1971 to prove the renormalizability of the model.

The major obstacle for describing weak interactions in terms of a gauge field theory stems from the fact that the vector bosons mediating weak interactions are not massless. Thus, it was not clear whether gauge symmetry is relevant at all in the case of weak interactions. Invocation of Higgs mechanism, as in the case of the Weinberg-Salam $SU(2) \times U(1)$ model, allows for implementation of gauge symmetry at the lagrangian level. However, such symmetry must be eventually broken in the sense that the ground-state solution to the problem does not respect the symmetry. The specific way of symmetry breaking, in which the lagrangian (hamiltonian) respects the symmetry but the ground-state solution does not, is called "spontaneous symmetry breaking." The Glashow-Salam-Weinberg (GSW) electroweak theory is a spontaneously broken $SU(2) \times U(1)$ gauge field theory.

The GSW electroweak theory predicts the existence of neutral weak interactions, the existence of the charm quark, and the existence of $W^{\pm}$ and Z^0, all of which have been substantiated in quantitative terms by experiments. Although the physics related to the Higgs sector remains elusive, there is little doubt that any better theory proposed in the foreseeable future must reproduce successes of the GSW electroweak theory.

13.1 Higgs Mechanism in an $SU(2) \times U(1)$ Gauge Theory

Consider an $SU(2) \times U(1)$ gauge theory. The gauge fields are to be denoted by $A^i_\lambda(x)$ and $B_\lambda(x)$, respectively. Following the procedure given earlier in Ch. 12 for QCD, we may write the lagrangian for gauge fields,

$$\mathcal{L}_{gauge} = -\frac{1}{4} F^i_{\mu\nu} F^i_{\mu\nu} - \frac{1}{4} B_{\mu\nu} B_{\mu\nu}, \tag{1}$$

where

$$F^i_{\mu\nu} \equiv \partial_\mu A^i_\nu - \partial_\nu A^i_\mu + g\epsilon^{ijk} A^j_\mu A^k_\nu, \tag{2a}$$

$$B_{\mu\nu} = \partial_\mu B_\nu - \partial_\nu B_\mu. \tag{2b}$$

A general gauge transformation is specified by

$$\frac{\tau}{2}\cdot\mathbf{A}_\mu(x)\to\frac{\tau}{2}\cdot\mathbf{A}'_\mu(x)=U(\theta)\{\frac{\tau}{2}\cdot\mathbf{A}_\mu(x)-\frac{i}{g}U^{-1}(\theta)\partial_\mu U(\theta)\}U^{-1}(\theta), \tag{3a}$$

$$B_\mu(x)\to B'_\mu(x)=B_\mu(x)-\frac{1}{g'}\partial_\mu\varphi(x) \tag{3b}$$

with

$$U(\theta)=exp[-i\theta(x)\cdot\frac{\tau}{2}]. \tag{3c}$$

It is a routine exercise to demonstrate that the lagrangian $\mathcal{L}_{gauge}$ is gauge invariant i.e., invariant under a local gauge transformation specified by Eqs. (3a)–(3c). Introduction of a mass term such as $-\frac{1}{2}m^2\mathbf{A}_\mu\cdot\mathbf{A}_\mu$ destroys gauge invariance so that weak boson masses can only be taken into account in a specific manner.

To incorporate weak boson masses, we introduce a pair of complex scalar fields,

$$\phi=\begin{pmatrix}\phi^+\\ \phi^0\end{pmatrix}, \tag{4}$$

which transforms like a doublet under $SU(2)$ and possesses a weak hypercharge $Y_W=1$ under $U(1)$. Thus, the gauge-invariant derivative for ϕ is given by

$$D_\mu=\partial_\mu-ig\frac{\tau}{2}\cdot\mathbf{A}_\mu(x)-i\frac{g'}{2}B_\mu(x). \tag{5}$$

We choose

$$\mathcal{L}_{scalar}=-[D_\mu\phi]^\dagger[D_\mu\phi]-V(\phi), \tag{6}$$

where

$$V(\phi)=\mu^2\phi^\dagger\phi+\lambda(\phi^\dagger\phi)^2, \tag{7}$$

with $\mu^2<0$ and $\lambda>0$. The lagrangian $\mathcal{L}$ is gauge invariant, i.e. invariant under the gauge transformation of Eqs. (3a)–(3c) and the equation given below:

$$\phi(x)\to\phi'(x)=U(\theta)\cdot exp(-i\frac{1}{2}Y_W\varphi(x))\phi(x). \tag{8}$$

However, the potential $V(\phi)$ has a minimum at

$$<\phi_0>=\begin{pmatrix}0\\ v/\sqrt{2}\end{pmatrix},\quad with\quad v=\sqrt{-\frac{\mu^2}{\lambda}}, \tag{9}$$

so that the physical ground state $<\phi_0>$ differs from the trivial ground state, i.e. $<\phi>=0$. Note that the explicit form for $<\phi_0>$ varies with the gauge. In the unitary gauge (U-gauge), $<\phi_0>$ assumes the form given by Eq. (9). In other gauges, Eq. (8) may be used to generate $<\phi'_0>$.

The situation is typical for the so-called "spontaneous symmetry breaking": The lagrangians $\mathcal{L}_{gauge}$ and $\mathcal{L}_{scalar}$ are invariant under an arbitrary gauge transformation but the ground-state solution to the problem varies with the gauge, i.e., breaks gauge symmetry. As we shall see shortly, such spontaneous breaking of gauge symmetry generates masses for weak bosons. The mechanism has been named "Higgs mechanism" (Higgs 1964; Anderson 1963; Englert et al. 1964; Kibble et al. 1964).

We rewrite Eq. (4) as follows,

$$\phi(x) = \begin{pmatrix} \phi^+(x) \\ \phi^0(x) \end{pmatrix} \equiv \exp[+\frac{i}{2v}\xi(x)\cdot\tau]\begin{pmatrix} 0 \\ \frac{1}{\sqrt{2}}(v+\eta(x)) \end{pmatrix}, \tag{10}$$

with $\xi(x)$ and $\eta(x)$ four real functions. Accordingly, we make a gauge transformation,

$$\begin{aligned} &U(\theta) = U(\xi/v) = exp[-\frac{i}{2v}\xi(x)\cdot\tau], \\ &\varphi(x) = 0, \end{aligned} \tag{11}$$

such that

$$\phi'(x) = U(\xi/v)\phi(x) = \begin{pmatrix} 0 \\ \frac{1}{\sqrt{2}}(v+\eta(x)) \end{pmatrix}. \tag{12}$$

For the sake of simplicity, we denote gauge fields as $\mathbf{A}_\mu(x)$ and $B_\mu(x)$ in this specific gauge (U-gauge). We introduce

$$W_\mu^\pm = \frac{1}{\sqrt{2}}(A_\mu^1 \pm iA_\mu^2), \tag{13a}$$

$$Z_\mu^0 = cos\theta_W A_\mu^3 - sin\theta_W B_\mu, \tag{13b}$$

$$A_\mu = sin\theta_W A_\mu^3 + cos\theta_W B_\mu, \tag{13c}$$

with

$$sin\theta_W = \frac{g'}{\sqrt{g^2+g'^2}}, \tag{14a}$$

$$cos\theta_W = \frac{g}{\sqrt{g^2+g'^2}}. \tag{14b}$$

Substituting Eq. (10) back into Eq. (6) and using Eqs. (13)–(14), we find, with $v^2 + \frac{\mu^2}{\lambda} = 0$,

$$\begin{aligned} \mathcal{L}_{scalar} =& \{-\frac{1}{2}\partial_\mu\eta\partial_\mu\eta - \frac{1}{2}(-2\mu^2)\eta^2 - \frac{\lambda}{4}(\eta^4 + 4v\eta^3)\} \\ &+ \frac{\mu^4}{4\lambda} \\ &- \frac{1}{8}\{v^2 + (2v\eta + \eta^2)\}\{2g^2 W_\mu^+ W_\mu^- + [g^2 + (g')^2] Z_\mu^0 Z_\mu^0\}. \end{aligned} \tag{15}$$

Accordingly, three gauge bosons become massive while the remaining one massless:

$$M_{W^\pm} = \frac{1}{2} g v, \tag{16a}$$

$$M_{Z^0} = \frac{1}{2}[g^2 + (g')^2]^{1/2} v = \frac{M_{W^\pm}}{cos\theta_W}, \tag{16b}$$

$$M_A = 0. \tag{16c}$$

The field $A_\mu(x)$ is indentified with the photon field while the massive gauge fields are indentified as weak bosons. What really happens is that, in the U-gauge, three degrees of freedom associated with $\phi(x)$ have been absored into $W_\mu^\pm$ and Z_μ^0 (as their longitudinal components) and the remaining one, namely $\eta(x)$, acquires a mass $-2\mu^2$.

The interaction terms in Eq. (15) read

$$\begin{aligned}\mathcal{L}_{\phi g}^{int} &= -\frac{\lambda}{4}(\eta^4 + 4v\eta^3) \\ &- \frac{1}{8}(2v\eta + \eta^2)\{2g^2 W_\mu^\pm W_\mu^- + [g^2 + (g')^2] Z_\mu^0 Z_\mu^0\}.\end{aligned} \tag{17}$$

The Feynman rules in the U-gauge yield Green's functions which are unrenormalizable. A generalized renormalizable gauge formulation of spontaneously broken gauge theories leads to the so-called "R_ξ gauge" (Fujikawa, Lee, and Sanda 1972), which is to be described in the Appendix at the end of this chapter.

The coupling of weak bosons to the photon field is of some interest. Substituting Eqs. (13a)–(13c) back into Eq. (1), we find

$$\begin{aligned}\mathcal{L}_{gauge} =& -\frac{1}{4}\{F_{\mu\nu}F_{\mu\nu} + Z_{\mu\nu}^0 Z_{\mu\nu}^0 + 2W_{\mu\nu}^+ W_{\mu\nu}^-\} \\ &- ei\{F_{\mu\nu}W_\mu^+ W_\nu^- + (W_{\mu\nu}^+ W_\mu^- - W_{\mu\nu}^- W_\mu^+)A_\nu\} \\ &- ei\, cot\theta_W \{Z_{\mu\nu}^0 W_\mu^+ W_\nu^- + (W_{\mu\nu}^+ W_\mu^- - W_{\mu\nu}^- W_\mu^+) Z_\nu^0\} \\ &- e^2\{W_\mu^+ W_\mu^- (A_\nu + cot\theta_W Z_\nu^0)^2 \\ &\quad - W_\mu^+ W_\nu^- (A_\mu + cot\theta_W Z_\mu^0)(A_\nu + cot\theta_W Z_\nu^0)\} \\ &- \frac{1}{2}(\frac{e}{sin\theta_W})^2 (W_\mu^+ W_\mu^- W_\nu^+ W_\nu^- - W_\mu^+ W_\mu^+ W_\nu^- W_\nu^-).\end{aligned} \tag{18}$$

Here we have used

$$e = g sin\theta_W; \tag{19}$$

$$\begin{aligned}F_{\mu\nu} &\equiv \partial_\mu A_\nu - \partial_\nu A_\mu, \\ Z_{\mu\nu}^0 &\equiv \partial_\mu Z_\nu^0 - \partial_\nu Z_\mu^0, \\ W_{\mu\nu}^\pm &\equiv \partial_\mu W_\nu^\pm - \partial_\nu W_\mu^\pm.\end{aligned} \tag{20}$$

Note that the weak bosons $W_\mu^\pm$ also couple to the photon field through a magnetic-moment coupling, namely $-eiF_{\mu\nu}W_\mu^+W_\nu^-$. In addition, there are "sea-gull" terms [$\propto WW\,AA$] which may also be of some interest.

To sum up, we wish to mention that we have succeeded in constructing an $SU(2)\times U(1)$ gauge theory in which three gauge bosons [$W_\mu^\pm$ and Z_μ^0] are massive while the remaining one [A_μ or the photon field] is massless. The lagrangians $\mathcal{L}_{gauge}$ [Eq. (1)] and $\mathcal{L}_{scalar}$ [Eq. (6)] are gauge invariant but the ground-state solution to the problem varies with the gauge [i.e., breaks gauge symmetry spontaneously]. The mechanism yields two important relations: $M_{Z^0} = M_{W^\pm}/cos\theta_W$ [Eq. (16b)] and $e = gsin\theta_W$ [Eq. (19)].

13.2 The $SU(2)\times U(1)$ Electroweak Theory with Two Generations of Fermions

We wish to consider an application of the $SU(2)\times U(1)$ gauge theory introduced earlier by identifying the four gague bosons with the observed weak bosons $\{W_\mu^\pm, Z_\mu^0\}$ and the photon $\{A_\mu\, or\, \gamma\}$. Specifically, we consider how quarks and leptons couple to these gauge bosons. For the sake of clarity, it is useful to study in some detail the case with two generations of fermions :

Leptons:

$$\begin{pmatrix}\nu_e\\ e^-\end{pmatrix}, \qquad \begin{pmatrix}\nu_\mu\\ \mu^-\end{pmatrix};$$

Quarks:

$$\begin{pmatrix}u\\ d\end{pmatrix}, \qquad \begin{pmatrix}c\\ s\end{pmatrix} \tag{21}$$

It is a straightforward task to generalize the scheme to incorporate the third generation of fermions.

First, we consider leptons. The $\mu - e$ university may be assumed so that the electroweak structure in the muon sector is identical with that in the electron sector. For a given Dirac field, we introduce the left-handed (L) and right-handed (R) components,

$$\psi_L \equiv \frac{1}{2}(1+\gamma_5)\psi, \qquad \psi_R \equiv \frac{1}{2}(1-\gamma_5)\psi. \tag{22}$$

In the electron sector, the physical fields relevant for weak interactions include e_L^-, e_R^-, and ν_{eL} only. A natural assignment is therefore given by

$$L = \begin{pmatrix}\nu_{eL}\\ e_L^-\end{pmatrix}, \quad SU(2) \quad doublet;$$
$$R = e_R^-, \quad SU(2) \quad singlet. \tag{23}$$

Note that, if ν_{eL} and e_L^- both were SU(2) singlets, then there would not be any coupling between e_L^- and $W^\pm$.

The fact that the photon remains massless upon spontaneous symmetry breaking implies that there remains an exact U(1) gauge symmetry with a quantum number which can be identified as the electric charge,[1]

$$Q = T_3^W + \frac{Y}{2}. \tag{24}$$

Accordingly, we have

$$Y_L = -1, \quad \text{and} \quad Y_R = -2 \tag{25}$$

The fermion lagrangian in the electron sector is therefore given by

$$\begin{aligned}\mathcal{L}_e = &- \bar{R}\gamma_\mu\{\partial_\mu + ig'B_\mu\}R \\ &- \bar{L}\gamma_\mu\{\partial_\mu - ig\frac{\tau}{2}\cdot \mathbf{A}_\mu + i\frac{g'}{2}B_\mu\}L.\end{aligned} \tag{26}$$

A little algebra yields

$$\begin{aligned}\mathcal{L}_e = &- \{\bar{e}\gamma_\mu\partial_\mu e + \bar{\nu}_L\gamma_\mu\partial_\mu\nu_L\} + \frac{g}{\sqrt{2}}\{i\bar{e}_L\gamma_\mu\nu_L W_\mu^- + h.c.\} \\ &+ \frac{e}{sin\theta_W cos\theta_W} Z_\mu^0\{i\bar{L}\frac{\tau_3}{2}\gamma_\mu L + sin^2\theta_W i\bar{e}\gamma_\mu e\} \\ &+ eA_\mu(-i)\bar{e}\gamma_\mu e.\end{aligned} \tag{27}$$

Next, we consider quarks. The standard assignment is given by

$$\begin{aligned}&\begin{pmatrix} u_L \\ d_{cL}\end{pmatrix}, \begin{pmatrix} c_L \\ s_{cL}\end{pmatrix} \; : \; SU(2)\ doublets,\ Y = \frac{1}{3}; \\ &u_R, c_R \; : \qquad SU(2)\ singlets,\ Y = \frac{4}{3}; \\ &d_R, s_R \; : \qquad SU(2)\ singlets,\ Y = -\frac{2}{3},\end{aligned} \tag{28}$$

where

$$\begin{aligned}d_c &= d\cos\theta_c + s\sin\theta_c, \\ s_c &= -d\sin\theta_c + s\cos\theta_c,\end{aligned} \tag{29}$$

with θ_c the Cabibbo angle. This yields

$$\begin{aligned}\mathcal{L}_Q = &- \{\bar{u}\gamma_\mu\partial_\mu u + \bar{d}\gamma_\mu\partial_\mu d + \bar{c}\gamma_\mu\partial_\mu c + \bar{s}\gamma_\mu\partial_\mu s\} \\ &+ \frac{g}{\sqrt{2}}\{i\bar{d}_{cL}\gamma_\mu u_L W_\mu^- + i\bar{s}_{cL}\gamma_\mu c_L W_\mu^- + h.c.\} \\ &+ \frac{e}{\sin\theta_W\cos\theta_W} Z_\mu^0\{\frac{i}{2}[\bar{u}_L\gamma_\mu u_L - \bar{d}_L\gamma_\mu d_L \\ &\quad + \bar{c}_L\gamma_\mu c_L - \bar{s}_L\gamma_\mu s_L] - \sin^2\theta_W J_\mu^{e.m.}\} \\ &+ eA_\mu J_\mu^{e.m.},\end{aligned} \tag{30}$$

[1] *Without loss of generality, it may be assumed that the conserved quantum number is* $Q = aT_3^W + bY$. *Application of this formula to the Higgs doublet, Eq.(4), yields* $a=1$ *and* $b=1/2$.

where the electromagnetic current for quarks is given by

$$J_\mu^{e.m.} = \frac{2}{3} i\bar{u}\gamma_\mu u - \frac{1}{3} i\bar{d}\gamma_\mu d + \frac{2}{3} i\bar{c}\gamma_\mu c - \frac{1}{3} i\bar{s}\gamma_\mu s. \tag{31}$$

Note that the neutral weak current, i.e., the current which couples to Z_μ^0, is flavor-conserving [i.e., $\triangle I = 0, \triangle I_3 = 0, \triangle S = 0$, and $\triangle C = 0$]. Introduction of the charm quark to make up another left-handed doublet allows one to avoid a sizable flavor-changing neutral weak current, so that consistency with experimental observations may be achieved. This is the so-called "GIM mechanism" (Glashow et al. 1970). It is worth mentioning that introduction of the charm quark in this context preceded the discovery of the ψ/J family (i.e., a family of quarkonium states consisting of a charm quark and an anticharm quark) in 1974.

The fermion-mass terms present some problem because the left-handed and right-handed components of a fermion belong to different representations of SU(2)×U(1). To preserve gauge symmetry, we may write, in the electron sector,

$$\begin{aligned}\mathcal{L}_e^m &= -G_e(\bar{R}\phi^\dagger L + \bar{L}\phi R) \\ &= -\frac{G_e v}{\sqrt{2}}\bar{e}e + ..., \quad \textit{in the } U - \textit{gauge},\end{aligned} \tag{32}$$

This may explain to some extent why $m(\nu_e) = 0$ [and $m(\nu_\mu) = 0$], provided that right-handed neutrinos do not exist. However, there are many masses which are known to differ from zero: $m_e, m_\mu, m_u, m_d, m_c$, and m_s for the first two generations. If the mass generation mechanism similar to Eq. (32) is used, there are six parameters, one for each nonzero mass. It is clear that such mass generation mechanism is *not* natural, but ideas for a better picture remain to be both speculative and qualitative.

13.3 Weak Interactions at Low Energies

The GSW SU(2)×U(1) electroweak theory is renormalizable, allowing for calculations of higher order graphs for a given physical process. Elements for discussing renormalizability of the GSW theory are similar to those given earlier in Ch. 10 for QED. Here we choose not to discuss the subject any further because such discussion is necessarily highly technically involved. Instead, we wish to consider mainly weak interactions at low energies, i.e., $E \ll M_{W^\pm}, M_{Z^0}$, where existing experimental data are all about. To this end, we shall consider three generations of fermions (cf. §.0.1.),
leptons:

$$\begin{pmatrix} \nu_e \\ e^- \end{pmatrix}, \quad \begin{pmatrix} \nu_\mu \\ \mu^- \end{pmatrix}, \quad \begin{pmatrix} \nu_\tau \\ \tau^- \end{pmatrix};$$

quarks:

$$\begin{pmatrix} u \\ d \end{pmatrix}, \quad \begin{pmatrix} c \\ s \end{pmatrix}, \quad \begin{pmatrix} t \\ b \end{pmatrix}. \tag{33}$$

We assume the $e-\mu-\tau$ universality for the sake of simplicity. In addition, the Cabibbo rotation in the case of two generations,

$$\begin{pmatrix} d_c \\ s_c \end{pmatrix} = \begin{pmatrix} cos\theta_c & sin\theta_c \\ -\sin\theta_c & \cos\theta_c \end{pmatrix} \begin{pmatrix} d \\ s \end{pmatrix}, \tag{34}$$

is to be replaced by a general Kobayashi-Maskawa rotation in the case of three generations (Kobayashi and Maskawa 1973),

$$\begin{pmatrix} d' \\ s' \\ b' \end{pmatrix} = \begin{pmatrix} c_1 & s_1c_3 & s_1s_3 \\ -s_1c_2 & c_1c_2c_3 - s_2s_3e^{i\delta} & c_1c_2s_3 + s_2c_3e^{i\delta} \\ -s_1s_2 & c_1s_2c_3 + c_2s_3e^{i\delta} & c_1s_2s_3 - c_2c_3e^{i\delta} \end{pmatrix} \begin{pmatrix} d \\ s \\ b \end{pmatrix}, \tag{35}$$

where $c_i \equiv cos\theta_i$, $s_i \equiv sin\theta_i$ $(i = 1, 2, 3)$, and δ is the CP-violating phase. There are many other parametrizations of the matrix, as briefly discussed in the 1988 publication of Particle Data Group. In the parametrization given above (and some others as well), the angle θ_1 may be identified with the Cabibbo angle θ_c.

Generalizing Eqs. (27) and (30) to the case of three generations, we obtain the weak-interaction lagrangian,

$$\begin{aligned} \mathcal{L}_W =& \frac{1}{2\sqrt{2}} \frac{e}{\sin\theta_W} \{\tilde{J}_\lambda^{(-)}(x) W_\lambda^{(-)}(x) + h.c.\} \\ &+ \frac{1}{2} \frac{e}{\sin\theta_W \cos\theta_W} Z_\lambda^0(x) \tilde{N}_\lambda(x), \end{aligned} \tag{36}$$

with $e = g\sin\theta_W = g'\cos\theta_W$ and

$$\tilde{J}_\lambda^{(-)} = \ell_\lambda^{(-)} + J_\lambda^{(-)}, \tag{37a}$$

$$\tilde{N}_\lambda = \ell_\lambda^{(0)} + N_\lambda. \tag{37b}$$

Here we have

$$\begin{aligned} \ell_\lambda^{(-)} &= i\bar{e}\gamma_\lambda(1+\gamma_5)\nu_e + i\bar{\mu}\gamma_\lambda(1+\gamma_5)\nu_\mu \\ &\quad + i\tau\gamma_\lambda(1+\gamma_5)\nu_\tau. \end{aligned} \tag{38a}$$

$$\begin{aligned} J_\lambda^{(-)} &= i\bar{d}'\gamma_\lambda(1+\gamma_5)u + i\bar{s}'\gamma_\lambda(1+\gamma_5)c \\ &\quad + i\bar{b}'\gamma_\lambda(1+\gamma_5)t. \end{aligned} \tag{38b}$$

$$\begin{aligned} \ell_\lambda^{(0)} &= \frac{i}{2}\{\bar{\nu}_e\gamma_\lambda(1+\gamma_5)\nu_e - \bar{e}\gamma_\lambda(1+\gamma_5)e\} + 2sin^2\theta_W i\bar{e}\gamma_\lambda e \\ &\quad + \{e \to \mu\} + \{e \to \tau\}. \end{aligned} \tag{38c}$$

$$\begin{aligned} N_\lambda =& \frac{i}{2}\{\bar{u}\gamma_\lambda(1+\gamma_5)u - \bar{d}\gamma_\lambda(1+\gamma_5)d \\ &+ \bar{c}\gamma_\lambda(1+\gamma_5)c - \bar{s}\gamma_\lambda(1+\gamma_5)s \\ &+ \bar{t}\gamma_\lambda(1+\gamma_5)t - \bar{b}\gamma_\lambda(1+\gamma_5)b\} \\ &- 2sin^2\theta_W J_\lambda^{e.m.}. \end{aligned} \tag{38d}$$

The hadronic electromagnetic current $J_\lambda^{e.m.}$ is given by

$$\begin{aligned} J^{e.m.} &= \frac{2}{3} i\bar{u}\gamma_\lambda u - \frac{1}{3} i\bar{d}\gamma_\lambda d + \frac{2}{3} i\bar{c}\gamma_\lambda c \\ &- \frac{1}{3} i\bar{s}\gamma_\lambda s + \frac{2}{3} i\bar{t}\gamma_\lambda t - \frac{1}{3} i\bar{b}\gamma_\lambda b. \end{aligned} \tag{39}$$

In cases where $W^\pm$ and Z^0 are not directly observed, we write the second order S-matrix as follows:

$$\begin{aligned} S^{(2)} &= -\frac{e^2}{8sin^2\theta_W} \int d^4x d^4y T(\tilde{J}_\lambda^{(-)}(x)\tilde{J}_\eta^{(+)}(y)) \\ &\cdot \frac{1}{(2\pi)^4} \int d^4k e^{-ik\cdot(x-y)} \frac{1}{i} \frac{\delta_{\lambda\eta}}{M_W^2 + k^2 - i\varepsilon} \\ &- \frac{e^2}{8sin^2\theta_W cos^2\theta_W} \int d^4x d^4y T(\tilde{N}_\lambda(x)\tilde{N}_\eta(y)) \\ &\cdot \frac{1}{(2\pi)^4} \int d^4k e^{-ik\cdot(x-y)} \frac{1}{i} \frac{\delta_{\lambda\eta}}{M_Z^2 + k^2 - i\varepsilon}. \end{aligned} \tag{40}$$

At energies well below M_{Z^0} and $M_{W^\pm}$, we may neglect k^2 as compared to M_{Z^0} or $M_{W^\pm}$ since the four-momentum k can be expressed as a simple linear combination of external momenta (by virtue of energy-momentum conservation at each vertex). Thus, the integration over d^4k yields $\delta^4(x-y)$. Define an effective lagrangian,

$$S^{(2)} \approx i \int d^4x \mathcal{L}_W^{eff}(x), \quad for \quad k \ll M_{W^\pm}, M_{Z^0}. \tag{41}$$

we find

$$\mathcal{L}_W^{eff}(x) = \frac{G_F}{\sqrt{2}} (\tilde{J}_\lambda^{(-)}(x)\tilde{J}_\lambda^{(+)}(x) + \tilde{N}_\lambda(x)\tilde{N}_\lambda(x)), \tag{42}$$

with the Fermi coupling constant G_F given by

$$\frac{G_F}{\sqrt{2}} = \frac{e^2}{8M_W^2 sin^2\theta_W} = \frac{e^2}{8M_Z^2 sin^2\theta_W cos^2\theta_W}. \tag{43}$$

A precise value for G_F can be extracted from studies of muon decay $\mu^- \to e^-\bar{\nu}_e\nu_\mu$ (Particle Data Group 1988; see Ch. 14):

$$G_F = (1.16637 \pm 0.00004) \times 10^{-5} GeV^{-2} c^4. \tag{44}$$

We may use Eqs. (37) and (38) to rewrite Eq. (42) as follows:

$$\begin{aligned} \mathcal{L}_W^{eff}(x) =& \frac{G_F}{\sqrt{2}} \{ [\ell_\lambda^{(-)}\ell_\lambda^{(+)} + \ell_\lambda^{(0)}\ell_\lambda^{(0)}] \\ &+ [\ell_\lambda^{(-)}J_\lambda^{(+)} + \ell_\lambda^{(+)}J_\lambda^{(-)} + 2\ell_\lambda^{(0)}N_\lambda] \\ &+ [J_\lambda^{(-)}J_\lambda^{(+)} + N_\lambda N_\lambda] \}, \end{aligned} \tag{45}$$

suggesting a classification of weak interactions into three distinct categories: (1) purely leptonic, (2) semileptonic, and (3) purely hadronic (or nonleptonic). Examples of these reactions are listed below. Formulae on decay rates, cross sections, and asymmetries are the subject of discussions in particle physics textbooks (e.g., Commins and Bucksbaum 1983); among them, most commonly used ones are listed in the 1988 publication of Particle Data Group and will be later reproduced in the next chapter. Note that we are already fully equipped to derive these formulae except that further discussion on such derivation will certainly divert our attention too much away from the focus of this book.

(a) Purely Leptonic Weak Interactions:

Examples of purely leptonic interactions which have been subject to experimental studies include:

$$muon\ \ decay: \ \mu^- \to e^- \bar{\nu}_e \nu_\mu, \tag{46a}$$

$$(\nu_\mu e)\ \ scattering: \ \nu_\mu + e^- \to \nu_\mu + e^-, \tag{46b}$$

$$(\nu_e e)\ \ scattering: \ \nu_e + e^- \to \nu_e + e^-, \tag{46c}$$

$$Weak\ \ interaction\ \ studies\ \ in\ \ e^+e^- \to \mu^+\mu^-. \tag{46d}$$

Note that only $W^\pm$ exchange is responsible for muon decay (46a) while only Z^0 exchange is allowed in the case of $(\nu_\mu e)$ scattering. However, both $W^\pm$ and Z^0 exchanges must be considered in the case of $(\nu_e e)$ scattering and both Z^0 and γ exchanges are involved in weak interaction studies associated with $e^+e^- \to \mu^+\mu^-$. Accordingly, it is not possible to further distinguish purely leptonic weak interactions into charged and neutral weak interactions.

(b) Semileptonic Weak Interactions:

Semileptonic weak interactions can be classified further into two distinct subcategories:

(b.1) semileptonic charged weak interactions

Examples of charged weak interactions include neutron β-decay $n \to p + e^- + \bar{\nu}_e$, β-decays of mesons such as $\pi^- \to \pi^0 + e^- + \bar{\nu}_e$ and $K^- \to \pi^0 + e + \bar{\nu}_e$, pion decay $\pi^+ \to \mu^+\nu_\mu$, nuclear β-decays such as $^{12}B \to ^{12}C + e^- + \nu_e$ and $^{12}N \to ^{12}C + e^+ + \nu_e$, muon capture in hydrogen $\mu^- + p \to \nu_\mu + n$, nuclear muon capture, and so on. Comparison of the $^{14}O\ \beta - decay$ rate with that for muon decay yields a value on the Cabibbo angle,

$$\theta_c = 0.210 \pm 0.025. \tag{47}$$

(b.2) semileptonic neutral weak interactions

Examples of semileptonic neutral weak interactions include

$$\begin{aligned}
&\nu(\bar{\nu}) + N \to \nu(\bar{\nu}) + X,\\
&\nu(\bar{\nu}) + p \to \nu(\bar{\nu}) + p,\\
&\nu(\bar{\nu}) + p \to \nu(\bar{\nu}) + N + \pi,\\
&\nu(\bar{\nu}) +^{12} C(g.s.) \to \nu(\bar{\nu}) +^{12} C^*(15.110),\\
&\vec{e} + p \to e + p,\\
&etc.
\end{aligned} \tag{48}$$

A "world average" value for the electroweak mixing parameter $sin^2\theta_W$ as obtained from neutral weak interaction experiments is given by

$$sin^2\theta_W = 0.230 \pm 0.005. \tag{49}$$

It is useful to mention that classification of semileptonic weak inteactions into exclusive ones such as $\nu_\mu + p \to \nu_\mu + p$ and inclusive ones such as $\nu_\mu + p \to \nu_\mu + X$ (with X denoting unobserved hadrons) derives mostly from the difference in theoretical treatments (rather than in physics contents).

(c) Nonleptonic Weak Interactions:

Among all weak interactions, nonleptonic or purely hadronic weak interactions are least understood since the last term in Eq. (45) must be augmented by QCD corrections before a qualitative description can even be perceived. Examples of such interactions include hadronic decays of hyperons such as $\Lambda^0 \to p + \pi^-$ and $\sum^- \to n + \pi^-$, $K_{2\pi}$ decays such as $K^\pm \to \pi^\pm + \pi^0$ and $K_s^0 \to \pi^+ + \pi^-$, $K_{3\pi}$ decays such as $K_L^0 \to \pi\pi\pi$, hadronic decays of heavy-quark systems, and flavor-conserving nonleptonic weak interactions. Note that only $W^\pm$ exchanges contribute to flavor-changing nonleptonic weak interactions while both $W^\pm$ and Z^0 exchanges enter the problem of flavor-conserving nonleptonic weak interactions. In any event, complications arising from importance of QCD corrections prevent us from drawing a definitive support toward the GSW electroweak theory on the basis of nonleptonic weak interaction studies.

In summary, experimental data on weak interactions at low energies are described very well by the GSW $SU(2) \times U(1)$ electroweak theory. An additional boost came from recent direct observations of the $W^\pm$ and Z^0 weak bosons (at the predicted masses). Results of analyses of experimental data, which support quantitatively the validity of the GSW electroweak theory, are summarized in the 1988 publication of Particle Data Group and some portion of it will be reproduced and discussed in the subsequent chapter. With the coming generation of e^+e^- colliders with the center-of-mass energies $\sqrt{s}$ greater than M_{Z^0} such as SLC at Stanford and LEP at CERN, it is expected that the theory will be subject to much more severe scrutiny. Owing to the fact that weak interaction studies over the last century constantly produced one surprise after another, one might suspect that the GSW $SU(2) \times U(1)$ electroweak theory may mark the beginning of an unfolding mystery, rather than the completion of a zigsaw puzzle.

Appendix

Feynman Rules in the R_ξ Gauge

The Glashow-Salam-Weinberg (GSW) $SU(2) \times U(1)$ electroweak theory is a spontaneously broken gauge theory. For $\mu < 0$, three of the four real fields associated with the complex weak-isodoublet scalar field are absorbed as the longitudinal polarization degrees of freedom, one for each of three gauge fields, while the fourth one becomes a massive real scalar Higgs field. The idea of the "Higgs mechanism", as a special case of spontaneous symmetry breaking, is usually explained in the context of a classical field theory, where the ground state of the classical field is found by minimizing the potential. To compute quantum corrections or to treat the problem of renormalization in general, we need to work with the quantized version of the theory.

To quantize a gauge field theory with a general multicomponent field ϕ_i and corresponding set of sources J_i, we may work with the partition functional $Z[J] = exp\{iW[J]\}$ and define a set of quantities

$$\Phi_i(x) = \frac{\delta W[J]}{\delta J_i(x)}. \tag{A1}$$

It may be shown that $\Phi_i(x)$ is the vacuum expectation value of ϕ_i in the presence of $J_i(x)$. Thus, for $J_i(x) = 0$, $\Phi_i(x) \mid_{J_i=0} = v_i$, which may differ from zero, is the vacuum expectation value of ϕ_i in the absence of sources. Following the standard exercise in classical mechanics, we may define a Legendre transformation $\Gamma[\Phi]$ by

$$\Gamma[\Phi] = W[\mathbf{J}] - \int d^4x J(x) \cdot \Phi(x), \tag{A2}$$

and a "superpotential" $\mathcal{V}$ by

$$\Gamma[\Phi = 0] = -(2\pi)^4 \delta^4(0) \mathcal{V}(\phi). \tag{A3}$$

It is then possible to demonstrate that $\mathcal{V}$ has properties strictly analogous to the classical potential $V(\phi_i)$. In other words, the simple classical analysis of $V(\phi_i)$ that describes the spontaneous symmetry breaking of Higgs mechanism (§.13.1) remains valid in the quantized version of the theory (Abers and Lee 1973).

Although the classical theory was presented in unitary gauge, or "U gauge", the quantum theory may be formulated in a variety of gauges. In addition to the U gauge, there are the R gauge (or the Landau gauge) and the 't Hooft-Feynman gauge, each of which is a special case of the so-called generalized renormalizable gauge, or the "R_ξ" gauge. The R_ξ gauge is characterized by continuous real parameters α, ξ, and η, which arise from choice of gauge-fixing terms to eliminate the mixing between gauge bosons and the unphysical scalar bosons (i.e., unobserved Higgs fields in the general gauge). The Faddeev-Popov ghost fields are then calculated accordingly (using the method introduced in Appendix A of Chapter 12). Although the Green's functions of the theory in general depend on the gauge-fixing parameters, the physical results, or the S-matrix elements, are gauge invariant (Fujikawa et al. 1972). In what follows, we wish to describe, without proof, Feynman rules in the R_ξ gauge.

In the R_ξ gauge, the various "particles" include the usual fermions, gauge bosons ($W^\pm$, Z^0, and γ), the Higgs scalar σ, three unphysical scalar bosons $s^\pm$ and χ, and an isotriplet of Faddeev-Popov ghosts. In the limit $\xi, \eta \to 0$, the unphysical scalar bosons disappear as a result of being absorbed as the longitudinal components of the gauge bosons. This is just the U gauge. On the other hand, for $\eta, \xi \to \infty$ (which gives rise to the R gauge), the unphysical scalar bosons become Goldstone bosons.

The propagators for vector mesons, Higgs, unphysical scalars, and ghost fields are given in the R_ξ gauge as follows:

(i) Vector mesons:

$$D_{\mu\nu}^{W^\pm}(k) = \frac{1}{i}[\delta_{\mu\nu} - \frac{k_\mu k_\nu}{k^2 + m_W^2/\xi}(1 - \frac{1}{\xi})]\frac{1}{k^2 + m_W^2 - i\epsilon}. \tag{A4}$$

$$D_{\mu\nu}^{Z}(k) = \frac{1}{i}[\delta_{\mu\nu} - \frac{k_\mu k_\nu}{k^2 + m_Z^2/\eta}(1 - \frac{1}{\eta})\frac{1}{k^2 + m_Z^2 - i\epsilon} \tag{A5}$$

$$D_{\mu\nu}^{\gamma}(k) = \frac{1}{i}[\delta_{\mu\nu} - \frac{k_\mu k_\nu}{k^2}(1 - \alpha)]\frac{1}{k^2 - i\epsilon}. \tag{A6}$$

(ii) Higgs boson (mass m_σ):

$$\sigma: \qquad D^\sigma(k) = \frac{1}{i}\frac{1}{k^2 + m_\sigma^2 - i\epsilon}. \tag{A7}$$

(iii) Unphysical scalar bosons:

$$s^\pm: \qquad D^{s^\pm}(k) = \frac{1}{i}\frac{1}{k^2 + m_W^2/\xi - i\epsilon}. \tag{A8}$$

$$\chi: \qquad D^{\chi}(k) = \frac{1}{i}\frac{1}{k^2 + m_Z^2/\eta - i\epsilon}. \tag{A9}$$

(iv) Fermion scalar ghosts:

$$D_\pm^g(k) = \frac{1}{i}\frac{1}{k^2 + m_W^2/\xi - i\epsilon}. \tag{A10}$$

$$D_0^g(k) = \frac{1}{i}\frac{1}{k^2 + m_Z^2/\eta - i\epsilon}. \tag{A11}$$

The gauge-fixing parameter α appearing in the photon propagator (A6) has no effect on S-matrix elements since $D^{\gamma}_{\mu\nu}(k)$ is always sandwiched between conserved currents. Or, the term $(k_\mu k_\nu/k^2)(1-\alpha)$ always yields zero in S-matrix elements. For $\xi, \eta \neq 0$, all other propagators vary as k^{-2} for large k^2, suggesting renormalizability by "naïve power counting".

Note that we may rewrite Eq. (A4) as follows:

$$D^{W^\pm}_{\mu\nu}(k) = \frac{1}{i}\frac{\delta_{\mu\nu} + (1/m_W^2)k_\mu k_\nu}{k^2 + m_W^2 - i\epsilon} - \frac{1}{i}\frac{k_\mu k_\nu}{m_W^2}\frac{1}{k^2 + m_W^2/\xi - i\epsilon}, \tag{A12}$$

where the second term on the right-hand side has a pole that is to be canceled in the S-matrix element by the pole in the $s^\pm$ propagator (A8). Eq. (A12) indicates that the R gauge is recovered in the limit $\xi, \eta \to \infty$.

$$\lim_{\xi\to\infty} D^{W^\pm}_{\mu\nu}(k) = \frac{1}{i}(\delta_{\mu\nu} - \frac{k_\mu k_\nu}{k^2})\frac{1}{k^2 + m_W^2 - i\epsilon}, \tag{A13}$$

which is the propagator in the R gauge. Similar results hold, of course, also for the Z^0 boson. For $\xi = 1$ and $\eta = 1$, we obtain the 't Hooft-Feynman gauge

$$D^{W^\pm}_{\mu\nu}(k) = \frac{1}{i}\frac{\delta_{\mu\nu}}{k^2 + m_W^2 - i\epsilon}. \tag{A14}$$

$$D^{s^\pm}(k) = \frac{1}{i}\frac{1}{k^2 + m_W^2 - i\epsilon}. \tag{A15}$$

Finally, we consider the limit as $\xi, \eta \to 0$, which must be taken with care (i.e., after S-matrix elements are computed) to avoid ambiguities. Formally, we have, in the U gauge,

$$\lim_{\xi\to 0} D^{W^\pm}_{\mu\nu}(k) = \frac{1}{i}\frac{\delta_{\mu\nu} + k_\mu k_\nu/m_W^2}{k^2 + m_W^2 - i\epsilon}, \tag{A16}$$

which is the propagator familiar for a massive vector meson. This gauge has the advantage that ghost and unphysical boson propagators vanish identically. However, for large k^2, $D^{W^\pm}_{\mu\nu}(k^2) = O(1)$ and renormalizability is not transparent by naïve power counting. Since the S-matrix is independent of ξ and η, it has become possible to prove renormalization for $\xi, \eta \neq 0$ and then to take the limit $\xi, \eta \to 0$.

We proceed to consider the vertex factors associated with the GSW electroweak theory. Using the (e^-, ν_e) to illustrate the couplings of fermions to gauge bosons, we use Eq. (27) in the text to obtain the following Feynman rules.

Electron-Photon Coupling:

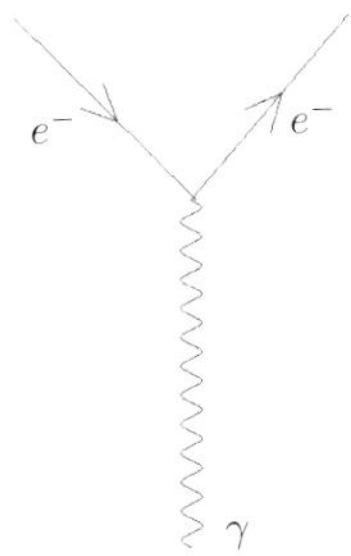

$$\Longleftrightarrow \qquad e\gamma_\mu \tag{A17a}$$

Electron-Z^0 Coupling:

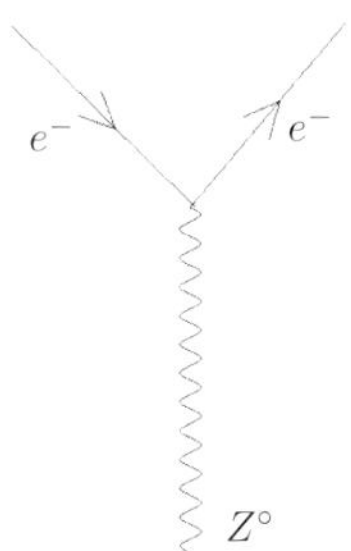

$$\Leftrightarrow \quad i\frac{ie}{sin\theta_W cos\theta_W}\{-\frac{1}{4}\gamma_\mu(1+\gamma_5)+sin^2\theta_W\gamma_\mu\} \tag{A17b}$$

Neutrino-Z^0 Coupling:

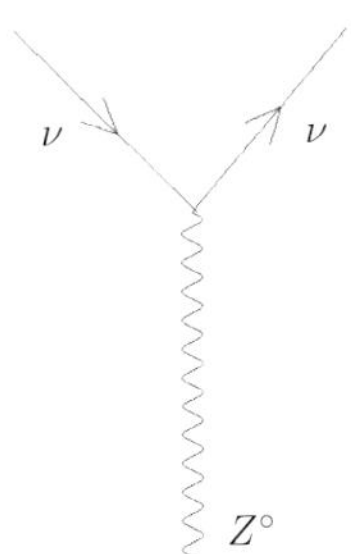

$$\Leftrightarrow \quad i\frac{ie}{sin\theta_W cos\theta_W}\cdot\frac{1}{4}\gamma_\mu(1+\gamma_5) \tag{A17c}$$

Fermion-$W^\pm$ Coupling:

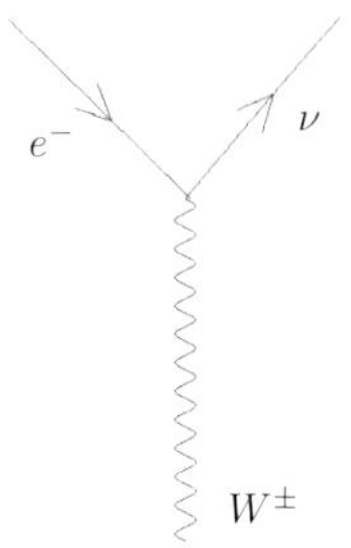

$$\Longleftrightarrow \quad i\frac{ig}{2\sqrt{2}}\gamma_\mu(1+\gamma_5) \tag{A17d}$$

Couplings of quarks to gauge bosons may be obtained using Eq. (30) [instead of Eq. (27)] in the text.

In view of Eq. (18) in the text, there are trilinear and quartilinear self couplings among gauge bosons. These couplings are illustrated in Fig. 1 with momenta and internal indices shown explicitly. We introduce

$$T_{\alpha\beta\gamma} = (p-q)_\beta\delta_{\alpha\mu} + (q-r)_\alpha\delta_{\mu\beta} + (r-p)_\mu\delta_{\beta\alpha}. \tag{A18a}$$

$$S_{\mu\nu,\lambda\rho} = 2\delta_{\mu\nu}\delta_{\lambda\rho} - \delta_{\mu\lambda}\delta_{\nu\rho} - \delta_{\mu\rho}\delta_{\nu\lambda}. \tag{A18b}$$

Using Eq. (18) given in the text, we assign to the trilinear $W^+W^-\gamma$ coupling the factor $e\,T_{\alpha\beta\gamma}$ and the trilinear $W^+W^-Z^0$ coupling the factor $e\,cos\theta_W T_{\alpha\beta\gamma}$. On the other hand, Feynman rules for the $WWWW$, $WW\gamma\gamma$, and $WWZZ$ couplings shown in Fig. 1 are given, respectively, by $-g^2S_{\mu\nu,\lambda\rho}$, $e^2S_{\mu\nu,\lambda\rho}$, $e^2cot^2\theta_W S_{\mu\nu,\lambda\rho}$.

These rules should be contrasted with the result obtained when the minimal substitution $\partial_\mu W_\nu^\pm \to (\partial_\mu \mp ieA_\mu)W_\nu^\pm$ is made in the Lagrangian for the old-fashioned intermediate boson theory. It is found that in the new theory an additional term $-ieW_\mu^+W_\nu^-F^{\mu\nu}$ arises, implying an additional contribution of $e\hbar/2m_Wc$ to the magnetic moment of W^-, over and above the "normal" magnetic moment of $\mu_{W^0} = e\hbar/2m_Wc$. The total W^- magnetic moment is thus expected to be

$$\mu_W = 2\mu_{W^0}. \tag{A19}$$

It is important to know that there are many more trilinear and quartilinear couplings as we include the Higgs boson σ and the three unphysical bosons. Feynman rules for these couplings can be inferred from the original Higgs lagrangian, Eq. (6) in the text. The generic forms for the trilinear couplings are illustrated in Fig. 2.
The couplings of the first type (LHS of Fig. 2) include $s^+s^-\gamma$, $s^+s^-Z^0$, $\sigma\chi Z^0$, σs^+W^-, χs^+W^-, σs^-W^+, and χs^-W^+ while those of the second type include γW^-s^+, $Z^0W^-s^+$, $W^+W^-\sigma$, and $Z^0Z^0\sigma$.

The generic form for the quartilinear couplings are illustrated by Fig. 3, where all possible combinations consistent with the interaction lagrangian must be taken into account.

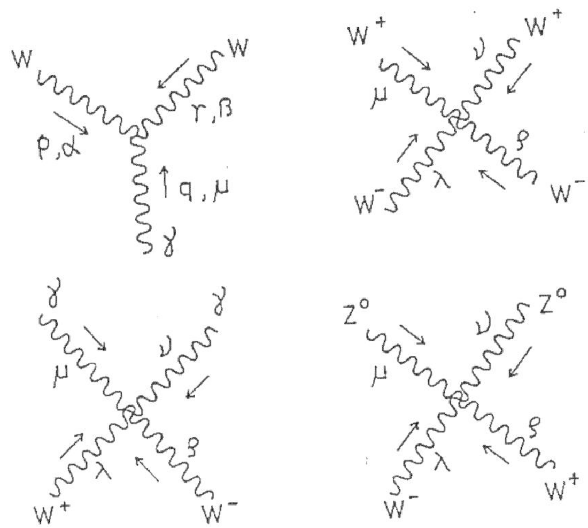

Figure 1: Trilinear and quartilinear couplings in the GSW $SU(2) \times U(1)$ electroweak theory. p, q, and r are momentum variables while α, μ, etc. are Lorentz indices.

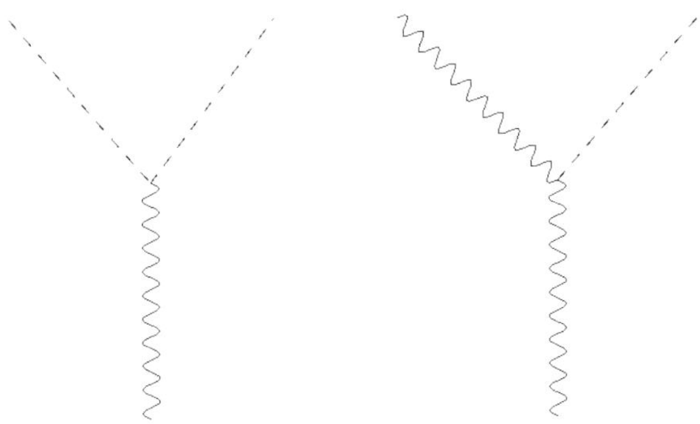

Figure 2: Trilinear couplings which involves the Higgs boson σ and the three unphysical scalar bosons.

In addition, there are of course couplings of fermions to the Higgs boson σ and any of the three unphysical scalar bosons $s^{\pm}$ and χ. The diagrams in the electron sector are illustrated in Fig. 4. Feynman rules may be inferred from Eq. (32) in the text.

Finally, there are three Faddeev-Popov ghost fields which are yet to be incorperated to complete the set of Feynman rules for the GSW electroweak theory. In view of the

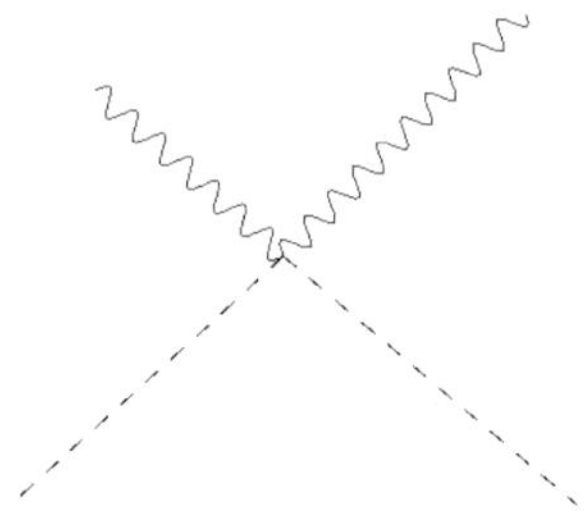

Figure 3: Quartilinear couplings which involves the Higgs boson σ and the three unphysical scalar bosons.

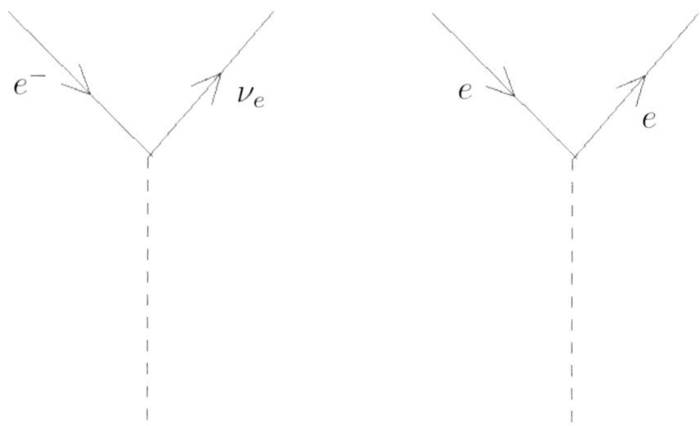

Figure 4: Couplings of fermions to the Higgs boson σ or the unphysical scalar boson.

large number of fields involved in the game, it is often convenient to focus on a specific higher-order loop calculation in order to decide how the various symmetries, such as gauge invariance, are maintained after all relevant diagrams are taken into account.[2]

[2] *As pointed out by Cheng and Tsai (1987), one must exercise care in obtaining Feynman rules in certain gauges such as the Coulomb gauge and the axial gauge. This is related to the fact that the method of path integrals suffers from ambiguities in these gauges.*

References

Weinberg, S., Phys. Rev. Lett. **19**, 1264 (1967); Salam, A., in *Elementary Particle Physics* (Nobel Symposium No. 8), ed. N. Svartholm (Almquist and Wiksell, Stockholm, 1968), p. 367; and Glashow, S. L., Nucl. Phys. **22**, 519 (1961); original references on the $SU(2)\times U(1)$ electroweak theory.

't Hooft, G., Nucl. Phys. **B33**, 173 (1971) and **B35**, 167 (1971) on the original proof of the renormalizability of the GSW electroweak theory.

Higgs, P.W., Phys. Lett. **12**, 132 (1964); Phys. Rev. Lett. **13**, 508 (1964); and Phys. Rev. **145**, 1156 (1966); Anderson, P.W., Phys. Rev. **130**, 439 (1963); Englert F. and Brout, R., Phys. Rev. Lett. **13**, 321 (1964); Englert, F., Brout, R., and Thiry, M.F., Nuovo Cimento **48**, 244 (1966); Guralnik, G.S., Hagen, C.R., and Kibble, T.W.B., Phys. Rev. Lett. **13**, 585 (1964); Kibble, T.W.B., Phys. Rev. **155**, 1554 (1967); on the Higgs mechanism.

Glashow, S.L., Iliopoulos, J., and Maiani, L., Phys. Rev. **D2**, 1285 (1970) on the GIM mechanism to suppress flavor-changing neutral weak currents.

Kobayashi, M., and Maskawa, T., Prog. Theor. Phys. Japan **49**, 652 (1973); Particle Data Group, *Review of Particle Properties*, Phys. Lett. **204B**, 1 (1988); on the Kobayashi-Maskawa matrix for three generations of fermions and other possible parametrizations.

Commins, E.D., and Bucksbaum, P.H., in *Weak Interactions of Leptons and Quarks* (Cambridge University Press, England, 1983) on an extensive review of weak interactions up to 1982.

Arnison, G., et al., Phys. Lett. **122B**, 103 (1983) and **126B**, 398 (1983); Banner, M., et al., Phys. Lett. **122B**, 476 (1983); Bagnaia, P., et al. Phys. Lett. **129B**, 130 (1983); on the experimental discovery of $W^{\pm}$ and Z^0 weak bosons.

Fujikawa, K., Lee, B.W., and Sanda, A.I., Phys. Rev. **D6**, 2923 (1972) on the R_ξ gauge.

H. Cheng and E.-C. Tsai, Chinese J. Phys. **25**, 95 (1987); on canonical quantization of non-abelian gauge field theories and Feynman rules.

Exercises: *Chapter 13*

1. Show that Eq. (18) follows from Eqs. (1) and (13a)–(13c). Note that $e = g\sin\theta_W$ arises naturally.

2. **(i)** Prove Eq. (30) from Eqs. (28) and (29).

(ii) Suppose that the charm quark were absent. s_{cL} must then be considered as an SU(2) singlet. Derive $\mathcal{L}_Q$ in this case. Is there any way to suppress flavor-changing neutral weak currents (other than the GIM mechanism)?

3. Consider $e^+e^- \to \mu^+\mu^-$ for $m_e, m_\mu \ll \sqrt{s} \ll M_{Z^0}$. (There are two Feynman diagrams: $e^+e^- \to \gamma \to \mu^+\mu^-$ and $e^+e^- \to Z^0 \to \mu^+\mu^-$.)

(i) Use Feynman rules to write down the T-matrix element.

(ii) Obtain $d\sigma/d\Omega$ with Ω the outgoing μ^+ (or μ^-)

(iii) Derive the charge asymmetry as defined by

$$\mathcal{A} \equiv \frac{\sigma(cos\theta > 0) - \sigma(cos\theta < 0)}{\sigma(cos\theta > 0) + (\sigma(cos\theta < 0)}.$$

4. Consider $e^+e^- \to W^+W^-$ at sufficiently high collider energies such as the proposed LEP II.

(i) Draw all Feynman diagrams relevant for the process.

(ii) Use Feynman rules to write down the T-matrix element.

(iii) Obtain $d\sigma/d\Omega$ with Ω the outgoing W^+ (or W^-) solid angle. Discuss your result.

Chapter 14. Experimental Tests: The Standard Model of the 20th Century

In this chapter, we wish to summarize briefly the status of the standard model as of 2000 A.D. To make our presentations up to date, we rely heavily on the latest biennual (once every two years) publication of Particle Data Group on "Review of Particle Properties". Since many relevant materials in this useful reference might be covered here because of the limitation in space and the level of our discussions, interested readers and students may consult this publication which is updated on a regular basis.

14.1 Quantum Chromodynamics

(1) QCD and the Parton Model

In §.12.2., we learned that quantum chromodynamics, or QCD, is asymptotically free. At sufficiently large Q^2 where the strong coupling $\alpha_s(Q^2)$ $(\equiv g^2/(4\pi))$ is sufficiently small compared to unity, a hadron must then look like a collection of non-interacting quarks, antiquarks, and gluons (or, in a collective term, "partons"), *at least to leading order in* $\alpha_s(Q^2)$. The parton picture of a hadron must be contrasted with its quark-model description at low Q^2, in which for example a baryon is considered as a system of three dressed, valence quarks confined to within a region specified by the baryon size and a meson is a quark-antiquark pair restricted to the meson volume. The two very different pictures of a single hadron, i.e. the parton-model description at large Q^2 and the valence quark-model language at small Q^2, may be made consistent with each other presumably by the asymptotically free nature of QCD. Nevertheless, reconciliation between the two very different pictures of a single hadron depends critically on how we specify the models and the subject is clearly beyond the scope of our present discussions.

In the parton-model language which is relevant at large Q^2, it is customary to work with the infinite-momentum frame in which the momentum of the hadron under study along a chosen $z-$direction (longitudinal direction) is infinitely large. Each parton will then carry a fraction x of the longitudinal momentum of the hadron. The information on parton's transverse momentum may be neglected at least in the first approximation. In addition, averaging over spins may also be carried out for the experiments which we shall mention in this chapter. Thus, a hadron, such as a proton, is characterized, at a given Q^2, by a set of distribution functions (or structure functions): $\{q_i(x), \bar{q}_i(x), G(x)\}$ with q_i, $\bar{q}_i$, and G the quark of flavor i, the antiquark of flavor i, and the gluon, respectively.

Suppose that we do an experiment on deep inelastic electron-proton scattering. In the infinite-momentum frame, the virtual photon strikes a quark or an antiquark of flavor i inside the target proton. Assuming that the electron-proton deep inelastic scattering (DIS) cross section is an incoherent sum of elementary electron-parton scattering cross sections, we find that the ep DIS cross section is proportional to a well-known structure function called $F_2(x, Q^2)$:

$$F_2(x, Q^2) = \sum_i e_i^2 \{q_i(x) + \bar{q}_i(x)\}, \tag{1}$$

where e_i is the electric charge of a quark of flavor i. Bjorken scaling says that $F_2(x, Q^2)$ depends only on x in the limit of large Q^2 and ν ($\equiv E_e - E'_e$) with x ($= Q^2/(2m_p\nu)$) held fixed.

As Q^2 is very large, we may choose to neglect quark-mass terms in QCD (at least for the light quarks). A massless quark can easily split into another quark of less energy and a gluon. Similarly, a massless gluon may also split into a quark-antiquark pair. Owing to energy-momentum conservation, these splitting processes occur collinearly (i.e., all particles involved line up along the same direction). In dimensional regularization, these collinear processes give rise to $\frac{1}{\epsilon}$ divergences. Thus, in massless QCD, we must absorb collinear divergences into a set of redefined structure functions, which depend explicitly on Q^2 causing violation of Bjorken scaling.

To describe the way in which scaling is broken in QCD, it is convenient to define the nonsinglet and singlet quark distributions:

$$\begin{aligned} F^{NS}(x, Q^2) &= q_i(x, Q^2) - q_j(x, Q^2), \\ F^S(x, Q^2) &= \sum_i \{q_i(x, Q^2) + \bar{q}_i(x, Q^2)\}. \end{aligned} \tag{2}$$

The nonsinglet structure functions have nonzero values of flavor quantum numbers such as isospin, strangeness, charm, or baryon number. The variation with Q^2 of these structure functions is described by the so-called Altarelli-Parisi evolution equations (Altarelli and Parisi, 1977).

$$\begin{aligned} Q^2 \frac{\partial F^{NS}}{\partial Q^2} &= \frac{\alpha_s(Q^2)}{2\pi} P^{qq} \circ F^{NS}, \\ Q^2 \frac{\partial}{\partial Q^2} \begin{pmatrix} F^S \\ G \end{pmatrix} &= \frac{\alpha_s(Q^2)}{2\pi} \begin{pmatrix} P^{qq} & 2N_f P^{qg} \\ P^{gq} & P^{gg} \end{pmatrix} \circ \begin{pmatrix} F^S \\ G \end{pmatrix}, \end{aligned} \tag{3}$$

where $G(x, Q^2)$ is the gluon distribution function and "$\circ$" denotes a convolution integral:

$$f \circ g = \int_x^1 \frac{dy}{y} f(y) g(\frac{x}{y}). \tag{4}$$

The splitting functions to leading order in α_s may be obtained by considering leading-order QCD corrections to DIS. Collinear divergences, which appear as $\frac{1}{\epsilon}$ singularities in the dimensional regularization scheme, must be subtracted and absorbed into the redefined structure functions, yielding explicit expressions for the splitting functions. This was done by Altarelli and Parisi (1977), who obtained

$$\begin{aligned} P^{qq} &= \frac{4}{3}(\frac{1+x^2}{1-x})_+ + 2\delta(1-x), \\ P^{qg} &= \frac{1}{2}(x^2 + (1-x)^2), \\ P^{gq} &= \frac{4}{3}(\frac{1+(1-x)^2}{x}), \\ P^{gg} &= 6(\frac{1-x}{x} + x(1-x) + (\frac{x}{1-x})_+ + \frac{11}{12}\delta(1-x)) - \frac{N_f}{3}\delta(1-x). \end{aligned} \tag{5}$$

Here $\frac{1}{(1-x)_+}$ is defined by

$$\int_0^1 dx \frac{f(x)}{(1-x)_+} = \int_0^1 dx \frac{f(x)-f(1)}{(1-x)}.$$

The evolution equations can easily be integrated to yield the expressions which determines the distribution functions at Q^2 in terms of those at a given Q_0^2, the latter being obtained by a global fit to the existing experimental data. Thus, one may assume that the structure function $F_2(x, Q^2)$ is determined by distribution functions which depend explicitly on Q^2:

$$F_2(x, Q^2) = \sum_i e_i^2 \{q_i(x, Q^2) + \bar{q}_i(x, Q^2)\}. \tag{6}$$

Note that Eq. (6) determines a specific subtraction scheme, i.e., the scheme in which how collinear divergences are absorbed into a set of redefined structure functions. The other schemes are of course possible; among them, the "modified minimum subtraction" scheme is often adopted.

The above results are for massless quarks. Algorithms may be obtained for inclusion of nonzero quark masses (Glück et al., 1982). In addition, "higher-twist" contributions of the form

$$F_i(x, Q^2) = F_i^{(LT)}(x, Q^2) + \frac{F_i^{(HT)}(x, Q^2)}{Q^2} + ..., \tag{7}$$

are also of numerical importance at low Q^2 (e.g., $< 10GeV^2$) or at x very close to unity.

We shall not discuss technical details concerning these corrections although the quality of the existing data already calls for suitable treatment of these higher-order corrections.

(2) Tests of QCD in Deep Inelastic Scattering

The original and still one of the most powerful quantitative tests of perturbative QCD comes from violation of Bjorken scaling observed in deep inelastic lepton-hadron scattering. The quality of contemporary experimental data already calls for proper inclusion of higher order corrections (Curci et al., 1980; Furmanski and Petronzio, 1980, 1982; Floratos, et al., 1981; Herrod and Wada, 1981). The current status of the experimental data has been reviewed by, for example, Voss (1987).

From the evolution equations (3), it is clear that a nonsinglet structure function offers in principle a better test of the theory since its Q^2 evolution is insensitive to the poorly known gluon distribution. However, such a measurement involves differences between cross sections, making a precision test less feasible. In practice, the most accurate measurements, involving singlet-dominated structure functions such as F_2, have resulted in strongly correlated measurements of the QCD renormalization scale $\Lambda_{\overline{MS}}$ in the modified minimum subtraction scheme[1] and the gluon distribution. The most accurate data currently available are from the BCDMS collaboration. The result obtained is,[2]

$$\Lambda = 230 \pm 20(stat.) \pm 60(sys.)MeV, \tag{8}$$

[1] *It is well-known that one must choose a renormalization scale, say Λ, in order to make sense out of massless QCD which is invariant under scale transformations.*

[2] *BCDMS collaboration: Benvenuti, A.C., et al., Phys. Lett.* **B195,** *97 (1987).*

which is consistent with the world average value of $\Lambda = (238 \pm 43)MeV$, (statistical and systematic uncertainty added in quadrature) obtained from the existing deep-inelastic experiments. Here and in the rest of the present chapter Λ is to be understood as the QCD scale parameter in the modified minimum subtraction scheme with four effective massless quarks.[3]

The impact on the measurement of α_s due to the higher order corrections may be estimated as follows. One may use the leading-order evolution equations in the data analysis in order to extract α_s, which is then compared with the result obtained by using the next-to-leading-order equations. In this way, BCDMS obtained $\alpha_s(5GeV) = 0.240$ in the leading-order approximation, whereas their next-to-leading-order fit yields $\alpha_s(5GeV) = 0.191$.

Typically, Λ is extracted from the data by parametrizing the parton distributions in simple analytic forms at some Q_0^2, evolving to higher Q^2 using the next-to-leading-order evolution equations, and fitting globally to the measured structure functions to obtain Λ. Thus an important task of such studies is the extraction of parton distributions at a fixed reference value of Q_0^2. These can then be evolved in Q^2 and used as input for phenomenological studies in hadron-hadron collisions. It is useful to have a simple analytic approximation to the parton distributions valid over a range of x and Q^2 values. Indeed, such parametrizations are available in the literature (Glück et al., 1982; Duke and Owens 1984; Eichten et al., 1984; Diemoz et al., 1987; Martin et al. 1988).

(3) Tests of QCD in High Energy Hadron Collisions

There are many ways in which perturbative QCD can be tested in high energy hadron colliders. The production of single large-transverse-momentum photons offers an interesting possibility, since the leading-order QCD subprocesses are $q\bar{q} \to \gamma g$, $qg \to \gamma q$, and $\bar{q}g \to \gamma \bar{q}$. Explicit expressions for the corresponding scattering amplitudes can be found, for example, in a review article of Owen (1987). If the parton distributions are taken from other processes and a value of Λ is assumed, then an absolute prediction is obtained. Conversely, the data can be used to extract information on parton distributions and a value for Λ. This is also one of the few hard scattering processes for which the next-to-leading-order corrections are known (Aurenche et al., 1984, 1988), so that a precision test is possible in principle. In practice, however, the uncertainties on the most accurate experimental data available to date are in the order of $20 - 30\%$, which is too large to limit the accuracy of the extracted value on α_s (or, equivalently, Λ). At this point, a value for Λ in the range $100 - 300MeV$ accounts for a wide range of data.

The production of hadrons with large transverse momentum in hadron-hadron collisions provides a direct probe of the scattering of quarks and gluons: $qq \to qq$, $qg \to qg$, $gg \to gg$, etc. The QCD prediction combines the parton distributions with the leading-order $2 \to 2$ parton hard scattering amplitudes. The present generation of $p\bar{p}$ colliders provide center-of-mass energies which are sufficiently high that these processes can be unambiguously identified in two-jet production at large transverse momentum. Corrected inclusive jet cross sections have been directly compared to the calculated parton cross sections, and the

[3] *We adopt the standard approximation in which quarks of mass greater than μ are neglected completely while quarks of mass less than μ are treated as massless. This means, for example, that the QCD parameter in the range of having five effective massless quarks, i.e., above the b-quark production threshold, is different from that with four effective massless quarks, although the two is related in a simple manner.*

agreement is impressive. Data are also available on the angular distribution of jets; they are also in agreement with QCD predictions.[4]

Many authors (Altarelli et al., 1978, 1979, 1984, 1985; etc.) have considered QCD corrections to Drell-Yan type cross sections, i.e., the lepton-pair production in hadron collisions by quark-antiquark annihilation into virtual photons, or W or Z bosons. The $O(\alpha_s)$ QCD corrections are sizable and approximately constant over the lepton-pair mass range probed by experiments, leading to a simple parametrization:

$$\sigma_{DY} = \sigma_{DY}^{(0)}[1 + \frac{\alpha_s(Q^2)}{2\pi}C + ...]. \tag{9}$$

It is interesting to note that the corresponding correction to W and Z production, as measured at $p\bar{p}$ colliders, has essentially the same theoretical form and is of order 30%. Total W and Z production cross sections to be measured soon will be accurate enough to be sensitive to such 30% QCD correction effects and can in principle offer another test of the theory. However, the calculation is still lacking on the complete $O(\alpha_s^2)$ QCD correction which is of potential importance in view of the large $O(\alpha_s)$ term.

Finally, QCD effects are also observable in the production of W and Z bosons with large transverse momentum (Halzen and Scott, 1978; Altarelli et al., 1984). There is good qualitative agreement, although the statistics is rather poor at present (UA1 collaboration, 1984; Bawa and Stirling, 1988).

(4) Tests of QCD in Heavy Quarkonium Decay

Potential model dependences often cancel out in the ratios of decay widths of heavy quarkonium states. Important examples of such ratios are

$$\frac{\Gamma(1^{--} \to ggg)}{\Gamma(1^{--} \to \mu^+\mu^-)}, \quad \frac{\Gamma(1^{--} \to \gamma gg)}{\Gamma(1^{--} \to ggg)}. \tag{10}$$

Thus, the ratios of partial decay widths allow for a determination of α_s at the heavy quark mass scale. The most precise data come from the decay widths of the $J^{PC} = 1^{--}$ J/ψ and Υ resonances. A summary of quarkonium decay rates can be found in a review article by, e.g., Kwong et al. (1987a).

However, the perturbative corrections to these ratios are rather large (Lepage and Mackenzie, 1981). They change the predictions by a factor of 1.64 and 0.77 respectively in the case of Υ decay. The corrections in the J/ψ case are even larger. In addition, relativistic corrections are unknown and could be substantial in the J/ψ case.

A recent analysis (Kwong et al., 1987b) of bottomonium decay-width ratios from CUSB, CLEO, and ARGUS[5] yields, upon neglecting the theoretical uncertainties,

$$\alpha_s(m_b) = 0.179 \pm 0.009. \tag{11}$$

[4] *UA1 collaboration: Arnison, G., et at., Phys. Lett.* **B177,** *244 (1986).*

[5] *Csorna, S.E., et al., Phys. Rev. Lett.* **56,** *1222 (1986); Schamberger, R.D., et al., Phys. Lett.* **138B,** *225 (1984); Albrecht, H., et al., Phys. Lett.* **B199,** *291 (1987).*

(5) Tests of QCD in e^+e^- Collisions

As a well-known result (Ch. 9), the total cross section for $e^+e^- \rightarrow$ *hadrons* may be obtained by multiplying the muon-pair cross section by the famous R-factor $R = 3\sum_q e_q^2$. The higher order QCD corrections to this quantity have been calculated, and the result may be expressed as follows:

$$= R^{(0)}[1 + \frac{\alpha_s}{\pi} + C_2(\frac{\alpha_s}{\pi})^2 + C_3(\frac{\alpha_s}{\pi})^3 + ...].$$

$$C_2 = (\frac{2}{3}\zeta(3) - \frac{11}{12})N_f + \frac{365}{24} - 11\zeta(3). \tag{12}$$

Numerically, $C_2 = 1.41$ in the modified minimum subtraction scheme. Recently C_3 has also been computed (Gorishny et al., 1988); numerically (for $N_f = 5$) $C_3 = 64.7$ in the modified minimum subtraction scheme. Quark mass effects are neglected in these calculations.

At the highest energies currently accessible, the corrections from QCD and Z exchange are comparable. A comparison of the theoretical prediction of Eq. (12) (corrected for the b-quark mass effect) with all the available data (including those from TRISTAN at $\sqrt{s} = 50\,GeV$) has been performed by the CELLO Collaboration.[6] The result is a correlated measurement of α_s and $sin^2\theta_W$. Fixing $sin^2\theta_W$ at the world-average value of 0.23, one obtains

$$\alpha_s(34GeV) = 0.132 \pm 0.016. \tag{13}$$

Two remarks are useful here. First, the principal advantage of determining α_s from R in e^+e^- annihilation is that there is no dependence on fragmentation models, jet algorithms, etc. Second, the order α_s^3 term in Eq. (12) is numerically twice as large as the order α_s^2 term, throwing the accuracy of the QCD prediction into serious doubt.

The traditional method of determining α_s in e^+e^- annihilation is from measuring quantities which are sensitive to the relative rate of two- and three-jet events. There are many possible choices of such "shape variables"; thrust (Farhi 1977), energy-energy correlations (Basham et al., 1978), planar triple-energy correlations (Csikor et al., 1985), average jet mass. etc. Calculations of these observables are free of infrared divergences, so that they can be reliably calculated in perturbation QCD. The starting point for calculating all these quantities is the simple "three-jet" cross section for $e^+e^- \rightarrow q\bar{q}g$ (which may be verified using the procedure introduced in Ch. 9 together with Feynman rules for QED and QCD):

$$\frac{1}{\sigma}\frac{d^2\sigma}{dx_1dx_2} = \frac{2\alpha_s}{3\pi}\frac{x_1^2 + x_2^2}{(1-x_1)(1-x_2)}, \tag{14}$$

where

$$x_i = \frac{2E_i}{\sqrt{s}}$$

are the center-of-mass energy fractions of the final-state (massless) quarks. A distribution in a "three-jet" variable, such as those mentioned above, is then obtained by integrating this differential cross section over an appropriate phase space region for a fixed value of the variable.

[6] *CELLO Collaboration: Behrend, H.J., et al., Phys. Lett.* **B183,** *400 (1987); and de Boer, W., SLAC-PUB-4428 (1987).*

A compilation of all the available data and a complete list of references can be found in, e.g., review papers by Stirling and Whalley (1987) and by Wu (1987). A "world average" is

$$\alpha_s(34GeV) = 0.14 \pm 0.02, \tag{15}$$

with the error being the spread between the different experiments including the fragmentation uncertainty, but not that due to the size of the higher order corrections, which might be somewhat larger than this error. Notice that this value of α_s is in agreement with the value obtained from the measurement of R described above.

(6) Conclusions

There are many other ways in which QCD can be tested in lepton-hadron deep inelastic scattering, hadron-hadron collisions, and electron-positron collisions. We may mention in particular the study of exclusive processes (form factors, elastic scattering, ...), the behavior of quarks and gluons in nuclei, the spin properties of the theory and the importance of polarized scattering data, QCD effects in hadron spectroscopy, etc.

The global consistency among the existing data in the framework of perturbative QCD suggest that QCD *describes* strong interactions among quarks. Nevertheless, there are still many important tests to be made before the role of QCD is firmly and uniquely established.

14.2 The Glashow-Salam-Weinberg $SU(2) \times U(1)$ Electroweak Theory

The standard Glashow-Salam-Weinberg (GSW) $SU(2) \times U(1)$ electroweak model has been introduced and elucidated in some detail in Ch. 13. Apart from the Higgs mass M_H, fermion masses, and generation mixing parameters, the theory has three parameters. A particularly useful set is:

(a) *the fine structure constant* $\alpha = 1/137.036$,[7] determined from electron magnetic moment anomaly $(g-2)$.

(b) *the Fermi coupling constant*, $G_F = 1.16637 \times 10^{-5}\ GeV^{-2}$, determined from the muon lifetime formula (including lepton mass and $O(\alpha)$ radiative corrections)[8]:

$$\begin{aligned}\tau_\mu^{-1} = \frac{G_F^2 m_\mu^5}{192\pi^3}&[1 + \frac{\alpha}{2\pi}(\frac{25}{4} - \pi^2)(1 + \frac{2\alpha}{3\pi}\ln\frac{m_\mu}{m_e})]\\ &\times [1 - \frac{8m_e^2}{m_\mu^2}] \times [1 + \frac{3}{5}\frac{m_\mu^2}{M_W^2}].\end{aligned} \tag{16}$$

and

[7] *α depends upon the energy scale under investigation. This standard value is good only at very low energies. For example, the value $1/128$ is appropriate at energies of order M_W.*

[8] *See Eq. (44), Ch. 13.*

(c) *the elecroweak mixing parameter* $sin^2\theta_W$, determined from neutral-current processes and the W and Z^0 masses. Note that the value of $sin^2\theta_W$ depends on the renormalization prescription. A very useful scheme (Sirlin 1980 and 1984) is to take the tree-level formula $sin^2\theta_W = 1 - M_W^2/M_Z^2$ as the definition of the renormalized $sin^2\theta_W$ to all orders in perturbation theory.[9] Alternatively, one may take M_Z rather than $sin^2\theta_W$ as the third fundamental parameter. This may become useful when a very precise value of M_Z is determined from experiments at SLC and LEP in the immediate future.

The quality of the existing data already demands that complete $O(\alpha)$ radiative corrections must be suitably taken into account. Accordingly, the tree-level expressions for M_W and M_Z, as given previously by Eqs. (16) and (43) in Ch. 13, are modified:

$$\begin{aligned} M_W &= \frac{A_0}{sin\theta_W(1-\triangle r)^{1/2}}, \\ M_Z &= \frac{M_W}{cos\theta_W}, \end{aligned} \tag{17}$$

where $A_0 = (\pi\alpha/\sqrt{2}G_F)^{1/2} = 37.281\, GeV$. The radiative correction parameter $\triangle r$ is predicted to be 0.0713 ± 0.0013 for $m_t = 45\, GeV$ and $M_H = 100\, GeV$, while $\triangle r \to 0$ for $m_t \sim 245\, GeV$. If M_Z is regarded as a fundamental parameter, then

$$sin^2\theta_W = \frac{1}{2}\{1 - [1 - \frac{4A_0^2}{M_Z^2(1-\triangle r)}]^{1/2}\} \tag{18}$$

is a derived parameter, and $M_W = M_Z\, cos\theta_W$.

(1) Formulae on cross sections and asymmetries

Following the 1988 publication of Particle Data Group, we shall first introduce the notations adopted by researchers working on different facets of the electroweak theory and then move on to list the basic formulae on cross sections and asymmetries which may easily be understood in terms of what we have learned in previous chapters.

It is convenient to write the four-fermion interactions relevant to ν-hadron (νH), νe, and e-hadron (eH) processes in a form that is valid in an arbitrary gauge theory (assuming massless left-handed neutrinos). One has

$$\begin{aligned} \mathcal{L}^{\nu H} &= \frac{G_F}{\sqrt{2}} i\bar{\nu}\gamma_\mu(1+\gamma_5)\nu \\ &\times \{\sum_i [\epsilon_L(i)\, i\bar{q}_i\gamma_\mu(1+\gamma_5)q_i + \epsilon_R(i)\, i\bar{q}_i\gamma_\mu(1-\gamma_5)q_i]\}, \end{aligned} \tag{19}$$

$$\mathcal{L}^{\nu e} = \frac{G_F}{\sqrt{2}} i\bar{\nu}_\mu\gamma_\mu(1+\gamma_5)\nu_\mu\, i\bar{e}\gamma_\mu(g_V^e + g_A^e\gamma_5)e, \tag{20}$$

[9] *An alternative is to use the modified minimal subtraction quantity* $sin^2\theta_W(\mu)$, *where* μ *is conveniently chosen to be* M_W *for electroweak processes. The two definitions are related by* $sin^2\theta_W(M_W) = C(m_t, M_H)sin^2\theta_W$, *where* $C = 0.9907$ *for* $m_t = 45GeV$, $M_H = 100GeV$.

$$\mathcal{L}^{eH} = \frac{G_F}{\sqrt{2}} \sum_i [C_{1i}\, i\bar{e}\gamma_\mu\gamma_5 e\, i\bar{q}_i\gamma_\mu q_i + C_{2i}\, i\bar{e}\gamma_\mu e\, i\bar{q}_i\gamma_\mu\gamma_5 q_i]. \tag{21}$$

Note that, for $\nu_e\bar{e}$ or $\bar{\nu}_e e$, the charged-current contribution must be included.

The standard model expressions for $\epsilon_{L,R}(i)$, $g^e_{V,A}$ and C_{ij} are summarized in the following table.

Table 1: Standard model expressions for the neutral-current parameters for ν-hadron, νe, and e-hadron processes. Entries are from Particle Data Group (1988).

Quantity	Standard Model Expression
$\epsilon_L(u)$	$\rho^{NC}_{\nu N}[\frac{1}{2} - \frac{2}{3}\kappa_{\nu N} sin^2\theta_W + \lambda_{uL}]$
$\epsilon_L(d)$	$\rho^{NC}_{\nu N}[-\frac{1}{2} + \frac{1}{3}\kappa_{\nu N} sin^2\theta_W + \lambda_{dL}]$
$\epsilon_R(u)$	$\rho^{NC}_{\nu N}[-\frac{2}{3}\kappa_{\nu N} sin^2\theta_W + \lambda_{uR}]$
$\epsilon_R(d)$	$\rho^{NC}_{\nu N}[\frac{1}{3}\kappa_{\nu N} sin^2\theta_W + \lambda_{dR}]$
g^e_V	$\rho_{\nu e}[-\frac{1}{2} + 2\kappa_{\nu e} sin^2\theta_W]$
g^e_A	$\rho_{\nu e}[-\frac{1}{2}]$
C_{1u}	$\rho'_{eq}[-\frac{1}{2} + \frac{4}{3}\kappa'_{eq} sin^2\theta_W]$
C_{1d}	$\rho'_{eq}[\frac{1}{2} - \frac{2}{3}\kappa'_{eq} sin^2\theta_W]$
C_{2u}	$\rho_{eq}[-\frac{1}{2} + 2\kappa_{eq} sin^2\theta_W]$
C_{2d}	$-C_{2u}$

Note that, if radiative corrections are ignored, $\rho = \kappa = 1$, $\lambda = 0$, as from Eqs. (45), (38a)-(38d), and (39) in Ch. 13. At $O(\alpha)$, $\rho^{NC}_{\nu N} = 1.00074$, $\kappa_{\nu N} = 0.9902$, $\lambda_{u_L} = -0.0031$, $\lambda_{d_L} = -0.0026$, and $\lambda_{u_R} = (1/2)\lambda_{d_R} = 3.5 \times 10^{-5}$ for $m_t = 45\, GeV$, $M_H = 100\, GeV$, $sin^2\theta_W = 0.23$, and $< Q^2 >= 20\, GeV^2$. For νe scattering, $\kappa_{\nu e} = 0.9897$ and $\rho_{\nu e} = 1.0054$ (at $< Q^2 >= 0$). For atomic parity violation, $\rho'_{eq} = 0.9793$ and $\kappa'_{eq} = 0.9948$. For the SLAC polarized electron experiment, $\rho'_{eq} = 0.970$, $\kappa'_{eq} = 0.993$, $\rho_{eq} = 0.993$, and $\kappa_{eq} = 1.03$ after incorporating additional QED corrections.

At present, the most precise determinations of $sin^2\theta_W$ are from deep inelastic neutrino scattering from isoscalar nuclear targets such as deuteron, carbon, or iron (the last being an approximate isoscalar target). The ratio $R_\nu = \sigma^{NC}_{\nu N}/\sigma^{CC}_{\nu N}$ of neutral- to charged-current cross sections has been measured to 1% accuracy by the CDHS and CHARM Collaborations,[10] so that it is imperative to obtain theoretical expressions for R_ν and $R_{\bar{\nu}}$ ($\equiv \sigma^{NC}_{\bar{\nu} N}/\sigma^{CC}_{\bar{\nu} N}$) as functions of $sin^2\theta_W$ to similar accuracy. Fortunately, most of the uncertainties arise from the strong interactions and neutrino spectra; they cancel in the ratio.

[10] *CDHS: Abramowicz, H., et al., Phys. Rev. Lett.* **57,** *298 (1986); CHARM: Allaby, J.V., et al., Phys. Lett.* **B177,** *446 (1986).*

Using the quark parton model, one finds, as a simple zeroth-order approximation,

$$R_\nu \equiv \sigma_{\nu N}^{NC}/\sigma_{\nu N}^{CC} = g_L^2 + g_R^2 r$$

$$R_{\bar{\nu}} \equiv \sigma_{\bar{\nu} N}^{NC}/\sigma_{\bar{\nu} N}^{CC} = g_L^2 + \frac{g_R^2}{r}, \tag{22}$$

where

$$g_L^2 \equiv \epsilon_L(u)^2 + \epsilon_L(d)^2 \simeq \frac{1}{2} - sin^2\theta_W + \frac{5}{9}sin^4\theta_W,$$

$$g_R^2 \equiv \epsilon_R(u)^2 + \epsilon_R(d)^2 \simeq \frac{5}{9} - sin^4\theta_W, \tag{23}$$

and $r \equiv \sigma_{\bar{\nu} N}^{CC}/\sigma_{\nu N}^{CC}$ is the ratio of $\bar{\nu}$ and ν charged-current cross sections, which can be measured directly. Note that, in practice, Eq. (22) must be corrected for quark mixing, the s and c seas, c-quark threshold effects (which mainly affect σ^{CC} - these turn out to be the largest theoretical uncertainty), nonisoscalar target effects, $W - Z$ propagator differences, and radiative corrections (which lower the extracted value of $sin^2\theta_W$ by ~ 0.009). Details of the neutrino spectra, experimental cuts, x and Q^2 dependence of structure functions, and longitudinal structure functions enter only at the level of these corrections and therefore lead to very small uncertainties (Amaldi *et al.* 1987). Altogether, the theoretical uncertainty is $\triangle sin^2\theta_W \sim \pm 0.005$, which would be very hard to improve in the forseeable future.

The laboratory cross section for $\nu_\mu e \to \nu_\mu e$ or $\bar{\nu}_\mu e \to \bar{\nu}_\mu e$ elastic scattering is given by

$$\frac{d\sigma_{\nu_\mu,\bar{\nu}_\mu}}{dy} = \frac{G_F^2 m_e E_\nu}{2\pi} \times [(g_V^e \pm g_A^e)^2 + (g_V^e \mp g_A^e)^2(1-y)^2 - (g_V^{e2} - g_A^{e2})\frac{y m_e}{E_\nu}], \tag{24}$$

where the upper (lower) sign refers to $\nu_\mu(\bar{\nu}_\mu)$, and $y \equiv E_e/E_\nu$ [which runs from 0 to $(1 + m_e/2E_\nu)^{-1}$] is the ratio of the kinetic energy of the recoil electron to the incident ν or $\bar{\nu}$ energy. For $E_\nu \gg m_e$ this yields a total cross section

$$\sigma = \frac{G_F^2 m_e E_\nu}{2\pi}[(g_V^e \pm g_A^e)^2 + \frac{1}{3}(g_V^e \mp g_A^e)^2]. \tag{25}$$

The most accurate leptonic measurements[11] of $sin^2\theta_W$ come from the ratio $R \equiv \sigma_{\nu_\mu e}/\sigma_{\bar{\nu}_\mu e}$ in which many of the systematic uncertainties cancel. Radiative corrections, which are small compared to the precision of present experiments, increase the extracted $sin^2\theta_W$ by $\simeq 0.002$. The cross section for $\nu_e e$ and $\bar{\nu}_e e$ may be obtained from Eq. (24) by replacing $g_{V,A}^e$ by $g_{V,A}^e + 1$, according to the Fierz reordering theorem (which enables one to take into account the charged-current contribution in a simple manner).

The SLAC polarized-electron experiment (Prescott et al., 1979) measured the parity-violating asymmetry

$$\mathcal{A} = \frac{\sigma_R - \sigma_L}{\sigma_R + \sigma_L}, \tag{26}$$

[11] *CHARM; Bergsma, F., et al., Phys. Lett.* **147B**, *481 (1984); and BNL E734: Ahrens, L.A., et al., Phys. Rev. Lett.* **54**, *18 (1985).*

where $\sigma_{R,L}$ is the cross section for the deep-inelastic scattering of a right- or left-handed electron: $e_{R,L}N \to eX$. In the quark parton model, we have

$$\frac{\mathcal{A}}{Q^2} = a_1 + a_2 \frac{1-(1-y)^2}{1+(1-y)^2}, \tag{27}$$

where $Q^2 > 0$ is the momentum transfer and y is the fractional energy transfer from the electron to the hadron. For the deuteron or other isoscalar target, one has, neglecting the s quark and antiquarks,

$$\begin{aligned} a_1 &= \frac{3G_F}{5\sqrt{2}\pi\alpha}(C_{1u} - \frac{1}{2}C_{1d}) \simeq \frac{3G_F}{5\sqrt{2}\pi\alpha}(-\frac{3}{4} + \frac{5}{3}sin^2\theta_W), \\ a_2 &= \frac{3G_F}{5\sqrt{2}\pi\alpha}(C_{2u} - \frac{1}{2}C_{2d}) \simeq \frac{9G_F}{5\sqrt{2}\pi\alpha}(sin^2\theta_W - \frac{1}{4}). \end{aligned} \tag{28}$$

Radiative corrections lower the extracted value of $sin^2\theta_W$ by ~ 0.005.

Experiments which detect atomic parity violation (Bouchiat and Pottier, 1986; Piketty 1986) are now quite precise, and the uncertainties associated with atomic wave functions are relatively small (especially for cesium). For heavy atoms one determines the "weak charge"

$$\begin{aligned} Q_W &= -2[C_{1u}(2Z+N) + C_{1d}(Z+2N)] \\ &\simeq Z(1-4sin^2\theta_W) - N. \end{aligned} \tag{29}$$

Radiative corrections increase the extracted $sin^2\theta_W$ by ~ 0.008.

The forward-backward asymmetry for $e^+e^- \to \ell\bar{\ell}$, $\ell = \mu$ or τ is defined as

$$\mathcal{A}_{FB} \equiv \frac{\sigma_F - \sigma_B}{\sigma_F + \sigma_B}, \tag{30}$$

where $\sigma_F(\sigma_B)$ is the cross section for ℓ^- to travel forward (backward) with respect to the e^- direction. The charge asymmetry $\mathcal{A}_{FB}$ and the ratio of the total cross section relative to pure QED, R, are given, from Problem 13.3. in Ch. 13, by

$$\begin{aligned} R &= F_1, \\ \mathcal{A}_{FB} &= \frac{3}{4}\frac{F_2}{F_1}, \end{aligned} \tag{31}$$

with

$$\begin{aligned} F_1 &= 1 - 2\chi_0 V^e V^\ell \cos\delta_R + \chi_0^2(V^{e2} + A^{e2})(V^{\ell 2} + A^{\ell 2}), \\ F_2 &= -2\chi_0 A^e A^\ell \cos\delta_R + 4\chi_0^2 A^e A^\ell V^e V^\ell, \end{aligned} \tag{32}$$

where

$$\begin{aligned} tan\delta_R &= \frac{M_Z\Gamma_Z}{M_Z^2 - s} \\ \chi_0 &= \frac{G_F}{2\sqrt{2}\pi\alpha}\frac{sM_Z^2}{[(M_Z^2 - s)^2 + M_Z^2\Gamma_Z^2]^{1/2}} \end{aligned} \tag{33}$$

and $\sqrt{s}$ is the CM energy. Eq. (32) is valid in the tree approximation. If QED radiative corrections are taken into account for the data, then the remaining electroweak corrections can be incorporated (Lynn and Stuart 1985; in an approximation adequate for existing PEP and PETRA data) by replacing χ_0 by $\chi(s) \equiv \chi_0(s)\alpha/\hat{\alpha}(s)$, where $\hat{\alpha}(s)$ is the running QED coupling. Numerically, $\alpha/\hat{\alpha}(s) \sim 1 - \Delta_r$ if Δ_r is evaluated for $m_t < 100\,GeV$.

At SLC and LEP, A_{FB} for $e^+e^- \to \ell^+\ell^-$ near the Z pole will be measured to high precision. Similarly, it is expected that the left-right asymmetry

$$\mathcal{A}_{LR} \equiv \frac{\sigma_L - \sigma_R}{\sigma_L + \sigma_R}, \tag{34}$$

where $\sigma_L(\sigma_R)$ is the cross section for a left(right)-handed incident electron, is to be measured very precisely at SLC and possibly at LEP. Neglecting terms of order $(\Gamma_Z/M_Z)^2$, one has, at tree level,

$$\begin{aligned} \mathcal{A}_{FB} &\simeq 3\eta_\ell \frac{\eta_e + \frac{1}{2}P_e}{1 + 2P_e\eta_e}, \\ \mathcal{A}_{LR} &\simeq 2\eta_e, \end{aligned} \tag{35}$$

where P_e is the initial e^- polarization and

$$\eta_i \equiv \frac{V^i A^i}{V^{i2} + A^{i2}}, \qquad i = e, \mu, \tau. \tag{36}$$

In Eq. (36), we have used the notations:

$$\begin{aligned} V^i &\equiv t_{3L}(i) - 2e_i sin^2\theta_W, \\ A^i &\equiv t_{3L}(i), \end{aligned} \tag{37}$$

with $t_{3L}(i)$ the 3rd component of the weak isospin for the fermion of flavor i and e_i its electric charge. The high-precision measurements will require careful application of both QED and electroweak radiative corrections to Eq. (35).

To complete our list on cross sections and asymmetries, we should also mention the width for gauge bosons to decay into massless fermions $f_1\bar{f}_2$, which is given by

$$\begin{aligned} \Gamma(W^+ \to e^+\nu_e) &= \frac{G_F M_W^3}{6\sqrt{2}\pi} \simeq 230 MeV \\ \Gamma(W^+ \to u_i\bar{d}_i) &= \frac{CG_F M_W^3}{6\sqrt{2}\pi} \mid V_{ij} \mid^2 \simeq 717 \mid V_{ij} \mid^2 MeV \\ \Gamma(Z \to \psi_i\bar{\psi}_i) &= \frac{CG_F M_Z^3}{6\sqrt{2}\pi}[V^{i2} + A^{i2}] \\ &\simeq \left\{ \begin{array}{ll} 170\,MeV(\nu\bar{\nu}), & 85.4\,MeV(e^+e^-) \\ 305\,MeV(u\bar{u}), & 394\,MeV(d\bar{d}). \end{array} \right. \end{aligned} \tag{38}$$

For leptons $C = 1$, while for quarks $C = 3[1 + \frac{\alpha_s(M_W)}{\pi}]$, where the factor 3 is due to color and the factor in parentheses is a QCD correction (Albert et al., 1980; Consoli et al., 1983; Güsken et al., 1985; Jegerlehner, 1986). The remaining corrections are negligible.

Assuming three fermion families, we obtain the total widths as follows:

$$\Gamma_Z \sim (2.58 - 2.55) GeV,$$
$$\Gamma_W \sim (2.52 - 2.12) GeV, \tag{39}$$

where the range arises from a variation of the top quark mass m_t, and the other fermion masses have been neglected. The data up to the year of 1988 are severely limited by the small number of the observed events:

$$\Gamma_{Z\,exp} < 6.5 GeV,$$
$$\Gamma_{W\,exp} < 5.6 GeV. \tag{40}$$

Nevertheless, results from Fermilab Tevatron ($p\bar{p}$ collider), Stanford SLC and CERN LEP (e^+e^- colliders) are piling up rapidly beginning at the summer of 1989, providing stringent experimental tests of Eq. (39).[12]

(2) Standard-model analyses of experimental results

The electroweak mixing parameter $sin^2\theta_W$ and equivalently, the Z^0 mass, M_Z, have been determined from the W and Z masses and from a variety of neutral-current processes covering a very wide Q^2 range. The results (Amaldi et al., 1987), as taken from Particle Data Group (1988) and shown in Table 2, are in impressive agreement with each other, indicating the quantitative success of the GSW electroweak theory. The best fit to all data yields $sin^2\theta_W = 0.230 \pm 0.0048$ which corresponds to $M_Z = 92.0 \pm 0.7\,GeV$, where the errors (as well as those given below for other neutral-current parameters) include full statistical, systematic, and theoretical uncertainties.

Note that the central values of all fits shown in Table 2 are obtained by assuming, in the determination of radiative corrections, that the top quark mass $m_t = 45\,GeV$ and the Higgs boson mass $M_H = 100\,GeV$. Whereas the first of the two errors shown is experimental, the second (in square brackets) is theoretical, as computed by assuming that (1) there are three fermion families and that (2) the top-quark mass and the Higgs boson mass are allowed to vary but subject to the arbitrarily chosen constraint $m_t < 100\,GeV$, and $M_H < 1\,TeV$. In the other cases the theoretical and experimental uncertainties are combined. When m_t is allowed to be totally arbitrary, the fits to all data yield $sin^2\theta_W = 0.229 \pm 0.007$ and $M_Z = 91.8 \pm 0.9\,GeV$. The existing e^+e^- data do not yield a useful determination of $sin^2\theta_W$: At PEP and PETRA energies, all values of $sin^2\theta_W$ from 0.1 to 0.4 give a good description of the existing data on the observed asymmetries.

The radiative corrections are sensitive to the isospin breaking associated with a large m_t. Consistency of the $sin^2\theta_W$ values derived from the various reactions requires $m_t < 180\,GeV$ at $90\,\%\,C.L.$ for $M_H \leq 100\,GeV$, with a slightly weaker limit for larger M_H. Similar limits hold for the mass splittings between fourth-generation quarks or leptons.

The measured values of M_W and M_Z are given in Table 3. They are in excellent agreement with the predictions of the standard model when full radiative corrections (to both the W and Z mass formulas and to deep inelastic scattering) are included, but in severe disagreement when the corrections are excluded.

---

[12] *See Eq. (41) below.*

Table 2: Determination of $sin^2\theta_W$ and M_Z (in GeV) from various reactions. Entries are from from Particle Data Group (1988).

Reaction	$sin^2\theta_W$	M_Z
Deep inelastic (isoscalar)	$0.233 \pm 0.003 \pm [0.005]$	$91.6 \pm 0.4 \pm [0.8]$
$\nu_\mu p \to \nu_\mu p$	0.210 ± 0.033	95.0 ± 5.2
$\bar{\nu}_\mu p \to \bar{\nu}_\mu p$	0.210 ± 0.033	95.0 ± 5.2
$\nu_\mu e \to \nu_\mu e$	$0.223 \pm 0.018 \pm [0.002]$	93.0 ± 2.7
$\bar{\nu}_\mu e \to \bar{\nu}_\mu e$	$0.223 \pm 0.018 \pm [0.002]$	93.0 ± 2.7
W, Z	$0.228 \pm 0.007 \pm [0.002]$	92.3 ± 1.1
Atomic parity violation	$0.209 \pm 0.018 \pm [0.014]$	95.1 ± 3.9
SLAC eD	$0.221 \pm 0.015 \pm [0.013]$	93.3 ± 2.7
μC	0.25 ± 0.08	89.6 ± 9.7
All data	0.230 ± 0.0048	92.0 ± 0.7

At the moment of this writing, the latest published data[13] on the mass and width of the Z^0 boson come from Fermilab Tevatron and Stanford SLC (August 1989):

$$\begin{aligned} M_Z &= 91.14 \pm 0.12\,GeV; \quad \Gamma_Z = 2.42^{+0.45}_{-0.35}\,GeV, \\ &\quad (Stanford\ SLC); \\ M_Z &= 90.9 \pm 0.3(stat + sys)\ \pm 0.2(scale)\,GeV; \\ &\qquad \Gamma_Z = 3.8 \pm 0.8 \pm 1.0\,GeV; \quad (Fermilab\ Tevatron). \end{aligned} \tag{41}$$

Note that the newly observed Z^0 mass values are in excellent agreement with the GSW model prediction with radiative corrections (the entry in Table 3). The observed widths also compare well with the prediction given previously in Eq. (39). The need to reduce the error of the theoretical prediction has become obvious.

In view of the fact that both Stanford SLC and CERN LEP, e^+e^- colliders capable of producing Z^0 copiously, already started producing extensive experimental results, it is clear that we may soon witness significant advances, or even another breakthrough, in the area of electroweak physics.

(3) Model-independent analyses of experimental results

The W and Z masses and neutral-current data can be used to set limits on possible deviations from the standard model. For example, the relation in Eq. (17) between M_W is modified if there are Higgs multiplets with weak isospin $> 1/2$ with significant vacuum expectation values. In order to calculate to higher orders in such theories one must select a set

[13] *Abrams, G.S., et al., Phys. Rev. Lett.* **63**, *2173 (1989); Abe, F., et al., Phys. Rev. Lett.* **63**, *720 (1989).*

Table 3: The W and Z masses (in GeV). The first uncertainties are mainly statistical and the second are energy calibration uncertainties that are 100% correlated between M_W and M_Z for each group. The last two rows are predictions of the GSW eletroweak theory, using $sin^2\theta_W$ determined from deep inelastic scattering, with and without radiative corrections, respectively. The entries are from Particle Data Group (1988).

Group	M_W	M_Z	
	UA2	$80.2 \pm 0.8 \pm 1.3$	$91.5 \pm 1.2 \pm 1.7$
	UA1	$83.5^{+1.1}_{-1.0} \pm 2.7$	$93.0 \pm 1.4 \pm 3.0$
UA1+	UA2 combined	80.9 ± 1.4	91.9 ± 1.8
Prediction	with radiative corrections	80.2 ± 1.1	91.6 ± 0.9
Prediction	without radiative corrections	75.9 ± 1.0	87.1 ± 0.7

of four fundamental renormalized parameters. It is convenient to take these as α, G_F, M_Z, and M_W, especially because M_W and M_Z are directly measurable. Then $sin^2\theta_W$ and ρ can be considered dependent parameters defined by

$$sin^2\theta_W \equiv \frac{A_0^2}{M_W^2}(1 - \triangle r), \tag{42}$$

and

$$\rho = M_W^2/(M_Z^2 \cos^2\theta_W). \tag{43}$$

As long as the new physics which yields $\rho \neq 1$ is a small perturbation which does not significantly affect the radiative corrections, ρ can be regarded as a phenomenological parameter which multiplies G_F in Eqs. (19)–(21) and (33).[14] The allowed regions in the $\rho - sin^2\theta_W$ plane are obtained and a global fit to all data (Amaldi et al. 1987) yields

$$\begin{aligned} sin^2\theta_W &= 0.229 \pm 0.0064, \\ \rho &= 0.998 \pm 0.0086, \end{aligned} \tag{44}$$

which is remarkably close to what we have obtained earlier in the context of the standard model. (It justifies to some extent the neglect of $\rho - 1$ due to radiative corrections).

Most of the parameters relevant to ν-hadron, νe, e-hadron, and e^+e^- processes are now determined uniquely and precisely from the data in "model independent" fits (i.e., fits which allow for an arbitrary electroweak gauge theory). The values for the parameters defined in Eq. (19)–(21) are given in Table 4 along with the predictions of the GSW electroweak theory

[14] *In addition, the expression for M_Z in Eq. (17) is divided by $\sqrt{\rho}$; the M_W formula is unchanged.*

Table 4: Values of the model-independent neutral-current parameters, compared with the GSW electroweak theory with $sin^2\theta_W = 0.230$. $\theta_i, i = L$ or R, is defined as $tan^{-1}[\epsilon_i(u)/\epsilon_i(d)]$. Entries are from Particle Data Group (1988).

Quantity	Experimental Value	Standard Model Prediction
$\epsilon_L(u)$	0.339 ± 0.017	0.345
$\epsilon_L(d)$	-0.429 ± 0.014	-0.427
$\epsilon_R(u)$	-0.172 ± 0.014	-0.152
$\epsilon_R(d)$	$-0.011^{+0.081}_{-0.057}$	0.076
g_L^2	0.2996 ± 0.0044	0.301
g_R^2	0.0298 ± 0.0038	0.029
θ_L	2.47 ± 0.004	2.46
θ_R	$4.65^{+0.48}_{-0.32}$	5.18
g_A^e	-0.498 ± 0.027	-0.503
g_V^e	-0.044 ± 0.036	-0.045
C_{1u}	-0.249 ± 0.071	-0.191
C_{1d}	0.381 ± 0.064	0.340
$C_{2u} - \frac{1}{2}C_{2d}$	0.19 ± 0.37	-0.039

with $sin^2\theta_W = 0.230$. Note that there is a second $g_{V,A}^e$ solution, given approximately by $g_V^e \leftrightarrow g_A^e$, which is eliminated by e^+e^- data under the assumption that the neutral current is dominated by the exchange of a single Z.

The agreement is excellent between the results from the model-independent fits and the standard model predictions with $sin^2\theta_W = 0.230$, indicating the amazing success of the GSW electroweak theory. Note that it is difficult to present the e^+e^- results in a model-independent way as Z-propagator effects are non-negligible at PETRA and PEP energies. However, assuming $e-\mu-\tau$ universality, the lepton asymmetries imply $A^e = -0.511 \pm 0.013$, in good agreement with the standard model prediction $-1/2$. The vector coupling is not well determined by existing e^+e^- data: values of V^e from -0.3 to 0.3 are allowed.

(4) Conclusions

In summary, it is clear that the existing experimental data on weak interactions are described remarkably well by the *GSW* $SU(2) \times U(1)$ electroweak theory. A major boost came from direct observations of the $W^\pm$ and Z^0 weak bosons at the predicted masses. With the coming generation of e^+e^- colliders with the center-of-mass energies $\sqrt{s}$ greater than M_{Z^0} (SLC and LEP), it is expected that the theory will be subject to much more severe scrutiny. Owing to the fact that weak interaction studies over the last century constantly produced one surprise after another, one might suspect that the GSW $SU(2) \times U(1)$ electroweak theory may mark the beginning of an unfolding mystery, rather than the completion of a zigsaw puzzle.

14.3 Concluding Remarks

Our presentations in §.14.1 and §.14.2.[15] should have made it clear why quantum chromodynamics (QCD) and the Glashow-Salam-Weinberg (GSW) electroweak theory altogether have assumed the status of the "Standard Model". Both QCD and the GSW electroweak theory are phrased in terms of renormalizable gauge field theories. Thus, we might wonder upon a number of basic questions, a few of which we wish to mention here from our own perspectives.

First of all, we do not understand the reason why gauge principle must be invoked in constructing a theory of elementary particles. This is against the so-called "operational point of view", since there is a gauge degree of freedom which is not directly observable.

Nor we understand the reason why the principle of renormalizability play such an essential role here. In fact, the statement posed by Dirac on that physics must be based on strict mathematics may be a serious challenge here, in view of all the divergences appearing in the present theory such as ultraviolet divergences, infrared divergences, and, even more in massless QCD, collinear divergences. It is indeed difficult to imagine why, in such mathematically ill-behaved theory, some finite parts make physical sense while a good chunk of infinite parts are only renormalizations of those entities which we do not understand, such as mass and electric charge. To some extent, all these divergences are tied to the fact that we are trying to use a local field theory to describe a "pointlike" particle, but we must pardon ourselves that the concept of a "pointlike" object does not make any physical sense at all. In this regard, we might hope that adoption of a better definition for "renormalization" and the possible close tie with the so-called "renormalization group equation" may turn the present mathematically ill-defined theory into the one that would please Dirac and many of us. But the quest for mathematical rigour is yet to be satisfied.

Of course, we may quit being too sophisticated and shall have no more digging into the bottom of it. However, we may still wonder whether strong interactions and electroweak interactions may be unified into a grand unified theory (GUT) with a gauge group having better looking than $SU(3) \times SU(2) \times U(1)$, since after all the running $\alpha_s(Q^2)$ must meet the slowly-changing electroweak $\alpha(Q^2)$ at some Q^2, the so called "grand unification scale". Georgi and Glashow (1974) led the efforts on this front by proposing the simplest $SU(5)$ grand unification scheme which predicts, among others, that a proton would decay into a system of mesons and leptons with a net zero baryon number. Experimental searches for proton decay or baryon number nonconservation set a limit of proton lifetime greater than $(10^{31} - 10^{32})$ years, contradicting the prediction of the simplest $SU(5)$ grand unified model. Of course, there is plenty of room to get life going again by mixing into the GUT scheme

[15] *The presentation in §.14.1. has much in common with "Review of Particle Properties", Particle Data Group, pp. 96–101 (Phys. Lett.* **204**, *1988), prepared April 1988 by Barnett, R.M., Hinchliffe, I., and Stirling, W.J. On the other hand, much of the material for the section on the GSW electroweak theory comes from pp. 102 - 106 of the same publication, prepared April 1988 by Langacker, P. The interested reader may consult the original references for further details which have been left out in our discussion.*

ideas of "supersymmetry" (fermion-boson symmetry), or cooking the pot more thoroughly by trying to put everything altogether including this time the feeble gravitational force in the framework of supersymmetric string models ("superstrings"). Of course, we might succeed along one of these routes for searching for a better theory, but the deeper meaning for why a gauge field theory has been so successful remains to be unraveled.

Closely tied to the the questions listed above, there are unsatisfactory features, or even failures, related to the Standard Model. For example, we have thus far failed to identify the existence of the Higgs particle predicted by the GSW electroweak theory. Searches for the top (t) quark have yet to turn up a positive signature. There are coupling constants, masses, etc., so many of them which we do not have any good idea about why they are there and why they take on specific values as they have had. To move forward, we need yet better ideas: ideas that allow us to go beyond the GSW electroweak theory, ideas that will help us to beat the highly nonlinear QCD thoroughly in order to get to the bottom of it, etc. We are stalled if, for example, we need to rely on a computer that is so many orders of magnitude more powerful than the existing ones in order to understand QCD in quantitative terms; but, fortunately, we all know, or believe, that we are still far from hitting the limit of our capability to understand.

Thus, it might be worth repeating the statement which we already used earlier in a slightly different context: If the history could be a guide, the Standard Model might just mark the beginning of another unfolding mystery, rather than completion of a seemingly-never-ending zigsaw puzzle that the Nature has given all of us to play.

References

Particle Data Group, *Review of Particle Properties*, Physics Letters **B204**, 1 (1988).

Altarelli, G., and Parisi, G., Nucl. Phys. B**126**, 298 (1977); Parisi, G., *Proc. 11th Rencontre de Moriond*, 1976, ed. J. Tran Thanh Van; on the QCD evolution equations.

Glück, M., Hoffmann, E., and Reya, E., Z. Phys. **C13.** 119 (1981); on the nonzero quark mass effect on the QCD evolution equations.

Curci, G., Furmanski, W., and Petronzio, R., Nucl. Phys. **B175,** 27 (1980); Furmanski, W. and Petronzio, R., Phys. Lett. **97B**, 437 (1980), and Z. Phys. **C11**, 293 (1982); Floratos, E.G., Kounnas, C., and Lacaze, R., Phys. Lett. **98B.** 89 (1981), Phys. Lett. **98B,** 285 (1981), and Nucl. Phys. **B192,** 417 (1981); and Herrod, R.T. and Wada, S., Phys. Lett. **96B,** 195 (1981), and Z. Phys. **C9,** 351 (1981); on higher order corrections to QCD.

Voss, R., talk at *the 1987 International Symposium on Lepton and Photon Interactions at High Energies*, Hamburg 1987; on the current status of the experimental data on deep inelastic scattering.

Glück, M., Hoffmann, E., and Reya, E., Z. Phys. **C13,** 119 (1982); Duke, D.W. and Owens, J.F., Phys. Rev. **D30,** 49 (1984); Eichten, E., Hinchliffe, I., Lane, K., and Quigg, C., Rev. Mod. Phys. **56,** 579 (1984); Erratum: **58**, 1065 (1986); Diemoz, M., Ferroni, F., Longo, E., and Martinelli, G., CERN preprint CERN-TH-4751-87 (1987); Martin, A.D., Roberts, R.G., and Stirling, W.J., Phys. Rev. **D37,** 1161 (1988); on parametrizations of parton distributions.

Owens, J.F., Rev. Mod. Phys. **59**, 465 (1987); on the production of single large-transverse-momentum photons in hadron colliders.

Aurenche, P., Baier, R., Douiri, A., Fontannaz, M., and Schiff, D., Phys. Lett. **140B**, 87 (1984); Nucl. Phys. **B297**, 661 (1988); on next-to-leading-order corrections in the production of single large-transverse-momentum photons in hadron colliders.

Altarelli, G., Ellis, R.K., and Martinelli, G., Nucl. Phys. **B143,** 521 (1978); Nucl. Phys. B**157**, 461 (1979); Altarelli, G., Ellis, R.K., Greco, M., and Martinelli, G., Nucl. Phys. B**246**, 12 (1984); Altarelli, G., Ellis, R.K., and Martinelli, G., Phys. Lett. **151B**, 457 (1985); Kubar-André, J. and Paige, F.E., Phys. Rev. D**19**, 221 (1979); Kubar-André, J., Le. Bellac, M., Meunier, J.L., and Plaut, G., Nucl. Phys. **B157**, 251 (1980); Dokshitzer, Yu. L., Dyakonov, D.I., and Troyan, S.I., Phys. Lett. **78B**, 290 (1978); Phys. Report **58**, 269 (1980); Parisi, G. and Petronzio, R., Nucl. Phys. **154**, 427 (1979); Curci, G., Greco, M., and Srivastava, Y., Phys. Rev. Lett. **43**, 434 (1979); Nucl. Phys. **B159**, 451 (1979); Kodaira, J., and Trentadue, L., Phys. Lett. **112B**, 66 (1982); **123B**, 335 (1983); Davies, C.T.H. and Sterling, W.J., Nucl. Phys. **B244**, 337 (1984); Davies, C.T.H., Webber, B.R., and Sterling, W.J., CERN preprint TH-3987/84; Collins, J., and Soper, D.E., Nucl. Phys. **B193**, 381 (1981); **B194**, 445 (1982); **B197**, 446 (1982); Collins, J., Soper, D.E., and Sterman, G., CERN preprint TH-3923/84; Collins, J., Invited talk at LAMPF Conference on Quarks and Gluons in Nuclei, February 1986 (unpublished); Bodwin, G.T., Brodsky, S.J., and Lepage, G.P., Phys. Rev. Lett. **47**, 1799 (1981); Bodwin, G.T., Phys. Rev. D**31**, 2616 (1985); W-Y. P. Hwang and L.S. Kisslinger, Phys. Rev. **D38**, 788 (1988); W-Y. P.

Hwang, talk at the 1988 LAMPF Workshop on "Nuclear and Particle Physics on the Light Cone" (World Scientific, Singapore, 1989); on QCD corrections to Drell-Yan lepton-pair production in hadron-hadron collisions.

Halzen, F., and Scott, W., Phys. Rev. **D18.** 3378 (1978); Altarelli, G., et al., Nucl. Phys. **B246,** 12 (1984); on QCD effects in the production of W and Z bosons with large transverse momentum.

UA1 collaboration: Arnison, G., et al., Phys. Lett. **139B,** 115 (1984); Bawa, A.C., and Stirling, W.J., Phys. Lett. **B203,** 172 (1988); on the data in the production of W and Z bosons with large transverse momentum.

Kwong, W., Quigg, C., and Rosner, J.L., Ann. Rev. Nucl. and Part. Sci. **37,** 325 (1987); on a summary of quarkonium decay rates. (1987a)

Lepage, P. and Mackenzie, P., Phys. Rev. Lett. **47,** 1244 (1981); on perturbative QCD corrections to quarkonium decay ratios.

Kwong, W., Mackenzie, P.B., Rosenfeld, R., and Rosner, J.L., University of Chicago preprint EFI 87-31 (1987); on a recent analysis of bottomonium decay-width ratios from CUSB, CLEO, and ARGUS. (1987b)

Gorishny, S.G., Kataev, A., and Larin, S.A., INR preprint, Moscow, 1988; for the calculation of $C_3 = 64.7$ in the modified minimum subtraction scheme. (See Eq. (12).)

Farhi, E. Phys. Rev. Lett. **39,** 1587 (1977); on the thrust variable in the production of jets in e^+e^- collisions.

Basham, C.L., Brown, L.S., Ellis, S.D., and Love, S.T., Phys. Rev. **D17,** 2298 (1978); on energy-energy correlations in the produiction of jets in e^+e^- collisions.

Csikor, F., et al., Phys. Rev. **D31,** 1025 (1985); on planar triple-energy correlations in the produiction of jets in e^+e^- collisions.

Stirling, W.J. and Whalley, M. R., Durham-RAL Database Publication RAL-87 -107 (1987); and Wu, S.L., in *"Proceedings of the 1987 International Symposium on Lepton and Photon Interactions at High Energies"* (Hamburg 1987); on a comprehensive review of the available data on jet productions in e^+e^- collisions.

Sirlin, A., Phys. Rev. **D22,** 971 (1980); **D29,** 89 (1984). Extensive references to other papers are given in "*Radiative Corrections in* $SU(2)_L \times U(1)$", eds. Lynn, B.W. and Wheater, J.F. (World Scientific, Singapore. 1984); on the renormalization prescription for $sin^2\theta_W$ in the GSW electroweak theory.

Prescott, C.Y., et al., Phys. Lett. **84B,** 524 (1979); on the SLAC polarized-electron experiment to measure the parity-violating asymmetry.

Bouchiat, M.A., and Pottier, L., *Science* **234,** 1203 (1986); and Piketty, C.A., *Weak and Electromagnetic Interactions in Nuclei,* ed. H. V. Klapdor (Springer-Verlag. Berlin, 1986), p. 603; for the reviews on the experiments measuring atomic parity violation.

Lynn, B.W. and Stuart, R.G., Nucl. Phys. **B253,** 216 (1985); and *Physics at LEP*, eds. Ellis, J. and Peccei, R., CERN 86-02, Vol. I; on electroweak corrections to the forward-backward asymmetry in $e^+e^- \to \ell\bar{\ell}$.

Amaldi, U., Böhm, A., Durkin, L.S., Langacker, P., Mann, A.K., Marciano, W.J., Sirlin, A., and Williams, H.H., Phys. Rev. **D36,** 1385 (1987); on analyses of the data related to the Glashow-Salam-Weinberg Electroweak Theory. Very similar conclusions are reached in an analysis by Costa, G., Ellis, J., Fogli, G.L., Nanopoulos, D.V., and Zwirner, F., CERN preprint CERN-TH, 4675/87 (1987).

UA1: Arnison, G., et al., Phys. Lett. **166B**, 484 (1986); UA2: Ansari, R., et al., Phys. Lett. **B186**, 440 (1987); on the W and Z boson masses shown in Table 3.

Albert, D., Marciano, W.J., Wyler, D., and Parsa, Z., Nucl. Phys. **B166**, 460 (1980); Consoli, M., Lo Presti, S., and Maiani, L. Nucl. Phys. **B223,** 474 (1983);G*ü*sken, S., K*ü*hn, J.H., and Zerwas, P.M., Phys. Lett. **155B,** 185 (1985); and Jegerlehner, F., Z. Phys. **C32,** 425 (1986); on QCD corrections to $Z \to \psi\bar{\psi}$ decays.

Georgi, H., and Glashow, S. L., Phys. Rev. Lett. **33**, 438 (1974); on the simplest $SU(5)$ grand unification scheme.

Exercises: *Chapter 14*

1. Assume that the electron-proton deep inelastic scattering (DIS) cross section is an incoherent sum of elementary electron-parton scattering cross sections,

$$\sigma(e + P \rightarrow e' + X) = \int_0^1 dx\, f_{i/P}(x)\, \sigma(e + i \rightarrow e' + i'),$$

where $f_{i/P}(x)$ is the probability of finding a parton of flavor i and longitudinal momentum fraction x inside the proton. Show that the *ep* DIS cross section is proportional to a well-known structure function called $F_2(x, Q^2)$:

$$F_2(x, Q^2) = \sum_i e_i^2 \{q_i(x) + \bar{q}_i(x)\},$$

where e_i is the electric charge of a quark of flavor i.

2. Draw the Feynman diagrams which describe $e^+e^- \rightarrow q\bar{q}g$. Use Feynman rules for QED (in Ch. 9) and for QCD (in Ch. 12) to obtain the T-matrix elements. As another major step, try to prove Eq. (14) quoted in the text.

3. As a simple exercise, show that entries in Table 1 follow from what we have obtained in §.13.3, in the absence of radiative corrections. Try to draw those Feynman diagrams which allow you to understand the need to introduce the factors in ρ, κ, and λ.

4. Prove Eq. (25) by starting from the appropriate T-matrix elements.

5. Prove Eq. (27), with Eq. (28), by starting from the appropriate T-matrix elements at the quark parton level. Consult Problem 13.3. as a reference.

6. Prove Eqs. (38) by starting from the appropriate T-matrix elements. Consult the Appendix in Chapter 13 on the Feynman rules for the GSW electroweak theory.

Chapter 15. The Standard Model

We begin by making a few pre-requite statements that require why the Standard Model (for all Centuries) should be there.

We recognize that the two pillars of the 20th Century, i.e., Einstein's relativity principle and the quantum principle, should be there. The smallest units of matter, such as electrons, neutrinos, and quarks, should also exist.

The gauge principle and the principle of renormalizability should also be there, for the reasons to be elucidated below.

For a unit of matter, we have to describe it *not only* at a particular space-time point *but also* its behavior at the nearby points. To accomplish this, the force description via the gauge principle is what is needed. Along this line, it is also unique to write the kinetic-energy term in the lagrangian.

The principle of renormalizability is a necessity. It applies to the Dirac field itself, the gauge field itself, the complex scalar field itself, and all the possible interaction terms included. In particular, the quartic term $(\phi^\dagger\phi)^2$ for the complex scalar field $\phi(x)$, in view of its renormalizability, must be there.

As for the "origin" of mass, we could write the beginning lagrangian such that the "mass" terms do not appear at all, before the spontaneous symmetry breaking (SSB) takes place - offering an interpretation of the "origin of mass". We should keep this sacred feature in the Standard Model, i.e., without the introduction of the notion of "mass".

Therefore, we may declare that we live in the quantum 4-dimensional Minkowski space-time with the force-fields gauge-group structure $SU_c(3) \times SU_L(2) \times U(1) \times SU_f(3)$ built-in from the very beginning. This overall background is evidenced by the 3 K cosmic microwave background (*CMB*), almost uniform owing to the masslessness of the photons. This overall background can see the quark world, which is protected by $SU_c(3) \times SU_L(2) \times U(1)$ (i.e., the (123) symmetry). The background can also see the lepton world, as protected by $SU_L(2) \times U(1) \times SU_f(3)$ or another (123) symmetry. The sizes in the lepton world are typically of atomic sizes. Meanwhile, the sizes (or the volumes of operation, or the objects) in the quark world are typically of nuclear sizes.

Besides *CMB*, our Universe also manufactured lots of neutrinos (including three flavors, and antineutrinos) in its early stage - forming cosmic background neutrinos ($CB\nu$), which may account for the 25% dark matter.

We call it as "the Standard Model of All Centuries", hoping to elevate the standard of philosophy to the level of Newton, Einstein, and Dirac. It is basically the theory that governs the point-like particles (quantum fields) living in the world with Einstein's relativity principle and the quantum principle. We imagine that the Standard Model should be there five hundred years or a thousand years from now, definitely longer than Newton's classic era of four hundred years. This means that the classic Newton's teachings should be replaced by the teachings based on the Standard Model of All Centuries.

Maybe at some point we should elevate the gauge principle and the principle of renormalizability to the status of the full principle, when we have had thorough studies of these two principles. At this point, we have too many unknowns out there.

1 Prelude

We save the last two chapters for presenting the Standard Model of All Centuries. Though very brief, the readers or our students have some basic ideas to pursue after. In this beginning of the 21st Century, the time is already mature to replace the classic Newton's teachings by those based on the Standard Model of All Centuries, starting out from Einstein's relativity principle and the quantum principle to describe the smallest objects of our world, electrons, quarks, photons, etc., to have a complete closing on the Standard Model.

To begin with, let's remind ourselves of a few facts. It is firmly established that there are three generations of quarks and three generations of leptons, at the level of the so-called "point-like Dirac particles". According to another newly established belief in Cosmology, the content of the current Universe would be 25 % in the dark matter while only 5% in the ordinary matter, the latter described by the "minimal Standard Model" (mSM). In this language, the dark-matter particles are supposed to be described by an extended Standard Model. Thus, there is certain urgent need for an "extended" Standard Model, which also accommodates those phenomena which are currently classified as "beyond the Standard Model".

In the beginning of this 21st Century, we should realize that we in fact are living in the quantum 4-dimensional Minkowski space-time with the force-fields gauge-group structure $SU_c(3) \times SU_L(2) \times U(1) \times SU_f(3)$ built-in from the very beginning. In this background, the quark world is accepted because of the (123) symmetry [i.e., under $SU_c(3) \times SU_L(2) \times U(1)$]. Meanwhile, the lepton world is accepted in view of the other (123) symmetry (i.e., under $SU_L(2) \times U(1) \times SU_f(3)$). This altogether gives us "our world" [1].

The observed 3 K cosmic microwave background (CMB) tells us that our World (our Universe) has the $SU_L(2) \times U(1)$ implemented with the Lorentz group as well. This is why we live in the quantum 4-dimensional Minkowski space-time (for the Lorentz group) with the force-fields gauge-group structure implemented at the same time — to make sure that the overall group, including the Lorentz group, is there as the background of our World.

Here $SU_f(3)$ stands for the $SU_{family}(3)$ family gauge theory, rather than $SU_{flavor}(3)$ which derives originally from the isospin symmetry (the latter *not* a gauge theory).

Among the detailed efforts to include the $SU_f(3)$ gauge group, we could mention those of T. Yanagida [2] and of Yue-Liang Wu [3]. In our language, the force-fields background, the quark world, and the lepton world are separate entities to begin with (they have their own individual ranges and other characteristics). Each of them are well-behaved, mathematically and physically. Thus, it is more natural that $SU_f(3)$ covers only the lepton world - making it free of Landau ghosts and making it asymptotically free. See Chapter 12 on QCD $SU_c(3)$. Thus, our $SU_f(3)$ does not cover the quark world, unlike [2, 3]. The applicability of $SU_f(3)$, as part of the overall gauge group, to the lepton world, or the quark world, or both, is the issue for clarification.

There are reasons why the $SU_f(3)$ is right in our efforts. Because it applies only in the lepton world, the left-handed leptons have to put together to form a $(3, 2)$ multiplet (3: $SU_f(3)$ triplet; 2: $SU_L(2)$ doublet); that would call for a $(3, 2)$ family Higgs multiplet; and then to make every $SU_f(3)$ gauge boson massive we require a $(3, 1)$ purely family Higgs multiplet. See Chapter 13 on the Higgs mechanism. It turns out that it is harmless to have the charged partner in $\Phi(3, 2)$ - they would not go through the spontaneous symmetry breaking (SSB). In terms of the degrees of freedom (DOF), it turns out to be just right. It accommodates three generations of neutrinos to oscillate among themselves. It is amazing, indeed!!

There is a story that could be understood in the 20th Century but was *not* in any of the 20th-Century field-theory textbooks. Namely, a real strange thing happens in the 4-dimensional Minkowski space-time [4, 1], a story which we should know but so far we have overlooked. A complex scalar field $\phi(x)$ should have the self-interaction $\lambda(\phi^\dagger\phi)^2$ with a dimensionless (positive) λ cannot be observed if being alone, since a positive λ means self-repulsive. See Section 2.5, Chapter 1 of this book. It is dimensionless so it is determined *globally* by the 4-dimensional Minkowski space-time, but *not individually* by the field itself. Maybe $\lambda = \frac{1}{8}$ but we can't prove it so far. Thus, the mutually "related" complex scalar fields $\Phi(1,2)$ (Standard-Model Higgs), $\Phi(3,2)$ (mixed family Higgs), and $\Phi(3,1)$ (purely family Higgs) come together to sing a chorus, an SSB chorus with the various force-fields gauge fields.

Thus, it *cannot exist alone* for the Standard-Model Higgs complex scalar field $\Phi(1,2)$. It requires the family Higgs $\Phi(3,2)$ and $\Phi(3,1)$ for the *co-existence*.

The logic is rather simple. We are given the (quantum) 4-dimensional Minkowski space-time. Maybe we are also given the various force fields, as evidenced by the 3 K cosmic microwave background (CMB) of our Universe. We have complex scalar fields, Dirac fields, and spin-one gauge fields at our disposal. The complex scalar fields have the same characteristics as the background, the 4-dimensional Minkowski space-time. But they are self-repulsive and cannot exist alone by itself. But they are "related", the SM Higgs $\Phi(1,2)$, mixed family Higgs $\Phi(3,2)$, and purely family Higgs $\Phi(3,1)$, "related" to lower their total energy and to make the SM and family gauge bosons massive. This makes up the so-called "background". The quark world, with the (123) symmetry, is accepted by this background. Further, the lepton world, with another (123) symmetry and with a much bigger scale (sizes), is also accepted by the same background. We are talking about the origin of fields or of point-like particles [1]. We should figure out a way to think about our world, even if temporarily we might not be on the right track.

In the 4-dimensional Minkowski space-time, there are 10 invariants (of Lorentz group), that leads to another invariant, the so-called "mass", which comes from $M^2 = E^2 - \vec{p}^{\,2}$ (E from the Hamiltonian H and $\vec{p}$ from the momentum operators). In the lepton world, the masses turn out to be off-diagonal, explaining, among others, neutrino oscillations. In reality, when we assert that we are living in the quantum 4-dimensional Minkowski space-time with the force-fields gauge-group structure $SU_c(3) \times SU_L(2) \times U(1) \times SU_f(3)$ built-in from the outset, the chirality operator and the mass operator are not diagonal - making this world nontrivial. For instance, when we write the Dirac equations and others, we are used to the mass eigen-states; we don't even know how to write the flavor eigen-states *directly* except in terms of linear combinations of the mass eigen-states.

In the attempt to write down the Standard Model [which is based on the gauge group $SU_c(3) \times SU_L(2) \times U(1) \times SU_f(3)$], we would try to write it as a whole - strong interactions, electroweak interactions, and family interactions as a whole, and not separately. This guarantees the overall self-consistency.

The **"basic units"** of motion, which are constructed from the right-handed Dirac components or left-handed Dirac components but have their group assignments *explicit* under the group $SU_c(3) \times SU_L(2) \times U(1) \times SU_f(3)$, would be more appropriate than the so-called "building blocks of matter". Moreover, each basic unit derives from one kinetic-energy term, $-\bar{\Psi}_R\gamma_\mu\partial_\mu\Psi_R$ or $-\bar{\Psi}_L\gamma_\mu\partial_\mu\Psi_L$, and from only one such term. This is the lesson learnt from classical mechanics - the linkage from the discrete system to the continuum system.

The gauge principle tells us that ∂_μ is to be replaced by some "gauge-invariant" derivative D_μ. It is clear that this would be an economical way to write down a theory in the globally consistent manner.

As said earlier, we may split the ordinary-matter world into the "quark" world and the "lepton" world, as the ranges of their "livings" are quite different from each other. At low enough temperature, the systems of quarks (and anti-quarks) are confined while systems in leptons could roam anywhere. Thus, we could try to decide which multiplet of the dark-matter group $SU_f(3)$ each basic unit should belong - for instance, we may rule out the direct coupling between the quark world and the dark-matter world, altogether. This is another way to argue that $SU_f(3)$ does not cover the quark world.

We know that the right-handed neutrinos do not appear in the minimal Standard Model - so, $(\nu_{\tau R}, \nu_{\mu R}, \nu_{eR})$ would make a perfect triplet under $SU_f(3)$. In a recent paper [5], we proposed to put $((\nu_\tau, \tau)_L, (\nu_\mu, \mu)_L, (\nu_e, e)_L)$ (*columns*) ($\equiv \Psi_L(3,2)$) as the $SU_f(3)$ triplet and $SU_L(2)$ doublet. The next question is how to assign right-handed τ_R, μ_R, and e_R under $SU_f(3)$. We could write down the mass term for the charged leptons if they are singlets or (τ_R, μ_R, e_R) is an $SU_f(3)$ triplet? The singlet assignment can be ruled out since there are undesirable crossed terms in, e.g., $\Psi_L(\bar{3},2)e_R\Phi(3,2)$. So, three of them would better form a triplet - $\Psi_R^C(3,1)$. In fact, this completes the list of "the basic units" in the lepton world.

Maybe we should say a few words more about "the basic units" or "the basic units of motion", since in the field theory things always take the continuum meaning and so we have to make the terms as sharp as possible. The "building blocks of matter" is in fact rather vague. The "generalized coordinates under some specific group" may be more appropriate - so "the basic units of motion" or simply "the basic units" would be the abbreviation. For example, the multiplet in the proposal of Hwang and Yan [5] has the specific values for all the groups and each entry such as τ_L can only appear once in the construction. Thus, the force-fields gauge-group structure $SU_c(3) \times SU_L(2) \times U(1) \times SU_f(3)$ has to be determined at the same time as Lorentz group - so we use the word "built-in at the very beginning" or "built-in at the outset".

So, there are only three members in the lepton world - two righted-handed triplets and one "composite" left-handed triplet under $SU_f(3)$. Everything in the quark world is singlet under $SU_f(3)$.

Since we wish to propose the $SU_f(3)$ family gauge theory as a way to understand why there are three generations, it requires all additional particles, i.e., (eight) gauge bosons and (four) residual family Higgs, very massive. In the proposal of Hwang and Yan [5], we would have one Standard-Model Higgs field $\Phi(1,2)$, one complex family Higgs triplet $\Phi(3,1)$, and another family triplet-doublet complex scalar fields $\Phi(3,2)$. Amazingly enough, two neutral complex triplets, $\Phi(3,1)$ and $\Phi^0(3,2)$ would indeed undergo the desired Higgs mechanism and the three charged scalar fields would remain massive, i.e., there is no spontaneous symmetry breaking (SSB) in the charged Higgs sector.

So far, we have decided on the basic units of motion - out of those left-handed and right-handed objects (quarks or leptons); the gauge group is chosen to be $SU_c(3) \times SU_L(2) \times$

$U(1) \times SU_f(3)$. As we said, we live in the quantum 4-dimensional Minkowski space-time with the force-fields gauge-group structure $SU_c(3) \times SU_L(2) \times U(1) \times SU_f(3)$ built-in from outset. This is the "background" of everything, i.e., the lepton world and the quark world. It is very interesting to ask why our World stops there (that is, why there is no more in our Universe). For notations, we just use this book [6].

In the quark world, the "basic units" of motion include the up-type right-handed quarks u_R, c_R, or t_R,

$$D_\mu = \partial_\mu - ig_c \frac{\lambda^a}{2} G^a_\mu - i\frac{2}{3} g' B_\mu, \tag{1}$$

and, for the rotated down-type right-handed quarks d'_R, s'_R, or b'_R,

$$D_\mu = \partial_\mu - ig_c \frac{\lambda^a}{2} G^a_\mu - i(-\frac{1}{3}) g' B_\mu. \tag{2}$$

On the left-handed particles, we have the basic units, for the $SU_L(2)$ quark doublets,

$$D_\mu = \partial_\mu - ig_c \frac{\lambda^a}{2} G^a_\mu - ig\frac{\vec{\tau}}{2} \cdot \vec{A}_\mu - i\frac{1}{6} g' B_\mu. \tag{3}$$

Note that six quark-mass terms require six dimensionless couplings to $\Phi(1,2)$, as usual. This competes all the dimensionless couplings in the quark world (i.e. nine dimensionless couplings altogether).

As said earlier, in the lepton world, we introduce the family triplets, $(\nu^R_\tau, \nu^R_\mu, \nu^R_e)$ (column) ($\equiv \Psi_R(3,1)$) and (τ_R, μ_R, e_R) (column) ($\equiv \Psi^C_R(3,1)$), under $SU_f(3)$. The "basic unit" of motion is, for $\Psi_R(3,1)$,

$$D_\mu = \partial_\mu - i\kappa \frac{\bar{\lambda}^a}{2} F^a_\mu. \tag{4}$$

Similarly for $\Psi^C_R(3,1)$.

And, for the left-handed $SU_f(3)$-triplet and $SU_L(2)$-doublet $((\nu^L_\tau, \tau^L), (\nu^L_\mu, \mu^L), (\nu^L_e, e^L)$ (all columns) ($\equiv \Psi_L(3,2)$), the basic unit of motion is

$$D_\mu = \partial_\mu - i\kappa \frac{\bar{\lambda}^a}{2} F^a_\mu - ig\frac{\vec{\tau}}{2} \cdot \vec{A}_\mu + i\frac{1}{2} g' B_\mu. \tag{5}$$

The generation of the various quark masses is through the Standard-Model way. For the charged leptons, we have to choose $\bar{\Psi}_L(\bar{3},2)\Psi^C_R(3,1)\Phi(1,2)$. Note that, in the U-gauge, the SM Higgs looks like $(0, \frac{1}{\sqrt{2}}(v + \eta(x)))$, only one degree of freedom [6]. Thus, the neutral neutrino triplet $\Psi_R(3,1)$ cannot be used in the above coupling, because of the charge conservation.

Moreover, there is an important point in the lepton world. In the $SU_c(3) \times SU_L(2) \times U(1) \times SU_f(3)$ Standard Model [7], the neutrino mass term assumes a new form:

$$i\frac{h}{2} \bar{\Psi}_L(3,2) \times \Psi_R(3,1) \cdot \tilde{\Phi}(3,2) + h.c., \tag{6}$$

where $\Psi(3,i)$ is the neutrino triplet (with the first label for $SU_f(3)$ and the second for $SU_L(2)$). The cross-dot (curl-dot) product is somewhat new, referring to the singlet combination of three triplets in $SU(3)$. The Higgs field $\tilde{\Phi}(3,2)$ is new in this effort [5], because it carries some nontrivial $SU_f(3)$ charge.

This off-diagonal neutrino mass term offers us a natural way to describe neutrino oscillations, since the neutral part of $\Phi(3,2)$ could each receive vacuum expectation values.

Note that, for charged leptons, the Standard-Model choice is $\bar{\Psi}_L(\bar{3},2)\Psi_R^C(3,1)\Phi(1,2) + c.c.$, which gives three leptons an equal mass. But, in view of that if (ϕ_1,ϕ_2) is an $SU(2)$ doublet then $(\phi_2^\dagger,-\phi_1^\dagger)$ is another doublet, we could form $\tilde{\Phi}^\dagger(3,2)$, or $\tilde{\Phi}(3,2)$ in simplified notation, from the doublet-triplet $\Phi(3,2)$. That explains the difference in notation between $\tilde{\Phi}(3,2)$ and $\Phi(3,2)$ - we keep $\Phi(3,2)$ and $\Phi(1,2)$ to the same level while the tilde for the other, a strange complex conjugate.

$$i\frac{h^C}{2}\bar{\Psi}_L(3,2)\times\Psi_R^C(3,1)\cdot\Phi(3,2) + h.c.. \tag{7}$$

Here vacuum expectation values of $\Phi^0(3,2)$ give rise to the imaginary off-diagonal (hermitian) elements in the 3×3 mass matrix, so removing the equal masses of the charged leptons.

Here the couplings h^C and h might be closely related to the coupling strength κ for the family gauge bosons [8]. They are dimensionless in the 4-dimensional Minkowski space-time; so, they should be determined *globally* by this space-time. In the lepton world, there are six dimensionless couplings, counting in g and g' (the electroweak couplings) twice since they also appear in the quark world.

In fact, it is important to realize that in the quark world or in the lepton world all the couplings are dimensionless - a signature that they are determined *globally* by the 4-dimensional Minkowski space-time, rather than by the details in the quark world or in the lepton world. All these are tied closely the 4-dimensional Minkowski space-time; if *not* the 4-dimensional Minkowski space-time, all which we have said are *not* there. That is why we argue that these couplings have to be determined *globally* by the 4-dimensional Minkowski space-time. *This aspect is tied to our confidence why the Standard Model may be close to the truth.*

As pointed out in an early (2013) version of the paper [7], we may imagine that, in the U-gauge, the Standard-Model Higgs $\Phi(1,2)$ looks like $(0, (v+\eta(x))/\sqrt{2})$ (*column*) and $\Phi^\dagger(\bar{3},2)\Phi(1,2)$ would pick out the neutral sector naturally. In fact, the mixed-quartic term $(\Phi^\dagger(\bar{3},2)\Phi(1,2))(\Phi^\dagger(1,2)\Phi(3,2))$ with a suitable sign, would modify the mass term for the $\Phi(3,2)$ field such that the neutral sector has SSB while the charged sector remains massive. This in fact happens naturally. This "projected-out" Higgs mechanism is what we need.

We may [7] write down the terms for potentials among the three Higgs fields, subject to (1) that they are renormalizable, and (2) that symmetries are only broken spontaneously (via the Higgs or induced Higgs mechanism). We may write [7]

$$\begin{aligned} V &= V_{SM} + V_1 + V_2 + V_3, &(8)\\ V_{SM} &= \mu^2\Phi^\dagger(1,2)\Phi(1,2) + \lambda(\Phi^\dagger(1,2)\Phi(1,2))^2, &(9)\\ V_1 &= M^2\Phi^\dagger(\bar{3},2)\Phi(3,2) + \lambda_1(\Phi^\dagger(\bar{3},2)\Phi(3,2))^2 \\ &\quad +\epsilon_1(\Phi^\dagger(\bar{3},2)\Phi(3,2))(\Phi^\dagger(1,2)\Phi(1,2)) + \eta_1(\Phi^\dagger(\bar{3},2)\Phi(1,2))(\Phi^\dagger(1,2)\Phi(3,2)) \\ &\quad +\epsilon_2(\Phi^\dagger(\bar{3},2)\Phi(3,2))(\Phi^\dagger(\bar{3},1)\Phi(3,1)) + \eta_2(\Phi^\dagger(\bar{3},2)\Phi(3,1))(\Phi^\dagger(\bar{3},1)\Phi(3,2)) \\ &\quad +(\delta_1 i\Phi^\dagger(3,2)\times\Phi(3,2)\cdot\Phi^\dagger(3,1) + h.c.), &(10)\end{aligned}$$

$$
\begin{aligned}
V_2 &= \mu_2^2\Phi^\dagger(\bar{3},1)\Phi(3,1) + \lambda_2(\Phi^\dagger(\bar{3},1)\Phi(3,1))^2 + (\delta_2 i\Phi^\dagger(3,1)\cdot\Phi(3,1)\times\Phi(3,1) + h.c.) \\
&\quad +\lambda_2'\Phi^\dagger(\bar{3},1)\Phi(3,1)\Phi^\dagger(1,2)\Phi(1,2), \qquad (11) \\
V_3 &= (\delta_3 i\Phi^\dagger(3,2)\cdot\Phi(3,2)\times(\Phi^\dagger(1,2)\Phi(3,2)) + h.c.) \\
&\quad +(\delta_4 i(\Phi^\dagger(3,2)\Phi(1,2))\cdot\Phi^\dagger(3,1)\times\Phi(3,1) + h.c.) \\
&\quad +\eta_3(\Phi^\dagger(\bar{3},2)\Phi(1,2)\Phi(3,1) + c.c.). \qquad (12)
\end{aligned}
$$

In obtaining the last expression, we maintain that every terms are naïvely renormalizable since the terms are of power four or less (in scalar fields). Note that the terms in $\delta_j i$ involve the so-called "$SU(3)$ operations", as before.

As shown earlier [8, 9], two triplets of complex scalar fields would make the eight family gauge bosons and four residual family Higgs particles all massive. In the $SU_f(3)$ family gauge theory treated alone, there are many ways of accomplishing such goal. In the present complicated case, the equivalence between two triplets is lost, but we discover that this modification would be in a nice way, to patch up that three neutrinos couple to independent vacuum expectation values (and the family Higgs). So, three kinds of neutrino oscillations happen independently.

The "principle" of renormalizability turns out to be very powerful. To the least, we may use it to define the calculability of the theory. It implies the existence of some branch of calculable mathematics (in quantum field theory).

But in the above lagrangian there would be three "ignition" terms, etc.; that is, there are too many "renormalizable" terms. We should distinguish those dimensionless terms from those those dimensional, thinking of the values of those dimensionless couplings as determined *globally* by the 4-dimensional Minkowski space-time. Remember that we have never paid serious attention to such "metaphysical" aspects (maybe we should).

2 The Origin of Mass

Before 2014, our understanding of mass generation for the building blocks of matter was still lacking, not mentioning the origin of mass; so, the same story for the mysterious Higgs mechanisms. The story has changed completely if we accept the recent proposal [10] on the origin of mass.

The Higgs lagrangian (i.e., Eqs. (8)–(12) above) is based on renormalizability. Physics-wise, we should think of its redundancy for example, if there is a spontaneous symmetry breaking (SSB), there are three different channels for switch-on, depending on which complex scalar field gets turn on. On the other hand, when the temperature is high enough, those mass terms, except the switch-on of ignition, would become negligible; or, before the generation of masses.

Thus, the real world should be like this: There is at most one ignition channel for spontaneous symmetry breaking (SSB). When the temperature is high enough, all the mass terms should be zero (if necessary, by comparison).

In the 4-dimensional Minkowski space-time, the complex scalar field $\phi(x)$ have one unique feature - that the renormalizable term $\lambda(\phi^\dagger\phi)^2$ is allowed (only in the 4-dimensional space-time) [1]. This self-interaction is repulsive, explaining why we cannot see it unless under exceptions. Further, this term is dimensionless, so the coupling λ is a pure number that is determined by the 4-dimensional Minkowski space-time, *not* by the complex scalar field itself [10, 1].

Thus, one important beginning point is the 4-dimensional Minkowski space-time. The other point is whether the force-fields gauge symmetry $SU_c(3) \times SU_L(2) \times U(1) \times SU_f(3)$ should also be a part of this beginning point. As all the future entities should have these gauge group assignments (similar to quantum numbers), our suggestion is "yes" - that gives the assertion that "we live in the 4-dimensional Minkowski space-time with the force-fields gauge group $SU_c(3) \times SU_L(2) \times U(1) \times SU_f(3)$ built-in from the outset" [1].

Let us re-iterate the origin of mass [10]:

Suppose that, before the spontaneous symmetry breaking (SSB), the Standard Model does not contain any parameter that is pertaining to "mass", but, after the SSB, all particles in the Standard Model acquire the mass terms as it should, we call it "the origin of mass". In [10], we show that this is indeed the case explaining the origin of mass. In this way, we sort of tie "the origin of mass" to the effects of the SSB, or the generalized Higgs mechanism.

In our opinion, this way of explaining the origin of mass is of fundamental importance. After all, we could not give some real meaning to the term "mass"; it makes sense to interpret masses as our proposal of the origin of mass [10]. According to [10], the masses would disappear altogether at the temperature high enough, if for example we are thinking of the early Universe. On other hand, when the matter system becomes so dense and so hot in the process of forming a "black hole", the mass parameters would lose its meaning altogether, making the interior inside the horizon of the Schwarzschild metric something which we can't understand so far.

Starting from the Higgs sector of the Standard Model [7], we have the Standard-Model Higgs $\Phi(1,2)$, the purely family Higgs $\Phi(3,1)$, and the mixed family Higgs $\Phi(3,2)$, with the first label for $SU_f(3)$ and the second for $SU_L(2)$. We need another triplet $\Phi(3,1)$ since all eight family gauge bosons have to be massive [8].

We could say that it is the most interesting to put the different but related complex scalar fields (Higgs) in the same pot, letting them interact together. Two fields ϕ_a and ϕ_b can interact "directly" via $(\phi_a^\dagger\phi_a)(\phi_b^\dagger\phi_b)$ but, if related, via $((\phi_a^\dagger\phi_b)(\phi_b^\dagger\phi_a))$. Here we always talk about those things which are renormalizable. Under the renormalizable game, these are closely linked with the quartic self coupling such as $\lambda(\phi_a^\dagger\phi_a)^2$.

Besides, in the 4-dimensional Minkowski space-time, these terms have dimensionless couplings, or pure numbers. So, λ depends on the 4-dimensional Minkowski space-time, *not* on the field $\phi(x)$ itself. This is a very interesting aspect. *Thus, when we put three complex scalar fields* $\Phi(1,2)$, $\Phi(3,2)$, *and* $\Phi(3,1)$ *in this pot of the 4-dimensional Minkowski space-time all the quartic terms, including those crossed, are playing pure-number games among themselves. Consequently, the masses for the various Higgs emerge as the result. This discovery is closely linked to the origin of fields (point-like particles) [1].*

In fact we are "playing" with "energies", in terms of the complex scalar fields. For two related complex scalar fields, the cross term helps to lower the total energy. For two unrelated complex scalar fields, the cross term does not exists.

In the U-gauge, we choose to have

$$\Phi(1,2) = (0, \frac{1}{\sqrt{2}}(v+\eta)),\ \Phi^0(3,2) = \frac{1}{\sqrt{2}}(u_1+\eta_1', u_2+\eta_2', u_3+\eta_3'),\ \Phi(3,1) = \frac{1}{\sqrt{2}}(w+\eta',0,0), \tag{13}$$

all in columns. The five components of the complex triplet $\Phi(3,1)$ get absorbed by the $SU_f(3)$ family gauge bosons and the neutral part of $\Phi(3,2)$ has three real parts left - together making all eight family gauge bosons massive.

Before the mixing, the masses of the various Higgs are given by, using the general Higgs lagrangian [cf. Eqs. (8)–(12)],

$$\begin{aligned}
\eta: &\quad (\mu^2/\lambda) + \tfrac{1}{4}(\epsilon_1+\eta_1)u_iu_i + \tfrac{\lambda_2'}{4}w^2,\\
\eta': &\quad (\mu_2^2/\lambda_2) + \tfrac{\epsilon_2}{4}u_iu_i + \tfrac{\eta_2}{4}u_1^2 + \tfrac{\lambda_2'}{4}v^2,\\
\eta_1': &\quad M^2 + \tfrac{1}{4}(\epsilon_1+\eta_1)v^2 + \tfrac{\epsilon_2}{4}w^2 + (\lambda_1 - term),\\
\eta_{2,3}': &\quad M^2 + \tfrac{1}{4}(\epsilon_1+\eta_1)v^2 + \tfrac{\epsilon_2}{4}w^2 + (\lambda_1 - term),\\
\phi_1: &\quad M^2 + \tfrac{1}{2}\epsilon_1v^2 + \tfrac{1}{2}\epsilon_2w^2 + \tfrac{1}{2}\eta_2w^2 + \tfrac{\lambda_1}{2}u_iu_i,\\
\phi_{2,3}: &\quad M^2 + \tfrac{1}{2}\epsilon_1v^2 + \tfrac{1}{2}\epsilon_2w^2 + \tfrac{\lambda_1}{2}u_iu_i.
\end{aligned} \tag{14}$$

The mixing term looks like, apart from some common factor:

$$2(\epsilon_1+\eta_1)u_i\eta_i'v\eta + 2\epsilon_2u_i\eta_i'w\eta' + 2\eta_2u_1\eta_1'w\eta' + 2\lambda_2'w\eta'v\eta. \tag{15}$$

There are the other mixings (such the mixing inside $\eta_{1,2,3}'$), to be discussed later in this paper. To work out on "the origin of mass", we would drop out all "mass" terms to begin with.

Let us try to fix the notations further. See Ch. 13, Ref. For η' going through SSB, we have the following terms, neglecting the small λ'_2 term,

$$\frac{\mu_2^2}{2}(\eta'+w)^2 + (\frac{\epsilon_2}{4}u_iu_i + \frac{\eta}{4}u_1^2)(\eta'+w)^2 + \frac{\lambda_2}{4}(\eta'+w)^4. \tag{16}$$

SSB means that all the linear terms add up to zero, resulting the change in sign of the mass term, $\frac{1}{2}(2\lambda_2w^2(\eta')^2)$. Note that, for real fields, a factor of $\frac{1}{2}$ should be factored out.

The same applies to other SSB fields, even though the original minus signs are generated by other fields. This applies for the SM Higgs η.

In other word, the three "related" scalar (Higgs) fields $\Phi(1,2)$, $\Phi(3,2)$, and $\Phi(3,1)$ should be "equivalent" among themselves. The spontaneous symmetry breaking (SSB) is happening for all of them, actively or passively. Thus, λ, λ_1, and λ_2 should be the same. λ_2' should be zero or vanishingly small.

All the mass terms, including those SSB-driving terms, should be absent, apart from one SSB-driving term. The scalar field (ϕ_a) could affect a different scalar field (ϕ_b), if they are triplets under $SU_f(3)$, etc. It is easy to see that only one SSB-driving term is enough for all the three Higgs fields – there may be several SSB's for the neutral fields - in our case, it works for all of them.

We are working with the lagrangian, i.e., the energy that should be bounded from below ("positive definiteness"). The combination $(\phi_a^\dagger\phi_a + \phi_b^\dagger\phi_2)^2 - 4(\phi_a^\dagger\phi_b)(\phi_b^\dagger\phi_a)$ would work very well. This in fact helps to fix $\epsilon_{1,2}$ and $\eta_{1,2}$. As we argued before, the three λ's are the same.

As for the SSB-driving term, we decide to keep the purely family term, $\mu_2^2\Phi^\dagger(3,1)\Phi(3,1)$. As we shall see that the symmetry breaking would occur much earlier (in the history of the early Universe), or at a much higher temperature.

Thus, we write

$$\begin{aligned} V_{Higgs} &= \mu_2^2\Phi^\dagger(3,1)\Phi(3,1) + \lambda(\Phi^\dagger(1,2)\Phi(1,2) + cos\theta_P\Phi^\dagger(3,2)\Phi(3,2))^2 \\ &+\lambda(-4cos\theta_P)(\Phi^\dagger(\bar{3},2)\Phi(1,2))(\Phi^\dagger(1,2)\Phi(3,2)) \\ &+\lambda(\Phi^\dagger(3,1)\Phi(3,1) + sin\theta_P\Phi^\dagger(3,2)\Phi(3,2))^2 \\ &+\lambda(-4sin\theta_P)(\Phi^\dagger(\bar{3},2)\Phi(3,1))(\Phi^\dagger(3,1)\Phi(3,2)). \end{aligned} \tag{17}$$

These are two prefect squares minus the other extremes, to guarantee the positive definiteness, when the minus μ_2^2 was left out. (θ_P may be referred to as "Pauchy's angle".) ϵ_1, η_1, and ϵ_2, η_2 are expressed in λ, a great simplification. Note that we only include the interference terms between those involving the same group, $SU_f(3)$ or $SU_L(2)$; thus $\lambda_2' = 0$ (or, it should be close to zero).

We propose that the Higgs potential specified just above, V_{Higgs}, provides the definition of our Standard Model. The dimensionless coupling λ is universal for the various Higgs fields $\Phi(1,2)$ (the SM Higgs), $\Phi(3,2)$ (the mixed family Higgs), and $\Phi(3,1)$ (purely family Higgs).

From the expressions of u_iu_i and v^2, we obtain

$$v^2(3cos^2\theta_P - 1) = sin\theta_P cos\theta_P w^2. \tag{18}$$

And the SSB-driven η' yields

$$w^2(1 - 2sin^2\theta_P) = -\frac{\mu_2^2}{\lambda} + (sin2\theta_P - tan\theta_P)v^2. \tag{19}$$

These two equations show that it is necessary to have the driving term, since $\mu_2^2 = 0$ implies that everything is zero. Also, $\theta = 45^\circ$ is the (lower) limit.

The mass squared of the SM Higgs η is $2\lambda cos\theta_P u_iu_i$, as known to be $(125\ GeV)^2$. The famous v^2 is the number divided by 2λ, or $(125\ GeV)^2/(2\lambda)$. Using PDG's for e, $sin^2\theta_W$, and the W-mass [11], we find $v^2 = 255\ GeV$. So, $\lambda = \frac{1}{8}$, a simple model indeed.

The ratio of the VEV to its Higgs mass is determined by 2λ, whether the channel is not ignited or not. We might choose the channel of η' (the purely family Higgs) or that of η (the SM Higgs) as the ignition channel, but three Higgs channels have different labels. The three Lorentz-invariant scalar fields have different internal structures, an amusing question for further investigation.

The mass squared of η' is $-2(\mu_2^2 - sin\theta_P u_1^2 + sin\theta_P(u_2^2 + u_3^2))$. The other condensates are $u_1^2 = cos\theta_P v^2 + sin\theta_P w^2$ and $u_{2,3}^2 = cos\theta_P v^2 - sin\theta_P w^2$ while the mass squared of η_1' is $2\lambda u_1^2$, those of $\eta_{2,3}'$ be $2\lambda u_{2,3}^2$. The mixings among η_i' themselves are neglected in this paper.

There is no SSB for the charged Higgs $\Phi^+(3,2)$. The mass squared of ϕ_1 is $\lambda(cos\theta_P v^2 - sin\theta_P w^2) + \frac{\lambda}{2}u_iu_i$ while $\phi_{2,3}$ be $\lambda(cos\theta_P v^2 + sin\theta_P w^2) + \frac{\lambda}{2}u_iu_i$. (Note that a factor of $\frac{1}{2}$ appears in the kinetic and mass terms when we simplify from the complex case to that of the real field; see Ch. 13 of [6].)

A further look of these equations tells that $3cos^2\theta_P - 1 > 0$ and $2sin^2\theta_P - 1 > 0$. A narrow range of θ_P is allowed (greater than 45° while less than 57.4°, which is determined by the group structure). For illustration, let us choose $cos\theta_0 = 0.6$ and work out the numbers as follows: (Note that $\lambda = \frac{1}{8}$ is used.)

$$\begin{aligned} & 6w^2 = v^2, \quad -\mu_2^2/\lambda = 0.32v^2; \\ \eta: \quad & 2\lambda cos\theta_0 u_i u_i = (125\,GeV)^2, \quad v^2 = (250\,GeV)^2; \\ \eta': \quad & mass^2 = (51.03\,GeV)^2, \quad w^2 = v^2/6; \\ \eta_1': \quad & mass^2 = (107\,GeV)^2, \quad u_1^2 = 0.7333v^2; \\ \eta_{2,3}': \quad & mass^2 = (85.4\,GeV)^2, \quad u_{2,3} = 0.4667v^2; \\ \phi_1: \quad & mass = 100.8\,GeV; \qquad \phi_{2,3}: mass = 110.6\,GeV. \end{aligned} \tag{20}$$

All numbers appear to be reasonable. In the above, $cos\theta_0$ is the only free parameter until one of the family Higgs particles $\eta_{1,2,3}'$ and η' is found experimentally. Since the new objects need to be accessed in the lepton world, it would be a challenge for our experimental colleagues.

As a footnote, our Standard Model predicts that the mass of the SM Higgs η is a half of the vacuum expectation value v - a prediction in the origin of mass [10].

As for the range of validity, $\frac{1}{3} \leq cos^2\theta_P \leq \frac{1}{2}$. The first limit refers to $w^2 = 0$ while the second for $\mu_2^2 = 0$.

We may fix up the various couplings, using our common senses. The cross-dot products would be similar to κ, the basic coupling of the family gauge bosons. The electroweak coupling g is 0.6300 while the strong QCD coupling $g_s = 3.545$; my first guess for κ would be about 0.1. The masses of the family gauge bosons would be estimated by using $\frac{1}{2}\kappa \cdot w$, so slightly less than $10\,GeV$. (In the numerical example with $cos\theta_P = 0.6$, we have $6w^2 = v^2$ or $w = 102\ GeV$. This gives $m = 5\ GeV$ as the estimate.) So, the range of the family forces, existing in the lepton world, would be 0.02 *fermi*.

Furthermore, the λ_2' term [i.e., the interference between the spaces (3, 1) and (1, 2)] could be neglected safely in these discussions. But similar discussions to justify our Standard Model could be presented. There is one interesting aspect that the scalar fields, when having same $SU_f(3)$ or $SU_L(2)$ group structure, should be dealt with together, like in this paper.

The term that ignites the SSB is chosen to be with η', the purely family Higgs. This in turn ignites EW SSB and others. It explains the origin of all the masses, in terms of the spontaneous symmetry breaking (SSB). SSB in $\Phi(3,2)$ is driven by $\Phi(3,1)$, while SSB in $\Phi(1,2)$ from the driven SSB by $\Phi(3,2)$, as well. The different, but related, scalar fields can accomplish so much, to our surprise.

There is more reason that we could be optimistic. Besides the $SU_c(3)$ protection over the quark world, we now have another $SU_f(3)$ to protect the lepton world - the remaining part of the world in the Standard Model. The (conceptual) troubling QED Landau ghost is no longer there since QED alone is only part of the story; it is all asymptotically free, via $SU(3)$, and free of the ghosts. Maybe everything can be put together elegantly as a final complete and consistent theory.

On the experimental side, the family Higgs η_1', $\eta_{2,3}'$, and $\phi_{1,2,3}^+$ couple to only cross-generation leptons, making them difficult to observe. On the other hand, the purely family Higgs η' is rather elusive (maybe hopeless to observe). It is an ideal candidate particle for the so-called dark matter, but it may be long-lived using the scale of the star or galactic formation. These all happen in the lepton world, if the Standard Model is valid.

In the theoretical processes of figuring out how to write the Standard Model, or what is the best way to write it down, which would be the first to spell out (i.e. the "background"), etc., it came out with the attempt to phrase "a very precise definition of the Standard Model" [12], an ongoing exercise. Then, we realized that the word "built-in from the outset" [1], or "built-in from the very beginning", should be used. Eventually, it came to our minds that the 3 K cosmic microwave background (CMB) is a perfect experimental evidence. Altogether, it is essential to figure out these steps in detail - it should eventually lead us in the process of understanding the Nature.

In an early paper, Hwang and Yan [5] attempted to put six objects in the multiplet $\Psi_L(3,2)$. We were rather reluctant to do so, despite some important breakthroughs [13]. One should remember that the processes of searching for the Standard Model is non-orthodox, varying from a physicist to another. To be honest, we were not sure that a specific Standard Model would survive, or survive for how long. It is just amazing that the entire body of particle physics, or the Nature, can be summarized by the Standard Model.

3 Leptons: Point-like Dirac Particles

We proceed to discuss some phenomena in the lepton world. The lepton world is governed by the gauge group $SU_L(2) \times U(1) \times SU_f(3)$. The individual lepton numbers could be violated though the overall lepton number is strictly conserved.

The weak interactions involving leptons are determined by $-\bar{\Psi}(3,2)\gamma_\mu D_\mu \Psi(3,2)$ [with D_μ given by Eq. (5)] plus two curl terms [Eqs. (6) and (7)].

For the basic processes such as the muon decay, $\mu^- \to e^- + \bar{\nu}_e + \nu_\mu$, we may write, symbolically, the transition amplitude [6]:

$$iT = \frac{G_F}{\sqrt{2}} \bar{u}_e(p')\gamma_\lambda(1+\gamma_5)v_e(k') \cdot \bar{u}_{\nu_\mu}(k)\gamma_\lambda(1+\gamma_5)u_\mu(p) + others, \tag{21}$$

but this might be incorrect, since $u(p)$, $v(k)$, etc. are, by definition, on the mass shells. Neutrino oscillations tell us $u_{\nu_\mu}(k) \equiv U_{\mu i}u_i(k)$, with the left-hand side defined by "$\equiv$"; similarly for the antineutrino $v_e(k)$.

In fact, our language here is only for the mass-shell Dirac spinors, not for something which oscillates. So, we should write $\sum_i U_{ei}u^i(k)$ for the electron-like neutrino, etc., since $u^i(k)$'s are the mass-eigenstates - that is how we set up the Dirac equations for.

Similarly, for the muon or the electron, they should be on mass shells in our language - that is the way which we represent the muon or the electron. So, this seems against the off-diagonal mass terms - in fact, this requires the overall consistency check at least.

Thus, we should write, for the muon decay,

$$\begin{aligned} iT &= \frac{G_F}{\sqrt{2}} \bar{u}_e(p')\gamma_\lambda(1+\gamma_5)U_{ei}v_i(k') \cdot \bar{u}_j(k)U^\dagger_{\mu j}\gamma_\lambda(1+\gamma_5)u_\mu(p) \\ &+ \frac{G'}{\sqrt{2}} \bar{u}_e(p')(1-\gamma_5) \times U_{\mu j}v_j(k') \cdot \bar{u}_i(k)U^\dagger_{ei}(1-\gamma_5) \times u_\mu(p). \end{aligned} \tag{22}$$

Here the second term is due to the charged family Higgs exchange $\phi_1^\dagger$. (Our η_1', or $\phi_1^\dagger$, refers to the τ channel, by our convention.) So, G' [$\propto h^{C2}/m(\phi_1^\dagger)^2$] is presumably much smaller than the Fermi coupling G_F. The Fierz reordering could be used for this G' term.

The important point is that all the Dirac spinors are on the mass shells, so that the expression can be further simplified and calculated. Thus, the differential cross sections, or the decay rates, could be obtained. Note that, to this leading order, the inference term between the dominant muon-decay amplitude and the cross-product charged family-Higgs exchange term should not be there.

The other important example is the $\mu \to e$ conversions, say, $\mu^-(p)+p(q) \to e^-(p')+p(q')$, which goes through the vacuum expectation value (VEV) via Eq. (6) - the neutrino mass term $ih\bar{\Psi}_L(3,2) \times \Psi_R(3,1) \cdot \Phi(3,2) + h.c.$. The amplitude is given by

$$\begin{aligned} iT = & \frac{1}{(2\pi^4)} \int d^4k \bar{u}_e(p') \cdot i\frac{1}{2\sqrt{2}} \frac{e}{sin\theta_W} \cdot i\gamma_\lambda(1+\gamma_5) \\ & \cdot \frac{1}{i} \sum_j U^\dagger_{ej} \frac{m_j - i\gamma \cdot k}{m_j^2+k^2-i\epsilon} \cdot i \cdot \frac{i}{2} h \cdot (-) \cdot \frac{1}{2}(1-\gamma_5) \cdot u_i \frac{1}{\sqrt{2}} \cdot \\ & \cdot \frac{1}{i} \sum_l \frac{m_l - i\gamma \cdot k}{m_l^2+k^2-i\epsilon} U_{\mu l} \cdot i\frac{1}{2\sqrt{2}} \frac{e}{sin\theta_W} \cdot 1\gamma_\eta(1+\gamma_5) \cdot u_\mu(p) \\ & \cdot \bar{u}(q') \cdot i\frac{1}{2\sqrt{2}} \frac{e}{sin\theta_W} \cdot i\gamma_\eta(1+\gamma_5) \cdot \frac{1}{i} \frac{m - i\gamma \cdot (q+k-p')}{m^2+(q+k-p')^2-i\epsilon} \\ & \cdot i\frac{1}{2\sqrt{2}} \frac{e}{sin\theta_W} \cdot i\gamma_\lambda(1+\gamma_5) \cdot u(q) \cdot \frac{1}{i} \frac{1}{m_W^2+(k-p')^2-i\epsilon} \cdot \frac{1}{i} \frac{1}{m_W^2+(k+p)^2-i\epsilon}, \end{aligned} \tag{23}$$

in the 't Hooft-Feynman gauge. This means that the amplitude is finite but not complete, but, as seen below, all are rather small.

Simplifying it further, we obtain

$$\begin{aligned} iT = & \left(\frac{1}{2\sqrt{2}} \frac{e}{sin\theta_W}\right)^4 \cdot i(-)\frac{hu_1}{2\sqrt{2}} \cdot (-i \cdot 2)^2 \\ & \cdot \frac{1}{(2\pi)^4} \int d^4k \bar{u}_e(p') \gamma_\lambda \sum_{j,l} U^\dagger_{ej} \gamma \cdot k m_l U_{\mu l} \gamma_\eta (1+\gamma_5) u_\mu(p) \\ & \cdot \bar{u}(q') \gamma_\eta \gamma \cdot (q+k-p') \gamma_\lambda (1+\gamma_5) u(q) \\ & \cdot \frac{1}{m_j^2+k^2-i\epsilon} \cdot \frac{1}{m_l^2+k^2-i\epsilon} \cdot \frac{1}{m^2+(q+k-p')^2-i\epsilon} \cdot \frac{1}{m_W^2+(k-p')^2-i\epsilon} \cdot \frac{1}{m_W^2+(k+p)^2-i\epsilon}. \end{aligned} \tag{24}$$

Comparing this amplitude to the dominant muon-decay amplitude, it is down by $u_1 m_l$ with the tiny neutrino mass m_l (and $u_1 = 107\,GeV$ in the above table) - about $10^{-9} GeV^2$. The remaining reduction is from the 2nd-order weak interaction - a factor of 10^{-5}. So, normally, we expect 10^{-28} smaller than the dominant modes.

We have discussed the $\mu \to e+\gamma$ before [5]. The decay rate is suppressed by the neutrino mass effect $u_1 m_l$ and further by gauge invariance (for a real photon). The mode turns out to be smaller than the current search limits by more than 20th order away (completely negligible).

To sum up, the modifications in the ordinary muon decays may be observed - it defines a new category of the high-precision experiments. The $\mu \to e$ conversions such as $\mu^- + p \to e^- + p$ and $\mu \to e + \gamma$ turn to be hopelessly small.

Nevertheless, in certain decays such as ${}^3H \to {}^3He + e^- + \bar{\nu}_e$, we have to rewrite the neutrino state as a sum of neutrino-mass eigen-states - as required by neutrino oscillations. These are another category of high-precision experiments.

In a short summary, we do expect to see some corrections in the ordinary muon decay $\mu^- \to e^- + \bar{\nu}_e + \nu_\mu$ but the branching ratios induced by the $\nu_\mu \to \nu_e$ conversion or vice versa, such as $p + \mu^- \to p + e^-$ or $\mu \to e + \gamma$, turn out to be rather small. In the muon decay or nuclear beta decays, such as ${}^3H \to {}^3He + e^- + \bar{\nu}_e$, the replacement, as required by neutrino oscillations, of flavor states by the mass eigen-states does produce some tiny observable effects which might be detected in the precision experiments of the next generation.

Besides the three Higgs, the primary prediction of our Standard Model is the existence of the force of a new kind - i.e., the family force mediated by the family gauge bosons. As said above, we could use $\frac{1}{2}\kappa w$ as an estimate of the mass(es) of the family gauge bosons. Our first guess is for some feeble force - $\kappa = 0.1$. The above numerical example corresponds to $w = 102\ GeV$, so as to the family gauge boson mass of $5\ GeV$.

The family gauge bosons would then be in the vicinity of $5\ GeV$ or nearby, or the range of 0.04 *fermi*. Or, 0.04×10^{-13} cm in the effective range, between leptons (such as two electrons or an electron-positron pair) is too short to be detected for the entire atomic physics or the entire chemistry.

The precision experiments such as $g-2$ would eventually detect the residual family effects, since the existing $g-2$ calculations [14] is so far the QED calculation and should be completed by inclusion of other effects with the emphasis on family gauge bosons. We are looking forward to prospects in this direction.

Of course, we need to examine the precision part of atomic physics when the story becomes clear; even though the effects are tiny, the evolutions usually come from the tiny effects to begin with.

4 The Standard Model of All Centuries

We are suggesting [1] that we live in the quantum 4-dimensional Minkowski space-time with the force-fields gauge-group structure $SU_c(3) \times SU_L(2) \times U(1) \times SU_f(3)$ built-in from the outset. The quark world, with its (123) symmetry, is accepted by this background. Meanwhile, the lepton world, with another (123) symmetry, is also accepted.

This is the simple statement of the "Standard Model". The observed $3\ K$ cosmic microwave background (CMB) provides the evidence of why the force-fields gauge-group structure $SU_c(3) \times SU_L(2) \times U(1) \times SU_f(3)$ was built-in at the very beginning as the Lorentz group.

This $SU_f(3)$ is the "family" gauge theory (i.e., a force-fields group), rather than the strong-interaction "flavor" $SU_f(3)$ - we have to distinguish, with care, the terminology if needed. The gauge group $SU_L(2) \times U(1) \times SU_f(3)$ protects the lepton world from the Landau ghost since the asymptotic free nature of $SU_f(3)$ would set in much earlier on as $Q^2 \to \infty$ or $r \to 0$.

Throughout the 20th Century, we did not realize an important property of the complex scalar fields - the importance of the dimensionless coupling $\lambda(\phi^\dagger\phi)^2$ in the 4-dimensional Minkowski space-time.

In the 4-dimensional Minkowski space-time, the complex scalar field $\phi(x)$ is described, generally and renormalizably, by

$$\mathcal{L} = -(\partial_\mu\phi)^\dagger(\partial_\mu\phi) - M^2\phi^\dagger\phi - \lambda(\phi^\dagger\phi)^2. \tag{25}$$

If $\lambda < 0$, the system collapses (i.e. unbounded from below). If $\lambda > 0$, it is self-repulsive so that the system cannot build up by itself. The interesting question is that λ is dimensionless in the 4-dimensional Minkowski space-time. A pure number that characterizes the 4-dimensional Minkowski space-time (maybe $\lambda = \frac{1}{8}$, but we need the proof), *not* by the complex scalar fields.

The force fields are described by the gauge fields in the group $SU_c(3) \times SU_L(2) \times U(1) \times SU_f(3)$. The longitudinal components are missing in the purely gauge-fields description so that complex scalar fields are needed in the context of the Higgs mechanisms. In the $SU_c(3) \times SU_L(2) \times U(1) \times SU_f(3)$ 4-dimensional Minkowski space-time (i.e., the "background" of our World), it is already so set up in the various force fields - using the complex scalar fields for the various gauge fields via the specific Higgs mechanisms to form the overall "background" of our World.

It yields, and only yields, three Higgs fields $\Phi(1,2)$, $\Phi(3,2)$, and $\Phi(3,1)$. These "related" Higgs fields, such as $\Phi(1,2)$ and $\Phi(3,2)$ with $SU_L(2)$ doublets, can barely overcome the "self-repulsive" nature and can *exist*. And so they are able to *exist* in this World.

Let's turn our attention to the lepton world, each of which is typically of atomic sizes: We know that they couple to the $SU_L(2) \times U(1)$ gauge sector (which makes them visible). To interpret the ordering via three generations, we proposed the force-fields nature of the $SU_f(3)$ gauge sector. Neutrino oscillations provide a direct proof that generations can switch among themselves. By introducing the $SU_f(3)$ gauge sector to the original $SU_L(2) \times U(1)$, the lepton world is free from the Landau ghost and is asymptotically free. So, the lepton world is accepted. Atoms, molecules, etc., are examples of the lepton world.

How about the quark world? It is already a perfect world since it couples to the $SU_c(3) \times SU_L(2) \times U(1)$ [i.e., the standard (123)] gauge sector. It exhibits the size-confinement effect, i.e., that the quark world exists only within a given volume; or, it exhibits the temperature effect, i.e., that it undergoes the phase transition (into something else). The quark world is so much different from the lepton world, that the two $SU(3)$ are so much different.

The natural size of the quark world, in view of the confinement, is much smaller than the lepton world. Using the common scales, it is comparing the fermi, or nuclear, scale (for the quark world) to the atomic scale (for the lepton world).

Thus, we could declare that we live in the quantum 4-dimensional Minkowski space-time with the force-fields gauge-group structure $SU_c(3) \times SU_L(2) \times U(1) SU_f(3)$ built-in from the outset [10, 1]. Neutrino oscillations, via Eq. (6), are there. The story of how the building blocks of matter, i.e., the basic units of matter, build up the entire matter world is complete - all the couplings in the lepton world, as well as in the quark world, are dimensionless in the 4-dimensional Minkowski space-time; they are pure numbers, determined *globally* by the 4-dimensional Minkowski space-time; *strange enough.*

5 Concluding Remarks

This world is, indeed, very special. It is based on the quantum 4-dimensional Minkowski space-time with the force-fields gauge-group structure $SU_c(3) \times SU_L(2) \times U(1) \times SU_f(3)$ built-in from the very beginning. We realize that each complex scalar field, by itself, would be self-repulsive and so *cannot exist* if alone and that two "related" complex scalar fields could interact attractively (and so exist) and they become the very-much-wanted longitudinal components of the gauge fields. The quark world, of nuclear sizes, would be accepted because of the (123) symmetry. Meanwhile, the lepton world, of atomic sizes, could be accepted in view of another (123) symmetry.

The 3 K cosmic microwave background (CMB) in our Universe is the evidence that our Universe is the quantum 4-dimensional Minkowski space-time *plus* the force-fields gauge-group structure built-in at the outset.

In this language, everything carries the group characteristic. For example, $\nu_{e,L}$ belongs to the same multiplet as e_L, so the same characteristic - thus, the case would be ruled out that the electron is a Dirac particle while the neutrino is a Majorana particle.

Back in 2011 [15], there were a lot of speculations on what may be beyond the (old) Standard Model. It was proposed that we could work with two working rules: "Dirac similarity principle", based on eighty years of experience, and "minimum Higgs hypothesis", from the last forty years of experience. Using these two working rules, the basic model [7] became rather unique in our choice - so, it is so much easier to check it against the experiments. To move forward in building up our knowledge, there are moments that we have to play conservatively - including the moments of using these two working rules.

At this point, the two working rules seem to be rather trivial.

We have to say that the phenomenon of three generations is one leading puzzle in particle physics; nowadays, it is safe to add that neutrino oscillations is another related puzzle. When we write everything together, the need for introducing the Higgs $\Phi(3,2)$ becomes rather clear [5]. As this world does not have another massless gauge bosons (unless confined like gluons), it is also clear that there is another pure family Higgs $\Phi(3,1)$ to complete the story [8]. Is there any other possibility? It seems that it is unique.

We have to admit that the natural consequence is the $SU_f(3)$ family gauge theory - to put everything together, it leads to the (extended) Standard Model [7], with a very natural generalized Higgs mechanism for the three Higgs. It is in accord with the "minimum Higgs hypothesis". These scalar fields are meant together and interact together, and nothing more.

In fact, the Standard-Model Higgs field $\Phi(1,2)$ alone *cannot exist*, due to its self-repulsiveness. So, $\Phi(3,2)$, $\Phi(3,1)$ those family Higgs fields should be there to make all three enough mutual attractions to overcome the individual self-repulsions. And all family gauge bosons are massive in the process.

It is also of importance to point out that there is no experimental evidence to the assertion that neutrinos are point-like Dirac particles - "Dirac similarity principle" should be in experimental checks. Of course, they are in the same multiplet, like in [5], so that they both are Dirac particles of the same nature. So, the two working principles are justified by our experiences for eighty or forty years. Surprising enough, it seems to work this way, at least so far.

Thus, we assume that our world is the quantum 4-dimensional Minkowski space-time with the $SU_c(3) \times SU_L(2) \times U(1) \times SU_f(3)$ force-fields gauge-group structure built in from the outset; in our world, the various Higgs "exist" such that all gauge bosons, except the photon, are either confined or massive - this provides the background to support the quark world as well as to support the lepton world. This is the origin of "point-like particles", or of "fields" [1].

To close up our discussions, there remain three philosophical questions in our minds: First, why are the complex scalar fields so special in the 4-dimensional Minkowski space-time? The 4-dimensional Minkowski space-time (not the complex scalar fields themselves) would be the reason to determine the value of λ ($= \frac{1}{8}$?). In the other dimensions (different from four), the existence of the self-interaction $\lambda(\phi^\dagger\phi)^2$ would not have the same status. Second, why do we need the ignition channel, which turns out to be in the purely family channel ($\mu_2^2 < 0$)? Third, the dimensional regularization, or others, might not give whole story regarding the (leading) ultraviolet divergences. In the beginning of this 21th century, we might have this ghost (infinite or divergence) story to resurface again [1, 16]. But this may be destined to be so, as our knowledge accumulates in the process.

We may add that, under two working hypotheses or under our Standard Model (in the 4-dimensional Minkowski space-time with the $SU_c(3) \times SU_L(2) \times U(1) \times SU_f(3)$ gauge-group structure built in from the outset), we should be able to close the Universe; that is, all the dark-matter particles and all the ordinary-matter particles are accounted for. Our Standard Model provides a description of the entire matter world - i.e., the 25% dark-matter world and the 5% ordinary-matter world.

In our language, the *a priori* world (the background) is the quantum 4-dimensional Minkowski space-time with force-fields gauge-group structure built-in from the outset. The quark world and the lepton world are something added on at a later stage, and in fact may be the added-on at a different stage - that makes the study of leptons a subject by itself [17]. So, this study would lead us to gain understanding of the dark matter particles, such as the (3,1) purely family Higgs and $SU_{family}(3)$ family gauge bosons, and cosmic background (CB) ν's.

We would be curious about how the dark-matter world looks like, though it is difficult to verify experimentally. The first question would be: The dark-matter world, 25 % of the current Universe (in comparison, only 5 % in the ordinary matter), would clusterize to form the dark-matter galaxies, maybe even before the ordinary-matter galaxies. The dark-matter galaxies would then play the hosts of (visible) ordinary-matter galaxies, like our own galaxy, the Milky Way. Note that a dark-matter galaxy is by our definition a galaxy that does not possess any ordinary strong and electromagnetic interactions (with our visible ordinary-matter world). Without any other significant information as the input, we tend to think that the dark-matter galaxies might take slightly longer periods (time) and at somewhat larger scales (size), in the structural formation.

These would be the scenarios if we don't know what the dark matter are. Everything comes from the most probable, pending that our existing knowledge is. That is, we should base on the Standard Model to figure out the final answers and the Standard Model should be the Standard Model of all centuries.

We may offer a few thoughts on this fundamental question. The Standard Model [7] is unique. It is dimensionless in the 4-dimensional Minkowski space-time - it is dimensionless in the lepton world, it is dimensionless in the quark world, and, except the "ignition" term, it is also dimensionless in the Higgs and gauge sector. "Dimensionless" means that all the couplings are determined *globally* by the 4-dimensional Minkowski space-time; they are the property of the 4-dimensional Minkowski space-time. Because of all these, the question of ultraviolet divergences should be looked upon in a different manner.

Thus, for the following centuries, it is unlikely that the Standard Model will be replaced by something else. When we say this, we have in mind the Newton's classic era of physics.

In the Standard Model, neutrinos and antineutrinos are the *only long-lived* dark-matter particles [18]. This means that the 25% dark matter are the bulks of neutrinos and antineutrinos; they have the masses, with the heaviest of $0.058\,eV$, so they will be clustered in lumps or in neutrino halos. But these neutrino halos could not form the gravitational "centers" so each of them has to have the *visual* gravitational center.

What are the *visual* gravitational centers? On this question, we should remind ourselves that, in our ordinary-matter world, those quarks can aggregate in no time, to hadrons, including nuclei, and the electrons serve to neutralize the charges also in no time. Then atoms, molecules, complex molecules, and so on. These serve as the seeds for the clusters, and then stars, and then galaxies, maybe in a time span of $1\,Gyr$ (i.e., the age of our young Universe). The aggregation caused by strong and electromagnetic forces is fast enough to help giving rise to galaxies in a time span of $1\,Gyr$. Thus, we should rule out the extra-heavy dark-matter particles as the gravitational centers.

6 The Episode

It is clear that there are so many unending stories, at this beginning of the 21st Century. I joined the efforts of my co-author, Ta-You Wu, to finish a book called "Relativistic Quantum Mechanics and Quantum Fields", but I didn't realize until these years that "Relativistic Quantum Mechanics and Quantum Fields" is *the language* for *the Standard Model of All Centuries*. It is the language to re-write the entire Newton's classic era. For the obvious reason, I'm satisfied.

References

[1] W-Y. Pauchy Hwang, The Universe, **3-1**, 3 (2015). "The Origin of Fields (Point-like Particles)". Note that we have adopted a slightly modified format for the last chapter of this textbook. The Standard Model of particle physics is what we believe to be the standard model in the 21st Century.

[2] T. Yanadida, Phys. Rev. **D20**, 2986 (1979).

[3] Yue-Liang Wu, Phys. Lett. **B 714**, 286 (2012).

[4] W-Y. Pauchy Hwang, The Universe, **3-2**, 11 (2015); "Why is Our Space-Time 4-Dimensional?".

[5] W-Y. Pauchy Hwang and Tung-Mow Yan, The Universe, **1-1**, 5 (2013).

[6] Ta-You Wu and W-Y. Pauchy Hwang, "Relatistic Quantum Mechanics and Quantum Fields" (World Scientific 1991); this is the early version of this book.

[7] W-Y. Pauchy Hwang, arXiv:1304.4705v2 [hep-ph] 25 Aug 2015; "The Standard Model".

[8] W-Y. Pauchy Hwang, Nucl. Phys. **A844**, 40c (2010); *ibid.*, International J. Mod. Phys. **A24**, 3366 (2009); *ibid.*, Intern. J. Mod. Phys. Conf. Series **1**, 5 (2011); *ibid.*, American Institute of Physics 978-0-7354-0687-2/09, pp. 25-30 (2009).

[9] W-Y. P. Hwang, Phys. Rev. **D32**, 824 (1985).

[10] W-Y. Pauchy Hwang, The Universe, **2-2**, 47 (2014). "The Origin of Mass".

[11] Particle Data Group, "Review of Particle Physics", J. Phys. G: Nucl. Part. Phys. **37**, 1 (2010); and its biennual publications.

[12] W-Y. Pauchy Hwang, arXiv:1409.6077 [hep-ph] 22 Sep 2014. "A precise definition of the Standard Model".

[13] W-Y. Pauchy Hwang, arXiv:1207.6443v1 [hep-ph] 27 Jul 2012; *ibid.*, Hypefine Interactions **215**, 105 (2013); *ibid.*, arXiv:1209.5488v1 [hep-ph] 25 Sep 2012.

[14] T. Kinoshita and W.B. Lindquist, Phys. Rev. Lett. **47**, 1573 (1981).

[15] W-Y. P. Hwang, arXiv:11070156v1 (hep-ph, 1 Jul 2011), Plenary talk given at the 10th International Conference on Low Energy Antiproton Physics (Vancouver, Canada, April 27 - May 1, 2011).

[16] W-Y. Pauchy Hwang, The Universe, **2-1**, 41 (2014).

[17] W-Y. Pauchy Hwang, arXiv:1409.6296v1 [hep-ph] 22 Sep 2014. "The Family Collider".

[18] W-Y. Pauchy Hwang, arXiv:1012.1082v4 [hep-ph] 13 Jan 2016; "Cosmology: Neutrinos as the *Only* Final Dark Matter".

Exercises: *Chapter 15*

1. Discuss the gauge principle. Why should it be listed as a "principle"?

2. Discuss the principle of renormalizability for the coupling system of one Dirac field, one $U(1)$ gauge field, and one complex scalar (Higgs) field, in the quantum 4-dimensional Minkowski space-time. In your discussions, try to go beyond a couple of non-renormalizable terms in the lagrangian.

3. Discuss the origin of "mass". The Standard Model discussed in this chapter provides an excellent example that, apart from the term for *SSB*, there is no mass term at all. But the Standard Model describes our world, or our Universe. This should be regarded as a clue of understanding the origin of "mass".

Chapter 16. Neutrinos in the Cosmos

Neutrinos (including three different flavors, and antineutrinos) are particles of the most mysterious. In the Standard Model, it is the *only* long-lived particle. Neutrino oscillations among the three different flavors does take place.

Neutrinos have masses those are so tiny among all particles (apart the massless photons). In other words, it should cluster under the influence of the gravitational force, forming the chunky neutrino halos, $1-1$ correspondence with respect to the visual ordinary-matter objects, such as the Earth, the Venus, the Sun, and the Stars.

In 1983, Eugene Commins wrote a textbook with the ending chapter on neutrinos. We admire Commins' courage for doing so, trying to do the same on this textbook.

Efforts are under way to detect ultra high energy (UHE) neutrinos, $\nu(UHE)$. However, the reaction $\nu(UHE) + \bar{\nu}(free) \rightarrow e^- + e^+$ has the threshold at $10^{13}\,eV$ on the UHE neutrino, assuming a neutrino of mass $0.058\,eV$, the upper bound as suggested by the recent experiments on neutrino oscillations. This serves the major hindrance for the ultra high energy neutrinos, as neutrino halos, arising from the clustering of cosmic background neutrinos (CBν's), are taken into account. Owing to the 25% dark matter as compared to 5% visual ordinary matter, it is of utmost importance to "verify experimentally" whether neutrino halos account for this 25% dark matter. According to the Standard Model realized at the beginning of the 21st Century, there are no other "natural" dark-matter candidates other than neutrino halos.

The behaviors of the associated neutrino halos near visual heavy objects, such as stars, planets, etc., need to be understood in further details. The neutrino halo of a given object, such as the Earth, has to stake up in the momentum space, due to Pauli's exclusion principle, giving rise to the Fermi-Dirac sphere. Thus, the "presumed" formation of a black hole of visual ordinary matter is stopped by the invisible "incompressible" dark-matter neutrino halo - "incompressible" due to Pauli's exclusion principle (i.e., a manifestation of the quantum principle).

Therefore, the mass effect, due to the tiny largest neutrino mass of $0.058\,eV$, writes the last chapter of the story on the Standard Model.

1 Prelude

Could the neutrinos or antineutrinos travel freely in our Universe? Could the interactions with the "believed" cosmic background $1.9\ K$ neutrinos be completely negligible? We know that the cosmic background neutrinos would be of $1.9\ K$ if uniformly distributed [1], but neutrinos are now known to have masses so that they should cluster and hence deviate from the uniform $1.9\ K$ distribution.

Newton's gravitational law states that the force between two objects of mass m_1 and m_2 at the distance r is

$$force = G_N \frac{m_1 m_2}{r^2}, \tag{1}$$

with the gravitational constant G_N. In Einstein's general relativity, the law becomes an equation in differential geometry.

Our first point is as follows: With such a tiny gravitational constant G_N, the objects in question must be macroscopic, that is, that each object would be (many) moles of the smallest units of matter. The masses m_1 and m_2 of the objects would be at least 10^{24} molecules. For planets or stars, it would be $10^{24} \cdot 10^{36}$ molecules (e.g., a star of five solar mass) so, it goes through many complexities if enumerating from the very simple. Thus, Newton's gravitational law is a macroscopic law, *not* a microscopic law. As a macroscopic law, it has to build up from all the involved complexities and the simplicity must be spelled out in terms of symmetries.

The second important point has to do with the presence of the 25% dark matter. Our view is that these dark-matter particles are cosmic background (CB) ν's (i.e., neutrinos and antineutrinos, to be abbreviated simply as "neutrinos"). In the ideal case, each visual ordinary-matter macroscopic object, such as the Earth or the Venus, should carry five times in weight the neutrino halo. The reason for that is from the simplicity of the Newton's gravitational law (from Einstein's general relativistic law). *To sum up, a re-scaling of G_N by a factor of six (five plus self) is needed, if the dark matter of five times can be consistently taken into account.*

We are so much accustomed to the unified view that we always put the gravitational force at the same level of the strong and electroweak forces. But it is difficult to ignore the obstacles posed by all the complexities of the macroscopic law - Newton's gravitational law should be one of them.

Nowadays it is firmly established that there are three generations of quarks and three generations of leptons, at the level of the so-called "point-like Dirac particles". According to another newly established belief in Cosmology, the content of the current Universe would be 25 % in the dark matter while only 5% in the ordinary matter, the latter described by the "minimal Standard Model" (mSM). In this language, the dark-matter particles are supposed to be described by a general Standard Model. Thus, there is certain urgent need to look for a general Standard Model.

Specifically, we suggest that we live in the quantum 4-dimensional Minkowski space-time with the force-fields gauge-group structure $SU_c(3) \times SU_L(2) \times U(1) \times SU_f(3)$ built-in from the outset. In this background, the quark world is accepted because of the (123) symmetry (i.e., under $SU_c(3) \times SU_L(2) \times U(1)$), while the lepton world is accepted in view of the other (123) symmetry (i.e., under $SU_L(2) \times U(1) \times SU_f(3)$). This gives us "our world" [2].

Here $SU_f(3)$ stands for the $SU_{family}(3)$ family gauge theory, rather than $SU_{flavor}(3)$ which derives from the isospin symmetry - the latter *not* a gauge theory.

Soon after the hot Big Bang, the particles of different kinds were manufactured with the neutrinos (abbreviated for "neutrinos and antineutrinos") as one of the central products. Even though they are oscillating from one kind into another, the neutrinos are basically "stable". In fact, neutrinos are only long-lived dark-matter particles [3], in the Standard Model. In view of the weak interactions which they participated, these neutrinos were decoupled slightly earlier than the 3 K cosmic microwave background (CMB), that is why it was cooled longer, expected at 1.9 K [1]. But these cosmic background neutrinos (CBν's), owing to the tiny mass, should cluster, into the neutrino halos, around the visual ordinary matter objects such as planets, stars, etc. The non-zero tiny masses help to re-write the story.

These CBν's are anticipated to form neutrino halos, forming the so-called dark-matter halos, around the visual ordinary-matter objects, such as planets, stars, etc. - all because of the Newton's gravitational force due to the fact that neutrinos have masses [3]. According to Newton's gravitational law, the object being exerted would be independent of the object's mass - so the neutrino's mass of $0.058\, eV$, though tiny, is irrelevant so long as the neutrino has mass. This is one mystery associated with the Newton's gravitational law.

The other important point is that the candidate for the 25% dark matter is neutrinos and antineutrinos, if we look at the early history of our Universe [4]. There is no other long-lived dark-matter particle, beside CBν's.

We may treat the eight major planets and the Sun having the same origin - having similar dark-matter (neutrino) halos. For a system that was formed very early in our Universe history, we believe that the possibility of the neutrino halos getting rip off would be rather small.

We would assume that the eight major planets and the Sun, and other large systems, all have their own neutrino halos. The reason is due to both that CBν's were there before the creation of the first stars and that neutrino halos need their own heavy centers. Basically, a neutrino halo is a feebly-interacting fermi gas, which was "attached" to the ordinary-matter chunk, when the latter was formed. It is the magic of the Newton's gravitational law.

The Earth also has the invisible neutrino halo, according to our view. But we have to go to the Venus or the Mercury to detect the feeble effect arising from their neutrino halos - basically, the effect to be detected is smaller than the feeble backgrounds which we could think of. We are not so sure if the Venus or the Mercury would fulfill our criterions. But we have to try.

Therefore, for the Venus or the Mercury, we assume that the neutrino halos would be (2 - 3) times the radius of the "mother" planet and, according to the 25% dark matter versus the 5% visual ordinary matter, its "weight" is five times the weight of the mother planet.

In what follows, we do estimates only for the Venus. There are different reasons why we keep the Mercury in the picture - since there may be no day-night difference (no self-rotation, or such effects small), no atmosphere presence, etc. Of course, the proximity to the Earth will make the "experiments" more feasible.

For the Venus, we have the following basic information:

$$R = 6,051\, Km, \quad M = 4.87 \times 10^{24} Kg, \quad T = 224.7\, days. \tag{2}$$

These numbers are quite close to those for the Earth, the twin brother ($R = 6378\, Km$, $M = 5.794 \times 10^{24} Kg$).

Assuming, for CBν's, that the five times of the mass distributed uniformly over 2.5 times over the radius, we obtain

$$\rho_{DM} = 1.68\, g/cm^3 = 5.61 \times 10^{32}\, (eV/c^2)/cm^3. \tag{3}$$

There would be a total of 10^{34} neutrinos of the largest mass $0.058\, eV$; divided by six (3 flavors, plus antiparticles), etc.

We'll use this number to get our final estimate.

For us the earthlings, the Mercury, the Venus, the Earth, and the Mars, why only on Earth are there living things? Soon or later, we hope that we could understand the origins of these planets. Why this difference? In fact, we might be able to understand these questions through the interactions of solar neutrinos shining for over billions of years - the messengers that could go very deep into these planets and yet change the species at the nuclear level. For these very reasons, we should spend time trying to investigate these questions - let's call it "Extraterrestrial Solar Neutrino Physics" [5].

As pointed out earlier [3], there is a huge neutrino halo associated with a galaxy. Locally, it would follow the distribution of the visual world, according to Newton's gravitational law. Thus, the Sun would be a local center of the CBν's, and, to be sensible, eight planets the eight centers, since the neutrino halos shouldn't be so huge because of the gravitational force (even for the neutrino halos as the relativistic Fermi gas).

PRL **113,** 121101 (2014) PHYSICAL RE\

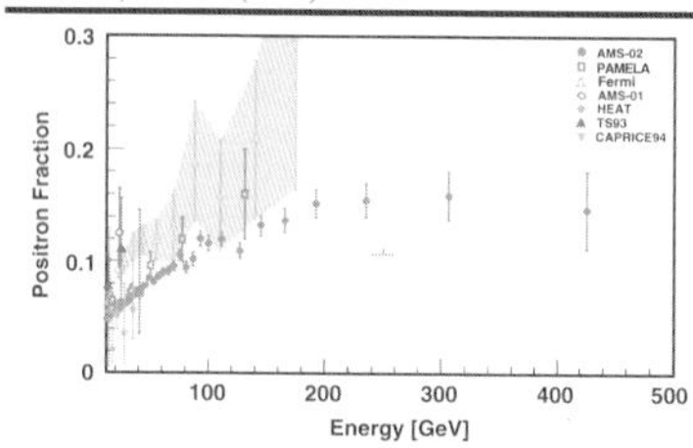

FIG. 3 (color). The positron fraction above 10 GeV, where it begins to increase. The present measurement extends the energy range to 500 GeV and demonstrates that, above ~200 GeV, the positron fraction is no longer increasing. Measurements from PAMELA [21] (the horizontal blue line is their lower limit), Fermi-LAT [22], and other experiments [17–20] are also shown.

Figure 1: The positron data from the AMS paper.

We can observe $\nu(Solar) + \bar{\nu}(CB; k_F) \rightarrow e^- + e^+$ using solar neutrinos, on the Space Station which are $400\,Km$ above the ground. The AMS experiments, as headed by Samuel Ting [6], provide the perfect example, with the results shown in the figure. Along the same line, it should be feasible to do Satellite experiment on the Venus or on the Mercury.

Thus, the existence of the neutrino mass implies that the CBν's cluster according to the local environments, according to Newton's gravitational law. Therefore, the detection of the CBν's might be feasible on the Earth, on the Venus, or on the Mercury. The existence of the CBν's is one of the fundamental issues that we, all physicists, are facing these days.

2 The Standard Model of All Centuries

In light of the Standard Model of all centuries [7, 2], the interest in the global behaviors of the CBν's is tremendous. The Standard Model [7] is basically the description of the *smallest* units of matter, such as electrons, neutrinos, quarks, etc. (a finite number of them), basing on the Einstein's relativity principle and the quantum principle. We name it as "of all centuries" since it could be there a thousand years from now. We imagine that this Standard Model would replace the thoughts of Newton's classic era, sooner or later.

In the beginning of the 21st Century, we could declare that we are in fact living in the quantum 4-dimensional Minkowski space-time with the force-fields gauge-group structure $SU_c(3) \times SU_L(2) \times U(1) \times SU_f(3)$ built-in at the very beginning, this defines the overall "background". The "background" can see the lepton world, of atomic sizes, which has the symmetry under $SU_L(2) \times U(1) \times SU_f(3)$. It can also see the quark world, of nuclear sizes, which has the well-known (123) symmetry. Altogether, it is called "the Standard Model" [7].

The force fields are realized by the gauge-group structure $SU_c(3) \times SU_L(2) \times U(1) \times SU_f(3)$ and are used to express the generalized gauge principle, as the changes from a space-time point x_μ to a neighboring point $x_\mu + \delta x_\mu$ need the forces to accomplish the goal. The various $\partial_\mu - D_\mu$ helps us for this moving and, meanwhile, it defines the kinetic energy uniquely.

The observed persistent existence of the 3 K cosmic microwave background (CMB) in our Universe is the *evidence* of that the force-fields gauge-group structure is "built-in from the very beginning". The CMB is rather uniform, to the level of one part in 10^5, due to the massless of the photons.

Theoretically, we need the family $SU_f(3)$ symmetry for the lepton world to make sure that the lepton world is asymptotically free and is free of Landau ghosts. Experimentally, it explains the three-generation problem and it also explains neutrino oscillations [7].

The "peculiar" characteristics are as follows: All the couplings in the lepton world are dimensionless in the 4-dimensional Minkowski space-time. All the couplings in the quark world are also dimensionless. Apart from the "ignition" term, all the couplings are dimensionless in the gauge and Higgs sector (forming the overall "background").

"Dimensionless-ness" means that it is the property of the 4-dimensional Minkowski space-time. It is truly "critical" in the quantum 4-dimensional Minkowski space-time as all these couplings would *not* be there if the dimension would be displaced by an infinitesimal amount from the integer four. So, we believe that these dimensionless couplings are determined *globally* by the 4-dimensional Minkowski space-time.

In this Standard Model [7], the neutrinos are the *only long-lived* dark-matter particles [4]. Yet, the theory excludes any additional particles from entering the theory thus, it is complete as a theory.

Thus, we should base on this Standard Model, in analyzing the various basic particle-physics problems.

3 $\nu(Solar) + \bar{\nu}(CB; k_F^\nu) \to e^- + e^+$ as the Evidence

First of all, we may make some crucial estimates on $\nu(Solar) + \bar{\nu}(CB; k_F^\nu) \to e^- + e^+$ on the Space Station 400 Km above the ground [6].

We wish to point out [8] that the reaction $\nu(Solar) + \bar{\nu}(CB) \to e^- + e^+$ cannot take place if the target antineutrino is *free* unless $\nu(Solar)$ is replaced by a neutrino with the energy near the threshold energy of $10^{13}\, eV$. So, the reaction proceeds if the target antineutrino comes from the surface of the Fermi-Dirac sphere of the neutrino halo of the Earth; that is, the energy of the antineutrino of $600\, GeV$, the Fermi energy of the system.

Thus, we try to reconstruct our two cases [8], viz.: (1) The target antineutrino is *free* such that the antineutrino could be set to at rest. And (2) the target antineutrino is of the Fermi energy $k_F \sim 600\,GeV$ while the beam neutrino comes from the Solar neutrinos.

On the other hand, we know that the largest mass of ν's is about $0.058\,eV$. We wish to clarify the importance of the Fermi-Dirac sphere of the neutrino halo [8].

The Conclusions: (1) In the first case, the neutrino mass of $0.058\,eV$ forces that the energy of the beam neutrino (i.e., the reaction threshold energy) has to be bigger than $10^{13}\,eV$ so to have $\nu(beam) + \bar{\nu}(free) \to e^- + e^+$ to take place. Otherwise, (2) $\nu(Solar) + \bar{\nu}(CB; k_F^\nu) \to e^- + e^+$, with MeV solar neutrino energies should be adequate (i.e., having enough energy) to take place, in view of the target antineutrino near the surface of the Fermi-Dirac sphere (of the Earth, or of the Venus).

The simple algebra on the kinematics is as follows [8]: We write, for the reaction $\nu + \bar{\nu}(free) \to e^- + e^+$,

$$\nu(p) + \bar{\nu}(p') \to e^-(p_e) + e^+(p_e'). \tag{4}$$

Using the *free* kinematics,

$$k_\mu = (\vec{k}, ik_0), \qquad k'_\mu = (0, im_\nu);$$
$$p_\mu = (\frac{\vec{k}}{2} + \vec{\delta}, iE), \qquad p'_\mu = (\tfrac{\vec{k}}{2} - \vec{\delta}, iE). \tag{5}$$

Let's compute the CM energy squared:

$$s \equiv (p_\mu + p'_\mu)^2, \tag{6}$$

or,

$$s = (k_\mu + k'_\mu)^2. \tag{7}$$

We obtain the threshold energy as follows:

$$k_0 = \frac{1}{m_\nu}(2m_e^2 + 2\delta^2 - m_\nu^2), \tag{8}$$

indicating that the threshold energy of $10^{13}\,eV$ if $m_\nu = 0.058\,eV$. Solar neutrinos cannot induce the reaction since it is far below the threshold energy, if the the target antineutrino were free. This would be the first impact of the nonzero neutrino mass.

But the target antineutrinos should be on the surface of the Fermi-Dirac sphere, of about $600\,GeV$, fairly close the threshold. If this fact could be verified experimentally in some way, we believe that this would be the fundamental discovery of physics.

We could use the head-on collision of $\nu(Solar) + \bar{\nu}(CB, k_F^\nu)$ as the sufficient example. That is,

$$k_\mu = (k\hat{z}, ik_0), \qquad k'_\mu = (-k'\hat{z}, ik'_0);$$
$$p_\mu = (q\hat{z} + \vec{\delta}, iE), \qquad p'_\mu = (q'\hat{z} - \vec{\delta}, iE'). \tag{9}$$

We could again consider the s-squared, noting that q and q' are much smaller than k'_0 but much larger k_0.

$$s \equiv (p_\mu + p'_\mu)^2 \approx -(m_e^2 + \delta^2)(\frac{q'}{q} + 2 + \frac{q}{q'}), \tag{10}$$

or,

$$s = (k_\mu + k'_\mu)^2 \approx -4kk'. \tag{11}$$

We obtain the threshold energy as follows:

$$4kk' = (m_e^2 + \delta^2)(\frac{q'}{q} + 2 + \frac{q}{q'}), \tag{12}$$

remembering that $k' \sim 200\,TeV$ and $k \sim 1MeV$.

The reaction $\nu_e(Solar) + \bar{\nu}_e \to e^- + e^+$ takes place because of the $W^\pm$ exchange, or, of the Z^0 s-pole creation. It might not be so small (as the ordinary weak processes) in view of the large energies involved in the process.

In fact, it is very interesting to work out the t-value:

$$t = (p_\mu - k_\mu)^2 \approx -m_e^2 - m_\nu^2 + (m_e^2 + \delta^2)\frac{k}{q} + m_\nu^2\frac{q}{k}. \tag{13}$$

The last term would be the biggest one.

On the other hand, we also have

$$t = (p'_\mu - k'_\mu)^2 \approx -m_e^2 - m_\nu^2 + (m_e^2 + \delta^2)\frac{k'}{q'} + m_\nu^2\frac{q'}{k'} + 4q'k'. \tag{14}$$

Here the third and the (last) fifth terms become the dominant contributions. These are the interesting plays of the large and small numbers. Note that it seems to yield the solution that $k << q' << q << k'$.

In the Standard Model [7], apart from the s-pole Z^0 creation, there is a cross-generation process, such as $\nu_\mu + \bar{\nu}_\mu \to e^- + e^+$, via the charged-Higgs exchange. We assume that this may be smaller than the $W^\pm$ exchange and the Z^0 s-pole creation.

To investigate the reaction $\nu(Solar) + \bar{\nu}(CB; k_F^\nu) \to e^- + e^+$ in particular, we have to be in a region where the neutrino halo (of the Earth) is still there but the ordinary-matter materials are absent - so, the natural place would be the high-attitude Satellites such as AMS [6]. The other option is deep in the ice at South Pole [9].

On the front of the Satellite experiment, we have the AMS experiment [6], etc., in which the height is such that the reaction $\nu(Solar) + \bar{\nu}(CB; k_F^\nu) \to e^- + e^+$ would be dominant. The excess positron spectra would be a direct check of the main idea, in that the Fermi-Dirac sphere of the neutrino halo (of the Earth) jacks up the neutrino energy and only those near surface Fermi energy can be excited. If not, the threshold of this reaction ($\sim 10TeV$) would forbid this from happening.

There is also the deep-inside-ice South-Pole experiment, namely, the *IceCube* Collaboration, with a couple of PeV ($1,000\,TeV$) detected [9]. But these neutrinos may be different from the peak or the maximum, i.e., different from the surface of the Fermi-Dirac sphere of the neutrino halo of the Earth, which we talk about in this chapter.

We could wonder where many ultrahigh energy neutrinos would come from. If the neutrino Fermi-Dirac sphere (of the Earth) does not exist, the free threshold of about $10\,TeV$ indicated above would cut off the neutrinos slightly higher than $10\,TeV$ - there should not be anything observed by *IceCube* near $1\,PeV$. As for the location, it might extend slightly beyond (i.e., higher than) the atmosphere, since the denser accumulates near the center while the neutrino halo is basically the Fermi-Dirac gas. Thus, there might be a high-height-enough region that is absent on the visual ordinary matter but is yet full of the dark matter (neutrinos), the location is ideal for the Satellite experiments, such as [6].

There are some other channels to detect the Fermi-Dirac sphere of the neutrino halo of the Earth, or, of the Venus. For instance, we could consider $\nu(Solar) + \bar{\nu}(CB; k_F^\nu) \rightarrow \gamma + \gamma$ against $\nu(Solar) + \bar{\nu}(free) \rightarrow \gamma + \gamma$, but it does not have the remarkable threshold behavior. Also, these experiments may have huge experimental backgrounds to worry about. The proposed experiment on $\nu(Solar) + \bar{\nu}(CB; k_F^\nu) \rightarrow e^- + e^+$, in view of the enormous threshold, may make our life much easier.

The reaction $\nu(Solar) + \bar{\nu}(CB) \rightarrow \gamma + \gamma$ can arise from a few box diagrams with the charged lepton and $W^\pm$ in the middle. It is slightly higher order but, based on the Standard Model [7], it would be the leading order for the annihilation of the $\nu(Solar) - \bar{\nu}(CB)$ pair if the Fermi-Dirac sphere was not there.

We think that the reaction $\nu(Solar) + \bar{\nu}(CB; K_F^\nu) \rightarrow e^- + e^+$ is rather unique - it happens if the Fermi-Dirac sphere of a neutrino halo of the Earth exists; it does not occur, otherwise. Note that it is of utmost importance to have a double check for the fundamental experiment like this.

In Figure 1, we show the famous plot of the ultrahigh energy positron spectra with the positron energy up to $300\,GeV$ and flatten out. We argue that, in addition to the positron near $300\,GeV$, there should be an electron of similar energy - the patten would be quite broad without the bumps (i.e., the resonance peaks). So, $\nu(Solar) + \bar{\nu}(CB; k_F^\nu) \rightarrow e^- + e^+$ discussed in this section serves to explain the outstanding puzzle.

The location of the Space Station, of about $400\,Km$ in height, well above the atmosphere, looks like that there is nothing there, except the Solar neutrinos, the lights, and relatively low abundance of cosmic rays (most of them from the Sun), that there should be no source for the positrons. Nevertheless, we argue that there is the Fermi-Dirac sphere of the neutrino halo of the Earth and the reaction $\nu(Solar) + \bar{\nu}(CB; k_F^\nu) \rightarrow e^- + e^+$ becomes the active source of these ultra high energy positrons.

We argue that the location of the Space Station, of about $400\,Km$ in height, is an ideal neutrino-only laboratory [6], abundant of the solar neutrinos (of $(1-10)\,MeV$) and abundant of the surface antineutrinos and neutrinos near the Fermi energy k_F^ν of the Fermi-Dirac sphere of the Earth. Of course, the location is bombarded by the cosmic rays (mostly from the Sun). The reaction $\nu(Solar) + \bar{\nu}(CB; k_F^\nu)$ provides the source of positrons - here the ultrahigh energy $\bar{\nu}(CB; k_F^\nu)$ is the necessary input. Otherwise, the non-neutrino dark-matter particles on the ultrahigh energy positron source seems to be rather *ad hoc.*

4 Invisible Neutrino Halos and Black Holes

Do neutrino halos stop the formation of black holes? That is, there is no black hole formed when the physics of neutrino halos is taken into account.

Neutrinos are fermions; they would be stacking up like fermi liquids since one state can accommodate only one neutrino.

In the limit of the very low temperature and very high densities, we have [10];

$$
\begin{aligned}
U &= \tfrac{3}{5} N \epsilon_F [1 + \tfrac{5}{12}\pi^2 (\tfrac{kT}{\epsilon_F})^2 + ...]; \\
P &= \tfrac{2}{3}\tfrac{U}{V} = \tfrac{2}{5}\tfrac{\epsilon_F}{v}[1 + \tfrac{5\pi^2}{12}(\tfrac{kT}{\epsilon_F})^2 + ...]; \\
\epsilon_F &= \tfrac{\hbar^2}{2m}(\tfrac{6\pi^2}{gv})^{2/3}; \qquad v \equiv \tfrac{V}{N}.
\end{aligned} \tag{15}
$$

In this limit, it is characterized by a large Fermi energy ϵ_F, of very small kinetic energy. That is, the very small mass is accompanied by a very large Fermi energy ϵ_F.

So, we need to apply these formulae to the neutrino gas in the case of the neutrino halo for the "visual" collapsed star - to see if the neutrino halo would stop the collapse of the star.

First of all, we note from the factor $\hbar^2/(2m)$ that it is an effect arising from the quantum principle and that it is also due to the non-zero mass in this limit. Secondly, it may be difficult to determine v, the volume per particle, in the same limit. But the white dwarf gives an example. It gives, for a dwarf star of a solar mass ($10^{33}\, gm$) [10],

$$\epsilon_F \approx \frac{\hbar^2}{2m_e}\frac{1}{v^{2/3}} \approx 20\, MeV. \tag{16}$$

Using this formula for the neutrino halo with $m_\nu = 0.058\, eV$,

$$\epsilon_F^\nu \approx \frac{\hbar^2}{2m_\nu}\frac{1}{v^{2/3}} \approx 200\, TeV. \tag{17}$$

Maybe we could generalize the uncertainty relation as follows:

$$(\epsilon_F v^{2/3}) \cdot m \geq \hbar^2. \tag{18}$$

In light of the tiny neutrino mass, there is some quantity "conjugate" to the mass quantity for the uncertainty relation. The "conjugate" quantity to the mass is "the Fermi energy times the occupied area per particle". According to our presentation, this uncertainty relation might make some sense though we don't know the unit volume v which may be the same as the electron case.

The Schwarzschild radius, or the horizon, is so tiny (by many orders) in this game while the neutrino halo has the minimum volume that is much bigger (by many orders) than the horizon (of the presumed black hole) (so that the pulling-back gravitational force exerted by the neutrino halo becomes overwhelming at the very last stage). The visual story is dictated by the invisible partner, the neutrino halo. The conclusion is that a black hole consisting of visual ordinary matter cannot be formed, because of the incompressible dark-matter neutrino halo.

Thus, the destiny of a visual ordinary-matter star plus its five times in weight invisible neutrino halo will be determined by the neutrino halo, as a degenerate (invisible) Fermi gas, well ahead of the presumed "black hole" of the visual ordinary-matter lump. A reasonable theory usually gives us a conclusion that is a little boring. We believe that the Standard Model would prevail.

5 Concluding Remarks

Neutrinos have been with us for nearly a whole century, since the early 1930's. Because the feeble nature of its interactions with the environments, we do not understand the neutrinos yet. We thought that we understand neutrinos in terms of the Standard Model [7] owing to the beauty, the consistency, and the completeness of the theory. Basically, the Standard Model is the description of the *smallest* units of matter on the basis of Einstein's relativity principle and the quantum principle. The Standard Model already covers *all* smallest units of matter (including electrons, neutrinos, quarks, etc., a finite number of them) since it is redundant on adding to it some new particle(s). It really is our world, and nothing more.

Basing on this Standard Model [7], cosmic background neutrinos and antineutrinos (CBν's) are the 25% dark matter, as contrast with the 5% ordinary matter in our Universe [4].

The neutrino halos (i.e. 25% dark matter) always accompany the 5% visual ordinary-matter objects, such as stars, planets, etc., such that the world behaves normally and that the collapse of a visual star into a black hole never happens.

Since the 5% ordinary matter are accompanied by the 25% invisible dark matter (in "incompressible" neutrino halos), the story of formation of black holes of visual ordinary matter is just a fiction.

In this chapter, we just try to "convert" those unknowns to those which we could understand. That is, we hope that we could base on the Standard Model [7] in trying to understand our Universe - our Universe is just the Standard Model and *vice versa.*

Notes: In the last two chapters, we do not offer the problem sets, except your personal discussions over some basic concepts, judging from the fact that there are lot's of challenging problems if you combine the current materials with those discussed in the previous chapters. A version of Chapter 16 was published partially in The Universe, **4-1**, 7 (2016).

References

[1] See, for example, Chapter 11, E.D. Commins and P.H. Bucksbaum, "Weak Interactions of Leptons and Quarks" (Cambride University Press, London et al., 1983).

[2] W-Y. Pauchy Hwang, The Universe, **3-1**, 3 (2015); "The Origin of Fields (point-like Particles)".

[3] W-Y. Pauchy Hwang, arXiv:1203.1407v2 [hep-ph] 13 Jan 2016; "Formation of partially dark-matter galaxies".

[4] W-Y. Pauchy Hwang, arXiv:1012.1082v4 [hep-ph] 13 Jan 2016; "Cosmology: Neutrinos as the Only Final Dark Matter".

[5] W-Y. Pauchy Hwang and Jen-Chieh Peng, arXiv:1003.4347v1 [hep-ph] March 2010; currently under revision.

[6] Samuel Ting, CERN Courier, **56-10**, 26 (2016); and the references therein.

[7] W-Y. Pauchy Hwang, arXiv:1304.4705v2 [hep-ph] 25 August 2015; "The Standard Model".

[8] W-Y. Pauchy Hwang, The Universe, **4-1**, 7 (2016); "Addendum to 'Neutrinos in the Cosmos': Neutrino Halos as the 25% Dark Matter".

[9] Francis Halzen and Spencer Klein, CERN Courier, **54-10**, 30 (2014); and the references therein.

[10] Kerson Huang, "Statistical Mechanics" (Copyright @ 1963, 1987, John Wiley & Sons, Inc.), Ch. 11, Fermi Systems.

Exercises: *Chapter 16*

1. The idea of a purely-neutrino collider: Imagine the Space Station at the height of about $400\,Km$ above the ground, circling the Earth. Solar neutrinos are coming to your direction all the time, until the Sun ceases to shine. The $CB\nu$, the five times the weight of the Earth, would be another part of the collider. $\nu(Solar) + \bar{\nu}(CB; k_F^\nu) \rightarrow e^- + e^+$ would not occur if there would not jack up energy to a certain k_F^ν, provided that the neutrino mass is less than $0.058\,eV$ (as suggested by some analysis). Try to explore this idea of a purely-neutrino collider.

2. Continue on Problem 1. Remember that the twin planet, the Venus, of the Earth could provides the best check of the purely-neutrino collider, if CB neutrinos are indeed the 25 % dark matter. So, the Venus was created for some purpose. Try to elucidate on how to expand the solar neutrino physics to the extraterrestrial solar neutrino physics ($etSNP$). After all, the Venus and the Earth have the similar visual weight so as to a similar weight of invisible (dark-matter) parts, such that the quantitative comparison between two twin planets is useful.

SUBJECT INDEX

NAME INDEX